Unitary Group Representations in Physics, Probability, and Numb... Theory

MATHEMATICS LECTURE NOTE SERIES

Volumes of the Series published from 1961 to 1973 are not officially numbered. The parenthetical numbers shown are designed to aid librarians and bibliographers to check the completeness of their holdings.

ISBN			
0-8053-5801-3	(1)	S. Lang	Algebraic Functions, 1965
0-8053-8703-X	(2)	J. Serre	Lie Algebras and Lie Groups, 1965 (3rd printing, with corrections, 1974)
0-8053-2327-9	(3)	P. J. Cohen	Set Theory and the Continuum Hypothesis, 1966 (4th printing, 1977)
0-8053-5808-0 0-8053-5809-9	(4)	S. Lang	Rapport sur la cohomologie des groupes, 1966
0-8053-8750-1 0-8053-8751-X	(5)	J. Serre	Algèbres de Lie semi-simples complexes, 1966
0-8053-0290-5 0-8053-0291-3	(6)	E. Artin and J. Tate	Class Field Theory, 1967 (2nd printing, 1974)
0-8053-0300-6 0-8053-0301-4	(7)	M. F. Atiyah	K-Theory, 1967
0-8053-2434-8 0-8053-2435-6	(8)	W. Feit	Characters of Finite Groups, 1967
0-8053-3555-4	(9)	Marvin J. Greenberg	Lectures on Algebraic Topology, 1966 (5th printing, with corrections, 1977)
0-8053-3757-1	(10)	Robin Hartshorne	Foundations of Projective Geometry, 1967 (3rd printing, 1978)
0-8053-0660-9	(11)	H. Bass	Algebraic K-Theory, 1968
0-8053-0668-4	(12)	M. Berger and M. Berger	Perspectives in Nonlinearity: An Introduction to Nonlinear Analysis, 1968
0-8053-5208-2 0-8053-5209-0	(13)	I. Kaplansky	Rings of Operators, 1968
0-8053-6690-3 0-8053-6691-1	(14)	I. G. MacDonald	Algebraic Geometry: Introduction to Schemes, 1968
0-8053-6698-9 0-8053-6699-7	(15)	G. W. Mackey	Induced Representation of Groups and Quantum Mechanics, 1968
0-8053-7710-7 0-8053-7711-5	(16)	R. S. Palais	Foundations of Global Non-linear Analysis, 1968
0-8053-7818-9 0-8053-7819-7	(17)	D. Passman	Permutation Groups, 1968
0-8053-8725-0	(18)	J. Serre	Abelian *l*-Adic Representations and Elliptic Curves, 1968
0-8053-0116-X	(19)	J. F. Adams	Lectures on Lie Groups, 1969
0-8053-0550-5	(20)	J. Barshay	Topics in Ring Theory, 1969
0-8053-1021-5	(21)	A. Borel	Linear Algebraic Groups, 1969
0-8053-1050-9	(22)	R. Bott	Lectures on K(X), 1969
0-8053-1430-X 0-8053-1431-8	(23)	A. Browder	Introduction to Function Algebras, 1969
		G. Choquet	Lectures on Analysis (3rd printing, 1976)
0-8053-6955-4	(24)		Volume I. Integration and Topological Vector Spaces, 1969
0-8053-6957-0	(25)		Volume II. Representation Theory, 1969
0-8053-6959-7	(26)		Volume III. Infinite Dimensional Measures and Problem Solutions, 1969
0-8053-2366-X 0-8053-2367-8	(27)	E. Dyer	Cohomology Theories, 1969

MATHEMATICS LECTURE NOTE SERIES *(continued)*

ISBN			
0-8053-2420-8 0-8053-2421-6	(28)	R. Ellis	Lectures on Topological Dynamics, 1969
0-8053-2570-0 0-8053-2571-9	(29)	J. Fogarty	Invariant Theory, 1969
0-8053-3080-1 0-8053-3081-X	(30)	William Fulton	Algebraic Curves: An Introduction to Algebraic Geometry, 1969 (4th printing, with corrections, 1978)
0-8053-3552-8 0-8053-3553-6	(31)	M. J. Greenberg	Lectures on Forms in Many Variables, 1969
0-8053-3940-X 0-8053-3941-8	(32)	R. Hermann	Fourier Analysis on Groups and Partial Wave Analysis, 1969
0-8053-4551-5	(33)	J. F. P. Hudson	Piecewise Linear Topology, 1969
0-8053-5212-0 0-8053-5213-9	(34)	K. M. Kapp and H. Schneider	Completely O-Simple Semigroups: An Abstract Treatment of the Lattice of Congruences, 1969
0-8053-5240-6 0-8053-5241-4	(35)	J. B. Keller and S. Antman, (eds.)	Bifurcation Theory and Nonlinear Eigenvalue Problems, 1969
		O. Loos	Symmetric Spaces
0-8053-6620-2 0-8053-6621-0	(36)		Volume I. General Theory, 1969
0-8053-6622-9 0-8053-6623-7	(37)		Volume II. Compact Spaces and Classification, 1969
0-8053-7024-2 0-8053-7025-0	(38)	H. Matsumura	Commutative Algebra, 1970
0-8053-7574-0 0-8053-7575-9	(39)	A. Ogg	Modular Forms and Dirichlet Series, 1969
0-8053-7812-X 0-8053-7813-8	(40)	W. Parry	Entropy and Generators in Ergodic Theory, 1969
0-8053-8350-6 0-8053-8351-4	(41)	W. Rudin	Function Theory in Polydiscs, 1969
0-8053-9100-2 0-8053-9101-0	(42)	S. Sternberg	Celestial Mechanics Part I, 1969
0-8053-9102-9	(43)	S. Sternberg	Celestial Mechanics Part II, 1969
0-8053-9254-8 0-8053-9255-6	(44)	M. E. Sweedler	Hopf Algebras, 1969
0-8053-3946-9 0-8053-3947-7	(45)	R. Hermann	Lectures in Mathematical Physics Volume I, 1970
0-8053-3942-6	(46)	R. Hermann	Lie Algebras and Quantum Mechanics, 1970
0-8053-8364-6 0-8053-8365-4	(47)	D. L. Russell	Optimization Theory, 1970
0-8053-7080-3 0-8053-7081-1	(48)	R. K. Miller	Nonlinear Volterra Integral Equations, 1971
0-8053-1875-5 0-8053-1876-3	(49)	J. L. Challifour	Generalized Functions and Fourier Analysis, 1972
0-8053-3952-3	(50)	R. Hermann	Lectures in Mathematical Physics Volume II, 1972
0-8053-2342-2 0-8053-2343-0	(51)	I. Kra	Automorphic Forms and Kleinian Groups, 1972
0-8053-8380-8 0-8053-8381-6	(52)	G. E. Sacks	Saturated Model Theory, 1972
0-8053-3103-4	(53)	A. M. Garsia	Martingale Inequalities: Seminar Notes on Recent Progress, 1973
0-8053-5664-3 0-8053-5665-1	(54)	T. Y. Lam	The Algebraic Theory of Quadratic Forms, 1973
0-8053-6702-0 0-8053-6703-9	55	George W. Mackey	Unitary Group Representations in Physics, Probability, and Number Theory, 1978

Other volumes in preparation

Unitary Group Representations in Physics, Probability, and Number Theory

George W. Mackey
Harvard University

1978

The Benjamin/Cummings Publishing Company, Inc.
Advanced Book Program
Reading, Massachusetts

London · Amsterdam · Don Mills, Ontario · Sydney · Tokyo

Other books by George W. Mackey, published by W. A. Benjamin, Inc., Advanced Book Program, Reading, Massachusetts 01867:

Induced Representations of Groups and Quantum Machanics, 1968
Mathematical Foundations of Quantum Mechanics, 1963
(3rd printing, 1977)

Library of Congress Cataloging in Publication Data

Mackey, George Whitelaw, 1916-
Unitary group representations in physics, probability, and number theory.

(Mathematics lecture notes series ; 55)
Bibliography: p.
Includes index.
1. Representations of groups. 2. Mathematical physics. 3. Probabilities. 4. Numbers, Theory of.
I. Title.
QA171.M173 512′2 78-23563
ISBN 0-8053-6702-0
ISBN 0-8053-6703-9 pbk.

Manufactured in the United States of America

ABCDEFGHIJK-MA-798

To

Alice

and

Ann

NOTE

Chicago Notes: Lectures delivered at the University of Chicago in the summer of 1955. Published in 1976, with an extensive appendix, by the University of Chicago Press.

Berkeley Notes:* Lectures delivered at the University of California, Berkeley, in the summer of 1965.

Oxford Notes: Lectures delivered at Oxford University during the academic year 1966-1967.
The present book, *Unitary Group Representations in Physics, Probability, and Number Theory,* is the published version of these Oxford Notes, with corrections as well as amended, improved text for "Automorphic Forms and Group Representations," and about 9,000 words of "Notes and References."

Quantum Mechanics Lecture Notes: Lectures delivered at Harvard University during the spring of 1960. Published in 1963 as *The Mathematical Foundations of Quantum Mechanics* by W. A. Benjamin, Inc.; third printing, 1977, by The Benjamin/Cummings Publishing Company, Reading, Massachusetts.

*The Berkeley Notes are superseded by the Oxford Notes; 25% of the latter constitute a verbatim transcription of 60% of the former.

Contents

Preface

This book contains the text of a set of mimeographed lecture notes written by the author and distributed in connection with a course given at Oxford University during the academic year 1966–1967. The course was designed to be an introduction to the modern theory of unitary group representations with emphasis on its wide range of applications and on the concept of an induced representation. In order to make the applications intelligible to mathematicians unfamiliar with quantum physics and advanced number theory, considerable space has been devoted to background material in these subjects. In particular, Sections 17 through 24 can be read as an introductory text on quantum mechanics and statistical mechanics—once one has a good understanding of the material in the first ten sections. Section 26 was designed to give mathematicians with little training in number theory some idea of the main course of development of the more systematic part of that subject.

There are, of course, limits to what one can do in less than sixty lectures (even when one writes more than was actually said). Something has to be sacrificed; in these lectures it was any attempt to give detailed proofs except in the very beginning. The theory of finite-dimensional representations of finite groups is a rather beautiful chapter in pure algebra which, unfortunately, is unfamiliar to many mathematicians. Sections 2 and 3 give a self-contained introduction with full proofs. It is hoped that a careful study of these sections will give the reader sufficient background to appreciate the description of how the theory extends to infinite-dimensional unitary representations of locally compact groups given in Sections 4 through 11. Although proofs are omitted, a serious effort has been made to motivate the theorems and explain their significance. In understanding Sections 9, 10, and 11, the reader may find it useful to specialize to finite groups, to try to find his own proofs, and thus to augment Sections 2 and 3. More or less full proofs of much of the material in Sections 4 through 11 will be found in the author's 1955 Chicago lecture notes. These lecture notes were published in book form by the University of Chicago Press, Chicago, Illinois, in 1976. Almost half of the book is devoted to a long appendix written in 1975 and designed to bring it up to date by summarizing some of the major advances made between 1955 and 1975.

The Oxford notes presented here have relatively little overlap with the Chicago notes and tend rather to supplement them. Indeed, except in the appendix, the Chicago notes have almost nothing to say about applications or about the theory of semi-simple Lie groups. The emphasis is all on detailed proofs of the basic theorems. The Oxford notes, on the contrary, omit most proofs, devote three sections to semi-simple Lie groups, and applications are heavily emphasized. While the appendix to the Chicago notes does discuss semi-simple Lie groups as well as applications to quantum mechanics and number theory, the treatment of the latter two topics is far more lengthy and detailed in the Oxford notes.

The situation is quite different with respect to the relationship between the Oxford notes and the mimeographed lecture notes entitled "Group representations and noncommunatative harmonic analysis with applications to analysis number theory and physics." The latter were distributed in connection with the author's lectures in the summer of 1965 at the University of California in Berkeley and were written by the author that summer. The Berkeley notes are best regarded as a condensed first draft of the Oxford notes and are more or less completely superseded by the latter when not actually contained in them. To be specific, Sections 5 through 11 of the Berkeley notes (with the exception of Section 8) appear verbatim as Sections 4 through 6 and 9 through 11 of the Oxford notes. Sections 7 and 8 of the Oxford notes are an expansion of Section 8 of the Berkeley notes. Moreover, Sections 12 and 13 of the Berkeley notes on connections with ergodic theory and probability appear verbatim as Sections 15 and 16 of the Oxford notes. Sections 3, 12, 13, 14, and 17 through 31 of the Oxford notes go far beyond what is to be found in the Berkeley notes. In particular, Sections 17 through 31 constitute a fivefold expansion of the material on applications to quantum mechanics and number theory that appears in the final two sections of the Berkeley notes.

While writing Section 14 of the Berkeley notes describing the applications to quantum mechanics, the author became impressed with how quickly one can explain the fundamentals of that subject once one has certain concepts and results to work with from the general theory of unitary group representations. He had already scheduled a course in the mathematical foundations of quantum mechanics for the fall semester at Harvard, and the Berkeley experience stimulated him to begin with an account of the theory of unitary group representations and to present quantum mechanics within the framework of that theory. He wrote out detailed lecture notes and while these were not typed or reproduced at the time a large part of their content is reproduced in Sections 2, 3, and 17 through 24 of the Oxford notes presented here. They go much further into the superstructure of quantum mechanics than the author's book *The Mathematical Foundations of Quantum Mechanics* [Benjamin/Cummings, Reading, Mass. (1963)] and integrate the whole subject quite thoroughly with the theory of unitary group representations. While they can be read independently of that book, they do not supersede it. The approach of the earlier book is different and more elementary. Moreover, more than half of it is devoted to topics not treated in the present work. On the other hand, the author's book *Induced Representations and Quantum Mechanics* [Benjamin/Cummings, Reading, Mass. (1968)] is based on lectures given at the Scuole Normale in Pisa during the Oxford spring vacation of 1967. These lectures were a condensation of a small part of the Oxford lectures with quantum mechanics as the only application considered.

Sections 12, 13, and 14 and 25 through 31 were composed especially for the Oxford lectures. Except for a part of Section 30, which has been completely rewritten to eliminate a serious error, Sections 27 through 31 describe the connections between unitary group representations, automorphic forms, and number theory as the author understood them in 1967. He would write this material quite differently today. However, both the subject and his understanding of it are in a state of rapid flux and this does not seem the time to attempt a new formulation.

Much less attention is given to the connections with ergodic theory and probability than to either quantum mechanics or number theory. The Berkeley treatment is simply repeated. The reader who would like to see a more extended treatment of this part of the subject may consult the published version of a set of ten lectures given by the author at Texas Christian University in Fort Worth in the summer of 1972. They appear in vol. 12 (1974) of *Advances in Mathematics* under the title "Ergodic theory and its significance for probability and statistical mechanics."

Another abbreviated version of the material in this book will be found in the notes for the author's 1970 C.I.M.E. lectures in Montecatini, Italy, [published as pages 223–330 of *Theory of Group Representations and Fourier Analysis*, C.I.M.E., Edizioni Cremonese, Rome (1971)]. They differ from the Berkeley lectures in being more condensed (eight hours of lectures instead of twelve) and in profiting a bit from new insights gained between 1965 and 1970.

The author's rather lengthy article "Harmonic Analysis as the Exploitation of Symmetry—An Historical Survey" (to appear in the Rice Institute Studies) supplements the present book in a completely different way than the Chicago lecture notes do. Almost three quarters of it is concerned with what was known before 1946, whereas the Chicago and Oxford lecture notes are both primarily concerned with the part of the theory that began to develop at a rapid rate only at the end of 1946. Its principal thesis is that the extensive modern applications of the theory of unitary group representation theory to number theory, physics, and probability are natural generalizations and extensions of much older applications of classical Fourier analysis to the very same subjects. In the course of developing that thesis, the author has sketched the historical background of much that is presented in this book—beginning in the mid-seventeenth century.

In addition to many minor corrections and the indicated revision of part of Section 30, the text of this book differs from the original Oxford lecture notes in that a few paragraphs (totalling about 9000 words) labeled "Notes and References" have been appended to many sections. They are designed in part to acquaint the reader with the broad outlines of developments that have taken place since the original lectures were given, but also contain historical remarks and bibliographical indications that could have been made earlier. They are neither very systematic nor very thorough and should by no means be construed as an attempt to write a detailed survey of developments in the last eleven years. The author simply read through each section and then wrote down what occurred to him at the time as things he might add to what had already been said. Whenever convenient, he has taken advantage of the existence of the appendix to his Chicago lecture notes and sent the reader to them for references and further details. He herewith apologizes to the (undoubtedly many) authors whose work he has unjustly neglected to mention.

The author's extremely agreeable and rewarding year at Oxford was made possible by an appointment as George Eastman visiting professor. He would like to take this opportunity to thank the administrators of the Eastman chair as well as the various individuals who in many ways helped to make this year a memorable one. Professor Michael Atiyah and the Master and Fellows of Balliol College deserve special mention. Thanks are also certainly due to Professor Jacob Feldman, who invited the author to Berkeley in the summer of 1965 and thereby stimulated him to begin to organize his thoughts about applications in a systematic way. Other acknowledgments are due to Henry Mitchell who worked hard editing the Berkeley notes, to E. Brian Davies who helped in a similar fashion with the Oxford notes, and to various individuals who pointed out errors and obscurities. Phillip Green in particular gave the author a long list of corrections to Sections 26 through 31.

George W. Mackey

1. Introduction

Let S be a set and let G be a group. A function $s,x \to sx$ from $S \times G$ to S will be said to convert S into a G-space if the following two conditions are satisfied:

(a) $(sx_1)x_2 = s(x_1x_2)$ for all $s \in S$, $x_1, x_2 \in G$;

(b) $se = s$ for all $s \in S$, where e is the identity of G.

A subset S_1 of the G-space S is said to be *invariant* if $sx \in S_1$ whenever $s \in S_1$ and $x \in G$. Clearly every subset of S is contained in a unique smallest invariant subset. We call this the invariant subset of S generated by the given subset. The invariant subset generated by a single point is called the *orbit* of that point. If the G-space S is such that S is itself an orbit, or equivalently if S and the empty set are the only invariant subsets, we say that S is a *transitive* G-space. In the general case every invariant subset is itself a G-space and every orbit is a transitive G-space. The various orbits are disjoint and the most general G-space is in an obvious sense a "direct sum" of transitive G-spaces.

At first sight transitive G-spaces would seem incapable of further analysis. However for any G-space S we may form the space $\mathfrak{F}(S)$ of all complex-valued functions on S and make $\mathfrak{F}(S)$ into a G-space by letting:

$$(fx)(s) = f(sx^{-1})$$

The G-space $\mathfrak{F}(S)$ is also a vector space and is one in such a way that each mapping

$$f \to fx^{-1} = V_x(f)$$

is a linear transformation. Even when S is transitive, $\mathfrak{F}(S)$ will usually not

---

George W. Mackey, Unitary Group Representations in Physics, Probability, and Number Theory

ISBN 0-8053-6702-0/0-8053-6703-9, pbk

only fail to be transitive but will have many invariant subsets that are *linear* subspaces of $\mathcal{F}(S)$. For example, let $S=G$ and let $s,x\to sx$ be group multiplication. Let f be any homomorphism of G into the group of all nonzero complex numbers. Then f is a member of $\mathcal{F}(S)$ and the set of all constant multiples of f is an invariant one-dimensional subspace of $\mathcal{F}(S)$. These subspaces for different homomorphisms f are linearly independent and when G is finite and commutative every element of $\mathcal{F}(S)=\mathcal{F}(G)$ is uniquely a sum of elements one from each one-dimensional invariant subspace. When G is infinite, one has a similar result provided that one allows infinite sums and replaces $\mathcal{F}(S)$ by the set of all functions that are sufficiently well-behaved in a suitable sense. If G is the group obtained from the group of additive real numbers by identifying two numbers when they differ by an integral multiple of 2π, then the only measurable functions f that are homomorphisms are those of the form:

$$x\to e^{inx} \quad \text{for } n=0,\pm 1,\pm 2,\ldots$$

and the Fourier expansion theorem states that every square-summable function on G is uniquely of the form

$$\sum_{n=-\infty}^{\infty} c_n e^{inx}$$

where convergence is in the mean.

A large part of the theory of group representations, especially in so far as it applies to analysis and physics, is concerned with generalizations of the Fourier expansion theorem in which $S\neq G$ and G is a separable locally compact group that need be neither compact nor commutative.

An example familiar in classical physics where $S\neq G$ and G is compact but not commutative is that in which S is a spherical surface in three space and G is the group of all rotations of S into itself. Here for each odd integer $2j+1$ we obtain a $(2j+1)$-dimensional subspace $\mathcal{H}_j$ of the space of all square-summable functions on S by taking the restrictions to S of those polynomials in three variables that are homogeneous of the jth degree and satisfy Laplace's equation. Each $\mathcal{H}_j$ is irreducible in the sense that it has no invariant subspaces except itself and zero, and every square-summable function on S is uniquely a sum:

$$f_1+f_2+\cdots$$

where $f_j\in\mathcal{H}_j$ and convergence is in the mean. The members of the $\mathcal{H}_j$ for $j=0,1,2\ldots$ are called *surface harmonics*.

When the group G is compact, one always has discrete sums as in the above two examples. However, when it is only locally compact, one usually has to deal with "continuous direct sums" or "direct integrals" of "infini-

ISBN 0-8053-6702-0/0-8053-6703-9, pbk

tesimal" invariant subspaces. A typical example is that in which $G=S$ and G is the group of all real numbers under addition. According to the theory of the Fourier transform every square-summable function f on G may be written uniquely in the form:

$$f(x)=\lim_{T\to\infty}\int_{-T}^{T} e^{ixt}\varphi(t)\,dt$$

where the limit exists in the mean and φ is another square-summable function called the Fourier transform of f.

Of course we need not confine attention to scalar-valued functions on the G-space S. It is also useful to consider the generalization of the above in which $\mathcal{F}(S)$ is replaced by various spaces of vector-valued functions on S. This generalization is especially interesting when the vector space in which the members of $\mathcal{F}(S)$ take their values varies from point to point of S—so that one has what is known as a vector bundle—and when the bundle itself is a G-space in a manner that we shall describe later.

In all cases there is a direct analogue of the mapping $x\to V_x$ defined above and this mapping is a homomorphism of G into the group of all one-to-one linear transformations of the given vector space into itself. Much of the theory can be developed without reference to the fact that the vector space in question is a function space, and we may handle many cases at once by developing a theory of abstract group representations. Here a group representation is by definition a homomorphism of a group into the group of all linear transformations of a vector space. Developing the theory in the abstract is not just a useful technique for handling a number of cases at once and avoiding irrelevant complications. In applications to quantum mechanics one needs the abstract theory itself—especially if one wants to deduce the basic equations of the subject from a minimum of ad hoc axioms.

We have described the theory of group representations as a generalization of Fourier analysis, and while true when suitably interpreted this simple statement does not do justice to the richness and extent of the subject. If A is a bounded linear operator on a Banach space, then

$$e^{At}=1+At+\frac{A^2t^2}{2!}+\cdots$$

exists for all real t and defines a bounded linear operator V_t^A. Moreover $t\to V_t^A$ defines a representation of the additive group of all real numbers and this representation uniquely determines the operator A. The correspondence $A\to V^A$ between single operators and group representations may be extended in various ways so as to permit A to be unbounded and when the Banach space is a Hilbert space we obtain a complete one-to-one correspondence between (not necessarily bounded) operators A that are

ISBN 0-8053-6702-0/0-8053-6703-9, pbk

skew-adjoint in the sense that $A^* = -A$ and continuous representations of the real line in which all the V_t are unitary. Multiplication by i converts skew-adjoint operators into self-adjoint operators and we obtain a natural one-to-one correspondence between self-adjoint operators and continuous unitary representations of the real line. This correspondence preserves all interesting properties and makes it possible to deduce the theory of self-adjoint operators—spectral theorem, multiplicity theory, and all—from the more general theory of unitary representations.

When G is a commutative group, the only irreducible representations are one-dimensional and they are completely described by the homomorphisms of G into the multiplicative group of all nonzero complex numbers. These homomorphisms are usually easy to determine and the interesting part of the theory is the part telling us how to describe general representations in terms of irreducible ones. On the other hand, when G is noncommutative the problem of finding the irreducible representations can be highly nontrivial even when G is finite. When G is finite, the decomposition theory is quite easy and the interest of the theory lies entirely in finding the irreducible representations. The problem of finding the irreducible representations of finite groups is a branch of abstract algebra that is very different in character from the problem of expanding a function in a Fourier series or "diagonalizing" a self-adjoint operator in a Hilbert space.

For a long time the phrase "theory of group representations" was used only to apply to the essentially algebraic part of the subject in which the emphasis is on finding the irreducible representations of noncommutative groups—especially the finite-dimensional ones. Indeed the phrase still has this connotation in the minds of many mathematicians. However, as we have indicated the theory we propose to outline in sections 4 through 11 is a thorough-going synthesis of the algebraic theory with Fourier analysis and spectral theory and includes considerable generalizations of all three subjects.

ISBN 0-8053-6702-0/0-8053-6703-9, pbk

2. Representations of finite commutative groups

If V is a one-dimensional representation of any group G, then we may write

$$V_x = \chi(x)I$$

where I is the identity operator and $x \to \chi(x)$ is a complex-valued function on G that is never zero. Since V is a representation:

$$\chi(xy) = \chi(x)\chi(y)$$

for all x and y in G. Such a function χ from G to the nonzero complex numbers is called a *character*. Conversely, if χ is a character of G and I is the identity operator in a complex vector space $\mathcal{H}$, then $x \to \chi(x)I$ will be a representation of G that is one-dimensional if and only if $\mathcal{H}$ is. In the special case in which G is finite and commutative, we shall see that every representation may be built up in a simple transparent way from representations $x \to \chi(x)I$, where χ is a character.

We shall begin by studying some general properties of characters. Note first that for any group G the set of all characters is itself a group under pointwise multiplication. When G is finite,

$$\chi(x^n) = \chi(x)^n = 1$$

for some n. Hence,

$$|\chi(x)| = 1$$

and

$$\overline{\chi(x)} = \chi(x)^{-1}$$

George W. Mackey, Unitary Group Representations in Physics, Probability, and Number Theory

ISBN 0-8053-6702-0/0-8053-6703-9, pbk

Now suppose that G is a cyclic group of order n with generator a. Then

$$\chi(a^b)=\chi(a)^b$$

so χ is completely determined by the complex number $\chi(a)$. Since $a^n=e$, $\chi(a)$ must be an nth root of unity, and any nth root of unity will do. Hence there are exactly as many characters of G as there are nth roots of unity, and the group of all characters is isomorphic to the group of all nth roots of unity and hence to G. Now let G_1 and G_2 be any two groups and let G be their direct product, that is, the set of all pairs x,y, where

$$(x_1,y_1)\cdot(x_2,y_2)=(x_1x_2,y_1y_2)$$

It follows at once from the definitions that the most general character χ of G may be constructed by choosing characters χ_1 and χ_2 of G_1 and G_2, respectively, and setting

$$\chi(x,y)=\chi_1(x)\cdot\chi_2(y)$$

Thus

$$\widehat{G_1\times G_2}\simeq\hat{G}_1\times\hat{G}_2$$

where $\hat{G}$ denotes the group of all characters of G and $\simeq$ denotes isomorphism. Now it is a theorem in elementary group theory that every finite commutative group is a direct product of cyclic groups, and it follows from this and the foregoing that G and $\hat{G}$ are isomorphic whenever G is finite and commutative. In particular, we see that G has just as many characters as it has elements, and that knowing the structure of G we can find all characters quite explicitly. It should be noted that the relationship between G and $\hat{G}$ is quite like that between a vector space and its dual. There is a canonical isomorphism between G and $\hat{\hat{G}}$, but one can only set up an isomorphism between G and $\hat{G}$ by choosing an arbitrary "basis" of cyclic subgroups of G.

Now let G be finite and commutative and let χ be any member of $\hat{G}$. Then, for each $y\in G$,

$$\sum_{x\in G}\chi(x)=\sum_{x\in G}\chi(xy)=\chi(y)\sum_{x\in G}\chi(x)$$

Thus

$$\{1-\chi(y)\}\sum_{x\in G}\chi(x)=0$$

and it follows that whenever $\chi\not\equiv 1$, then

$$\sum_{x\in G}\chi(x)=0$$

ISBN 0-8053-6702-0/0-8053-6703-9, pbk

Now let χ_1 and χ_2 be two distinct characters. Then

$$\chi_1(x)/\chi_2(x) \equiv \chi_1(x)\overline{\chi_2(x)} \not\equiv 1$$

Hence

$$\sum_{x\in G} \chi_1(x)\overline{\chi_2(x)} = 0 \quad \text{whenever } \chi_1 \neq \chi_2 \tag{*}$$

The various characters on G are mutually orthogonal functions. It follows at once that they are linearly independent, and since their number is equal to the dimension of the vector space of all functions on G, we conclude that the characters form a basis for the vector space of all complex-valued functions on G.

With these simple facts about characters at our disposal, it is easy to analyze the structure of the most general representation of a finite commutative group. Let V be a representation of the finite commutative group G in the complex vector space $\mathcal{H}(V)$. For each character χ on G, let P_χ denote the operator:

$$\frac{1}{o(G)} \sum_{x\in G} \overline{\chi(x)}\, V_x$$

where $o(G)$ is the number of elements in the group G. Then using the orthogonality relation (*) and the defining identity for characters, it is easy to prove the following statements:

(1) $P_\chi^2 = P_\chi$ for all $\chi \in \hat{G}$;

(2) $P_{\chi_1} P_{\chi_2} = P_{\chi_2} P_{\chi_1} = 0$ whenever $\chi_1 \neq \chi_2$;

(3) $\Sigma_{\chi\in\hat{G}} P_\chi = I$ the identity operator;

(4) $V_x P_\chi = \chi(x) P_\chi$ for all x and χ.

Indeed:

$$\begin{aligned}
P_{\chi_1} P_{\chi_2} &= \frac{1}{o(G)^2} \sum_{x\in G, y\in G} \overline{\chi_1(x)}\;\overline{\chi_2(y)}\, V_{xy} \\
&= \frac{1}{o(G)^2} \sum_{x\in G, t\in G} \overline{\chi_1(x)}\; \overline{\chi}_2(x^{-1}t)\, V_t \\
&= \frac{1}{o(G)^2} \Big(\sum_{x\in G} \overline{\chi_1(x)}\, \chi_2(x) \Big) \Big(\sum_{t\in G} \overline{\chi_2(t)}\, V_t \Big) \\
&= \begin{cases} 0 & \text{if } \chi_1 \neq \chi_2 \\ \dfrac{1}{o(G)} \displaystyle\sum_{t\in G} \overline{\chi_2(t)}\, V_t = P_{\chi_2} & \text{if } \chi_1 = \chi_2 \end{cases}
\end{aligned}$$

ISBN 0-8053-6702-0/0-8053-6703-9, pbk

So (1) and (2) are proved. Also,

$$\sum_{\chi\in\hat{G}} P_\chi = \frac{1}{o(G)} \sum_{\substack{x\in G \\ \chi\in\hat{G}}} \overline{\chi(x)}\, V_x$$

$$= \frac{1}{o(G)} \sum_{x\in G} V_x \left(\sum_{\chi\in\hat{G}} \chi(x) \right)$$

and the inner sum is zero whenever $x \neq e$ because $\chi \to \chi(x)$ is a member of $\hat{\hat{G}}$. Thus

$$\sum_{\chi\in\hat{G}} P_\chi = \frac{1}{o(G)} V_e \sum_{\chi\in\hat{G}} 1 = V_e = I$$

and (3) is proved. Finally,

$$V_x P_\chi = \frac{1}{o(G)} \sum_{y\in G} \overline{\chi(y)}\, V_x V_y$$

$$= \frac{1}{o(G)} \sum_{y\in G} \overline{\chi}(yx^{-1}) V_y$$

$$= \chi(x) \frac{1}{o(G)} \sum_{y\in G} \overline{\chi}(y) V_y$$

$$= \chi(x) P_\chi$$

Let $\mathcal{H}_\chi$ be the range of P_χ. It follows at once from (1), (2), and (3) that every vector in $\mathcal{H}$ may be written uniquely as a sum of vectors one from each $\mathcal{H}_\chi$ and it follows from (4) that for every vector φ in $\mathcal{H}_\chi$,

$$V_x(\varphi) = \chi(x)\varphi$$

Each $\mathcal{H}_\chi$ thus defines a subrepresentation V^χ of V and this representation is of the form $x \to \chi(x)I$, where I is the identity operator in a vector space. Moreover, in an obvious sense (which we shall make explicit below), V is the direct sum over χ of the subrepresentations V^χ. This particular direct sum decomposition of V has the following important property: If $\mathcal{H}'$ is any invariant subspace of $\mathcal{H}$ then $\mathcal{H}'$ is the direct sum of its intersections with the $\mathcal{H}_\chi$. In fact, for each $\varphi \in \mathcal{H}'$,

$$P_\chi(\varphi) = \frac{1}{o(G)} \sum_{\chi\in\hat{G}} \overline{\chi(x)}\, V_x(\varphi)$$

ISBN 0-8053-6702-0/0-8053-6703-9, pbk

and hence is also in $\mathcal{H}'$. Thus,

$$\varphi = \sum_{\chi \in \hat{G}} P_\chi(\varphi)$$

is a sum of elements from the $\mathcal{H}_\chi \cap \mathcal{H}'$. It follows that to determine the most general invariant subspace $\mathcal{S}$ of $\mathcal{H}$, it suffices to determine the most general invariant subspace of each $\mathcal{H}_\chi$. But every subspace of each $\mathcal{H}_\chi$ is invariant. Thus we have a complete over view of all possible invariant subspaces of $\mathcal{H}$ and all possible direct sum decompositions of V. To obtain a decomposition of V as a direct sum of irreducible subrepresentations, we have only to choose a basis in each $\mathcal{H}_\chi$ and we may choose this basis arbitrarily. This partial nonuniqueness of the decomposition into irreducibles is a feature of the general case. The strictly unique decomposition into the V^χ is called the central decomposition or the canonical decomposition into "disjoint" "primary" parts. We shall give precise definitions below.

Notes and References

Characters and Fourier analysis on finite commutative groups occurred (at least implicitly) as a key tool in number theory throughout the nineteenth century. The word "character" is due to Gauss and was used by him in his *Disquisitiones Arithmeticae* published in 1801 to denote what is essentially a character of order 2 on a certain finite commutative group (also introduced by Gauss) that plays a central role in the theory of binary quadratic forms. Of course, group theory did not exist as such in 1801 and characters of finite commutative groups as we now understand them were first defined in 1878 and 1881 by Dedekind and Weber, respectively. Weber was the first to define them for *abstract* finite commutative groups.

ISBN 0-8053-6702-0/0-8053-6703-9, pbk

3. *Characters and the decomposition of representations of finite groups that are not necessarily commutative*

In this section we shall find that there is a fairly complete generalization of the results of the preceding section to the case where G is finite but not necessarily commutative. However, this generalization is more complicated in several respects. One has a generalization of the notion of character and it may be used in the same way to effect the same sort of decomposition of an arbitrary representation. However, the notion is less easy to define, the problem of finding all the characters of a given group is a highly nontrivial one and the structure of the representations induced in the $\mathcal{H}_\chi$ is less transparent.

To begin with, it will be useful to introduce some definitions. If U and V are both representations of the same group, we shall say that U and V are *equivalent* if there exists a linear transformation from $\mathcal{H}(U)$ to $\mathcal{H}(V)$ that is one-to-one, has all of $\mathcal{H}(V)$ for its range and is such that

$$TU_xT^{-1} = V_x \quad \text{for all } x \in G$$

Here and in the sequel $\mathcal{H}(W)$ will denote the space in which the W_x act. By the *direct sum* of two representations U and V, we shall mean the unique representation $U \oplus V$ whose space is the direct sum of the spaces $\mathcal{H}(U)$ and $\mathcal{H}(V)$ and which is defined by the equation

$$(U \oplus V)_x(\theta_1, \theta_2) = (U_x(\theta_1), V_x(\theta_2))$$

for all $x \in G$, $\theta_1 \in \mathcal{H}(U)$, and $\theta_2 \in \mathcal{H}(V)$. If U is a representation of G and

George W. Mackey, Unitary Group Representations in Physics, Probability, and Number Theory

ISBN 0-8053-6702-0/0-8053-6703-9, pbk

$\mathcal{H}'$ is a subspace of $\mathcal{H}(U)$, we call this subspace *invariant* if $U_x(\varphi) \in \mathcal{H}'$ for all $x \in G$ and $\varphi \in \mathcal{H}'$. By replacing each U_x by its restriction to $\mathcal{H}'$, given $\mathcal{H}'$ invariant, we obtain a new representation whose space is $\mathcal{H}'$ and which we call the *subrepresentation* defined by $\mathcal{H}'$. Given an invariant subspace $\mathcal{H}'$ of $\mathcal{H}(U)$, let $\mathcal{H}(U)/\mathcal{H}'$ denote the corresponding "quotient space," that is, the vector space of all classes mod $\mathcal{H}'$ where two vectors are put in the same class whenever their difference lies in $\mathcal{H}'$. Let $\bar{\varphi}$ denote the equivalence class to which φ belongs. Then we may unambiguously define $\bar{U}_x(\bar{\varphi})$ to be $\overline{U_x(\varphi)}$ since $\bar{\varphi}_1 = \bar{\varphi}_2$ implies $\varphi_1 = \varphi_2 \in \mathcal{H}'$, and this implies

$$U_x(\varphi_1) - U_x(\varphi_2) \in \mathcal{H}'$$

for all $x \in G$. In this way U_x defines a linear transformation $\bar{U}_x$ in $\mathcal{H}(U)/\mathcal{H}'$ and $x \to \bar{U}_x$ is easily seen to be a representation of G. It is called the *quotient representation* of G defined by the invariant subspace $\mathcal{H}'$. Suppose that the invariant subspace $\mathcal{H}'$ admits a complementary subspace that is also invariant; that is an invariant subspace $\mathcal{H}''$ such that

$$\mathcal{H}' \cap \mathcal{H}'' = \{0\} \quad \text{and} \quad \mathcal{H}' + \mathcal{H}'' = \mathcal{H}(U)$$

Then $\bar{\varphi} \neq 0$ for all $\varphi \in \mathcal{H}''$ and the mapping $\varphi \to \bar{\varphi}$ sets up an equivalence between the quotient representation defined by $\mathcal{H}'$ and the subrepresentation defined by $\mathcal{H}''$. It follows at once that any two complementary subrepresentations of the same subrepresentation are equivalent. We shall deal mainly with representations in which every subrepresentation has a complementary subrepresentation and hence need not concern ourselves to any great extent with quotient representations. A representation with no proper subrepresentations is said to be *irreducible*.

Consider the identity

$$TU_xT^{-1} = V_x$$

defining the notion of equivalence. It may also be written in the form

$$TU_x = V_xT$$

and then makes sense whether or not T has an inverse. A linear transformation T from $\mathcal{H}(U)$ to $\mathcal{H}(V)$ that satisfies this identity is called an *intertwining operator*. The set of all intertwining operators is a vector space, which we shall denote by $\mathcal{R}(U, V)$; we shall call its vector space dimension the *intertwining number* of U and V. When $U = V$, $\mathcal{R}(U, U)$ is a linear algebra that we shall call the *commuting algebra* of U.

When an intertwining operator is one-to-one and onto, it sets up an equivalence between the representations concerned. Let us now see what we can conclude about two representations when we know only that there exists a nonzero intertwining operator. Let T be any such intertwining operator, let N_T be the null-space of T, and let R_T be its range. Then N_T is

ISBN 0-8053-6702-0/0-8053-6703-9, pbk

an invariant subspace of $\mathcal{H}(U)$. In fact, if $T(\varphi)=0$, then

$$TU_x(\varphi)=V_xT(\varphi)=0$$

so that each U_x carries N_T into N_T. Similarly, R_T is an invariant subspace of $\mathcal{H}(V)$. Suppose that N_T has a complementary invariant subspace N_T'. Then T restricted to N_T' will be one-to-one, and T will set up an equivalence between the subrepresentation of U defined by N_T' and the subrepresentation of V defined by R_T. If N_T' does not exist, it can still be shown that T sets up an equivalence between the quotient representation defined by N_T' and the subrepresentation defined by R_T. Working backwards it is easy to see that the existence of an equivalence between a quotient representation of U and a subrepresentation of V implies the existence of a nonzero intertwining operator between U and V. Thus, $\mathcal{R}(U,V)$ reduces to zero if and only if no quotient representation of U is equivalent to any subrepresentation of V. When G is finite, it can be shown that any subrepresentation of U has a complementary subrepresentation. Hence, in this case, $\mathcal{R}(U,V)$ reduces to zero if and only if no subrepresentation of U is equivalent to any subrepresentation of V. When either and hence both of these conditions hold, we shall say that U and V are *disjoint*. The representations V^χ of Section 2 are clearly mutually disjoint. Moreover, each V^χ is *primary* in the sense that it cannot be written as a direct sum of two disjoint representations. The decomposition

$$V=\sum_\chi V^\chi$$

is clearly the unique decomposition of V into disjoint primary parts.

When U and V are irreducible, N_T and R_T are either 0 or the whole space. Thus either $\mathcal{R}(U,V)=0$ or U and V are equivalent. If U and V are equivalent, then $\mathcal{R}(U,V)$ is isomorphic with $\mathcal{R}(U,U)$ and every nonzero member of $\mathcal{R}(U,U)$ is one-to-one and onto. It follows that $R(U,U)$ is a division ring whenever U is irreducible; this result is known as Schur's lemma. When U is both irreducible and finite-dimensional, $R(U,U)$ will be a finite-dimensional division ring over the complex numbers and hence must be just the ring of all complex numbers. To see this in a more elementary way, let T be any nonzero element of $\mathcal{R}(U,U)$ and let h be any complex number such that $T-hI$ has a zero determinant. Then $T-hI$ will have a nonzero null-space and is a member of $\mathcal{R}(U,U)$ so that $T-hI=0$ and $T=hI$. Thus $\mathcal{R}(U,U)$ consists just of constant multiples of the identity. Let $i(U,V)=\dim\mathcal{R}(U,V)=$intertwining number of U and V. It follows immediately from the definitions that

$$i(U_1\oplus U_2,V)=i(U_1,V)+i(U_2,V)$$

and

$$i(U,V_1\oplus V_2)=i(U,V_1)+i(U,V_2)$$

ISBN 0-8053-6702-0/0-8053-6703-9, pbk

Thus, if

$$U=L^1\oplus L^2\oplus\cdots\oplus L^n$$

where the L^j are finite-dimensional and irreducible and M is any irreducible representation of G then

$$i(M,U)=\sum_j i(M,L^j)=\text{number of } j \text{ with } L^j\simeq M$$

The truth of the following statement is an immediate consequence. Let $L^1\oplus\cdots\oplus L^n$ be equivalent to $M^1\oplus\cdots\oplus M^m$, where the L^j and M^k are finite-dimensional and irreducible. Then $m=n$ and there is a permutation π of the integers from 1 to n such that

$$L^j\simeq M^{\pi(j)}\quad\text{for } j=1,\ldots,n$$

The number $i(L^j,U)$ is equal to the number of summands of U equivalent to L^j in any decomposition of U as a direct sum of irreducibles, and is called the *multiplicity* with which L^j occurs in U. If U and V are both direct sums of finite-dimensional irreducibles, then

$$i(U,V)=\sum i(L,U)i(L,V)$$

where L varies over the equivalence classes of irreducible representations of G.

When G is finite, it is easy to show that every finite-dimensional representation of G is equivalent to a direct sum of irreducible representations. Indeed let U be finite-dimensional and let $\varphi,\psi\rightarrow[\varphi,\psi]$ be any positive-definite Hermitean inner product in $\mathcal{H}(U)$, that is,

$$\begin{aligned}&[\varphi\cdot\psi]=\overline{[\psi,\varphi]}\\&[\varphi,\varphi]>0\quad\text{except when } \varphi=0\\&\varphi\rightarrow[\varphi,\psi]\quad\text{is linear}\end{aligned}$$

Let

$$[\varphi,\psi]^{\sim}=\frac{1}{o(G)}\sum_{x\in G}[U_x(\varphi),U_x(\psi)]$$

Then $\varphi,\psi\rightarrow[\varphi,\psi]^{\sim}$ is also a positive-definite Hermitean inner product and will have the further property that each U_x will be *unitary* in the sense that

$$[U_x(\varphi),U_x(\psi)]^{\sim}=[\varphi,\psi]^{\sim}$$

for all x, φ, and ψ. In other words, when G is finite every finite-dimensional representation of G is unitary with respect to some inner product.

ISBN 0-8053-6702-0/0-8053-6703-9, pbk

Now let U be a unitary finite-dimensional representation of any group G and let $\mathcal{H}_1$ be any invariant subspace of $\mathcal{H}(U)$. Let $\mathcal{H}_1^{\perp}$ be the "orthogonal complement" of $\mathcal{H}_1$; that is, the set of all θ such that $[\varphi,\theta]=0$ for all $\varphi\in\mathcal{H}_1$. It follows from elementary vector space theory that $\mathcal{H}_1^{\perp}$ is a complementary subspace, and we shall show now that it is invariant. Indeed, if $\theta\in\mathcal{H}_1^{\perp}$ and $x\in G$, then

$$[U_x(\theta),\varphi]=[\theta,U_{x^{-1}}(\varphi)]=0$$

for all $\varphi\in\mathcal{H}_1$ since $U_{x^{-1}}(\varphi)\in\mathcal{H}_1$ whenever $\varphi\in\mathcal{H}_1$. Thus $U_x(\theta)\in\mathcal{H}_1^{\perp}$ for all $x\in G$ and all $\theta\in\mathcal{H}_1^{\perp}$. Thus $\mathcal{H}_1^{\perp}$ is invariant.

Putting the above two conclusions together we see that for finite-dimensional representations of finite groups every subrepresentation has a complementary subrepresentation. It follows immediately by induction that every finite-dimensional representation of a finite group is equivalent to a direct sum of irreducible representations. We have already seen that this decomposition is unique to within equivalence. Thus the finite-dimensional representations of a finite group are all known to within equivalence as soon as one knows all irreducible finite-dimensional representations to within equivalence.

In Section 2 we saw that for a finite commutative group there are only a finite number of inequivalent irreducible representations, and we described the most general infinite-dimensional representation in terms of these. We shall now generalize these results to the noncommutative case. Let G be any finite group. Let $\mathcal{F}$ be the vector space of all complex valued functions on G. For each $x\in G$ let R_x be the operator $f\to f_x$, where

$$f_x(y)=f(yx)$$

Then $x\to R_x$ is a representation of G called the *right regular representation.* Now let V be any representation of G and let φ be any element in $\mathcal{H}(V)$. For each $f\in\mathcal{F}$ let

$$T_\varphi(f)=\sum_{x\in G}f(x)V_{x^{-1}}(\varphi)$$

Then T_φ is a linear transformation from $\mathcal{F}$ to $\mathcal{H}(V)$. Moreover,

$$\begin{aligned}T_\varphi R_y(f)&=\sum_{x\in G}f(xy)V_{x^{-1}}(\varphi)\\&=\sum_{x\in G}f(x)V_{yx^{-1}}(\varphi)\\&=V_yT_\varphi(f)\end{aligned}$$

ISBN 0-8053-6702-0/0-8053-6703-9, pbk

Thus T_φ intertwines R and V. Hence T sets up an equivalence between a subrepresentation of R and a subrepresentation of V whose space contains φ. We can at once draw the following conclusions:

(a) Every element φ in the space of any representation V of a finite group is a finite sum of elements each contained in a finite-dimensional irreducible invariant subspace.

(b) Every irreducible representation of a finite group is equivalent to a subrepresentation of the regular representation.

(c) The number of inequivalent irreducible representations of a finite group does not exceed the order of the group.

(d) The dimension of an irreducible representation of a finite group does not exceed one less than the order of the group.

Now let T be any intertwining operator for R and V. Then,

$$T(f) = \sum_{x \in G} f(x) T(\delta_x)$$

where

$$\delta_x(y) = \begin{cases} 0 & \text{if } y \neq x \\ 1 & \text{if } y = x \end{cases}$$

because T is linear and

$$f = \sum_{x \in G} f(x) \delta_x$$

But

$$T(\delta_x) = T(R_{x^{-1}} \delta_e) = V_{x^{-1}} T(\delta_e)$$

Thus $T = T_\varphi$, where $\varphi = T(\delta_e)$. In other words, the operators T_φ defined above constitute just the vector space of all intertwining operators for R and V. If $T_\varphi = 0$ for some φ, then

$$\sum_{x \in G} f(x) V_{x^{-1}}(\varphi) = 0$$

for all f and in particular for $f = \delta_e$. Thus $\varphi = 0$. It follows that $\mathcal{R}(R, V)$ is isomorphic to $\mathcal{H}(V)$ and hence that

$$i(R, V) = \dim \mathcal{H}(V)$$

In particular, we conclude that the right regular representation contains each irreducible representation with multiplicity equal to its dimension.

Now the dimension of the regular representation is just $o(G)$ and is also equal to the sum of the dimensions of all the irreducible components. Since the representation with dimension d occurs d times, its equivalence class contributes d^2 to the full dimension of R. Thus we have the following remarkable fact: Let $L^1,\ldots,L^r$ be the inequivalent irreducible representations of the finite group G. Let d_j be the dimension of L^j. Then

$$d_1^2+d_2^2+\cdots+d_r^2=o(G)$$

When G is commutative all d_j are 1, and this result reduces to the result already noted, that

$$o(G)=o(\hat{G})$$

Suppose that all of the irreducible representations of the finite group G are one-dimensional. By the above there must exist $o(G)$ characters $\chi_1,\ldots,\chi_r$, where $r=o(G)$, and they must constitute a basis for the set of all functions on G. Thus the mapping

$$x\to\{\chi_1(x),\ldots,\chi_r(x)\}$$

must be one-to-one. Since this is a homomorphism of G into a commutative group, G must be commutative. In other words, every finite noncommutative group must have at least one irreducible representation whose dimension exceeds one.

The smallest noncommutative group is the six-element group of all permutations of three objects By the result proved above, $d_1^2+d_2^2+\cdots=6$. Hence $d_j\leqslant 2$ for all j since $3^2=9>6$. On the other hand, $2^2+2^2=8>6$ and at least one $d_j>1$. Thus there is just one possibility apart from order:

$$d_1=1,\quad d_2=1,\quad d_3=2$$

In other words, we have proved that the so called "symmetric group" on three objects must have just three inequivalent irreducible representations —two of dimension one and one of dimension two. Similar arguments yield information about the number and dimensions of the irreducible representations of other finite groups of low orders. There is just one group of order seven and it is commutative. However, there are two noncommutative groups of order eight. Since $2^2+2^2=8$ and every group has a one-dimensional representation, there must be at most one twodimensional irreducible representation, and since $3^2=9>8$, there must be at least one two-dimensional irreducible representation and no higher dimensional one. Thus both noncommutative groups of order eight have one two-dimensional irreducible representation and four one-dimensional ones. Both the groups of order nine are commutative but there is a noncommutative

ISBN 0-8053-6702-0/0-8053-6703-9, pbk

group of order ten. Here are the nontrivial ways of writing 10 as a sum of squares:

$$10=3^2+1^2$$
$$10=2^2+2^2+1^2+1^2$$
$$10=2^2+1^2+1^2+1^2+1^2+1^2+1^2$$

The last possibility can be ruled out because if there were six characters there would be a quotient group of order six and six does not divide ten. The first possibility can only be ruled out by using a theorem we have not proved to the effect that the dimension of an irreducible representation must divide the order of the group. Doing so we find that the noncommutative group of order ten has two one-dimensional representations and two irreducible two-dimensional ones. We leave it to the reader to show there are just two possibilities for the three noncommutative groups of order twelve:

$$12=3^2+1^2+1^2+1^2$$
$$12=2^2+2^2+1^2+1^2+1^2+1^2$$

Let V be a finite-dimensional representation of the finite group G. We wish to study the structure of the commuting algebra $\mathfrak{R}(V,V)$. We know that we may write V as a direct sum of irreducibles. Let us collect together the equivalent ones and write

$$V \simeq V^1 \oplus \cdots \oplus V^r$$

where each V^j is a direct sum of equivalent irreducibles and the irreducibles of the different V^j are inequivalent. If T is any member of $\mathfrak{R}(V,V)$, let T_j be the operator that takes φ into the jth component of $T(\varphi)$ and let T_j^i be the restriction of T_j to the space of V^i. Then T_j^i is an intertwining operator for V^i and V^j and hence must be zero unless $i=j$. In other words, T maps $\mathcal{H}(V^i)$ into $\mathcal{H}(V^i)$ for all i. It follows at once that $\mathfrak{R}(V,V)$ is the direct sum of the algebras $\mathfrak{R}(V^i,V^i)$. Thus we need only study $\mathfrak{R}(W,W)$ in the special case where W is the direct sum of k replicas of the same irreducible representation L. Writing

$$W=L\oplus L\oplus \cdots \oplus L$$

let $\mathcal{H}_j$ be the subspace of $\mathcal{H}(W)$ consisting of all k-tuples $(\varphi_1,\ldots,\varphi_k)$ such that $\varphi_i=0$ for $i\neq j$. Then we can define T_j^i as above. T is uniquely defined by the matrix T_j^i and is in $\mathfrak{R}(W,W)$ if and only if each T_j^i is in $\mathfrak{R}(L,L)$. We verify at once that $\mathfrak{R}(W,W)$ is isomorphic to the ring of all $k\times k$ matrices with elements from $\mathfrak{R}(L,L)$. Since L is irreducible it follows from Schur's lemma that $\mathfrak{R}(L,L)$ is just the ring of all complex numbers. Thus $\mathfrak{R}(W,W)$ is isomorphic with the ring of all $k\times k$ complex matrices.

ISBN 0-8053-6702-0/0-8053-6703-9, pbk

Putting the two results together we draw the following conclusion: If V is a finite-dimensional representation of the finite group G, then $\mathfrak{R}(V,V)$ is a direct sum of full matrix algebras over the complex numbers. There is one summand for each equivalence class of irreducible representations that occurs, and the multiplicity with which it occurs is the number of rows and columns in the corresponding matrix algebra.

Now consider the special case in which V is the right regular representation. We have already seen that setting

$$T_\varphi(f)=\sum_{x\in G} f(x)W_{x^{-1}}(\varphi)$$

yields a one-to-one linear map $\varphi\rightarrow T_\varphi$ of $\mathcal{H}(W)$ onto $\mathfrak{R}(R,W)$, where R is the right regular representation and W is arbitrary. Setting $R=W$, we see that there is a one-to-one linear map of the space $\mathcal{F}(G)$ of all complex-valued functions on G onto $\mathfrak{R}(R,R)$ which takes $\varphi\in\mathcal{F}(G)$ into the operator:

$$f\rightarrow f' \quad \text{where } f'(y)=\sum_{x\in G} f(x)\varphi(yx^{-1})$$
$$=\sum_{t\in G}\varphi(yt^{-1})f(t)$$

In this way $\mathcal{F}(G)$ itself becomes an algebra. Let us see what the multiplication law is. Since

$$\varphi=\sum_{x\in G}\varphi(x)\delta_x$$

where

$$\delta_x(y)=\begin{cases}0, & x\neq y\\ 1, & x=y\end{cases}$$

it will suffice to find out what δ_{z_1} multiplied by δ_{z_2} is. Now the operator $\mathfrak{R}(R,R)$ defined by δ_z takes f into f', where

$$f'(y)=\sum_{x\in G} f(x)\delta_z(yx^{-1})$$
$$=f(z^{-1}y)$$

Thus δ_{z_1} multiplied by δ_{z_2} is $\delta_{z_1z_2}$. In other words, the multiplication in $\mathcal{F}(G)$ that we obtain by identifying it with $\mathfrak{R}(R,R)$ is identical with the one we get by regarding φ as

$$\sum_{x\in G}\varphi(x)\delta_x$$

ISBN 0-8053-6702-0/0-8053-6703-9, pbk

and multiplying the δ_x by the group multiplication law. $\mathcal{F}(G)$ with this multiplication is called the *group algebra* of G. If we denote the multiplication by $*$, we see that

$$(\varphi_1 * \varphi_2)(x) = \sum_{y \in G} \varphi_1(xy^{-1})\varphi_2(y)$$

This operation is sometimes called *convolution*.

The result obtained above about the structure of commuting algebras, combined with the fact that $\mathcal{F}(G)$ is isomorphic to the commuting algebra of the right regular representation, tells us that $\mathcal{F}(G)$ is the direct sum of a finite number of full matrix algebras, one for each equivalence class of irreducible representations of G. The dimensions of these direct summands are of course the squares of the dimensions of the corresponding irreducible representations.

Consider the center of the group algebra $\mathcal{F}(G)$. Clearly it consists of all finite linear combinations of the identity elements in the constituent matrix algebras. Hence its dimension is the number of inequivalent irreducible representations of G. On the other hand, it is clear that $f \in \mathcal{F}(G)$ is in the center of the group algebra of G if and only if

$$f * \delta_x = \delta_x * f \quad \text{for all } x \in G$$

But this means that

$$f(xy) = f(xy)$$

or

$$f(xyx^{-1}) = f(y) \quad \text{for all } x, y \in G$$

In other words, f is in the center of $\mathcal{F}(G)$ if and only if f is a constant on the conjugate classes of G. (We remind the reader that y_1 and y_2 are in the same conjugate class if and only if there exists $x \in G$ with $xy_1x^{-1} = y_2$.) It follows that the dimension of the center of the group algebra must also be equal to the number of conjugate classes in G. Thus we arrive at the conclusion that the number of inequivalent irreducible representations of a finite group G is exactly equal to the number of conjugate classes. Of course, when G is commutative the conjugate classes all contain one element so that the number of irreducible representations is equal to the number of elements of G as we well know already.

Let us now look at the identity elements in the full matrix direct summands of $\mathcal{F}(G)$. These constitute a basis for the center of $\mathcal{F}(G)$ and equivalently for the "class functions" in G, that is, the functions that are constant on the conjugate classes. Let us call them $X_1, \ldots, X_r$. It follows

ISBN 0-8053-6702-0/0-8053-6703-9, pbk

immediately from the definitions that $X_i * X_j = 0$ if $i \neq j$ and $X_i * X_i = X_i$. Moreover, $X_1 + \cdots + X_r$ is the identity δ_e of the group algebra. When G is commutative the functions X_j are just the characters divided by the constant $o(G)$, and our definition of the P_χ may be written in terms of the correspondence $X = \chi / o(G)$ as follows:

$$P_\chi = \sum_{x \in G} \overline{X(x)}\, V_x$$

This definition makes sense in general and leads to the same results. Let V be an arbitrary (possibly infinite-dimensional) representation of the finite group and let $X_1, \ldots, X_r$ be the identities of the full matrix components of the group algebra. For each j let

$$P_{X_j} = \sum_{x \in G} \overline{X_j(x)}\, V_x$$

To establish the properties of the operators P_{X_j} note first that for any $f \in \mathcal{F}(G)$ we may set

$$V_f = \sum_{x \in G} f(x) V_x$$

and that

$$\begin{aligned} V_{f*g} &= \sum_{x,y \in G} f(xy^{-1}) g(y) V_x \\ &= \sum_{x,y \in G} f(x) g(y) V_x V_y \\ &= \left(\sum_{x \in G} f(x) V_x \right) \left(\sum_{y \in G} g(y) V_y \right) \\ &= V_f V_g \end{aligned}$$

Since $f \rightarrow V_f$ is obviously linear, it is a homomorphism of $\mathcal{F}(G)$ into the algebra of all linear transformations on $\mathcal{H}(V)$. Now P_{X_j} is just $V_{\overline{X}_j}$ and the identities

$$\begin{aligned} &X_i * X_j = 0, \quad i \neq j \\ &X_i * X_i = X_i \\ &X_1 + \cdots + X_r = \delta_e \end{aligned}$$

imply at once the corresponding identities for the $\overline{X}_j$. Thus we conclude

ISBN 0-8053-6702-0/0-8053-6703-9, pbk

that:

(a) $P_{X_j}^2 = P_{X_j}$,

(b) $P_{X_i}P_{X_j} = P_{X_j}P_{X_i} = 0$ for $i \neq j$,

(c) $P_{X_1} + \cdots + P_{X_r} = I$, the identity operator.

Since $\overline{X_j}$ is in the center of $\mathfrak{F}(G)$, we also deduce that $\delta_x \overline{X_j} = \overline{X_j}\delta_x$ and hence that

$$V_x P_{X_j} = P_{X_j} V_x \quad \text{for all } x \text{ and } j$$

Let $\mathcal{H}_{X_j}$ be the range of P_{X_j}. Just as in the commutative case we deduce that $\mathcal{H}(V)$ is the direct sum of the invariant subspaces $\mathcal{H}_{X_j}$ and let V^{X_j} denote the subrepresentation of V defined by $\mathcal{H}_{X_j}$.

In order to study the structure of the components V^{X_j}, it is useful to introduce two new definitions. For each f in $\mathfrak{F}(G)$, we let

$$\tilde{f}(x) = \overline{f(x^{-1})}$$

and for each f and g in $\mathfrak{F}(G)$ we let

$$(f,g) = \sum_{x \in G} f(x)\overline{g(x)}$$

Then

$$(f*g)^{\sim} = \tilde{g}*\tilde{f}$$

as a simple calculation shows and $\tilde{f}$ is in the center of the group ring whenever f is. Moreover,

$$\begin{aligned}(f*\tilde{f})(e) &= \sum_{y \in G} f(y^{-1})\tilde{f}(y) \\ &= \sum_{y \in G} f(y)\overline{f(y)} \\ &> 0 \quad \text{unless } f = 0\end{aligned}$$

Thus $f*\tilde{f}$ is never zero unless $f=0$. On the other hand, it is easy to see that the X_j may be characterized as the only idempotents in the center of $\mathfrak{F}(G)$ that are not sums of other idempotents. Since $f \to \tilde{f}$ is an isomorphism, we see each $\tilde{X}_j$ must coincide with some X_i. But if $\tilde{X}_j = X_i$ where $i \neq j$, then $\tilde{X}_j * X_j = 0$, and we have seen that this is impossible. It follows that $\tilde{X}_j = X_j$ for all j. We can also conclude that the X_j are orthogonal with respect to

ISBN 0-8053-6702-0/0-8053-6703-9, pbk

our inner product because

$$0=\sum_{y\in G} X_i(xy^{-1})X_j(y) \quad \text{for all } x\in G, i\neq j$$

Thus

$$0=\sum_{y\in G} X_i(y^{-1})X_j(y)=\sum_{y\in G} \overline{X_i(y)}\,X_j(y), \quad i\neq j$$

Returning to the V^{X_j} and the $\mathcal{H}_{X_j}$ let us consider the intertwining operator T_φ for φ in $\mathcal{H}_{X_j}$. It will intertwine R with V^{X_j}. We show now that it is zero on the subspace of $\mathcal{F}(G)$ consisting of all f with $f*X_j=0$. Indeed,

$$\begin{aligned} P_{X_j}T_\varphi(f) &= V_{\overline{X}_j}\left[\sum_{x\in G} f(x)V_{x^{-1}}(\varphi)\right] \\ &= V_{\overline{X}_j}V_{f^0}(\varphi) \\ &= V_{\overline{X}_j * f^0}(\varphi) \end{aligned}$$

where $f^0(x)=f(x^{-1})$. Now as $f*X_j=0$, so $X_j^0*f^0=0$ and $X_j^0=\overline{X}_j$. Hence

$$P_{X_j}T\varphi(f)=0$$

Now a complement of the subspace on which $f*X_j=0$ is the subspace on which $f*X_j=f$. Let us call this subspace $\mathcal{F}_j$. $\mathcal{F}(G)$ is the direct sum of the $\mathcal{F}_j$ and the $\mathcal{F}_j$ are the invariant subspaces of $\mathcal{F}(G)$ on which R reduces to a direct sum of replicas of the same irreducible. Thus T_φ restricted to $\mathcal{F}_j$ has the same range as T_φ and we conclude that the subrepresentation of V^{X_j} generated by φ is equivalent to a subrepresentation of the subrepresentation of R defined by $\mathcal{F}_j$. It follows that all irreducible subrepresentations of V^{X_j} are equivalent to each other and to all irreducible subrepresentations of the subrepresentation of R defined by $\mathcal{F}_j$. Hence no irreducible subrepresentation of V^{X_j} is equivalent to any irreducible subrepresentation of V^{X_k} if $k\neq j$. We conclude that

$$V=V^{X_1}\oplus\cdots\oplus V^{X_r}$$

is the unique decomposition of V into disjoint primary subrepresentations. Exactly as in the commutative case one can show that every invariant subspace of $\mathcal{H}(V)$ is a direct sum of invariant subspaces of the $\mathcal{H}_{X_j}$.

In the commutative case every subspace of $\mathcal{H}_{X_j}$ was invariant. This is false in the noncommutative case but it is still true that the set of all invariant subspaces of $\mathcal{H}_{X_j}$ has the same structure as the set of all subspaces. To see this, and for other purposes as well, it is useful to

ISBN 0-8053-6702-0/0-8053-6703-9, pbk

introduce the notion of tensor product. Let $\mathcal{H}_1$ and $\mathcal{H}_2$ be two vector spaces, and let $\mathcal{H}_1^*$ and $\mathcal{H}_2^*$ denote their algebraic duals. For each pair φ, ψ with $\varphi \in \mathcal{H}_1$ and $\psi \in \mathcal{H}_2$, let $\varphi \times \psi$ be the bilinear functional on $\mathcal{H}_1^* \times \mathcal{H}_2^*$ taking l_1, l_2 into $l_1(\varphi) l_2(\psi)$. Let $\mathcal{H}_1 \times \mathcal{H}_2$ denote the set of all bilinear functional of the form

$$c_1(\varphi_1 \times \psi_1) + c_2(\varphi_2 \times \psi_2) + \cdots + c_n(\varphi_n \times \psi_n)$$

$\mathcal{H}_1 \times \mathcal{H}_2$ is called the tensor product of $\mathcal{H}_1$ and $\mathcal{H}_2$. Let L be an irreducible representation of G and let $\mathcal{H}$ be an arbitrary vector space. For each $x \in G$ and each $B \in \mathcal{H}(L) \times \mathcal{H}$, let $M_x(B)$ be the bilinear form

$$M_x(B)(l_1, l_2) = B(L_{x^{-1}}^*(l_1), l_2)$$

so that

$$M_x(\varphi \times \psi) = L_x(\varphi) \times \psi$$

Then M is a representation of G whose space is $\mathcal{H}(L) \times \mathcal{H}$. For each subspace $\mathcal{H}'$ of $\mathcal{H}$, we obtain an invariant subspace of $\mathcal{H}(L) \times \mathcal{H}$ by taking all finite sums $\Sigma_r c_r(\varphi_r \times \psi_r)$ where each $\psi_j \in \mathcal{H}'$, and is not hard to show that every invariant subspace is of this form. In particular, if $\{\psi_\alpha\}$ is a Hamel basis for $\mathcal{H}$ and $\mathcal{H}_\alpha$ is the set of all $\varphi \times \psi_\alpha$ where $\varphi \in \mathcal{H}(L)$, then $\mathcal{H}_\alpha$ is an irreducible invariant subspace and the corresponding representation is equivalent to L. Moreover, it is easy to see that each $\theta \in \mathcal{H}(L) \times \mathcal{H}$ may be written uniquely in the form

$$\theta = \theta_{\alpha_1} + \cdots + \theta_{\alpha_n}$$

where $\theta_{\alpha_j} \in \mathcal{H}_{\alpha_j}$. Thus in an obvious sense M is a "restricted" direct sum of as many replicas of L as $\{\psi_\alpha\}$ has elements. Here "restricted" means that each element of the direct sum has only a finite number of nonzero components. Conversely, such a restricted direct sum can always be realized in the form $x \rightarrow U_x$, where $\mathcal{H}(U) = \mathcal{H}(L) \times \mathcal{H}$, $U_x(\varphi \times \psi) = L_x(\varphi) \times \psi$, and the dimension of $\mathcal{H}$ is the number of summands. Finally, a simple transfinite induction just like that used in showing the existence of a Hamel basis tells us that each V^{X_j} is a restricted direct sum of equivalent irreducibles. Hence V^{X_j} may be written in the form

$$V_x^{X_j}(\varphi \times \psi) = L_x^j(\varphi) \times \psi$$

and the invariant subspaces correspond one-to-one with the set of all subspaces of $\mathcal{H}$.

Given any representation of any group the set of all invariant subspaces is a "partially ordered set" in a natural way. We remind the reader that a

ISBN 0-8053-6702-0/0-8053-6703-9, pbk

partially ordered set is a set $\mathcal{S}$ together with a relation $a \leqslant b$ such that the following axioms are satisfied:

(1) $a \leqslant a$ for all $a \in \mathcal{S}$;

(2) $a \leqslant b$ and $b \leqslant c$ implies $a \leqslant c$;

(3) $a \leqslant b$ and $b \leqslant a$ implies $a = b$.

If $a \leqslant b$ means a is a subset of b, then the following are examples of partially ordered sets:

(a) The set of all subsets of a set A.

(b) The set of all subspaces of a vector space V.

(c) The set of all invariant subspaces of the space $\mathcal{H}(V)$ of some group representation V.

If $\mathcal{S}_1$ and $\mathcal{S}_2$ are two partially ordered sets, we define a new partially ordered set called the direct sum $\mathcal{S}_1 \oplus \mathcal{S}_2$ by letting the elements of $\mathcal{S}_1 + \mathcal{S}_2$ be the pairs (x,y) such that $x \in \mathcal{S}_1$ and $y \in \mathcal{S}_2$, and setting

$$(x_1,y_1) \leqslant (x_2,y_2) \quad \text{when } x_1 \leqslant x_2 \quad \text{and } y_1 \leqslant y_2$$

Let $\mathcal{L}(V)$ denote the partially ordered set of all invariant subspaces of $\mathcal{H}(V)$, where V is a group representation. It is not in general true that:

$$\mathcal{L}(V_1 \oplus V_2) \simeq \mathcal{L}(V_1) \oplus \mathcal{L}(V_2)$$

However, this is so when every invariant subspace of $V_1 \oplus V_2$ is the direct sum of an invariant subspace of $\mathcal{H}(V_1)$ and an invariant subspace of $\mathcal{H}(V_2)$. Now let V be an arbitrary representation of the finite group G and let $V^{X_1}, \ldots, V^{X_r}$ be the subrepresentations defined above. It follows from what has been said that

$$\mathcal{L}(V) = \mathcal{L}(V^{X_1}) \oplus \cdots \oplus \mathcal{L}(V^{X_r})$$

and that each $\mathcal{L}(V^{X_j})$ is isomorphic to the partially ordered set of all subspaces of some vector space. Clearly no $\mathcal{L}(V^{X_j})$ can be further decomposed as a direct sum. We thus have a quite complete analysis of the invariant subspace structure of any representation of a finite group. The irreducible constituents correspond one-to-one to the irreducible representations of the group just as do the simple constituents of the commuting algebra.

Another corollary of our analysis is that to obtain the most general representation of a finite group G it suffices to determine the (finitely-many) equivalence classes of irreducible representations of G. This latter

ISBN 0-8053-6702-0/0-8053-6703-9, pbk

problem is trivial when G is commutative. However, when G is noncommutative, it is nontrivial and is in fact one of the main problems of the theory. The problem is trivial when G is commutative because every finite commutative group is a direct product of cyclic groups—it is easy to determine the irreducible representations of cyclic groups— and there is a simple rule for finding the irreducible representations of a direct product of two commutative groups when one knows the irreducible representations of the factors. There is no such simple structure theorem for finite noncommutative groups. The groups that cannot be written as a direct product of smaller groups constitute an immense unsurveyable class. However, it is useful to know even in the noncommutative case that there is still a simple rule for finding the irreducible representations for $G_1 \times G_2$ when one knows them for $G_1 \times G_2$. Let L and M be representations of G_1 and G_2, respectively. We define a new representation $L \times M$ called the tensor (or Kronecker) product of L and M whose space is $\mathcal{H}(L) \times \mathcal{H}(M)$ by setting

$$\{(L \times M)_{x_1,x_2}(B)\}(l_1,l_2) = B\big(L^*_{x_1}(l_1), M^*_{x_2}(l_2)\big)$$

This means in particular that

$$(L \times M)_{x_1,x_2}(\varphi \times \psi) = L_{x_1}(\varphi) \times M_{x_2}(\psi)$$

It is not hard to show that $L \times M$ is irreducible whenever L and M are and that every irreducible representation of $G_1 \times G_2$ is equivalent to a representation of the form $L \times M$. To prove the latter one first shows that the representation V of $G_1 \times G_2$ restricted to $G_1 \times e$ is a direct sum of replicas of the same irreducible and observes that each $V_{e,y}$ is in the commuting ring of $x \to V_{x,e}$. We leave the details to the reader as an exercise.

We have observed that the minimal idempotents $X_1, \ldots, X_r$ in the center of the group algebra of G reduce in the commutative case to the characters divided by $o(G)$ and that the functions $o(G)X_j$ have other properties in common with the characters of commutative groups. We shall now generalize the notion of character and relate the $o(G)X_j$ to the generalized or noncommutative characters. For any finite-dimensional representation V of a group G let $\chi^V(x)$ be the complex-valued function

$$x \to \operatorname{Tr}(V_x)$$

Since

$$\operatorname{Tr}(TAT^{-1}) = \operatorname{Tr} A$$

it follows at once that equivalent representations have identical characters. Also, if χ is any character in the sense of Section 2 and $x \to \chi(x)I$ is the

ISBN 0-8053-6702-0/0-8053-6703-9, pbk

corresponding one-dimensional representation, then χ is just the character of the representation. In other words, the characters in the sense of Section 2 are just the characters of one-dimensional representations. We shall refer to them from now on as *one-dimensional characters* and use the unqualified term character to mean the new more general notion we have just defined. It is obvious that

$$\chi^{V_1+V_2}(x)=\chi^{V_1}(x)+\chi^{V_2}(x)$$

and hence that every character is a linear combination with positive integral coefficients of characters of irreducible representations—or so-called *irreducible characters*. Notice that

$$\mathrm{Tr}(V_{xyx^{-1}})=\mathrm{Tr}(V_x V_y V_{x^{-1}})$$

Hence for every character χ

$$\chi(xyx^{-1})=\chi(y)$$

That is, every character is a constant on the conjugate classes. Now let $\mathfrak{F}_0$ be any irreducible invariant subspace of the regular representation and let $f_1,\ldots,f_r$ be any orthonormal basis for $\mathfrak{F}_0$. Let χ be the character for the corresponding irreducible representation of G. Then

$$\chi(x)=\sum_{i=1}^{r}(R_x f_i,f_i)$$

But

$$(R_x f_i,f_i)=\sum_{y\in G} f_i(yx)\overline{f_i}(y)$$

Hence χ is a linear combination of left translations of elements of $\mathfrak{F}_0$. But $\mathfrak{F}_0$ lies in one of the $\mathfrak{F}_j$, where $\mathfrak{F}_j$ is the set of all f with $f*X_j=f$ for some j. Thus χ is in $\mathfrak{F}_j$. Since χ is in the center of $\mathfrak{F}(G)$, χ must be a constant multiple of X_j. In other words, the irreducible characters are just the functions c_jX_j, where the c_j are constants to be determined. We have already seen that the X_j are orthogonal and it follows that the χ_j also are. If L^j is an irreducible representation whose character is χ_j and

$$V=n_1L_1\oplus n_2L_2\oplus\cdots\oplus n_rL_r$$

where

$$nW=W\oplus\cdots\oplus W \quad n \text{ times}$$

ISBN 0-8053-6702-0/0-8053-6703-9, pbk

then

$$\chi^V(x) = n_1\chi_1(x) + \cdots + n_r\chi_r(x)$$

and the n_j are uniquely determined by χ^V because of the orthogonality. Thus V is determined to within equivalence by its character, the function χ^V. To determine n_j we compute

$$n_j = \frac{(\chi^V, \chi_j)}{(\chi_j, \chi_j)}$$

We shall show below that

$$(\chi_j, \chi_j) = o(G)$$

and this will enable us to determine the c_j.

One verifies at once from the definitions that the trace of the tensor product of two operators is the product of their traces and hence that

$$\chi^{L\times M}(x,y) = \chi^L(x)\chi^M(y)$$

We remark in passing that this implies that we obtain the irreducible characters of the direct product of two groups from those of the groups themselves in exactly the same way whether the group is commutative or not. Now let L and M be two irreducible representations of the same group G. Then $L \times M$ is an irreducible representation of $G \times G$ whose character is

$$(x,y) \to \chi^L(x)\chi^M(y)$$

Restricting x,y to be in the diagonal subgroup $\tilde{G}$ consisting of all x,y with $x = y$, we obtain a representation of $\tilde{G}$ and hence a representation of G. This new representation of G need no longer be irreducible and usually will not be. It is called the (inner) Kronecker product $L \times M$ of L and M. Its character will be the function

$$x \to \chi^L(x)\chi^M(x)$$

that is the product of the functions χ^L and χ^M. It follows that the product of two irreducible characters is again a character and hence a linear combination of irreducible characters with positive integral coefficients. Unlike the commutative case the product of two irreducible characters need not be irreducible and the irreducible characters need not form a group.

ISBN 0-8053-6702-0/0-8053-6703-9, pbk

On the other hand, for each finite group G we do have a problem analogous to that of determining the structure of the character group in the commutative case. This is the problem of determining the nonnegative integers n_{ijk} in the formula

$$\chi_i \chi_j = \sum_k n_{ijk} \chi_k$$

or, equivalently, the problem of determining the reduction into irreducibles of each inner Kronecker product $L \otimes M$. Let us now make one important general observation about this reduction problem; namely, that $L \otimes M$ contains the trivial representation at most once and contains it once if and only if $\overline{\chi^L} = \chi^M$. In this connection let us note that if V is any finite-dimensional representation of G, then

$$x \to V^*_{x^{-1}}$$

is also a representation whose character is the complex conjugate of that of V. We denote this representation by $\overline{V}$ and call it the *adjoint* of V. Now recall that the space of $L \otimes M$ consists of all bilinear functional on $\mathcal{H}(L)^* \times \mathcal{H}(M)^*$ and that the most general bilinear functional may be put into the form

$$B(l_1, l_2) = l_2(A(l_1))$$

where A is a linear transformation from $\mathcal{H}(L)^*$ to $\mathcal{H}(M)$. Now it follows at once from the definitions that A will be an intertwining operator for $\overline{L}$ and M if and only if B is left fixed by all $(L \otimes M)_x$. Thus the number of times that $L \otimes M$ contains the identity is equal to $i(\overline{L}, M)$. It follows that when both are irreducible we have $i(\overline{L}, M) = 0$ or 1 according as $\overline{\chi^L} = \chi^M$. It follows in particular that $L \otimes \overline{L}$ contains the identity just once whenever L is irreducible. Thus for each irreducible character χ_j we have

$$\chi_j \overline{\chi} = 1 + \sum n_k \chi_k$$

where the χ_k range over the irreducible characters different from 1. Thus, as announced previously,

$$\sum_{x \in G} |\chi_j(x)|^2 = o(G)$$

Using this fact we can determine the c_j. Indeed, since $X_j * X_j = X_j$, it follows that

$$\sum_{y \in G} |X_j(y)|^2 = X_j(e)$$

ISBN 0-8053-6702-0/0-8053-6703-9, pbk

Hence

$$\sum_{y \in G} \frac{|\chi_j(y)|^2}{c_j^2} = \frac{\chi_j(e)}{c_j}$$

Hence,

$$\frac{o(G)}{c_j^2} = \frac{d_j}{c_j}$$

and

$$c_j = \frac{o(G)}{d_j}$$

where d_j is the dimension of the representation whose character is χ_j. Thus,

$$\chi_j(x) = \frac{o(G)X_j(x)}{d_j}$$

The fact that the inner Kronecker product of two irreducible representations need not be irreducible is a special case of a much more general fact. If L is any irreducible representation of G and H is a subgroup of G, then the restriction of L to H need not be irreducible and the problem arises of finding its irreducible constituents and their multiplicities. Given a group G one is far from finished when one knows all the equivalence classes of irreducible representations of G. One wants, in addition, to know how to decompose all inner Kronecker products and how the restrictions to the various subgroups decompose. These questions play an important part in many applications of the theory—in particular, in applications to quantum mechanics.

Notes and References

The big step from one-dimensional characters of finite commutative groups to higher dimensional characters of noncommutative finite groups was taken by G. Frobenius in 1896. Frobenius was led to his discovery by a problem posed by Dedekind. The exact story is complicated but interesting and will be found in a series of three articles by Thomas Hawkins published between 1971 and 1974 in *Archive for History of Exact Sciences*. Frobenius's theory was extensively developed in the next quarter century by many mathematicians, especially himself, I. Schur, and W. Burnside. A comprehensive modern work on the subject that has become a standard reference is the book of C. W. Curtis and I. Reiner published in 1962 and entitled *Representation Theory of Finite Groups and Associative Algebras*.

ISBN 0-8053-6702-0/0-8053-6703-9, pbk

4. Preliminaries concerning infinite groups and measure spaces

In extending the theory of group representations from finite to infinite groups, one needs an appropriate substitute for the process of summing a function over the elements of the group. For countable discrete groups such as the additive group of all the integers, one has only to replace a finite sum by an infinite series. However, in dealing with the additive group of the real line the functions f for which $\Sigma_{x\in G}f(x)$ exist are surely not very interesting. Fortunately, there is a general notion applicable to a wide class of groups that reduces to summation over the group when the group is discrete and to taking $\int_{-\infty}^{\infty} f(x)dx$ when the group is the additive group of the real line. The class of groups for which the notion exists is the class of locally compact topological groups. We remind the reader that a topological group is a group that is at the same time a T_1 topological space in such a fashion that $x,y \rightarrow xy^{-1}$ is a continuous function. A topological group is locally compact if it contains an open set whose closure is compact. For example, the additive group of the real line is locally compact and so is any finite direct product of locally compact groups. Thus the additive group of a finite-dimensional real or complex vector space is locally compact. An example of a noncommutative locally compact group is the group of all $n\times n$ nonsingular matrices with real (or complex) elements. Any closed subgroup of a locally compact group is also locally compact. In particular, the group of rotations in three-space and the Lorentz group are locally compact—the rotation group is even compact. All of these examples have the additional property of being *separable* in the sense that there exists a countable basis for the open sets. In any topological space, one defines the Borel sets to be the members of the smallest family of sets containing all of the open sets and closed under the

George W. Mackey, Unitary Group Representations in Physics, Probability, and Number Theory

ISBN 0-8053-6702-0/0-8053-6703-9, pbk

following three operations: (1) taking complements, (2) taking countable unions, and (3) taking countable intersections. Of course (3) is implied by (1) and (2). More generally, one defines a *Borel space* to be a set S together with any distinguished family of subsets closed under these three operations. By a *measure* on a Borel space S we shall always mean a function μ, defined on all Borel sets, taking values in the interval $[0,\infty]$, which is:

(i) completely additive in the sense that

$$\mu(E_1 \cup E_2 \cup E_3 \cup \cdots) = \mu(E_1) + \mu(E_2) + \cdots$$

whenever the E_j are mutually disjoint and

(ii) is σ-finite in the sense that we may find Borel sets $E_1, E_2, \cdots$ such that $S = E_1 \cup E_2 \cup \cdots$ and $\mu(E_j) < \infty$ for $j = 1, 2, \cdots$.

A right-invariant Haar measure in a topological group is a measure $\mu \neq 0$ with the property that $\mu(Ex) = \mu(E)$ for all Borel sets E and all group elements x. One defines a left-invariant Haar measure similarly. The importance of local compactness in the theory of group representations stems from the fact that every separable locally compact group admits a right- (left-) invariant Haar measure and this measure is as unique as it could possibly be. If μ_1 and μ_2 are both right- (left-) invariant Haar measures for the same group, then there exists a positive real number λ such that $\mu_1(E) \equiv \lambda \mu_2(E)$. Unfortunately, a left-invariant Haar measure need not be right-invariant. If it is, the group is said to be unimodular. One can extend the theorem on the existence and uniqueness of Haar measure to nonseparable locally compact groups provided that certain refinements in the definition making a distinction between Borel and Baire sets are made. We shall consider only separable groups and will not concern ourselves with these refinements.

It is easy to see that Haar measure in any countable discrete group is such that $\mu(E)$ is a fixed constant times the number of elements in E. Integration of a function on the group with respect to μ is then simply the sum of all the values of the function multiplied by this same constant. Similarly, Haar measure on the real line is such that $\mu([a,b]) = \lambda(b-a)$, where λ is a fixed constant, and integration with respect to this measure is just λ times integration with respect to Lebesgue measure. The question naturally arises as to whether or not there exist groups that are not locally compact and that admit a Haar measure. We shall show that barring certain measure-theoretic pathology the answer is no. To give this answer we shall need certain measure-theoretic notions which we shall also need below. We proceed to develop these.

We define a function f from the Borel space S_1 to the Borel space S_2 to be a *Borel function* if $f^{-1}(E)$ is a Borel subset of S_1 whenever E is a Borel subset of S_2. We observe that when S_2 is the Borel space associated with a topological space, then f is a Borel function if and only if $f^{-1}(\mathcal{O})$ is a Borel

ISBN 0-8053-6702-0/0-8053-6703-9, pbk

set for every open subset $\mathcal{O}$ of S_2. By an *isomorphism* of the Borel space S_1 on the Borel space S_2 we shall mean a one-to-one Borel function f whose domain is S_1 and whose range is S_2 such that f^{-1} is also a Borel function. Equivalently, an isomorphism is a one-to-one map f from S_1 to S_2 such that E is a Borel subset of S_1 if and only if $f(E)$ is a Borel subset of S_2. Note the close analogy between Borel spaces and topological spaces. In each case the notion is defined as a set together with a distinguished family of subsets and the axioms in the two cases differ only slightly: Countable unions replace arbitrary unions, countable intersections replace finite intersections, and complementation is added. In this analogy a Borel function is the analog of a continuous one and an isomorphism is the analog of a homeomorphism. Pursuing the analogy further, one defines subspaces of Borel spaces, product spaces, and so forth. The definitions are the obvious ones and we leave their exact formulation to the reader.

There is one important respect, however, in which the theory of Borel spaces is vastly different from the theory of topological spaces. There are many, many fewer isomorphism classes in the former. One knows that Euclidean n-space is not homeomorphic to Euclidean m-space if $m \neq n$. Moreover, one knows that the closed subsets of Euclidean n-space for any n constitute a tremendous variety of nonhomeomorphic topological spaces. In contrast to this one has the following rather surprising theorem about Borel spaces: Let M_1 and M_2 be arbitrary complete separable metric spaces. Let E_1 and E_2 be Borel subsets of M_1 and M_2, respectively, and give E_i the Borel structure it inherits from M_i. Then E_1 and E_2 are isomorphic as Borel spaces whenever their cardinal numbers are the same. Moreover, their cardinal numbers will be the same whenever these cardinal numbers exceed $\aleph_o$. This means in particular that every nondiscrete separable locally compact group is isomorphic as a Borel space to the unit interval on the real line. Guided by these considerations, we define a Borel space to be *standard* if it is isomorphic as a Borel space to a Borel subset of a separable complete metric space. It is clear from the above that a large share of the Borel spaces which arise naturally in analysis will be standard. However, there is a slightly wider class of Borel spaces that has to be considered. The range E of a Borel function from a standard Borel space to a complete separable metric space M need not be a Borel set. The Borel spaces defined by such subsets and their isomorphic images are called *analytic* Borel spaces. For many purposes analytic Borel spaces are as good as standard ones because of the truth of the following theorem: Every analytic Borel space S is *metrically standard* in the sense that given any measure μ on S there exists a Borel subset N with $\mu(N)=0$ such that $S-N$ is standard.

Let μ be a measure in the Borel space S and let f be a nonnegative real-valued Borel function. We obtain a new measure μ_f by setting $\mu_f(E)=$

ISBN 0-8053-6702-0/0-8053-6703-9, pbk

$\int_E f(x)\,d\mu(x)$, and this measure clearly has the property that $\mu(E)=0$ implies $\mu_f(E)=0$. Conversely, the Radon–Nikodym theorem asserts that any measure ν such that $\mu(E)=0$ implies $\nu(E)=0$ is of the form μ_f for some f. We sometimes call f the density of ν with respect to μ and sometimes call it the Radon–Nikodym derivative and write it as $d\nu/d\mu$. When $d\nu/d\mu$ is never zero, then $\nu(E)=0$ implies $\mu(E)=0$ as well and also that $d\mu/d\nu=1/(d\nu/d\mu)$. Two measures μ and ν having the same sets of measure zero, so that both $d\mu/d\nu$ and $d\nu/d\mu$ exist, are said to belong to the same class. We shall usually use the symbol C_μ to denote the measure class to which μ belongs. We note that every measure class contains a member ν such that $\nu(S)<\infty$.

We are now almost ready to state the theorem about the necessary local compactness of groups with invariant measures. Note first that a separable locally compact group G is not only a standard Borel space but is also a *Borel group* in the sense that the function $x,y\rightarrow xy^{-1}$ is a Borel function from $G\times G$ to G. Note also that any two right-invariant Haar measures are in the same class and so are any two left-invariant Haar measures. It is not hard to show that the left and right Haar measures are in the same class, that this measure class is both right- and left-invariant, and is the only measure class in G that is either right- or left-invariant. Now let G be any analytic Borel group and let C be a nontrivial measure class in G that is either left- or right-invariant. The theorem states that C is both left- and right-invariant and that there is a unique separable locally compact topology in G with the following properties:

(1) G is a topological group with respect to this topology;

(2) the given Borel structure is that associated with the topology;

(3) C is the Haar measure class for G as a locally compact group.

This theorem tells us that the separable locally compact groups are just the analytic Borel groups that admit an invariant measure class and therefore are the natural ones on which to build a theory that makes use of integration over the group.

In Section 1 we defined the notion of G space and showed how each G space was associated with a representation of G. In the sequel we shall be interested only in the case in which G is a separable locally compact group and our space is a bundle whose base is a Borel space equipped with an invariant measure class. In this section we leave bundles aside and discuss only Borel G spaces. Let G be a separable locally compact group and let S be a G space. If S is also a Borel space, we shall say that S is a *Borel G space* if the mapping $s,x\rightarrow sx$ is a Borel function. Let μ be a measure in the Borel G space S. If $\mu(Ex)=\mu(E)$ for all x in G and all Borel subsets E of G, we shall say that μ is an *invariant* measure. If the class C_μ of μ is

ISBN 0-8053-6702-0/0-8053-6703-9, pbk

invariant in the sense that for all $x \in G, E \rightarrow \mu(Ex)$ is in the class whenever μ is, we shall say that μ is quasi-invariant. An invariant measure class may or may not have invariant members.

Now let G be separable and locally compact as above and let μ be an invariant measure in the analytic Borel G space S. Let $\mathcal{L}^2(S,\mu)$ denote the space of all complex-valued Borel functions f on S such that $\int |f(x)|^2 \, d\mu(x) < \infty$. If we identify functions that are almost everywhere equal, this becomes a Hilbert space with scalar product $(f,g) = \int f(x)\overline{g(x)}\, d\mu(x)$, and this Hilbert space is always separable in the sense that it admits a countable dense subset or equivalently a countable orthonormal basis. We remind the reader that a (complex) Hilbert space $\mathcal{H}$ is a vector space over the complex numbers together with a complex-valued function $f,g \rightarrow (f \cdot g)$ on $\mathcal{H} \times \mathcal{H}$ having the following properties:

(i) $(f \cdot g) = \overline{(g \cdot f)}$ for all f and g on $\mathcal{H}$;

(ii) for each fixed $g, f \rightarrow (f \cdot g)$ is linear;

(iii) $(f \cdot f) > 0$ if $f \neq 0$;

(iv) $\mathcal{H}$ is complete under the metric $f,g \rightarrow \|f - g\|$, where $\|f\| = \sqrt{(f,f)}$.

For each $x \in G$ let U_x be the linear transformation that maps f in $\mathcal{L}^2(S,\mu)$ into the right translate $y \rightarrow f(yx)$. It is clear that $x \rightarrow U_x$ is a representation of G and (because of the invariance of the measure μ) that each U_x is a *unitary* operator in the sense of being one-to-one, linear, onto, and norm preserving. It is easy to see that for each f and g in $\mathcal{L}^2(S,\mu)$, the function $x \rightarrow (U_x(f), g)$ is a Borel function. In the rest of these lectures we will be concerned mainly with representations of separable locally compact groups by unitary operators in separable Hilbert spaces. We shall speak of these as unitary representations and it will always be understood that the Hilbert space is separable and that the functions $x \rightarrow (U_x(f), g)$ are Borel functions. It is a surprising and useful fact that this weak "measurability" condition on $x \rightarrow U_x$ implies the following strong continuity condition: For each f in $\mathcal{H}(U)$ the function $x \rightarrow U_x(f)$ is a continuous function from the group to the Hilbert space.

The notions of equivalence, irreducibility, and so forth described in Section 3 for representations of finite groups in abstract vector spaces need to be modified slightly for unitary representations of locally compact groups. In defining equivalence we demand that the operator T which sets up the equivalence be *unitary*, and we demand that an intertwining operator be continuous (or, equivalently, *bounded* in the sense that $\|T(\varphi)\| / \|\varphi\|$ is bounded as φ varies over the nonzero elements in the Hilbert space). We also consider only closed invariant subspaces and say that a unitary representation is irreducible only if it admits no proper closed invariant subspaces. Given a closed invariant subspace $\mathcal{H}_1$, let $\mathcal{H}_1^{\perp}$

ISBN 0-8053-6702-0/0-8053-6703-9, pbk

be the set of all φ in $\mathcal{H}(U)$ with $(\varphi,\theta)=0$ for all θ in $\mathcal{H}_1$. Then $\mathcal{H}_1^\perp$ is also invariant and closed and U is the direct sum of the corresponding subrepresentations. In this case, then, every quotient representation has a canonical realization as a subrepresentation. The range of an intertwining operator need not be closed, but its closure is also an invariant subspace and Schur's lemma takes the following form. If T is an intertwining operator for U and V, then T defines a one-to-one intertwining operator from the subrepresentation of U defined by the orthogonal complement of the null space of T to the subrepresentation of V defined by the closure of the range of T. Moreover, these two subrepresentations are actually equivalent. It follows, in particular, that if U and V are equivalent in the weak sense of admitting a one-to-one intertwining operator with a dense range, then U and V are actually equivalent in the strong sense (i.e., are unitarily equivalent). As in Section 3, U and V are said to be *disjoint* if $R(U,V)=0$ or equivalently if no subrepresentation of U is equivalent to any subrepresentation of V. Let $U^1,U^2,U^3,\ldots$ be a countable sequence of unitary representations of G. We define the infinite direct sum $U=U^1\oplus U^2\oplus\cdots$ as follows: $\mathcal{H}(U)$ is the set of all sequences. $\varphi_1,\varphi_2,\ldots$ with $\varphi_j\in\mathcal{H}(U^j)$ such that $\|\varphi_1\|^2+\|\varphi_2\|^2+\cdots<\infty$, the product and the vector space operations being defined in the obvious way. For each $x\in G$ and $\varphi_1,\varphi_2,\ldots\in\mathcal{H}(U)$, we set $U_x(\varphi_1,\varphi_2,\ldots)=(U_x(\varphi_1),U_x(\varphi_2),\ldots$.

Any separable locally compact group $G\neq(e)$ has at least one unitary representation besides the trivial one-dimensional one that takes each x into the identity. This is the so-called *right regular representation* obtained from the general construction defined above by taking $S=G$ and μ to be Haar measure. Of course G acts on $S=G$ by right multiplication.

The construction of a unitary representation of G from an invariant measure μ in a G-space S may be generalized in two ways. We may replace μ by a quasi-invariant measure and compensate for the noninvariance by the appropriate use of Radon–Nikodym derivatives. Also, we may adopt the vector bundle notion of Section 1 to a Hilbert space, Borel space context. We shall give details in later lectures and see that this generalized construction comes near to being universal.

ISBN 0-8053-6702-0/0-8053-6703-9, pbk

5. Compact groups and the Peter–Weyl theorem

Let the separable locally compact group G be compact. Then its left and right Haar measures are the same and every Haar measure μ is such that $\mu(G)<\infty$. Conversely, for any noncompact locally compact group G, $\mu(G)=\infty$ for every left- or right-invariant Haar measure μ. The compact groups are thus "more finite than others" in both a topological and a measure theoretic sense. It turns out that the theory of the unitary representations of compact groups differs from the theory for finite groups in only minor points. The main result is the celebrated Peter–Weyl theorem, which may be formulated as follows: The right (left) regular representation of any (separable) compact group may be decomposed as a direct sum of irreducible finite-dimensional representations. Each irreducible representation occurs with a finite multiplicity equal to the dimension of the representation and every irreducible representation of the group is equivalent to a direct summand of the regular representation. It follows from the separability of $\mathcal{L}^2(G,\mu)$ that there can be only countably many inequivalent irreducible representations of G. We define the character χ_L of each representation L just as in Section 3 and let $\hat{G}$ denote the countable set of all irreducible characters. Using $\hat{G}$ we can analyze the general unitary representation U of G just as we analyzed the general representation of a finite group in Section 3. Choose the Haar measure μ so that $\mu(G)=1$, and for each $\chi\in\hat{G}$ let $P_\chi=d_\chi\int\bar{\chi}(x)U_x\,d\mu(x)$, where d_χ is the dimension of the irreducible representation whose character is χ. This integral exists because of the boundedness of χ and U and the finiteness of $\mu(G)$. Let $\mathcal{H}_\chi$ be the range of P_χ. It is not hard to prove that the $\mathcal{H}_\chi$ are mutually orthogonal invariant subspaces of $\mathcal{H}(U)$ whose closed linear span is $\mathcal{H}(U)$, and that the subrepresentation defined by $\mathcal{H}_\chi$ is a direct sum (in many ways) of irreducible representations all mutually equivalent and having character χ. The decomposition of $\mathcal{H}(U)$ effected by the $\mathcal{H}_\chi$ is the unique decomposition into disjoint primaries.

George W. Mackey, Unitary Group Representations in Physics, Probability, and Number Theory

ISBN 0-8053-6702-0/0-8053-6703-9, pbk

6. *Locally compact commutative groups*

When we abandon the last vestiges of finiteness by considering groups having no *finite* invariant measure, the decomposition theory of group representations becomes distinctly more difficult. Indeed, when specialized to the case in which the group is the additive group of the real line, it becomes equivalent to the union of the spectral theorem and the Hahn–Hellinger unitary equivalence theory for self-adjoint operators. Among the difficulties that arise are the following:

(1) There exist reducible unitary representations having no irreducible subrepresentations. In decomposing such unitary representations one is forced to use integrals instead of discrete sums.

(2) There exist infinite-dimensional irreducible unitary representations. Indeed, there are many interesting groups where the only finite-dimensional irreducible unitary representation is the trivial one. One cannot form $\mathrm{Tr}(U_x)$ when U_x is infinite-dimensional and unitary, and characters must be defined in a round about way when they can be defined at all.

(3) There exist unitary representations that are primary in the sense that they cannot be written as the direct sum of two disjoint parts, and yet are not direct sums or even direct integrals of replicas of the same irreducible. The study of these "nontype I" primaries is more or less equivalent to the Von Neumann–Murray theory of dimension functions in operator rings.

In the case of commutative groups only the first difficulty arises and it is useful to study this case separately. We begin by examining more closely the representation theory of compact commutative groups, which we obtain by specializing the results of Section 5.

ISBN 0-8053-6702-0/0-8053-6703-9, pbk

George W. Mackey, Unitary Group Representations in Physics, Probability, and Number Theory

As in the finite case any locally compact commutative group has only one-dimensional irreducible unitary representations and these are all of the form $x \to \chi(x)I$, where χ is a complex-valued function with the following properties:

(i) χ is continuous;

(ii) $|\chi(x)| \equiv 1$;

(iii) $\chi(xy) = \chi(x)\chi(y)$.

Conversely, any complex-valued function with these properties defines an irreducible one-dimensional unitary representation. As in the finite case, we call such functions *characters* and note that they form a group under multiplication. Again, as in the finite case, this group is called the character group and is denoted by the symbol $\hat{G}$. Unlike the finite case, G need not be isomorphic to $\hat{G}$. In fact, it follows from the Peter–Weyl theorem that $\hat{G}$ is countable whenever G is compact and separable. The infinite case differs from the finite case also in that there is no simple rule for determining $\hat{G}$ for a given G. On the other hand, we can get quite a bit of insight into the possibilities that exist for G and $\hat{G}$ by way of the Pontryagin duality theory. Let G be any separable compact commutative group. It folllows from the Peter–Weyl theorem that the characters χ of G exist in sufficient numbers to "separate the points of G"; that is, if $\chi(x) = \chi(y)$ for all χ, then $x = y$. On the other hand, note that for each fixed x the function $\chi \to \chi(x)$ is a character of $\hat{G}$, that is, a member of $\hat{\hat{G}}$. Since $\hat{G}$ is countable, we can give it the discrete topology in which every subset is open, and it will be separable and locally compact. With this topology every function is continuous. Hence $\chi \to \chi(x)$ satisfies (i) as well as (ii) and (iii). If we set $f_x(\chi) = \chi(x)$, we see that $x \to f_x$ is an isomorphism of G into $\hat{\hat{G}}$. Now, given any countable commutative group D, we may make it separable and locally compact by giving it the discrete topology, and we can form its character group D. It is not hard to show that the members of $\hat{D}$ separate the points of D and that $\hat{D}$ becomes a separable compact group if we give it the weakest topology with respect to which $\chi \to \chi(d)$ is continuous for all d in D. Giving $\hat{\hat{G}}$ its topology as the character group of the discrete group $\hat{G}$, the mapping $x \to f_x$ becomes a one-to-one homomorphic mapping of the compact group G into the compact group $\hat{\hat{G}}$. As shown by Pontryagin, this mapping is continuous, onto, and has a continuous inverse. Thus we may identify G with $\hat{\hat{G}}$ and recognize that the relationship between G and $\hat{G}$ is a reciprocal one. Each is the character group of the other. For this reason one often speaks of G and $\hat{G}$ as dual groups and sometimes refers to $\hat{G}$ as the dual group of G rather than as the character group of G. Note in particular that one obtains the most general separable compact commutative group by taking the most general countable commutative group D giving it the discrete topology and forming $\hat{D}$.

ISBN 0-8053-6702-0/0-8053-6703-9, pbk

Thus classifying the separable compact commutative groups is a purely algebraic problem.

The simplest infinite countable commutative group is the infinite cyclic group; that is, the additive group of all the integers. If χ is a character on this group, then $\chi(n)=(\chi(1))^n$, and $\chi(1)$ can be any complex number of modulus one. The character group of the integers is thus isomorphic to the group of rotations in the plane or equivalently to the quotient group of the additive group of the reals by the subgroup of all integral multiples of 2π. Conversely, then, the character group of the latter group is the infinite cyclic group and the characters are the functions $x \rightarrow e^{2\pi inx}$ for $n=0, \pm 1, \pm 2, \ldots$.

Let G be any separable compact commutative group and choose Haar measure so that $\mu(G)=1$. It follows from the Peter–Weyl theorem that the characters $\chi \in \hat{G}$ are mutually orthogonal and that every member f of $\mathcal{L}^2(G,\mu)$ may be written uniquely in the form $f(x)=\sum_{\chi \in \hat{G}} C_\chi \chi(x)$, where convergence is in the $\mathcal{L}^2(G,\mu)$ metric, that is, "in the mean." Moreover, $C_\chi = \int_G f(x)\bar{\chi}(x)\,d\mu(x)$. In the particular case where G is the real numbers modulo integral multiples of 2π, the members of $\mathcal{L}^2(G,\mu)$ may be regarded as complex-valued functions with period 2π. The above formulas then reduce to the following: $f(x)=\sum C_n e^{inx}$ and $C_n=(1/2\pi)\int_0^{2\pi} f(x)e^{-inx}\,dx$ that is, to the formula expressing the fact that any square summable periodic function on the line may be expanded as a Fourier series, and to the formula for the Fourier coefficients. Thus the theory of the regular representation of a separable compact commutative group includes the $\mathcal{L}^2$ theory of Fourier series as a very special case. We obtain the Fourier series for functions of k variables by letting G be the direct product of k replicas of the real numbers modulo 2π; that is, by letting $\hat{G}$ be the group of k-tuples of integers. Possibilities for $\hat{G}$ are sufficiently numerous to provide a great variety of further and less familiar specializations. A particularly interesting one is obtained by letting $\hat{G}$ be the quotient group obtained from the additive group of the rational numbers by identifying two rationals whenever they differ by an integer. Every member of this group is represented by a rational in the interval $0 \leqslant x < 1$, and, we may add, by applying ordinary addition and reducing modulo 1. Let a be the member of $\hat{\hat{G}}$ $(=G)$ that takes r into $e^{2\pi iar}$ and let Z be the subgroup of G generated by a. It is not hard to show that Z is an infinite cyclic group whose closure is G. Thus, complex-valued functions on Z that are uniformly continuous in the topology which Z inherits from G have unique continuous extensions to G and thus define functions in $\mathcal{L}^2(G,\mu)$ to which we may apply the Peter–Weyl theorem. Such functions have expansions of the form $f(n)=\sum C_r e^{2\pi irn}$, where r varies over the rationals in $0 \leqslant r < 1$ and are of frequent occurrence in number theory. Indeed, a number of important formulas in number theory may be interpreted as just such Fourier expansions.

ISBN 0-8053-6702-0/0-8053-6703-9, pbk

Recall that every rational may be written uniquely as a finite sum $r_{p_1}+r_{p_2}+\cdots+r_{p_k}$, where the p_j are distinct primes and the denominator of r_{p_j} is some power of p_j. Moreover, for each prime p the set of all $r\in\hat{G}$ whose denominator is a power of p is a subgroup $\hat{G}_p$ of $\hat{G}$. It is nearly obvious that G is naturally isomorphic to the direct product of all of the compact groups $\hat{\hat{G}}_p$ and that the map taking each element of Z into its $\hat{\hat{G}}_p$ component is a dense imbedding. The multiplication in Z has a unique continuous extension to $\hat{\hat{G}}_p$ that makes $\hat{\hat{G}}_p$ into a compact ring. This ring is called the ring of p-adic integers. It is an integral domain and the field of quotients may be topologized so as to be locally compact by giving it the unique topology for which each coset of G_p is open and has the topology of $\hat{\hat{G}}_p$. This locally compact field is called the field of p-adic numbers. Its additive group and the multiplicative group of its nonzero elements are both locally compact noncompact commutative groups that have the additional property of being totally disconnected. The p-adic integers and the p-adic numbers play a central role in modern number theory.

The construction permitting one to imbed an integral domain in a field has an immediate generalization permitting one to imbed a group with no elements of finite order in one that is completely divisible in the sense that $x\rightarrow x^n$ is onto ($x\rightarrow nx$ if the group is written additively). Since each $\hat{\hat{G}}_p$ has no elements of finite order, this is also true of $G=\Pi_p\hat{\hat{G}}_p$. The resulting divisible extension may be topologized as before so that G is a compact open subgroup and the result is a group called the non-Archimedean component of the adele group of the rationals. The adele group itself is the direct product of this group with the additive group of the real line. It also plays an important role in modern number theory. We shall discuss the adele groups of more general algebraic number fields and other number theoretical objects in a later section.

For each prime p let $\rho_p(n)$ be the function on the integers that is one if p divides n and is zero otherwise. Then the so-called *strongly additive* number-theoretic functions are functions of the form $\Sigma_p c_p\rho_p(n)$. The work of Kac and others on strongly additive functions is based on the observation that the ρ_p have properties reminiscent of those of the independent functions of probability theory. It is therefore of interest to observe that the ρ_p are uniformly continuous in the G topology and are therefore extendible to continuous functions on G. Moreover, $\rho_p(x)$ depends only on the $\hat{\hat{G}}_p$ component of x. It follows that the ρ_p as functions on G are in fact independent. It is almost certainly true that the results of Kac and his collaborators can be deduced from this observation and the known properties of independent functions.

The above encounter with the prime numbers may seem due to our special choice of the group $\hat{G}$. Actually, the primes occur naturally in a fairly general setting. Let D be any countable commutative group. Let D_t

ISBN 0-8053-6702-0/0-8053-6703-9, pbk

be the set of all elements of D some power of which is the identity. If $D=D_t$, the group D is said to be a *torsion group* and if $D_t=\{e\}$, D is said to be *torsion free*. In any event the quotient group D/D_t is always torsion free and we have a natural analysis of any D into its torsion part D_t and its torsion free part D/D_t. Of course D is not uniquely determined by D_t and D/D_t, but these two related groups are important invariants. Now let D be any torsion group. It is a fairly easy theorem that every element of D is uniquely expressible in the form $x_{p_1}\cdot x_{p_2}\cdots x_{p_k}$, where the p_j are distinct primes and the order of x_{p_j} is a power of p_j. Moreover, for each prime p the set of all $x\in D$ such that $x^{p^r}=e$ for some r is a subgroup D_p. It follows that $\hat{D}_t$ is the direct product of the compact groups $\hat{D}_p$. Our theorem on the dual of the rationals mod one is thus a special case of this quite general result. Those compact groups that are the duals of torsion groups have a simple direct characterization. They are totally disconnected compact groups. In the general case the connected component of the identity in $\hat{D}$ is the "annihilator" of D_t, that is, the set of all $\chi\in\hat{D}$ with $\chi(x)=1$ for all $x\in D_t$. It follows that $\hat{D}$ is connected if and only if D is torsion free. While the problem of classifying the torsion free D's is still open, as is the problem of finding the most general D with a given D_t and D/D_t, the problem of finding the most general countable torsion group is completely solved by a theorem known as Ulm's theorem. This theorem is too complicated to state here and we refer the reader to Kaplansky's monograph on abelian groups for a statement and proof.

So much for compact groups. That noncompact commutative groups have a more complicated representation theory can be seen by looking at the regular representation of a countable discrete group. This can be studied by applying Pontryagin duality to turn the Peter–Weyl theorem upside down, so to speak. Consider the formulas generalizing those of Fourier analysis:

$$f(x)=\sum_{\chi\in\hat{G}} C_\chi\chi(x)$$

$$C_\chi=\int f(x)\overline{\chi(x)}\,d\mu(x)$$

These may be given a more symmetrical form if we let $\hat{\mu}$ be the Haar measure in $\hat{G}$ that assigns the measure one to each one-element set, and write C_χ as $C(\chi)$ using functional instead of subscript notation. The equations then become

$$f(x)=\int_{\hat{G}} C(\chi)\chi(x)\,d\hat{\mu}(\chi)$$

$$C(\chi)=\int_G f(x)\overline{\chi(x)}\,d\mu(x)$$

ISBN 0-8053-6702-0/0-8053-6703-9, pbk

and we see that C is obtained from f just as f is obtained from C. The set of Fourier coefficients is just a function on the dual group and the relationship between the two functions is perfectly symmetrical. Now, it follows from the general theory of orthogonal functions that

$$\int_G |f(x)|^2 d\mu(x) = \sum_{\chi \in \hat{G}} |C_\chi|^2$$

$$= \int_{\hat{G}} |C(\chi)|^2 d\hat{\mu}(\chi)$$

and that every function C on $\hat{G}$ with $\Sigma|C(\chi)|^2 < \infty$ is the C associated with some f in $\mathcal{L}^2(G,\mu)$. Thus the mapping $C \to f$, where

$$f(x) = \int_{\hat{G}} C(\chi)\chi(x)\, d\hat{\mu}(\chi)$$

is a unitary mapping of $\mathcal{L}^2(G,\mu)$ on $\mathcal{L}^2(\hat{G},\hat{\mu})$ whose inverse is the unitary mapping $f \to C$, where

$$C(\chi) = \int_G f(x)\, \overline{\chi(x)}\, d\mu(x)$$

Let us compute the effect on f of translating C by χ_o. We have

$$f'(x) = \int_{\hat{G}} C(\chi\chi_o)\chi(x)\, d\hat{\mu}(\chi)$$

$$= \int C(\chi)\chi \cdot \chi_o^{-1}(x)\, d\hat{\mu}(\chi)$$

$$= \bar{\chi}_0(x) \int C(\chi)\chi(x)\, d\hat{\mu}(\chi)$$

$$= \bar{\chi}_0(x) f(x)$$

Thus the regular representation of $\hat{G}$ is equivalent to the representation defined as follows. Its space is $\mathcal{L}^2(G,\mu)$ and $L_{\chi_o} f(x) = \bar{\chi}_o(x) f(x)$. If E is any Borel subset of G, then the subspace $\mathcal{H}_E$ of all f with $f(x)=0$ for $x \in G-E$ is obviously a closed invariant subspace. Conversely, it is not hard to show that the $\mathcal{H}_E$ are the only closed invariant subspaces. Thus the lattice of closed invariant subspaces of $\mathcal{L}^2(\hat{G},\hat{\mu})$ is isomorphic to the Boolean algebra of all Borel sets modulo null sets in G. This lattice never has minimal elements unless there are points of positive measure in G and this is the case only when G is discrete. Of course G is both compact and discrete only when it is finite. Thus, when $\hat{G}$ is discrete and infinite its regular

ISBN 0-8053-6702-0/0-8053-6703-9, pbk

representation has no irreducible invariant subspaces. On the other hand, in realizing this representation by multiplication operators in $\mathfrak{L}^2(G,\mu)$ we have in effect written it as a continuous direct sum or direct integral of "infinitisimal" irreducible representations, and we have made the structure of its system of subrepresentations quite clear. We shall show that a complete analysis of the most general unitary representation of the most general separable locally compact commutative group may be given along these lines.

To begin with let us observe that the Pontryagin duality may be extended to locally compact commutative groups that are neither compact nor discrete. Given any such G, we define $\hat{G}$ just as before and we make $\hat{G}$ into a topological group by declaring a sequence of characters χ_n to converge to a character χ whenever $\chi_n(x)$ converges uniformly to $\chi(x)$ for x ranging in any compact subset of G. This definition causes $\hat{G}$ to be discrete when G is compact and compact when $\hat{G}$ is discrete and thus to agree with our previous definition. As before, let $f_x(\chi)=\chi(x)$ for each $x\in G$. Then $f_x\in\hat{\hat{G}}$ and $x\to f_x$ is a homomorphism of G into $\hat{\hat{G}}$. The general duality theorem of Pontryagin and Van Kampen says that this mapping is an isomorphism of G with $\hat{\hat{G}}$ that is also a topological homomorphism. Thus G and $\hat{\hat{G}}$ may be identified and locally compact commutative groups occur in dual pairs. The most familiar example of a locally compact commutative group that is neither discrete nor compact is the additive group of the real line. For each fixed real number $y, x\to e^{ixy}$ is a character, and every character may be so obtained. The topologies came out right and the additive group of the real line is isomorphic to its own character group just as in the case of a finite commutative group. More generally, if G is the additive group of a finite-dimensional real vector space, then the general character is $x\to e^{il(x)}$, where l is a member of the dual vector space. The additive group of a finite-dimensional vector space over the p-adic numbers is another example of a locally compact commutative group that is isomorphic to its own dual.

Let μ be a Haar measure on G, and for each $f\in\mathfrak{L}^1(G,\mu)\cap\mathfrak{L}^2(G,\mu)$ let $\hat{f}(\chi)=\int f(x)\overline{\chi(x)}\,d\mu(x)$. It can be shown that $\hat{f}$ is in $\mathfrak{L}^2(\hat{G},\nu)$ for any choice of the Haar measure ν in $\hat{G}$ and that there is a unique choice $\hat{\mu}$ such that $\int|f(x)|^2\,d\mu(x)=\int|\hat{f}(\chi)|^2\,d\hat{\mu}(\chi)$ for all f. It can be shown, moreover, that $f\to\hat{f}$ maps the dense subspace $\mathfrak{L}^1(G,\mu)\cap\mathfrak{L}^2(G,\mu)$ of $\mathfrak{L}^2(G,\mu)$ onto a dense subspace of $\mathfrak{L}^2(\hat{G},\hat{\mu})$ and hence has a unique extension to a unitary map of $\mathfrak{L}^2(G,\mu)$ on $\mathfrak{L}^2(G,\mu)$. That $f\to\hat{f}$ has these properties is called the Plancherel theorem (for commutative groups) after Plancherel who proved it when G is the additive group of the real line. When G is the additive group of the real line and $\chi_y(x)=e^{iyx}$, then $\hat{f}(\chi_y)=\int_{-\infty}^{+\infty}e^{-iyx}f(x)\,dx$ and is the classical Fourier transform of f. One speaks of $f\to\hat{f}$ as the generalized Fourier transform. It includes as special cases not only the classical Fourier

ISBN 0-8053-6702-0/0-8053-6703-9, pbk

transform, but also the transformation that takes a periodic function into its Fourier coefficients and the transformation taking a sequence of Fourier coefficients into the corresponding periodic function.

The generalized Fourier transform sets up an equivalence between the regular representation of G and the representation of G on $\mathcal{L}^2(\hat{G},\hat{\mu})$ defined by setting $L_x g(\chi) = \chi(x) g(\chi)$. Now for each Borel subset E of $\hat{G}$ let P_E be the projection operator that takes $g \in \mathcal{L}^2(\hat{G},\hat{\mu})$ into $g\varphi_E$, where φ_E is the characteristic function of E, that is, the function that is one in E and zero in $\hat{G} - E$. One verifies at once that $E \rightarrow P_E$ is a projection valued measure on $\hat{G}$ in the sense of the following definition: A projection valued measure on a Borel space S is a function $E \rightarrow P_E$ defined on the Borel sets of S and having the following properties:

(i) Its values are all projection operators in a separable Hilbert space $\mathcal{H}(P)$.

(ii) $P_o = 0, P_S = I$.

(iii) $P_{E \cap F} = P_E P_F$ for all E and F.

(iv) $P_E = P_{E_1} + P_{E_1} + P_{E_2} + \cdots$ whenever $E = E_1 + E_2 + \cdots$ and $E_i \cap E_j = 0$ for $i \neq j$.

This notion will play a fundamental role in much of what follows and merits close study. When S has only countably many elements and every point is a Borel set, then every subset of S is a Borel set and P is uniquely defined by its values at the one-point sets. Indeed, $P_E = \Sigma_{s \in E} P_{\{s\}}$. Conversely, if we write $\mathcal{H}$ as a direct sum of orthogonal subspaces labeled by the elements of S, there exists a unique projection valued measure P such that $P_{\{s\}}$ is the projection on $\mathcal{H}_s$. It follows, in particular, that the possible unitary representations of the separable compact commutative group G correspond one-to-one to the projection valued measures on $\hat{G}$ in such a way that when P corresponds to V, then $P_{\{\chi\}}$ is the P_χ of Section 5. Given P we can reconstruct V from S using the identity $(V_x f \cdot f) = \Sigma_{\hat{G}} \chi(x)(P_\chi f \cdot f)$. Moreover, $E \rightarrow (P_E(f) \cdot f)$ is a measure on $\hat{G}$ and we may write

$$(V_x(f) \cdot f) = \int_{\hat{G}} \chi(x)\, d(P_\chi f \cdot f)$$

In other words, the unitary representations of G and the projection valued measures on $\hat{G}$ correspond one-to-one in such a fashion that when V and P are corresponding elements, then

$$(V_x(f) \cdot f) = \int \chi(x)\, d(P_\chi f \cdot f)$$

for all f in $\mathcal{H}(V)$.

ISBN 0-8053-6702-0/0-8053-6703-9, pbk

Now let G be noncompact and let P be the projection valued measure on $\hat{G}$ defined above using the regular representation R of G. For each $f\in\mathfrak{L}^2(\hat{G},\mu)$ the measure $E\rightarrow(P_E f\cdot f)$ is the measure $E\rightarrow\int_E|f|^2 d\hat{\mu}$, and $\int_{\hat{G}}\chi(x)\,d(P_\chi f\cdot f)$ is $\int_{\hat{G}}|f(\chi)|^2\chi(x)\,d\mu(\chi)$, which is $(R_x(f)\cdot f)$. Thus the regular representation of any G is defined by a certain projection valued measure on $\hat{G}$ in just the way that the general unitary representation of a compact group is. However, when G is not compact, the associated projection valued measure is zero for all one point sets and one has no analog of the P_χ. This suggests that more general representations of G might correspond to more general projection valued measures on $\hat{G}$, and this is in fact the case. Indeed, we can prove the following theorem: Given any unitary representation V of the separable locally compact group G, there is a unique projection valued measure P on $\hat{G}$ such that for all $f\in\mathcal{H}(V)$ we have $(V_x(f)\cdot f)=\int\chi(x)\,d(P_\chi f\cdot f)$. Moreover, every projection valued measure on $\hat{G}$ arises in this way from a unique V. We shall refer to this theorem as the *spectral theorem for commutative groups*. The special case in which G is the additive group of the real line is equivalent via Stone's theorem to the spectral theorem for self-adjoint operators. Indeed, then $\hat{G}$ is just the additive group of the real line and the projection valued measure in $\hat{G}$ corresponding to $x\rightarrow e^{iHx}$ is just the projection valued measure such that $(Hf\cdot f)=\int x\,d(P_x f\cdot f)$. It very often happens in passing from the discrete to the continuous that one is able to proceed only by replacing point functions by set functions in the manner just illustrated.

We have just seen that the determination of the unitary representations of G is equivalent to the determination of the projection valued measures on $\hat{G}$. This fact, while useful in itself, does not solve the problem of determining the unitary representations of G. It merely reduces it to another problem. This other problem, or rather the part that remains after $\hat{G}$ is known, has an easy solution when $\hat{G}$ is countable. In this case a projection valued measure on $\hat{G}$ is completely determined by its values on the points of $\hat{G}$ and is hence completely determined to within equivalence by giving the dimension of each $P_{\{\chi\}}$. [In any case, we define P and P' to be equivalent if there exists a unitary operator W such that $P_E = WP_E W^{-1}$ for all Borel subsets E of the S in question. It is clear from the foregoing that V and V' are equivalent if and only if the associated projection valued measures are equivalent.] Thus (in the case of compact G) the general P on $\hat{G}$ is determined by the general function from $\hat{G}$ to the possible dimensions, that is, by assigning a "multiplicity" to each $\chi\in\hat{G}$.

Is there an analog of this result where $\hat{G}$ is not countable? We shall see that there is a subtle but fairly complete analog. This more general result is in no way dependent upon the group structure of $\hat{G}$, and is valid for quite general Borel spaces. We turn now to the details.

Let S be any analytic Borel space and let μ be a measure in S. Form the separable Hilbert space $\mathfrak{L}^2(S,\mu)$, and for each Borel subset E of S let P_E^μ

denote the operation of multiplication by the function φ_E that is one in E and zero on its complement. $E \to P_E^\mu$ is clearly a projection valued measure on S. Concerning these projection valued measures P^μ, it is possible to prove the following theorems:

(i) For fixed μ, any bounded linear operator that commutes with all P_E^μ is multiplication by a bounded Borel function. In particular,

(i′) if A and B are any two operators on $\mathcal{L}^2(S,\mu)$ that commute with all P_E^μ, then $AB = BA$;

(i″) any projection that commutes with all P_E^μ is of the form P_F^μ for some F.

(ii) A projection valued measure P having either property (i′) or (i″) has the other property as well and is equivalent to some P^μ.

(iii) P^{μ_1} and P^{μ_2} are equivalent if and only if μ_1 and μ_2 belong to the same measure class.

When S is countable, P has properties (i) and (ii) if and only if every $P_{\{s\}}$ is as one-dimensional or zero-dimensional. When $S = \hat{G}$, this means that the corresponding representation of G has no irreducible subrepresentation occurring with multiplicity greater than one. Accordingly, we make the definition: A projection valued measure with either, and hence both, of the properties (i′) and (ii′) above will be said to be *multiplicity free*. The theorem quoted above tells us that $\mu \to P^\mu$ sets up a one-to-one correspondence between the measure classes in S and the equivalence classes of multiplicity free projection valued measures.

When S is countable, two measures belong to the same measure class if and only if they give positive measure to the same points. Thus the measure classes in S, and hence the multiplicity free projection valued measures on S, correspond in this case to the different subsets of S. This is a reflection of the fact that when $\hat{G}$ is countable the multiplicity free representations of G correspond one-to-one to the subsets of $\hat{G}$. In general, however, a measure class is a more complicated kind of object and is not defined by a subset. On the other hand, there is a sense in which the measure classes on a general S may be regarded as generalized subsets. In particular, they form a Boolean algebra in a natural way.

An example of how much measure classes can differ from subsets may be helpful. Let S be the Euclidean plane and let μ be the measure that assigns to each Borel subset the Lebesgue measure of its intersection with the unit square $0 \leqslant x \leqslant 1$, $0 \leqslant y \leqslant 1$. Let v be the measure that assigns to each Borel subset the (linear) Lebesgue measure of its intersection with the line segment $y = \frac{1}{2}$, $\frac{1}{3} \leqslant x \leqslant \frac{2}{3}$. Then the measure class of μ seems to be described by the subset of S consisting of all points in the unit square. But then what subset describes the quite different measure class of $\mu + v$? The

ISBN 0-8053-6702-0/0-8053-6703-9, pbk

projection valued measures P^{μ} and $P^{\mu+v}$ are inequivalent and describe inequivalent multiplicity free unitary representations of the additive group of the plane. Yet, naively, each representation seems to be the "direct integral" of all characters of the form $a,b \to e^{i(ax+by)}$ with x,y in the unit square.

Having shown that the unitary representations of G correspond one-to-one to the projection valued measures on $\hat{G}$ and that the equivalence classes of multiplicity free projection valued measures on $\hat{G}$ correspond one-to-one to the measure classes on $\hat{G}$, it follows that we have a one-to-one correspondence between the measure classes in $\hat{G}$ and the members of a certain class of unitary representations of G. It is natural to call these the multiplicity free representations of G and to ask whether the correspondence can be set up directly. The answer is yes and the correspondence is easy to describe. Let v be any measure on $\hat{G}$ and form $\mathfrak{L}^2(\hat{G},v)$. For each $x \in G$ let $L_x^v f(\chi) = \chi(x) f(\chi)$. Just as in the case of the regular representation (where v was Haar measure), one can show that L^v is a representation and that its associated projection valued measure is P^v. In particular, we see that the regular representation of the separable locally compact group G is that particular multiplicity free representation whose measure class is that of Haar measure in $\hat{G}$. More generally, L^v may be regarded as a sort of "continuous direct sum" or "direct integral" of one-dimensional irreducible representations with respect to the measure v.

Having thus described the possible multiplicity free projection valued measures, and hence the multiplicity free unitary representations of a separable locally compact group, what can we say about the representations and projection valued measures that are *not* multiplicity free? We shall give the answer in terms of the representations themselves—leaving it to the reader to formulate the obvious analog for projection valued measures. Actually, a large part of the answer is valid for noncommutative groups and we shall so formulate it. Let G be any separable locally compact group and let us define a unitary representation V of G to be multiplicity free if $R(V,V)$ is commutative. Let n and m be cardinal numbers $\leqslant \infty$ and let V^1 and V^2 be multiplicity free unitary representations of G. One can prove that $V^1 \oplus V^1 \oplus \cdots n$ terms is equivalent to $V^2 \oplus V^2 \oplus \cdots m$ terms if and only if $n = m$ and V^1 and V^2 are equivalent. This suggests that we define a unitary representation to be of uniform multiplicity n wherever it is equivalent to a direct sum of n replicas of the same multiplicity free representation, and tells us that for fixed n the possible representations of uniform multiplicity n correspond one-to-one in a natural way to the possible multiplicity free representations. Let us say that a unitary representation of G is of type I if it is equivalent to a (discrete, but possibly infinite) direct sum of multiplicity free representations. The study of type I representations is completely reducible to the study of multiplicity free representations by the following theorem: Let V

be any type *I* representation. Then $\mathcal{H}(V)$ has a unique sequence of closed invariant subspaces $\mathcal{H}_\infty, \mathcal{H}_1, \mathcal{H}_2, \ldots$ with the following properties:

(1) If $\mathcal{H}_j$ is not trivial, it defines a subrepresentation of V that is uniformily of multiplicity j.

(2) The subrepresentations defined by distinct $\mathcal{H}_j$ are *disjoint*.

When B is commutative, one can prove that every unitary representation is of type *I* and that L^μ and L^ν are disjoint if and only if μ and ν are "mutually singular" in the sense that each is supported by a null set of the other. It follows that a complete set of invariants for a unitary representation of a separable locally compact commutative group G is a sequence $C_\infty, C_1, C_2, \ldots$ of mutually singular measure classes in $\hat{G}$. (Of course, some of these measure classes may be trivial.)

The noncompact locally compact commutative groups include not only the countable discrete groups and the additive groups of finite-dimensional vector spaces, but also direct products of these with each other and with compact groups. In addition, they include the additive groups of vector spaces over the p-adic numbers. There is a sense, however, in which we can almost describe the most general example in terms of countable discrete groups. It can be proved that every locally compact commutative group is uniquely a direct product of a finite-dimensional vector group and a group G^1 having a compact subgroup H whose quotient G^1/H is discrete. Except for the extension problem (which of course is not trivial), G^1 is described by the two countable discrete groups $\hat{H}$ and G^1/H.

ISBN 0-8053-6702-0/0-8053-6703-9 pbk

7. *Direct integrals of unitary representations in the general case*

If we are to have any hope of extending the theory of Section 6 to separable locally compact groups that are neither compact nor commutative, we must find a substitute for the construction allowing us to pass from a measure μ on $\hat{G}$ to a unitary representation V^{μ} of G. That is, we must find a way of defining the direct integral of a family of irreducible unitary representations with respect to a measure when these representations are not one-dimensional. Actually we shall have need of a notion of direct integral of unitary representations that are not necessarily irreducible. We base our definition on the auxiliary notion of a "Hilbert bundle" —a notion for which we shall have other uses later. Let S be an analytic Borel space. By a *Hilbert bundle over S* or a *Hilbert bundle with base S* we shall mean an assignment $\mathcal{H}: s \rightarrow \mathcal{H}_s$ of a separable Hilbert space to each $s \in S$. The set of all pairs (s,φ) with $\varphi \in \mathcal{H}_s$ will be denoted by $S \triangle \mathcal{H}$ and called the *space of the bundle*. By a *cross section* of our bundle we shall mean an assignment $s \rightarrow \varphi_s$ of a member of $\mathcal{H}_s$ to each $s \in S$. Of course a cross section may be throught of as a certain kind of subset of $S \triangle \mathcal{H}$. If φ is a cross section and (s_o, θ_o) a point of $S \triangle \mathcal{H}$, we may form the scalar product $(\theta_o, \varphi_{s_o})$. In this way every cross section defines a complex-valued function on $S \triangle \mathcal{H}$. By a *Borel Hilbert Bundle* we shall mean a Hilbert bundle together with an analytic Borel structure in $S \triangle \mathcal{H}$ such that the following conditions are satisfied:

(1) Let $\pi(s,\varphi) = s$. Then $E \subseteq S$ is a Borel set if and only if $\pi^{-1}(E)$ is a Borel set in $S \triangle \mathcal{H}$.

George W. Mackey, Unitary Group Representations in Physics, Probability, and Number Theory

ISBN 0-8053-6702-0/0-8053-6703-9, pbk

(2) There exist countably many cross sections $\varphi^1, \varphi^2, \ldots$ such that

(a) the corresponding complex-valued functions on $S \triangle H$ are Borel functions,

(b) these Borel functions separate points in the sense that no two distinct points (s_i, θ_i) of $S \triangle H$ assign the same values to all φ^j unless $\theta_1 = \theta_2 = 0$, and

(c) $s \to (\varphi^i(s), \varphi^j(s))$ is a Borel function for all i and j.

As an example, let $\mathcal{H}_0$ be any fixed Hilbert space, let $\mathcal{H}_s = \mathcal{H}_0$ for all s so that $S \triangle H$ is in fact $S \times \mathcal{H}_0$, and give $S \times \mathcal{H}_0$ the product Borel structure. In this case we obtain our countable family of cross sections by choosing a countable separating family $f_1, f_2, \ldots$ of complex-valued Borel functions on S, a complete orthonormal set $\theta_1, \theta_2, \ldots$ in $\mathcal{H}_0$, and taking the cross sections $s \to f_j(s)\theta_k$.

As another example, let S be countable and let $s \to \mathcal{H}_s$ be arbitrary and let $S \triangle H$ be given the Borel structure generated by sets of the form $s \times E$ where $s \in S$ and E is a Borel subset of $\mathcal{H}_s$.

When the function on $S \triangle \mathcal{H}$ defined by a cross section is a Borel function we shall call it a *Borel cross section*. It can be shown that the set of all Borel cross sections is a vector space under the obvious operations. Now let μ be a measure on S. We shall say that the Borel cross section $s \to \varphi_s$ is *square-summable* with respect to μ if

$$\int_S (\varphi_s, \phi_s)\, d\mu(s) < \infty$$

It can be shown that the sum of two square-summable cross sections is again square-summable and that

$$\int_S (\varphi_s, \psi_s)\, d\mu(s)$$

exists whenever $s \to \varphi_s$ and $s \to \psi_s$ are square-summable. Let $\mathcal{L}^2(S, \mu, \mathcal{H})$ denote the set of all equivalence classes of square-summable cross sections where two cross sections φ and ψ are in the same equivalence class if $\psi_s = \varphi_s$ for almost all $s \in S$. Then it can be shown that $\mathcal{L}^2(S, \mu, \mathcal{H})$ is a separable Hilbert space with respect to the indicated addition and the scalar product,

$$(\varphi, \psi) = \int_S (\varphi_s, \psi_s)\, d\mu(s)$$

We shall call it the direct integral of the $\mathcal{H}_s$ with respect to μ and denote it by the symbol $\int_S \mathcal{H}_s\, d\mu(s)$. In the special case in which S is countable and μ is the counting measure, it reduces to the direct sum of the $\mathcal{H}_s$. In the

ISBN 0-8053-6702-0/0-8053-6703-9, pbk

special case where $\mathcal{H}_s = C =$ one-dimensional Hilbert space for all $s \in S$, then $\int_S \mathcal{H}_s \, d\mu(s)$ reduces to $\mathcal{L}^2(S,\mu)$.

Given any direct integral $\int_S \mathcal{H}_s \, d\mu(s)$, we may define a projection-valued measure P as follows. For each Borel subset E of S, P_E is the projection taking φ into φ', where $\varphi'_s = \varphi_s$ for all $s \in E$ and $\varphi'(s) = 0$ for all $s \notin E$. The projection-valued measure P determines the system $(S, \mathcal{H}, \mu)$ in its most essential features. More precisely, let P^1 and P^2 be the projection-valued measures associated with $(S, \mathcal{H}^1, \mu^1)$ and $(S, \mathcal{H}^2, \mu^2)$, respectively. Suppose that W is a unitary operator setting up an equivalence between P^1 and P^2. Then it is obvious that μ^1 and μ^2 have the same subsets of measure zero. Moreover, it can be shown that there exists a Borel subset N of S and for all $s \in S - N$ a unitary operator W_s of $\mathcal{H}_s^1$ onto $\mathcal{H}_s^2$ such that the following conditions are satisfied:

(a) $\mu^1(N) = \mu^2(N) = 0$;

(b) $s, \varphi_s \to s, W_s(\varphi_s)$ is a Borel isomorphism of $(S-N) \triangle \mathcal{H}^1$ on $(S-N) \triangle \mathcal{H}^2$;

(c) the map $\varphi \to \varphi'$, where $\varphi'_s = W_s \varphi_s$ defines a unitary map of $\int_S \mathcal{H}_s' \, d\mu^1(x)$ on $\int_S \mathcal{H}_s^2 \, d\mu^2(s)$, which coincides with W.

Conversely, if P is any projection-valued measure on the analytic Borel space S, there exists a Hilbert bundle $s \to \mathcal{H}_s$ and a measure μ on S such that $P_E = 0$ if and only if $\mu(E) = 0$ and such that P is equivalent to the projection-valued measure associated with $\int_S \mathcal{H}_s \, d\mu(s)$. Thus in a sense a projection-valued measure on a space S *is* a direct integral decomposition of the Hilbert space of the projection-valued measure.

Now let $\mathcal{H}^\mu = \int_S \mathcal{H}_s \, d\mu(s)$ be a direct integral of Hilbert spaces and for each $s \in S$ let there be given a unitary representation L^s of a separable locally compact group G. Then $S \triangle \mathcal{H}$ becomes a G-space if we set

$$(s, \varphi)x = (s, L^s_{x^{-1}}(\varphi))$$

If it is a Borel G-space, we shall say that $s \to L^s$ is a Borel function. For each cross section φ_s and each $\varphi \in G$, let

$$\{L_x(\varphi)\}_s = L^s_x(\varphi_s)$$

Then L_x defines a unitary operator in $\int_S \mathcal{H}_s \, d\mu(s)$ and $x \to L_x$ is a unitary representation of G. We call L the *direct integral* of the L^s with respect to μ, and write $L = \int_S L^s \, d\mu(s)$. When G is commutative, $S = \hat{G}$, and L^χ is $x \to \chi(x)I$, it is easy to see that $\int_G L^\chi \, d\mu(\chi)$ is just the V^μ defined in the last section. In general, we can prove that replacing μ by a measure μ' in the same measure class replaces L by an equivalent representation. Thus it makes sense to talk about a direct integral with respect to a measure class. It can also be proved that the $\int_S L^s \, d\mu(s)$ is equivalent to $\int_S M^s \, d\mu(s)$ whenever L^s and M^s are equivalent for almost all $s \in S$. If we are given a

ISBN 0-8053-6702-0/0-8053-6703-9, pbk

measure μ in an analytic Borel space S and for each s an equivalence class of unitary representations of G, there may or may not exist a way of choosing an L^s in each equivalence class and a Borel structure in the corresponding Hilbert bundle so that $\int_S L^s \, d\mu(s)$ is defined. However, if a way of choosing does exist, the equivalence class of the result is independent of the arbitrary choices involved. When the choices can be made we shall say that $s \to \textcircled{L^s}$ is μ-integrable. Here $\textcircled{L^s}$ denotes the equivalence class of L^s.

Now let G be any separable locally compact group, and let $\hat{G}$ denote the set of all equivalence classes of irreducible unitary representations of G. When G is commutative, $\hat{G}$ is a separable locally compact commutative group and we know what we mean by a Borel subset of $\hat{G}$. When G is not commutative, $\hat{G}$ is certainly not a group and at the moment we have not even given it a topology. One can define a topology on $\hat{G}$ but in general this topology will not be a well-behaved one. In particular, it need not be Hausdorff or even T_1. On the other hand, we need only the underlying Borel structure, and it is possible to introduce a suitable Borel structure directly without the aid of an intermediate topology. Moreover, in many cases in which the topology is not Hausdorff, the Borel structure comes out to be standard. To define it we first write

$$\hat{G} = \hat{G}^{\infty} \cup \hat{G}^1 \cup \hat{G}^2 \cup \cdots$$

where $\hat{G}^j$ denotes the set of all equivalence classes of j-dimensional irreducible representations. We shall introduce a Borel structure into each $\hat{G}^j$ and then define a subset of G to be a Borel set if each intersection with $\hat{G}^j$ is a Borel set. For j finite we may identify $\hat{G}^j$ with the set of all characters of j-dimensional irreducible representations and then introduce the weakest Borel structure with the property that $\chi \to \chi(x)$ is a Borel function for all $x \in G$. In other words, we demand that for each Borel subset B of the complex numbers and each $x \in G$ the set of all χ with $\chi(x) \in B$ shall be a Borel set. We then take the elements of the σ field generated by these sets as the Borel subsets of $\hat{G}^j$. It can be shown without great difficulty that $\hat{G}^j$ is a standard Borel space for each finite j.

When $j = \infty$, the members of $\hat{G}^j$ cannot be defined by characters and we must define the Borel structure of $\hat{G}^{\infty}$ in an indirect manner. Let $\mathcal{H}_{\infty}$ be the concrete Hilbert space of Hilbert, that is, the space of all infinite sequences of complex numbers for which $\sum_{j=1}^{\infty} |a_j|^2 < \infty$. Let $\hat{G}^{c,\infty}$ denote the set of all unitary representations of G in the space $\mathcal{H}_{\infty}$ and where we do not identify equivalent representations. Then for each $L \in \hat{G}^{c,\infty}$ and each $x \in G$, the operator L_x is defined by a matrix $\|a_{ij}^L(x)\|$ with infinitely many rows and columns. We give $\hat{G}^{c,\infty}$ the weakest Borel structure for which all of the functions $L \to a_{ij}^L(x)$ are Borel functions. It is not hard to prove that this definition makes $\hat{G}^{c,\infty}$ into a standard Borel space. To give $\hat{G}^{\infty}$ a Borel structure, we consider the mapping π that takes each $L \in \hat{G}^{c,\infty}$

ISBN 0-8053-6702-0/0-8053-6703-9, pbk

into the equivalence class to which it belongs. π is a mapping of $\hat{G}^{c,\infty}$ onto $\hat{G}^{\infty}$. We give $\hat{G}^{\infty}$ a Borel structure by declaring the Borel subsets E of $\hat{G}^{\infty}$ to be those for which $\pi^{-1}(E)$ is a Borel subset of $\hat{G}^{c,\infty}$. Unfortunately, $\hat{G}^{\infty}$ is not always a standard Borel space or even an analytic one. Thus $\hat{G}$ need not be analytic or standard. On the other hand, there are many important groups for which $\hat{G}$ is standard. We shall have more to say about this point later.

Now let μ be any measure on the Borel space $\hat{G}$. If there exists a Borel set N such that $\hat{G}-N$ is standard and $\mu(N)=0$, we shall say that μ is a standard measure. Let μ be standard and let N be as above. Let $S=\hat{G}-N$. Then the identity map takes each $s\in S$ into an equivalence class of irreducible unitary representations of G and it can be shown that the assignment is μ-integrable. In other words, for each $s\in S$ we may choose an L^s in the equivalence class and form $\int_S L^s\,d\mu(s)$. The resulting representation is independent to within equivalence of the choice of S and the L^s. We shall call it V^{μ}. It reduces to the V^{μ} defined earlier when G is commutative. Moreover, it follows from what has already been said that V^{μ_1} and V^{μ_2} are equivalent whenever μ_1 and μ_2 belong to the same class. It is now natural to suppose that every V^{μ} must be multiplicity free and that $\mu\rightarrow V^{\mu}$ must set up a one-to-one correspondence between all standard measure classes in G and all equivalence classes of multiplicity free representations of G. If this were true and if it were also true that every unitary representation of G is a direct sum of multiplicity free representations, then we would have a theory of the unitary representations of G fully as complete as and quite analogous to that developed for commutative groups in the last section. Unfortunately, neither of these things is true in general. It can happen that V^{μ} is not multiplicity free—indeed that $\mathfrak{R}(V^{\mu},V^{\mu})$ has a trivial center. Moreover, whenever V^{μ} is not multiplicity free it is not even an infinite direct sum of multiplicity free representations. Finally, it can happen that V^{μ} and $V^{\mu'}$ are equivalent even when μ and μ' are mutually singular; in other words, the same representation can be written as a direct integral of irreducibles in two completely different ways.

While these facts may seem rather discouraging, all is not lost. First of all, it is at least true that every multiplicity free representation is equivalent to one of the form V^{μ}—where the class of μ is uniquely determined. Thus one does have a natural one-to-one correspondence between the multiplicity free representations and certain standard measure classes in $\hat{G}$. The problem is to state which measure classes occur. Moreover, the theorem stated at the end of the last section allows us to reduce the study of all Type I representations to that of multiplicity free representations and hence to the measure class problem. Finally, there are many interesting noncompact noncommutative groups for which none of these difficulties appear. For these groups every measure class is standard, every V^{μ} is multiplicity free and every unitary representation is of Type I. We shall give further details in the next section.

ISBN 0-8053-6702-0/0-8053-6703-9, pbk

8. Primary representations and the von Neumann–Murray theory of factors

In decomposing a unitary representation of a compact group into its irreducible constituents there are two stages—with a different kind of uniqueness attached to each stage. In the first stage one looks at the projections

$$P_\chi^V = \int_G V_{x^{-1}} \chi(x)\, d\mu(x)$$

attached to the irreducible characters of G. Then

$$\mathcal{H}(V) = \sum_{\chi \in \hat{G}} \oplus \mathcal{H}_\chi^V$$

where $\mathcal{H}_\chi^V = \text{range } P_\chi^V$, and each $\mathcal{H}_\chi^V$ is invariant. These subspaces $\mathcal{H}_\chi^V$ are canonically determined. However, the subrepresentations V^χ which they determine are not irreducible. Each can be further decomposed as a direct sum of equivalent irreducibles. However, the decomposition is not unique in the same sense. The invariant subspaces can be chosen in many ways. All that is unique is the number of constituents and the equivalence class to which they belong. In our discussion of unitary representations of commutative groups we have introduced a different spatially unique decomposition. Let us see how it relates to the above. For each χ let $n(\chi)$ be the number of summands in the decomposition of V^χ into irreducibles, that is, the multiplicity with which the irreducible representation of character χ occurs in V. For each $m = \infty, 1, 2, \ldots$ let $\mathcal{H}_m^V$ be the direct sum of

George W. Mackey, Unitary Group Representations in Physics, Probability, and Number Theory

ISBN 0-8053-6702-0/0-8053-6703-9, pbk

all $\mathcal{H}_\chi^V$ with $n(\chi)=m$. Then

$$\mathcal{H}(V)=\mathcal{H}_\infty^V \oplus \mathcal{H}_1^V \oplus \mathcal{H}_2^V \oplus \cdots$$

is the discrete spatially unique decomposition that played a central role in our discussion of the commutative case. Note that it can be regarded as an earlier stage than either of the two discussed above. The $\mathcal{H}_\chi$ decomposition occurs by further subdividing each $\mathcal{H}_n$. As remarked in Section 7 there is an $\mathcal{H}_n$ decomposition only for certain representations, the so-called Type I representations, and this interferes with what at first sight could be an almost perfect extension of the commutative theory to the general one. However, there is a close analog of the $\mathcal{H}_\chi$ decomposition even in the general case. Moreover, a study of it gives considerable insight into what goes wrong and into what a Type I representation really is.

Let V be an arbitrary unitary representation of the separable locally compact group G. Let P be any projection in $\mathfrak{R}(V,V)$ and let V^P denote the subrepresentation whose space is the range of P. As we have seen, every subrepresentation of V is of the form V^P for some P in $\mathfrak{R}(V,V)$. Moreover, it is clear that V is equivalent to the direct sum of V^P and V^{1-P}. We now raise the following question. For what P is it true that V^P and V^{1-P} are *disjoint*? If V^P and V^{1-P} are disjoint and T is any bounded operator in $\mathfrak{R}(V,V)$, then T must carry the range of P into itself; otherwise $(1-P)$ times its restriction to the range of P would define a nonzero intertwining operator for V^P and V^{1-P}. Thus $PT=TP$ and P must be in $C\mathfrak{R}(V,V)$, the *center* of $\mathfrak{R}(V,V)$. Conversely, if P is in $C\mathfrak{R}(V,V)$, it is easy to show that V^P and V^{1-P} are disjoint if and only if P is in the center of $\mathfrak{R}(V,V)$. Now let $\mathfrak{B}_V$ denote the set of all projections in $C\mathfrak{R}(V,V)$. $\mathfrak{B}_V$ becomes a partially ordered set if we let $P_1 \leqslant P_2$ whenever $P_1P_2=P_1$, that is, whenever the range of P_1 is contained in the range of P_2. $\mathfrak{B}_V$ may or may not contain *atoms*; that is elements with no properly smaller elements except zero. If it does not we shall say that V has a *continuous center*. If their ranges are mutually orthogonal, there can be at most countable many of them. Corespondingly, we have a direct sum decomposition of V, namely,

$$V \simeq (V^{P_1} + V^{P_2} + \cdots) + V^{\bar{P}}$$

where $\bar{P}=1-P_1-P_2-\cdots$ and $V^{\bar{P}}$ has a continuous center. The V^{P_j} are of course mutually disjoint and each is *primary* in the sense that it cannot be written as a direct sum of two disjoint representations. In the special case in which G is compact, the P_j are just the P_χ for the various irreducible characters and $\bar{P}$ is always zero. On the other hand, when G is arbitrary and V is multiplicity free, then $C\mathfrak{R}(V,V)=\mathfrak{R}(V,V)$ and $\mathfrak{B}_V$ is isomorphic to the partially ordered set of all subrepresentations. When V

ISBN 0-8053-6702-0/0-8053-6703-9, pbk

has no irreducible subrepresentation, $\mathcal{B}_V$ is isomorphic as we have seen to the set of all Borel sets mod null sets of the unit interval. This can be shown to be true of $\mathcal{B}_V$ whenever V has a continuous center. More generally, one has the following: Let $\mathcal{B}$ be any complete Boolean algebra of projections, that is, any family of mutually commuting projections that is closed under the following operations:

$$\begin{aligned} P_1, P_2 &\to P_1 P_2 \\ P &\to 1 - P \\ P_1, P_2, \ldots &\to P_1 + P_2 + \cdots \quad \text{whenever } P_i P_j = P_j P_i = 0 \end{aligned}$$

Suppose that $\mathcal{B}$ has no atoms. Then $\mathcal{B}$ is isomorphic to the partially ordered set of all Borel subsets of the unit interval—two subsets being identified if they differ by a set of Lebesgue measure zero. In particular, any two atom-free complete Boolean algebras of projections are isomorphic. It follows immediately that any complete Boolean algebra of projections is the range of some projection-valued measure Q defined on a standard Borel space S.

Now suppose that our unitary representation V has a continuous center $\mathcal{B}_V$. Choose a projection-valued measure P on a standard Borel space S so that $\mathcal{B}_V$ is the range of P. As we have seen we may write $\mathcal{H}(V)$ as $\int_S \mathcal{H}_s \, d\mu(s)$ in such a way that P is the projection-valued measure associated with the direct integral. Using the fact that the P_E all lie in $\mathcal{R}(V,V)$, one may show further that it is possible to assign a unitary representation L^s to each $s \in S$ whose space is $\mathcal{H}_s$ and such that $\int_S L^s \, d\mu(s)$ exists and is equivalent to V. This much follows simply from the fact that each P_E is in $\mathcal{R}(V,V)$. Using the fact that each P_E is in $C\mathcal{R}(V,V)$, one can show further that the L^s can be chosen so as to be mutually disjoint (they are uniquely determined only up to μ-null sets). Finally, using the fact that the range of $E \to P_E$ is the whole of $\mathcal{B}_V$, one can show that almost all of the L^s are primary (and hence so that the choices can be made so that all the L^s are primary). This decomposition of V as a direct integral of disjoint primaries is essentially independent of the arbitrary choice of S and μ. If these are made in some other way leading to the formula

$$V = \int_{S_1} M^{s_1} \, d\nu(s_1)$$

then there exist Borel subsets N and N_1 of S and S_1, respectively, such that $\mu(N) = \nu(N_1) = 0$ and a Borel isomorphism θ of $S - N$ onto $S_1 - N_1$ such that $M^{\theta(s)}$ and L^s are equivalent for all s and such that $\mu(E) = 0$ if and only if $\nu(\theta(E)) = 0$.

It is important to notice that the theorems stated above do not imply that there is essentially only one way of writing a given unitary representa-

ISBN 0-8053-6702-0/0-8053-6703-9, pbk

tion as a direct integral of disjoint primaries. It is possible to have quite a different decomposition if we do not demand that the range of the associated projection-valued measure lies in the center of the commuting algebra. We shall give further details below. The essentially unique one in which the range of the associated projection-valued measure is $\mathfrak{B}_V$ will be called the central decomposition.

Putting together an analysis of the subrepresentations $V^{\bar{P}}$ and $V^{1-\bar{P}}$ and remembering that a discrete direct sum is a special kind of direct integral, we see that if V is an arbitrary unitary representation of an arbitrary separable locally compact group, then we have a canonical decomposition (the central one) of V as a direct integral of mutually disjoint primary representations. It is clear from the definitions involved that any subrepresentation of V is a direct integral of subrepresentations of the primary representations occurring in the central decomposition. Our next task then is to study the structure and decomposibility of primary representations.

Let V be an arbitrary primary representation of the separable locally compact group G. If V is of Type I, that is, if it is a direct sum of multiplicity free representations, then it follows from our general analysis of such that it must be of uniform multiplicity k for some k. That is, it must be of the form kM where M is multiplicity free. But M must also be primary for if $M = M_1 \oplus M_2$ where M_1 and M_2 are disjoint, then

$$kM = kM_1 \oplus kM_2$$

and kM_1 and kM_2 are also disjoint. But a representation M cannot be both primary and multiplicity free unless $C\mathfrak{R}(M,M) = \mathfrak{R}(M,M)$ and $\mathfrak{R}(M,M)$ is one-dimensional, that is, unless M is irreducible. Thus a Type I primary representation must be a direct sum of equivalent irreducible representations. Conversely, of course every direct sum of mutually equivalent irreducibles is primary. Since every unitary representation of G is of Type I when G is either compact or locally compact and commutative, it follows that every primary unitary representation of such a group is a direct sum of equivalent irreducibles. Actually, as we have already indicated, there are many separable locally compact groups that are neither compact nor abelian for which every unitary representation is of Type I. For these, every primary unitary representation is a direct sum of equivalent irreducibles and the theory for commutative groups has a complete extension along the lines indicated at the end of the last section. Actually one can show that a unitary representation V is of Type I if and only if "almost all" of the primary constituents in the central decomposition are of Type I. Thus the circumstance preventing a more or less complete carry-over of the commutative theory to the general case is the existence of primary representations that are not of Type I. Let us see what those are like.

To begin with let us record some obvious properties of Type I primary representations. Let $V^1 = k_1 M^1$ and $V^2 = k_2 M^2$ where k_1 and k_2 are

ISBN 0-8053-6702-0/0-8053-6703-9, pbk

positive integers or ∞ and M^1 and M^2 are irreducible. Then either M^1 and M^2 are inequivalent, in which case V^1 and V^2 are disjoint, or M^1 and M^2 are equivalent, in which case one of V^1 and V^2 is equivalent to a subrepresentation of the other. Clearly having M^1 and M^2 equivalent is an equivalence relation and we make the following definition: Two Type I primary representations are *quasi-equivalent* if they are not disjoint. It is obvious that the direct sum of two Type I primary representations is again primary if and only if they are quasi-equivalent, and that every subrepresentation of a Type I primary representation is a primary representation.

Now it is an interesting (but not obvious) fact that all of these statements are true for general primary representations. Let us define two primary representations V^1 and V^2 to be quasi-equivalent if V^1 and V^2 are not disjoint. Then one can prove the following:

(a) If V^1 and V^2 are quasi-equivalent, then either V^1 is equivalent to a subrepresentation of V^2 or vice versa.

(b) If V^1 and V^2 are quasi-equivalent, and V^2 and V^3 are quasi-equivalent, then V^1 and V^3 are quasi-equivalent.

(c) If V^1 and V^2 are quasi-equivalent, then $V^1 \oplus V^2$ is also primary. Indeed if $V^1, V^2, \ldots$ are quasi-equivalent, then $V^1 \oplus V^2 \oplus \cdots$ is primary.

(d) Every subrepresentation of a primary representation is primary.

It follows from the above that equivalence classes within a given quasi-equivalence class form an *ordered semigroup*. Here the semigroup operator is the taking of direct sums and the ordering is that of being (equivalent to) a subrepresentation. Moreover, when the primary representation is of Type I, this ordered semigroup is isomorphic to that of the positive integers $+\infty$ under addition. Indeed, two members of the same quasi-equivalence class are equivalent if and only if their unique irreducible occurs with the same multiplicity, and the possible multiplicities are the positive integers and $+\infty$.

Now it can be proved quite generally that for every quasi-equivalence class of primary representations one and only one of the following occurs:

(a) The members are of Type I so that the ordered semigroup is that of the positive integers and $+\infty$ under addition.

(b) The ordered semigroup is isomorphic to the additive semigroup of all positive real numbers and ∞. In particular, there is no least member of the quasi-equivalence class.

(c) The ordered semigroup has just one member, ∞. In other words, any two members of the quasi-equivalence class are actually equivalent.

A primary representation for which (b) holds is said to be of Type II and one for which (c) holds is said to be of Type III. Quite simple examples

ISBN 0-8053-6702-0/0-8053-6703-9, pbk

show that Type II primary representations can exist. Let G be a countable discrete group with the property that every element but the identity has an infinite number of conjugates. For example, G might be the group of all one-to-one transformations of the rational numbers onto themselves of the form $t \to at+b$, where $a \neq 0$. It can be shown that the regular representation of G is primary and of Type II. It is more complicated to construct Type III primary representations but it can be proved that any group that has a Type II primary representation also has one of Type III, and vice versa.

One might be tempted to suppose that a Type II primary representation is always a direct integral of equivalent irreducibles, but this is not at all true. In fact, a direct integral of equivalent irreducibles is always of Type I. Just as a given infinite-dimensional Hilbert space may be realized alternatively either as $\mathfrak{L}^2[0,1]$ or as a space of square-summable sequences, so can a given Type I primary representation be written alternatively as a direct sum or as a direct integral of equivalent irreducibles. One can perhaps gain more insight into the nature of a Type II primary representations by considering the following facts. If V is of Type II, there will be a subrepresentation V' unique to within equivalence such that $V'+V' \simeq V$. It is consistent with our terminology: $2V=V+V$ to denote V' by $\frac{1}{2}V$. By induction we define $(1/2^n)V$ and hence rV, where r is any rational number whose denominator is a power of 2. But if h is any real number, we may write $h=\sum_{j=1}^{\infty} r_j$, where the r_j are dyadic rationals, and then define hV to be $r_1V \oplus r_2V \oplus \cdots$. Thus hV is defined for all $0<h\leqslant\infty$. If V is not the largest member of its quasi-equivalence class, it can be shown that hV are all inequivalent and that every representation quasi-equivalent to V is equivalent to some hV. Thus the members of the quasi-equivalence class, except for the biggest, can be regarded as consisting of all real multiples of a single reference member. It is as though we had a peculiar kind of "ideal" irreducible representation that could occur with fractional multiplicities as well as integral ones. At any rate it apparently does not pay to attempt to analyze Type II (or Type III) primaries any further. One can show that any unitary representation V can be written as a direct integral of irreducibles. [One simply chooses a maximal Boolean algebra of projections in $\mathfrak{R}(V,V)$ by Zorn's lemma and deals with it as one dealt with the set of all projections in $C\mathfrak{R}(V,V)$ in finding the central decomposition.] However, when V is primary of Type II, the result is wildly nonunique. It is possible to write $V \simeq \int L^s\,d\mu(s)$ and $V \simeq \int M^t\,d\nu(t)$ in such a way that L^s is inequivalent to any M^t, and so that no L^s is equivalent to any L^{s_1} with $s \neq s_1$, and no M^t is equivalent to any M^{t_1} with $t \neq t_1$.

It seems to be necessary to regard a quasi-equivalence class of a Type II or III primary representation as on the same footing with an irreducible representation. To determine all the unitary representations of a given G (to within equivalence), one must determine not only the equivalence classes of irreducibles but also the quasi-equivalence classes of Type II and

ISBN 0-8053-6702-0/0-8053-6703-9, pbk

Type III primaries. The general representation may be written in a canonical fashion as a direct integral of primaries—one from each equivalence class. However, one does not know how to describe the measure classes that occur and one does not have nearly so satisfactory a theory as one does in the commutative case. To make matters still worse, there is no case in which (a) there exist non-Type I primaries *and* (b) it is possible to classify the irreducible unitary representations and thus "actually find" $\hat{G}$. There is even a theorem indicating that (a) and (b) cannot both be expected to be true for the same group. If one can in any "sufficiently explicit" sense actually classify the irreducible unitary representations of G, one expects to be able to find "canonical forms" in $\hat{G}^{c,\infty}$, that is, one expects to be able to find a subset A of $\hat{G}^{c,\infty}$ that meets each equivalence class just once. Of course such an A will always exist by the axiom of choice if we put no conditions on it. However, the above considerations suggest that it be an analytic set at least and if this is so it can be proved that $\hat{G}$ is standard. On the other hand, it has been proved by J. Glimm that $\hat{G}$ will be standard if it is analytic and that it is analytic and hence standard if and only if every representation of G is of Type I. Thus when non-Type I primaries exist, the structure of $\hat{G}$ is such as to cast doubt on the possibility of finding $\hat{G}$ in any "explicit" sense. It seems that separable locally compact groups fall sharply into two classes consisting of "good" groups and "bad" groups. For the good groups $\hat{G}$ is standard and every representation is of Type I. For the bad groups, to say that $\hat{G}$ is not analytic is putting it mildly. It is not even countably separated, that is, no countable family of Borel functions can separate points. Moreover, there will always exist primary representations of Type II and Type III. It is customary to refer to the good groups as *Type I groups*. For Type I groups one has a more or less perfect generalization of the decomposition theory that is valid in the commutative case. Every unitary representation is a direct sum of multiplicity free representations and so can be written canonically as a direct sum of disjoint representations of uniform multiplicity. The mapping $\mu \to V^{\mu}$ sets up a one-to-one correspondence between all measure classes in $\hat{G}$ and all equivalence classes of multiplicity free unitary representations.

Clearly it is important to know which groups are Type I groups and which are not. Unfortunately, there is no general theorem stating that a group G is of Type I if and only if it has some easily verifiable property. We must content ourselves at present with special results and a number of sufficient conditions. For example, one knows the following:

(a) Every compact group is of Type I.

(b) Every locally compact commutative group is of Type I.

(c) Every semi-simple Lie group is of Type I.

ISBN 0-8053-6702-0/0-8053-6703-9, pbk

(d) Every real "algebraic" group is of Type I.

(e) A countable infinite discrete group G is of Type I if and only if it has a commutative normal subgroup N such that G/N is finite.

(f) Every connected nilpotent Lie group is of Type I.

(g) There exist connected Lie groups that are not of Type I.

Curiously enough, rather slight changes in the definition of a group can change it from being a Type I Lie group to a non-Type I Lie group. Also, there is an example of a non-Type I Lie group whose regular representation is of Type I.

In a later section we shall give a general method for determining $\hat{G}$ for certain G, and as a by-product we shall be able to show that these G are Type I groups. This method applies to some of the groups of greatest interest in quantum mechanics; namely, the group of rigid motions in Euclidean space and the so-called Poincaré group generated by the Lorentz group and the space–time translations.

It is not hard to show that $G_1 \times G_2$ is a Type I group if and only if both G_1 and G_2 are of Type I. Moreover, when G_1 and G_2 are of Type I, we have

$$\widehat{G_1 \times G_2} = \hat{G}_1 \times \hat{G}_2$$

in the sense that every unitary irreducible representation of $G_1 \times G_2$ is of the form $L \times M$, where L is an irreducible representation of G_1 and M is an irreducible representation of G_2.

We close this section with a brief account of the relationship between the notion of primary representations and the von Neumann–Murray theory of operator rings.

Let $\mathcal{H}$ be a Hilbert space. If $\mathcal{Q}$ is any family of bounded linear operators on $\mathcal{H}$, we let $\mathcal{Q}'$ denote the set of all bounded linear operators B such that $BA = AB$ for all $A \in \mathcal{Q}$. Then $\mathcal{Q}'' \supseteq \mathcal{Q}$ and $\mathcal{Q}_1 \supseteq \mathcal{Q}_2$ implies $\mathcal{Q}_1' \supseteq \mathcal{Q}_2'$. Thus $\mathcal{Q}''' \subseteq \mathcal{Q}'$ and $\mathcal{Q}''' = (\mathcal{Q}')'' \supseteq \mathcal{Q}'$. Thus $\mathcal{Q}''' = \mathcal{Q}'$, and we conclude that $\mathcal{Q} = \mathcal{Q}''$ if and only if $\mathcal{Q} = \mathcal{Q}_1'$ for some $\mathcal{Q}_1$. Clearly $\mathcal{Q}'$ is always closed under addition and multiplication and contains the complex multiples of the identity, that is, $\mathcal{Q}'$ is a subalgebra of the algebra of all bounded operators of $\mathcal{H}$. Thus it is only for subalgebras that we may hope to have $\mathcal{Q}'' = \mathcal{Q}$. Now if $\mathcal{Q}$ is *self-adjoint* in the sense that $A^* \in \mathcal{Q}$ whenever $A \in \mathcal{Q}$, then $\mathcal{Q}'$ is also self-adjoint. When $\mathcal{H}$ is finite-dimensional, *every* self-adjoint subalgebra $\mathcal{Q}$ that contains the identity is such that $\mathcal{Q} = \mathcal{Q}''$. This is false when $\mathcal{H}$ is infinite-dimensional, but it is clear that the self-adjoint algebras such that $\mathcal{Q} = \mathcal{Q}''$ must play a central role and be somewhat more like subalgebras in the finite-dimensional case than those

ISBN 0-8053-6702-0/0-8053-6703-9, pbk

for which $\mathcal{A} \subset \mathcal{A}''$. They may be characterized as those self-adjoint subalgebras that contain the identity and are closed in the so-called weak operator topology. This is the weakest topology under which $A \to (A\varphi, \psi)$ is continuous for every φ and ψ in $\mathcal{H}$. von Neumann, who initiated their study, called them *rings of operators*. This term is out of date and they are now variously termed *W*-algebras* and *von Neumann algebras*.

When $\mathcal{H}$ is finite-dimensional, one has a very simple structure theorem for the von Neumann algebras associated with $\mathcal{H}$. If $\mathcal{A}$ is any such, one can form its center and let $P_1, \ldots, P_r$ be the minimal projections in this center. For each j let $\mathcal{H}_j$ be the range of P_j. Each $A \in \mathcal{A}$ takes $\mathcal{H}_j$ into itself and the set of all operators in $\mathcal{H}_j$ obtained in this fashion is a von Neumann algebra $\mathcal{A}_j$. Clearly $\mathcal{A}$ is the direct sum of the $\mathcal{A}_j$. Moreover, it is easy to see that each $\mathcal{A}_j$ has a one-dimensional center and so cannot be further decomposed in this way. Now let $\mathcal{H}_1$ and $\mathcal{H}_2$ be two Hilbert spaces and let $\mathcal{B}$ denote the set of all bounded linear operators on $\mathcal{H}_1 \times \mathcal{H}_2$ of the form $T \times I$, where T is a bounded linear operator in $\mathcal{H}_1$. It can be shown that $\mathcal{B}'$ is the set of all bounded linear operators of the form $I \times T'$, where T' is a bounded linear operator in $\mathcal{H}_2$, and hence that $\mathcal{B}'' = \mathcal{B}$ so that $\mathcal{B}$ is a von Neumann algebra. The center of $\mathcal{B}$ is $\mathcal{B} \cap \mathcal{B}'$ and is clearly one-dimensional. In the finite-dimensional case it can be shown that every von Neumann algebra with a one-dimensional center can be obtained from a suitable "factorization" $\mathcal{H} = \mathcal{H}_1 \otimes \mathcal{H}_2$ of the underlying Hilbert space. Accordingly one calls a von Neumann algebra in a finite-dimensional Hilbert space a *factor*. We see then that a von Neumann algebra on a finite-dimensional Hilbert space is uniquely a direct sum of factors and that every factor is isomorphic to the algebra of all operators on a space whose dimension divides the order of the given one.

In their pioneering work in the 1936 Annals of Mathematics entitled "On rings of operators," von Neumann and Murray set out to see to what extent these results could be generalized to von Neumann algebras in infinite-dimensional Hilbert spaces. Laying aside the direct sum problem (which von Neumann dealt with later by himself), they concentrated on factors and asked whether every factor could be obtained from a factorization of the Hilbert space as in the finite-dimensional case. They showed that it could not and did so as follows. Consider the set of all projections in a factor. If a factor comes from a factorization, one can assign a dimension to every projection, this dimension being the dimension of the range of the projection P' such that the given projection is $P' \times I$. When $\mathcal{H}_2$ is infinite-dimensional, $P' \times I$ will have an infinite-dimensional range even when that of P' is finite-dimensional. This positive real-valued function can be shown to be the only one (up to a constant multiple) having certain simple properties. They were then able to show that a function with these formal properties exists on the projections of an arbitrary factor, that it is unique up to a multiplicative constant and that there are only the following five

ISBN 0-8053-6702-0/0-8053-6703-9, pbk

possibilities for its range:

I_n $d, 2d, \ldots, nd$ for some positive integer n and some positive real d

I_∞ $d, 2d, 3d, \ldots$ for some positive real d

II_1 $0 \leqslant x \leqslant a$ for some positive real a

II_∞ $0 \leqslant x \leqslant \infty$

III_∞ ∞

The factor is correspondingly said to be of type I_n, I_∞, II_1, II_∞, or III_∞. They showed that all of the first four types actually exist and that a factor comes from a factorization if and only if it is of Type I_n or Type I_∞. (In a later paper von Neumann produced an example of a factor of Type III_∞.)

To see the connection of all this with primary group representations, let us notice that for any group representation V the commuting ring $\mathfrak{R}(V, V)$ is necessarily a von Neumann algebra, being the $\mathcal{Q}'$ for a self-adjoint family of bounded operators. Moreover, it can be shown that for every von Neumann algebra $\mathcal{Q}$ there exists a countable group U of unitary operators such that $U' = \mathcal{Q}'' = \mathcal{Q}$. Thus every von Neumann algebra occurs as an $\mathfrak{R}(V, V)$. Now the definition of a factor is such that $\mathfrak{R}(V, V)$ is a factor if and only if V is primary, and for this reason primary representations are often called factor representations. Thus the study of von Neumann algebras that are factors is exactly the same thing as the study of the commuting algebras of primary group representations.

Let V be a primary representation of E. Then each projection P in $\mathfrak{R}(V, V)$ defines a subrepresentation V^P quasi-equivalent to V. Let $a(P)$ denote the member of $0 < x \leqslant \infty$ assigned to V^P by the fact that the quasi-equivalence class of V defines a semigroup isomorphic to a sub semigroup of $0 < x \leqslant \infty$. Of course the function $P \to a(P)$ is uniquely determined only up to a multiplicative constant. Translating the defining properties of a into properties expressible in terms of projections in $\mathfrak{R}(V, V)$, one verifies at once that $P \to a(P)$ is the von Neumann–Murray dimension function. $\mathfrak{R}(V, V)$ is a factor of Type II_∞ if V is of Type II and V is the infinite member of its quasi-equivalence class. It is a factor of Type II_1 if V is of Type II and is a finite member of its quasi-equivalence class. Similar statements can be made for Type I and Type III. The von Neumann–Murray theory and the theory of primary representations are equivalent in the sense that either can be deduced from the other. Of course the von Neumann–Murray theory came first. On the other hand, when one does everything from the representation theory point of view, one gets a unified treatment of the von Neumann–Murray theory and the multiplicity part of the unitary equivalence theory of self-adjoint operators. This is carried out in detail in Chapter 1 of the author's 1955 Chicago lecture notes.

ISBN 0-8053-6702-0/0-8053-6703-9, pbk

9. *Hilbert G-bundles, systems of imprimitivity, and induced representations*

We have given a fairly thorough discussion of how one reduces the problem of finding the most general unitary representation of a separable locally compact group G to that of finding its most general irreducible representation. However, except for the special case of one-dimensional representations, we have said nothing about how one determines the irreducible representations. Indeed, we have given no examples of irreducible representations of dimension greater than one and no explicit examples of representations of any sort except for the regular representations and the one-dimensional ones. In this section we shall study a general method for constructing irreducible representations and for studying the relationship between the unitary representations of a group and those of its closed subgroups. Let G be a separable locally compact group, let S be a standard Borel G-space, and let μ be a measure in S that is invariant under the G action. As remarked in Section 4, we obtain a representation V of G in the Hilbert space $\mathcal{L}^2(S,\mu)$ by setting $(V_x f)(s)=f(sx)$. The method of construction that we propose to study is a straightforward generalization of this one in which $\mathcal{L}^2(S,\mu)$ is replaced by $\mathcal{L}^2(S,\mu,\mathcal{H})$ for some Hilbert bundle $\mathcal{H}$ over S and in which G acts upon $S\Delta\mathcal{H}$ and not just upon S. The details are as follows.

By an *isomorphism* of the Hilbert bundle $S,\mathcal{H}$ with the Hilbert bundle $S',\mathcal{H}'$, we shall mean a Borel isomorphism α of $S\Delta\mathcal{H}$ on $S'\Delta\mathcal{H}'$ such that for each $s\in S$ the restriction of α to $s\times\mathcal{H}_s$ has some $t\times\mathcal{H}'_t$ for its range and is unitary when regarded as a map of $\mathcal{H}_s$ on $\mathcal{H}'_t$. The map carrying s into t is clearly an isomorphism of S with S'. We shall call it the associated

George W. Mackey, Unitary Group Representations in Physics, Probability, and Number Theory

ISBN 0-8053-6702-0/0-8053-6703-9, pbk

base isomorphism and denote it by α^{π}. By an *automorphism* of the Hilbert bundle $S,\mathcal{H}$, we shall mean an isomorphism of $S,\mathcal{H}$ with itself.

Now let G be a separable locally compact group and let $S,\mathcal{H}$ be a Hilbert bundle such that $S\Delta\mathcal{H}$ is also a G-space. If $s,\varphi\to(s,\varphi)x$ is an automorphism of $S\Delta\mathcal{H}$ for each $x\in G$, we shall say that $S,\mathcal{H}$ is a Hilbert G-bundle. For each fixed x we may write $(s,\varphi)x=t,L^{s}_{x^{-1}}(\varphi)$, where t depends only on s and x and may be written sx, and $L^{s}_{x^{-1}}$ is a unitary transformation from $\mathcal{H}_s$ to $\mathcal{H}_{sx}$. Clearly, $s,x\to sx$ makes S into a G-space. We shall call S the base G-space. Suppose now that we are given an invariant measure μ in S. We shall construct a unitary representation U of G generalizing our earlier construction for the special one in which all $\mathcal{H}_s$ were identical and one-dimensional and all L_x the identity.

Let f be a cross section. If f is Borel and $\int(f(s)\cdot f(s))\,d\mu(s)<\infty$, we shall call f a square integrable cross section. The set of all square integrable cross sections is a Hilbert space under the obvious addition and the inner product $f:g=\int_S(f(s),g(s))\,d\mu(s)$ provided we do not distinguish between cross sections that are almost everywhere equal. Let us denote this Hilbert space by $\mathcal{L}^2(S,\mu,\mathcal{H})$. For each $x\in G$ we obtain a unitary operator U_x in $\mathcal{L}^2(S,\mu,\mathcal{H})$ by letting $(U_xf)(s)=L^s_xf(sx)$. It is easy to see that $x\to U_x$ is a unitary representation of G. We call it the *representation induced* by the action of G on $S,\mu,\mathcal{H}$.

Let $S,\mathcal{H}$ and $S',\mathcal{H}'$ be two Hilbert G-bundles and let α be an isomorphism between $S,\mathcal{H}$ and $S',\mathcal{H}'$ as Hilbert bundles. We shall call it a G-bundle isomorphism if it commutes with the action of G, that is, if $(s,\varphi)x\alpha=(s,\varphi)\alpha x$ for all x in G. Now let there be given invariant measures μ and μ' on S and S', respectively. A G-bundle isomorphism α of $S,\mathcal{H}$ on $S',\mathcal{H}'$ will be said to be measure preserving if $\mu'(\alpha^{\pi}(E))=\mu(E)$ for all Borel subsets E of S. A measure preserving G-bundle isomorphism of $S,\mathcal{H}$ on $S',\mathcal{H}'$ defines a unitary map W of $\mathcal{L}^2(S,\mu,\mathcal{H})$ on $\mathcal{L}^2(S',\mu',\mathcal{H}')$, namely, the obvious analog of the U_x defined above with $s,\varphi\to(s,\varphi)\alpha$ replacing $s,\varphi\to(s,\varphi)x$. One verifies at once that W sets up a unitary equivalence between the representations of G induced by its actions on $S,\mathcal{H},\mu$ and $S',\mathcal{H}',\mu'$, respectively.

To survey the unitary representations of a given G defined by its actions on Hilbert bundles, we must survey the possible Hilbert G-bundles and their invariant measures, and by the above remark we need not distinguish between isomorphic G bundles. Let μ be an invariant measure in S, where $S,\mathcal{H}$ is a Hilbert G-bundle, and suppose that there exists an invariant Borel set E in S such that $\mu(E)\neq 0$ and $\mu(S-E)\neq 0$. We obtain two new Hilbert G-bundles over E and $S-E$, respectively, by restricting $s\to\mathcal{H}_s$ to these subsets, and the restrictions of μ to E and $S-E$ are nontrivial invariant measures. Given these two "sub-bundles," the original Hilbert G-bundle can be reconstructed by an obvious direct sum procedure. Moreover, the unitary representation of G induced by the original bundle

ISBN 0-8053-6702-0/0-8053-6703-9, pbk

is obviously equivalent to the direct sum of the unitary representations induced by the two sub-bundles. Clearly, we shall want to concentrate our attention on those systems $S, \mathcal{H}, \mu$ for which such a decomposition is impossible—especially if we are interested in constructing irreducible representations. Presumably we can then pass to the general case via summation and integration. When no such E exists, the action of G on S is said to be *ergodic* or *metrically transitive* with respect to the measure μ. Alternatively, one says that the invariant measure μ is ergodic or metrically transitive with respect to the given action. In Section 2 we defined a G-space to be *transitive* if there are no invariant subsets except the empty set and the whole space. Clearly, a transitive action is ergodic with respect to every invariant measure and one might suppose at first that all ergodic actions are essentially transitive in the sense that one can achieve transitivity by throwing away an invariant set of measure zero. However, this is false and the subject of ergodic theory may be regarded as owing its very existence to the fact that there exist ergodic actions—even of the additive group of the integers—in which every orbit has measure zero. We shall have more to say about this later. In the meantime we merely remark that while the strictly ergodic case has to be faced, the transitive one is susceptible of a much more complete analysis and, moreover, is the only one that arises in many interesting situations.

Given our separable locally compact group G, then, we proceed to study the Hilbert G-bundles and their invariant measures in the special case in which the action of G on S is transitive. Given such a G-bundle $S, \mathcal{H}$, let s_0 be any point in S and let K_{s_0} denote the subgroup of G consisting of all x with $s_0 x = s_0$. Then, for each $x \in K_{s_0}$, $L_x^{s_0}$ is a unitary map of $\mathcal{H}^{s_0}$ on $\mathcal{H}^{s_0}$ and $x \rightarrow L_x^{s_0}$ is a unitary representation of the subgroup K_{s_0}. In this way every Hilbert G-bundle over the transitive G-space S is invariantly associated with a unitary representation of the subgroup K_{s_0}. Now it is an extremely interesting and useful fact that the whole Hilbert G-bundle is determined to within isomorphism by the equivalence class of this representation of K_{s_0}. To be precise, let $\mathcal{H}$ and $\mathcal{H}'$ be any two Hilbert G-bundles over S and let L and L' be the unitary representations of K_{s_0} defined by restricting the actions to $\mathcal{H}^{s_0}$. Then $\mathcal{H}$ and $\mathcal{H}'$ define isomorphic G-bundles over S if and only if L and L' are equivalent representations of K_{s_0}. Moreover, every unitary representation of K_{s_0} is associated in this way with a Hilbert G-bundle over S. These facts are quite easy to prove and imply that the problem of finding the most general Hilbert G-bundle over a given transitive standard G-space S is completely equivalent to the problem of finding the most general unitary representation of K_{s_0}.

Now, to define a unitary representation of G induced by a Hilbert G-bundle, we must not only have the G-bundle but also an invariant measure μ in S. Another useful fact in the transitive case is that if an

ISBN 0-8053-6702-0/0-8053-6703-9, pbk

invariant measure exists at all in S this invariant measure is unique up to a multiplicative constant. Moreover, changing the constant does not change the unitary equivalence class of the induced representation. In other words, given a standard transitive G-space S one has a well-defined equivalence class of unitary representations of G for each equivalence class of unitary representations of K_{s_0}. One usually omits the phrase "equivalence class" and speaks of the representation of G induced by the representation L of K_{s_0}. Indeed, one can omit the reference to Hilbert G-bundles altogether and define an explicit representation U^L of G for each unitary representation L of K_{s_0} in such a manner that the equivalence class of U^L is that induced by the equivalence class of L. Here are the details. Consider the mapping $x \rightarrow s_0 x$ of G onto S. It is clear that $s_0 x = s_0 y$ if and only if $s_0 x y^{-1} = s_0$, so that $xy^{-1} \in K_{s_0}$. Thus $s_0 x = s_0 y$ if and only if $K_{s_0} x = K_{s_0} y$, that is, if and only if x and y belong to the same right K_{s_0} coset. In other words, we have a natural one-to-one mapping of the space G/K_{s_0} of all right K_{s_0} cosets onto the space S. Using this mapping we may carry the invariant measure μ on S into a measure $\tilde{\mu}$ on the space G/K_{s_0} of right cosets. Starting with a unitary representation L of K_{s_0} and the measure $\tilde{\mu}$, we construct a Hilbert space as follows. Consider the vector-space of all Borel functions f from G to $\mathcal{H}(L)$ that satisfy the following identity:

$$* \qquad f(kx) = L_k(f(x)) \text{ for all } k \in K_{s_0} \text{ and all } x \in G$$

Notice that the set of functions satisfying the identity are invariant under right translation; and notice also that if f satisfies the identity *, then $(f(x) \cdot f(x))$ is a constant on the right K_{s_0} cosets. Indeed, $(f(kx) \cdot f(kx)) = (L_k f(x) \cdot L_k f(x)) = (f(x) \cdot f(x))$ by unitarity. We give G/K_{s_0} the Borel structure of S by means of the above mapping and it is not hard to see that $(f(x) \cdot f(x))$ defines a function on G/K_{s_0} that is a Borel function. Thus it makes sense to consider $\int_{G/K_{s_0}} (f(x) \cdot f(x))\, d\tilde{\mu}(x)$. And it is clear that it is invariant under right translation of the function f. We now form the Hilbert space of all Borel functions that satisfy the identity * and the following inequality:

$$** \qquad \int_{G/K_{s_0}} (f(x) \cdot f(x))\, d\tilde{\mu}(x) < \infty,$$

where one identifies functions equal almost everywhere, and defines an inner product by $f:g = \int_{G/K_{s_0}} (f(x) \cdot g(x))\, d\tilde{\mu}(x)$. For each $x \in G$ let $U_x^L f(y) = f(yx)$. Then $x \rightarrow U_x^L$ is a unitary representation of G. We call it the unitary representation of G *induced* by the unitary representation L of K_{s_0}. This is consistent with our earlier usage because the representation of G induced by the action of G on $S, \mathcal{H}, \mu$, where μ is an invariant measure in the transitive Borel G-space S and $S, \mathcal{H}$ is a Hilbert G-bundle over S is equivalent to U^L, where L is the unitary representation of K_{s_0} invariantly associated with the bundle.

Let us now ask which subgroups K of G arise as K_{s_0}'s for suitable transitive G-spaces S. It can be proved that K_{s_0} is always a closed subgroup

ISBN 0-8053-6702-0/0-8053-6703-9, pbk

of G. Moreover, given any closed subgroup K of G we may convert the space G/K of all right K cosets into a standard Borel G-space in the following manner. Let $\theta(x)$ be the right coset K_x to which x belongs. Then a subset E of G/K is defined to be a Borel set wherever $\theta^{-1}(E)$ is a Borel subset of G. Moreover, we define $\theta(x)y=\theta(xy)$. This does it. Since the subgroup leaving $\theta(e)$ fixed is just K we see that every closed K arises. It is easy to see that S_1 and S_2 are isomorphic as Borel G-spaces in the sense that there exists a Borel isomorphism g of S_1 on S_2 such that $g(sx)=g(s)x$ for all s and x if and only if, for each $s_1 \in S_1$ and each $s_2 \in S_2$, the subgroups K_{s_1} and K_{s_2} are conjugate (i.e., $\exists y \in G$ with $yK_{s_1}y^{-1}=K_{s_2}$). In other words, to determine the most general transitive standard Borel G-space is the same as to determine the most general closed subgroup—except that conjugate subgroups determine isomorphic Borel G-spaces. (Note that changing from s_0 to s_1 changes K_{s_0} to a conjugate subgroup.)

Unfortunately there are closed subgroups K of separable locally compact groups G such that the transitive G-space $S=G/K$ does not admit any invariant measure. It thus appears at first sight that the process described above for passing from unitary representations of a closed subgroup to unitary representations of the whole group only applies to special closed subgroups. However, only a slight modification is necessary in order to make it apply to all closed subgroups. Returning to our definition of the representation induced by a triple $\mu, S, \mathcal{H}$, where μ is an invariant measure in S, let us suppose that μ is only quasi-invariant. The definition goes through as before except that U_x need no longer be unitary because $E \rightarrow Ex$ need not be measure preserving. However, since μ is quasi-invariant, the measures μ and $E \rightarrow \mu(Ex)=\mu_x(E)$ do belong to the same measure class and we can form the Radon–Nikodym derivative ρ_x of μ_x with respect to μ. Let us change our definition of U_x to read: $U_x f(s)=\sqrt{\rho_x(s)}\, L_x f(sx)$. This definition agrees with the old when μ is invariant so that $\rho_x(s)\equiv 1$. Moreover, the factor $\sqrt{\rho_x(s)}$ just compensates for the noninvariance of the measure μ in such a fashion as to make the U_x unitary. It is not hard to verify that $x \rightarrow U_x$ is a unitary representation of G. Now wherever one has one quasi-invariant measure one immediately has many others because all measures in the class of a quasi-invariant measure are themselves quasi-invariant. It is therefore important to observe that changing μ to another measure in the same class does not change the equivalence class of the representation U. If $\rho=d\mu/d\mu'$, then the linear operator $f \rightarrow \sqrt{\rho}\, f$ is a unitary map of $\mathcal{L}^2(S,\mathcal{H},\mu')$ on $\mathcal{L}^2(S,\mathcal{H},\mu)$, which sets up an equivalence between the U defined by μ' and the U defined by μ.

In the special case in which the action of G on S is transitive so that S may be identified with G/K for some closed subgroup K of G, there will always exist at least one quasi-invariant measure in $S=G/K$. To see this let ν be any finite measure in the measure class of Haar measure and let

ISBN 0-8053-6702-0/0-8053-6703-9, pbk

$\theta(x) = Kx$. Then $\tilde{\nu}$ where $\tilde{\nu}(E) = \nu(\theta^{-1}(E))$ is a finite quasi-invariant measure on G/K. It is easy to see that the class of $\tilde{\nu}$ is independent of the choice of ν and it can in fact be proved that every quasi-invariant measure G/K is in the class of the $\tilde{\nu}$. Thus a standard transitive G-space always admits a unique invariant measure class.

It follows now that our construction for passing from a unitary representation L of a closed subgroup K of G may be carried out for any closed subgroup K whether or not G/K actually has an invariant measure. G/K will always have quasi-invariant measures and the construction carried out using one of these is independent of the one we choose. To within equivalence at least we may speak without ambiguity of *the* unitary representation U^L of G induced by any unitary representation L of any closed subgroup K of G. In the special case in which L is taken to be the one-dimensional identity of K, U^L is just the representation of G in the space $\mathfrak{L}^2(S,\mu)$ defined by setting $(U_x f)(s) = f(sx)\sqrt{\rho_x(s)}$, and where K is the identity subgroup, so that S may be identified with G, we obtain the regular representation. The representations U^L play a central role in the theory of group representations. Many of the known irreducible representations of the more complicated groups are of the form U^L, where L is a one-dimensional representation of a proper subgroup and many important applications of the theory involve a study of the decomposition theory of U^L for L's such that U^L is not irreducible. To have a complete theory of the unitary representations of a separable locally compact group G one needs to know not just the unitary representations of G but also the unitary representations L of each closed subgroup K and the structure of U^L for each L. It is easy to show that inducing commutes with direct sums in the sense that $U^{L \oplus M}$ is always equivalent to $U^L \oplus U^M$. Thus in studying the structure of the various U^L, one may concentrate on irreducible L's. In particular, we see that U^L cannot be irreducible unless L is. As an example in which fairly complete answers can be given, let us consider the commutative case. Let χ_0 be a one-dimensional character of the closed subgroup K of the separable locally compact commutative group G. It can be shown that U^{χ_0} is a multiplicity free representation of G and that the measure class in $\hat{G}$ that defines it may be constructed as follows. Let $K^\perp$ be the closed subgroup of $\hat{G}$ consisting of all χ with $\chi(x) = 1$ for all x in K. Let ν be a Haar measure for $K^\perp$ and let ν' be the measure in $\hat{G}$ such that $\nu'(E) = \nu(K^\perp \cap E)$ for all Borel subsets E of $\hat{G}$. Finally, let $\chi_0\nu'$ be the measure $E \rightarrow \nu'(E\chi_0)$, the translate of ν' by χ_0. The measure class of U^{χ_0} is just $\chi_0\nu'$. Notice that this measure class is supported by the coset $\chi_0 K^\perp$ and this is precisely the set of all characters which coincide with χ_0 when restricted to K. When G is compact as well as commutative, our result reduces to the statement that U^{χ_0} is the direct sum of all of those irreducible representations of G which coincide with χ_0 when restricted to K. This result is just the specialization to the commutative case of a

ISBN 0-8053-6702-0/0-8053-6703-9, pbk

theorem about not necessarily commutative compact groups known as the Frobenius reciprocity theorem. Namely, let L be any irreducible unitary representation of the closed subgroup K of the separable compact group G and let M be any irreducible unitary representation of G. Then the multiplicity with which M occurs in the decomposition of U^L is exactly equal to the multiplicity with which L occurs in the decomposition of the restriction of M to K. Except in rather special cases (such as when G is commutative), the restriction of an irreducible representation to a subgroup will no longer be irreducible and parallel to the question of finding the decomposition of U^L for all irreducible representations L of the closed subgroups K, we have the question of finding the decomposition of the restriction of M to K for all irreducible representations M of G. The Frobenius reciprocity theorem tells us that these questions are closely related—indeed, that for fixed K a solution of either problem for all representations implies a solution of the other. One can generalize the Frobenius reciprocity theorem to groups that are neither compact nor commutative; but the statement is a little complicated and will not be given here.

Our statement about the structure of U^{χ_0} when G is commutative is essentially the Frobenius reciprocity theorem for locally compact commutative groups and in this case the duality between G and $\hat{G}$ leads to certain refinements. It follows from what has already been said that every character χ_0 of K is the restriction to K of at least one character of G. Thus as an algebraic object $\hat{K}$ coincides with the quotient group $\hat{G}/K^{\perp}$. It is possible to prove that the natural mapping of $\hat{G}/K^{\perp}$ on $\hat{K}$ is actually a topological group isomorphism so that we may identify $\hat{K}$ with the topological quotient group $\hat{G}/K^{\perp}$. In addition, we note that $K \subseteq K^{\perp\perp}$ and here that $(K^{\perp\perp})^{\perp} \subseteq K^{\perp} \subseteq (K^{\perp})^{\perp\perp}$. Since $K^{(\perp\perp)\perp} = (K^{\perp})^{\perp\perp} = K^{\perp\perp\perp}$ we see that $K^{\perp\perp\perp} = K^{\perp}$. Actually it is easy to prove that $K = K^{\perp\perp}$ if and only if K is closed. Thus every closed K is $K_1^{\perp}$ for some closed subgroup of the dual and $K \rightarrow K^{\perp}$ sets up a one-to-one inclusion inverting correspondence between closed subgroups of G and $\hat{G}$, respectively, such that $\hat{G}/K^{\perp}$ and K are dual pairs and G/K and $K^{\perp}$ are dual pairs. These facts play an important role in applications to number theory. We note in particular that whenever K is discrete and G/K is compact (as when K is the subgroup of the real line consisting of all the integers), then $\hat{K}$ is compact and $\hat{G}/\hat{K}$ is discrete from which it follows that $K^{\perp}$ is discrete and $\hat{G}/K^{\perp}$ is compact.

Returning to the general case, let K_1 and K_2 be two closed subgroups of the separable locally compact group G and suppose that $K_1 \subseteq K_2$. Let L be a unitary representation of K_1. In addition to the induced representation U^L of G, we may also form another unitary representation of G as follows. First use the inducing process to define U^L as a unitary representation of K_2. (We distinguish these two meanings of U^L by subscripts: ${}_GU^L$, ${}_{K_2}U^L$.)

ISBN 0-8053-6702-0/0-8053-6703-9, pbk

Then form the representation U^M of G induced by the representation $M = {}_{K_2}U^L$ of K_2. It is a useful theorem that the representations U^M and ${}_GU^L$ are always equivalent. In an awkward but less ambiguous symbolism

$$ {}_GU^L \simeq {}_GU^{({}_{K_2}U^L)} $$

We shall refer to this theorem as the chain rule or the theorem on inducing in stages. If we take $K_1 = \{e\}, K_2 = K$ to be arbitrary and L to be the one-dimensional identity representation of K_1, we see that ${}_{K_2}U^L$ is the regular representation of K_2 and ${}_GU^L$ is the regular representation of G. Thus we have the corollary that U^L is equivalent to the regular representation of G whenever L is equivalent to the regular representation of K.

Let G be as above and let $S, \mathcal{H}$ be a Hilbert G-bundle. Let μ be a measure in S that is quasi-invariant under the action of G and let U be the unitary representation of G induced by the action of G in $S, \mu, \mathcal{H}$. For each Borel subset E of S let P_E be the projection operator that takes the cross section f in $\mathcal{L}^2(S, \mathcal{H}, \mu)$ into $\varphi_E f$. As usual φ_E is the function that is one on E and zero outside of E. Then $E \to P_E$ is a projection valued measure on S and we verify at once that

$$ U_x^{-1} P_E U_x = P_{Ex^{-1}} \qquad (\dagger) $$

for all $x \in G$ and all Borel sets $E \subseteq S$.

If U is any unitary representation of G and S is a standard Borel G-space, we call a projection valued measure on S that satisfies the identity (†) a *system of imprimitivity for U based on S*. We shall call the system of imprimitivity defined above for the induced representation U the *associated system of imprimitivity*. When the action of G on S is transitive so that the Hilbert G-bundle $S, \mathcal{H}$ is completely determined by the pair K, L, where L is a unitary representation of the closed subgroup K of G, we may describe the associated system of imprimitivity directly without introducing the bundle $s \to \mathcal{H}_s$. Recall that the Hilbert space $\mathcal{H}(U^L)$ consists of functions f from G to $\mathcal{H}(L)$ satisfying the identity $f(kx) = L_k f(x)$ for all $k \in K, x \in G$. The associated system of imprimitivity P is then such that $(P_E f)(x) = \varphi_E(\bar{x}) f(x)$, where $\bar{x}$ is the right coset Kx.

The importance of the associated system of imprimitivity lies in the fact that it essentially determines the structure of U as an induced representation. The result is easiest to state and also most useful when S is a transitive G-space. We confine ourselves to a discussion of this case. Let K be a closed subgroup of the separable locally compact group G. Let U be a unitary representation of G and let P be a system of imprimitivity for U based on the G-space G/K. Then there exists a unitary representation L of K and a unitary operator W from $\mathcal{H}(U)$ to $\mathcal{H}(U^L)$ such that $WU_xW^{-1} = U_x^L$ and $WP_EW^{-1} = P_E^L$ for all $x \in G$ and all Borel $E \subseteq G/K$. Here U^L is the representation of G induced by the representation L of K and P^L is the

ISBN 0-8053-6702-0/0-8053-6703-9, pbk

associated system of imprimitivity. The equivalence class of L is uniquely determined by the pair U, P and the commuting algebra of L is isomorphic to the subalgebra of $\mathcal{R}(U, U)$ consisting of all T that commute with all P_E. We shall refer to this collection of statements as the imprimitivity theorem. As we shall see below it has many applications.

Let S be a standard Borel space and let $s \rightarrow L^s$ assign a unitary representation to each $s \in S$. Let $\mathcal{H}$ denote the assignment $s \rightarrow \mathcal{H}(L^s)$. If $S \Delta \mathcal{H}$ may be given a Borel structure making $S, \mathcal{H}$ into a Hilbert bundle, it will become a Hilbert G-bundle if we set $(s, \varphi)x = s, L^s_{x^{-1}}(\varphi)$. When this is possible we shall say that $s \rightarrow L^s$ is a Borel function. The Hilbert G-bundles we obtain in this way are just those for which the action of G on S is trivial. In particular, any measure μ in S is quasi-invariant and leads to the construction of an induced representation. It turns out that the equivalence class of this induced representation depends only upon the class to which μ belongs and is independent of the choice of the Borel structure in $S \Delta \mathcal{H}$. When S is countable, it reduces to the direct sum of the L^s over all s for which $\mu(s) \neq 0$. In general we call it the *direct integral* of the L^s with respect to μ and write it as $\int_S L^s\, d\mu(s)$. The representation L alluded to at the end of Section 8 is just $\int_{\hat{G}} L^s\, d\mu(s)$, where L^s is a member of the equivalence class that *is* the point s.

ISBN 0-8053-6702-0/0-8053-6703-9, pbk

10. *The unitary representations of semi-direct products*

We introduced Section 9 by remarking that we had so far failed to present any examples of irreducible unitary representations except one-dimensional ones. We shall now show how the results of Section 9 may be used not only to construct many such examples, but to give us a satisfying explicit determination of all irreducible unitary representations of the members of an interesting and important family of noncommutative groups.

Let G be a separable locally compact group. If G is not commutative, it may be almost commutative (meta-abelian) in the sense that it has a closed commutative normal subgroup N such that the quotient group G/N is commutative. Meta-abelian groups are especially easy to analyze when G is a "semi-direct product" of N and G/N in the sense that G admits a second closed subgroup H (necessarily isomorphic to G/N) such that every element x of G is uniquely a product nh, where $n \in N$ and $h \in H$. For example, the group of all permutations of 3 objects is a semi-direct product of a cyclic group of order 3 and a cyclic group of order 2, and the group of all distance-preserving, orientation-preserving one-to-one transformations in the plane is a semi-direct product of the commutative group of all translations and the commutative group of all rotations about some fixed point. The theory we shall give does not depend upon the commutativity of H and reduces the problem of finding the irreducible representations of G to that of finding the irreducible representations of H and certain of its subgroups. This is useful even if we do not immediately know the irreducible representations of H and the relevant subgroups because these may also be semi-direct products to which the theory applies and we may proceed inductively. For example the group of all permutations of 4

ISBN 0-8053-6702-0/0-8053-6703-9, pbk

George W. Mackey, Unitary Group Representations in Physics, Probability, and Number Theory

objects is a semi-direct product of a commutative group of order 4 and the group of all permutations of three objects. Even if H is not a semi-direct product, the reduction can be useful because H and the relevant subgroups are "smaller" and perhaps simpler than G and other methods may be applicable.

The notion of semi-direct product itself does not depend upon the commutativity of either N or H and may be defined quite generally as follows. The separable locally compact group G is said to be a semi-direct product of the closed normal subgroup N and the closed subgroup H if every x in G is uniquely representable in the form nh, where $n \in N$ and $h \in H$. Given the representability, $G = NH$, uniqueness holds if and only if $N \cap H = \{e\}$. Notice now that

$$\begin{aligned}(n_1h_1)(n_2h_2) &= n_1h_1n_2h_1^{-1}h_1h_2 \\ &= \left[n_1\left(h_1n_2h_1^{-1}\right)\right](h_1h_2)\end{aligned}$$

and that $h_1n_2h_1^{-1}$ is in N because N is normal. Thus, if we know N and H and the mapping $n,h \to hnh^{-1}$ of $N \times H$ into N, we can reconstruct G. Note also that for each fixed h, $n \to hnh^{-1}$ is automorphism α_h of N and that $h \to \alpha_h$ is a homomorphism of H into the group of all automorphisms of N. Conversely, suppose that we are given two separable locally compact groups N and H and a homomorphism $h \to \alpha_h$ of H into the group of all automorphisms of N. We may then construct a new group G that is a semi-direct product of subgroups isomorphic to N and H as follows. We let the points of G be pairs n,h with $n \in N$ and $h \in H$ and we set $(n_1,h_1)(n_2,h_2) = n_1\alpha_{h_1}(n_2), h_1h_2$. It is readily verified that G is a group and that this group is a locally compact topological group in the product topology whenever $\alpha_h(n)$ is continuous in both variables. The subset of all n,h with $h = e$ is a closed subgroup isomorphic to N, the subset of all n,h with $n = e$ is a closed subgroup isomorphic to H, and G is a semi-direct product of these two subgroups. In obvious senses of the words, we see that the notion of semi-direct product may be defined either "internally" or "externally." A group of linear transformations is always a group of automorphisms of the additive group of the vector space in which they act, and one can form the semi-direct product of these two groups using the identity homomorphism. This semi-direct product is isomorphic to the group generated by the translations and the linear transformations. For example, the inhomogeneous Lorentz group, or Poincaré group, is a semi-direct product of the space–time translations and the Lorentz transformations.

Let G be a semi-direct product of the separable locally compact groups N and H with respect to a homomorphism $h \to \alpha_h$ of H into the group of automorphisms of N. We wish to study the unitary representations of G. Let V be any such, and let A and B be its restrictions to N and H,

ISBN 0-8053-6702-0/0-8053-6703-9, pbk

respectively. That is, let $A_n = V_{n,e}$ and let $B_h = V_{e,n}$. Since

$$V_{(n,h)} = V_{(n,e)(e,h)} = V_{n,e}V_{e,h} = A_n B_h$$

we see that V is completely known as soon as the unitary representations A and B of N and H are known. Of course, these representations cannot be chosen arbitrarily, but must be so selected that

$$V_{(n_1,h_1)(n_2,h_2)} = V_{n_1,h_1}V_{n_2,h_2}$$

for all n_1, h_1, n_2, h_2. Since

$$(n_1,h_1)(n_2,h_2) = n_1\alpha_{h_1}(n_2), h_1h_2$$

this condition is just

$$A_{n_1\alpha_{h_1}(n_2)}B_{h_1h_2} = A_{n_1}B_{h_1}A_{n_2}B_{h_2}$$

and since A and B are both representations, it reduces to

$$A_{n_1}A_{\alpha_{h_1}(n_2)}B_{h_1}B_{h_2} = A_{n_1}B_{h_1}A_{n_2}B_{h_2}$$

Canceling A_{n_1} and B_{h_2} and dropping superfluous subscripts, the condition becomes the very simple identity: $A_{\alpha_h(n)}B_h = B_hA_n$. This may be suggestively rewritten in the form

$$B_hA_nB_h^{-1} = A_{\alpha_h(n)} \tag{*}$$

The identity (*) reminds one of the identity defining a system of imprimitivity. It actually becomes this if we suppose that N is commutative and apply the spectral theorem to replace A by the projection valued measure that defines it.

To be specific, note first that each α_h defines a dual automorphism α_h^* of $\hat{N}$. Indeed, for each $h \in H$ and each $\chi \in N$, $n \to \chi(\alpha_h(n))$ is again a character. If we call it $\alpha_h^*(\chi)$, we verify without difficulty that $\chi \to \alpha_h^*(\chi)$ is an automorphism of $\hat{N}$ and that $h \to \alpha_{h^{-1}}^*$ is a homomorphism. Thus if we set $\chi h = \alpha_{h^{-1}}^*(\chi)$ we find that $\hat{N}$ becomes a standard Borel H-space. Now, by the spectral theorem each unitary representation A of N is associated with a unique projection valued measure P^A on $\hat{N}$ and P^A determines A. It thus makes sense to ask: Under what conditions on P^A does A satisfy the identity (*) with a given B? A straightforward calculation shows that A satisfies (*) if and only if the projection valued measure P^A on N satisfies

$$B_hP_E^AB_{h^{-1}} = P_{Eh^{-1}}^A \tag{**}$$

ISBN 0-8053-6702-0/0-8053-6703-9, pbk

and (**) says nothing more nor less than: P^A is a system of imprimitivity for B based on $\hat{N}$. It follows that finding the most general pair A, B such that $n, h \to A_n B_h$ is a unitary representation of G, and hence finding the most general unitary representation of G is the same thing as finding the most general pair P, B consisting of a unitary representation B of H and a system of imprimitivity P for B based on the Borel G-space $\hat{N}$.

Our main theorem is a consequence of this remark and the imprimitivity theorem described in Section 9. Let V be any primary unitary representation of G. It is not hard to show then that P^A must have the following two properties.

(i) It is uniformly k-dimensional for some k.

(ii) The defining measure class in $\hat{N}$ must be invariant and ergodic under the action of H.

We remark that when N is compact this is just a complicated way of saying that A must be a direct sum of *all* the characters in a *single orbit*, each one occuring with the same multiplicity k. (Here "all" is the translation of invariant and "single orbit" is the translation of ergodic.) Now, we have already classified ergodic invariant measure classes into those that are strictly ergodic and those that are essentially transitive. Let us say that our primary representation V is of *transitive type* (relative to N) if the ergodic invariant measure class associated with P^A is essentially transitive; that is, gives positive measure to an *orbit* $\mathcal{O}$ of H in $\hat{N}$. In this event $P^A_E = P^A_{\mathcal{O} \cap E}$ for all E and we may suppose that P^A is defined only in the orbit $\mathcal{O}$ and hence is a transitive system of imprimitivity for B based on $\mathcal{O}$. It follows from the imprimitivity theorem that $B = U^L$ for some unitary representation L of the subgroup H_χ of H leaving some $\chi \in \mathcal{O}$ fixed. Moreover,

$$\begin{aligned}\mathfrak{R}(V,V) &= \mathfrak{R}(B,B) \cap \mathfrak{R}(A,A)\\ &= \mathfrak{R}(B,B) \cap \mathfrak{R}(P^A,P^A)\end{aligned}$$

and we have remarked that this is isomorphic to $\mathfrak{R}(L,L)$. Thus L must be primary and of the same type as V and V will be irreducible if and only if L is irreducible. Conversely, let χ be any member of $\hat{N}$, let $\mathcal{O}$ be its orbit under H, let H_χ be the subgroup of all h with $[\chi]h = \chi$, and let L be any primary unitary repsentation of H_χ. Let B^L be the representation of H induced by L and let P^L be the associated system of imprimitivity transferred from H/H_χ to $\mathcal{O}$ by the map $x \to [\chi]x$. Defining $P^A_E = P^L_{E \cap \mathcal{O}}$, we obtain a system of imprimitivity for B^L based on $\hat{N}$ and hence a primary representation V of G of transitive type whose orbit is $\mathcal{O}$. Thus every orbit $\mathcal{O}$ occurs and for each $\chi \in \mathcal{O}$ every primary representation L of H_χ occurs. It is not hard to see that the representation V defined by P^L and B^L may

ISBN 0-8053-6702-0/0-8053-6703-9, pbk

be defined directly from χ and L as follows. Let NH_χ be the subgroup of G consisting of all nh with $n \in N$, $h \in H_\chi$. Then we obtain a unitary representation χL of NH_χ by setting $(\chi L)_{n,h} = \chi(n)L_h$. V is just the unitary representation $U^{\chi L}$ of G induced by χL.

Expanded and augmented here and there, the above considerations lead to a proof of the following theorem: Let N, H, and G be as above. Let J be a subset of $\hat{N}$ that meets each H orbit once and only once. For each $\chi \in J$ let H_χ be the closed subgroup of all h with $\alpha_h^*(\chi) = \chi$. Let L be any primary unitary representation of H_χ and let χL be the unitary representation $n,h \to \chi(n)L_h$ of NH_χ. Then the induced representation $U^{\chi L}$ of G is primary and of transitive type. Every primary representation of G of transitive type is equivalent to $U^{\chi L}$ for some unique pair χ, L and $U^{\chi L}$ is irreducible, Type I, Type II, or Type III according as L is irreducible, Type I, Type II, or Type III.

Notice that whenever H is commutative, then every H_χ is commutative and every orbit of $\hat{N}$ gives us as many distinct irreducible representations as H_χ has characters. These irreducible representations will all be of the form U^L for one-dimensional L but will have dimension equal to the number of elements in $\mathcal{O}_\chi$.

In the general case there will be primary representations and irreducible representations that are not of transitive type and the theorem quoted above tells us nothing about them. Indeed, every strictly ergodic invariant measure class in $\hat{N}$ is associated with a large family of them. On the other hand, it may happen that no such measure classes exist; then our theorem gives complete results. In fact, the following theorem is true: If the orbit cross section J can be chosen so as to be an analytic subset of $\hat{N}$, then every primary representation of G is of transitive type and hence of the form $U^{\chi L}$ described above. When J can be so chosen we shall say that N is *regularly imbedded* in G, or that our semi-direct product is regular. Notice, in particular, that when the semi-direct product is regular and every H_χ is of Type I, then G itself is of Type I. Many solvable Lie groups can be proven to be Type I groups in this way. In particular, a regular semi-direct product of two commutative groups is always a Type I group in which every irreducible unitary representation is induced by a one-dimensional representation of a subgroup.

Example I: Let N be the cyclic group of order 3; let H be the cyclic group of order 2; and let $\alpha_h(x) = x^{-1}$ if $h \neq e$. Then G is the unique noncommutative group of order 6 and is isomorphic to the group of all permutations on 3 objects. $\hat{N}$ is also the cyclic group of order 3 and $\alpha_h^*(x) = x^{-1}$ if $h \neq e$. Thus there are two orbits in $\hat{N}$: the one containing the identity alone and the one containing χ and χ^{-1}, where χ is not the identity. The subgroup leaving e fixed is H. The two characters of H define one-dimensional (irreducible) representations of G via $nh \to \chi(h)$. [Quite generally, $nh \to L_h$ is an irreducible representation of G whenever L is an irreducible representation of H and these are just the representations associated with the orbit $\{e\}$.] The subgroup H_χ is $\{e\}$ so there is a unique irreducible representation associated with the orbit of χ. It is the induced representation U^χ and is two dimensional since G/N has two elements. Every irreducible representation of G is equivalent to one of these. If we replace N by an arbitrary cyclic group obtaining the general "dihedral group," we get very similar results. If χ_0 is a generator for $\hat{N}$ and the order n of χ_0 is odd, then the orbits of $\chi_0, \chi_0^2, \cdots \chi_0^{(n-1)/2}$ are all distinct and the $U^{\chi_0^j}$ are inequivalent and two dimensional. The only other irreducible representations are the two one-dimensional ones defined by the characters of H. If the order n of χ_0 is even, then $\{\chi_0^{n/2}\}$ is an orbit with just one member and $H_{\chi_0^{n/2}} = H$. Thus in this case there are four one-dimensional representations. The rest are all two dimensional and are of the form $U^{\chi_0^j}$ with $j = 1, 2, \ldots, (n-2)/2$.

Example II: N is the group of all translations in the Euclidean plane, H is the group of all rotations about 0,0, and $\alpha_h(x,y) = x',y'$, where x',y' is x,y rotated through the angle representing the rotation h. For each pair of real numbers a,b let $\chi^{a,b}(x,y) = e^{i(ax+by)}$. Then the $\chi^{a,b}$ constitute just exactly the members of $\hat{N}$. Moreover, if α_h is a rotation through the angle θ in the x,y plane, then α_h^* is a rotation through the angle θ in the a,b plane. Hence $\chi^{a,b}$ and χ^{a_1,b_1} are in the same orbit if and only if $a^2 + b^2 = a_1^2 + b_1^2$. Hence the set of all points $r,0$ with $r \geqslant 0$ is a set cutting each orbit just once. Since this set is a Borel set, our semi-direct product is regular. Now for each $r > 0$, $H_{\chi^{r,0}}$ contains only the identity. Hence $U^{\chi^{r,0}}$ is an irreducible infinite-dimensional unitary representation of G and is the only one associated with the orbit $a^2 + b^2 = r^2$. The $U^{\chi^{r,0}}$ are all inequivalent and except for the trivial one-dimensional representations defined by the characters of H they are all of the irreducible unitary representations of G.

Example III: Let N be the additive group of all real numbers and let H be the multiplicative group of all nonzero real numbers. Let $\alpha_h(n)$ be the result of multiplying n by h^2. Then N is isomorphic to $\hat{N}$ in such a way that $\alpha_h^*(n)$ is also multiplication by h^2. There are clearly just three orbits: the set $\mathcal{O}^+$ of all positive real numbers, the set $\mathcal{O}^-$ of all negative real numbers, and $\{0\}$. The subgroup H_1 of H leaving 1 fixed is the two element group $1, -1$. Let χ_0 and χ_1 be the two characters of H_1. Then $x,h \to e^{ix}\chi_j(h)$ is a one-dimensional representation L_j of NH_1 for $j = 1, 2$. Similarly $x,h \to e^{-ix}\chi_j(h)$ is a one-dimensional representation M_j of NH_1 for $j = 1, 2$. The four infinite-dimensional representations $U^{L_1}, U^{L_2}, U^{M_1}, U^{M_2}$ are all inequivalent and constitue all of the irreducible unitary representations of G except for the one-dimensional ones defined by the characters of H.

11. Unitary representations of general group extensions

If G is an arbitrary separable locally compact group, one can ask whether it has closed normal subgroups. If it has such a subgroup N, one can form the quotient group G/N and, to some extent, reduce the study of the properties of G to the study of the properties of N and G/N. The simplest case is that in which G is isomorphic to $N \times G/N$, and the next simplest is that in which G is a semi-direct product of N and a closed subgroup H isomorphic to G/N. We have just seen how the unitary representation theory of G can be reduced to that of N and of certain selected subgroups of G/N whenever N is commutative and G is a semi-direct product. It is natural to ask to what extent one can still do this when N is an arbitrary closed normal subgroup of G. We shall indicate briefly how, with one important change, substantially the whole theory can be carried over—not to the case of completely general N—but to the case of an arbitrary N of Type I.

Let N be of Type I and let $\hat{N}$ denote the set of all equivalence classes of irreducible unitary representations of N. For each x in G and each irreducible unitary representation L of N, let L^x denote the representation $n \to L_{xnx^{-1}}$. Then if x is in N, $L_n^x = L_{xnx^{-1}} = L_x L_n L_x^{-1}$ so L and L^x are equivalent. It follows that the equivalence class (L^x) of L^x depends only upon the image of x in G/N and hence that $(L)^z$ is well defined whenever (L) is in $\hat{N}$ and z is in G/N. It is not hard to show that (L), $z \to (L)^z$ converts $\hat{N}$ into a standard Borel G/N-space. Just as above we may single out certain primary unitary representations of G as being of *transitive type* with respect to N, and as before it can be shown that no other kind occurs if $\hat{N}$ admits an analytic cross section for the G/N orbits. For each (L) in $\hat{N}$ we define $(G/N)_{(L)}$ to be the (closed) subgroup of G/N consisting of all

George W. Mackey, Unitary Group Representations in Physics, Probability, and Number Theory

z with $(L)^z=(L)$, and one might suppose that the primary unitary representations of G that are of transitive type and are associated with the orbits of (L) would correspond one-to-one to the primary representations of $(G/N)_{(L)}$. However, this is not so in general. It turns out instead that the primary unitary representations of G associated with the orbit of (L) correspond one-to-one to the primary unitary "projective" representations of G/N associated with a certain essentially uniquely defined "multiplier" for G/N.

A few words of explanation are in order. By definition, a *projective unitary representation* of a group G is a map $x \to U_x$ of G into the unitary operators of some Hilbert space $\mathcal{H}(U)$ in which the familiar identity $U_{xy}=U_xU_y$ is replaced by the weaker one $U_{xy}=\sigma(x,y)U_xU_y$, where $x,y\to\sigma(x,y)$ is a function from $G\times G$ to the complex numbers of modulus one. The reason for the word "projective" is this: A unitary operator defines an automorphism of the "projective geometry" of all closed subspaces of the Hilbert space in which it acts. However, if c is a complex number of modulus one, then U and cU determine the same automorphism, and U is determined by the automorphism only up to multiplication by a complex number of modulus one. Thus if we have a homomorphism of the group G into the group of automorphisms of this projective geometry, it may be described by assigning a unitary operator U_x to each x in G; but, because of the indicated ambiguity, we cannot conclude that $U_{xy}=U_xU_y$ but only that U_{xy} is some complex number of modulus one (depending on x and y) multiplied by U_xU_y. Of course, one can always hope to readjust the arbitrary constants in the U_x so that $\sigma(x,y)\equiv 1$. Unfortunately, this hope is not always realizable. If we replace U_x by $U'_x=U_x/g(x)$, then

$$\begin{aligned} U'_{xy} &= U_{xy}/g(xy) \\ &= (\sigma(x,y)/g(xy))\,U_xU_y \\ &= (\sigma(x,y)g(x)g(y)/g(xy))\,U'_xU'_y \end{aligned}$$

Thus σ can be eliminated if and only if g can be chosen so that $\sigma(x,y)=g(xy)/(g(x)g(y))$. It turns out that there are multipliers that cannot be thrown into this form and thus that cannot be eliminated by readjusting the arbitrary constants.

It follows immediately from the associative law that every multiplier satisfies the identity

$$\sigma(x,yz)\sigma(y,z)=\sigma(xy,z)\sigma(x,y) \qquad (*)$$

Conversely, any Borel function σ from $G\times G$ to the complex numbers of modulus one that satisfies (*) can be shown to be a multiplier for certain projective representations of G. It is clear that the set of all Borel solutions of (*) is a commutative group under multiplication and that the subset of those of the form $g(xy)/(g(x)g(y))$ is a subgroup. The quotient group is called the multiplier group M_G of G. Its elements correspond one-to-one to the essentially different multipliers of G. By essentially different we mean not changeable one into the other by changing the arbitrary constants. When G is the additive group of the real line, it can be shown that M_G contains only the identity. In this case every projective unitary representation can be reduced to an ordinary one. On the other hand, when G is the rotation group in three dimensions, it can be shown that M_G is of order 2. There is a multiplier that is not eliminatable but essentially only one.

Let us call a projective unitary representation whose multiplier is σ a *σ-representation*. For each fixed G and σ one can develop a theory of σ-representations of G that is analogous in almost all respects to the theory of ordinary representations developed in the preceding sections. The direct sum of two σ-representations is again such and one can define irreducibility, disjointness, primary representation, type, and so forth in a completely analogous fashion. The chief concept that does not extend is that of Kronecker product. The Kronecker product of two σ-representations is a σ^2-representation. Changing σ to another multiplier σ' such that σ/σ' is of the form $g(xy)/(g(x)g(y))$ does not change the theory in any important respect. The mapping $L \rightarrow L'$, where $L'_x = L_x/g(x)$, sets up a one-to-one correspondence between σ-representations and σ'-representations that preserves all significant operations and properties.

These remarks having been made, let us return to the primary representations of G of transitive type associated with the orbit of (L) in $\hat{N}$. One can show that (L) determines a well-defined member of the multiplier group of $(G/N)_{(L)}$. Let σ be any multiplier belonging to this equivalence class in $M_{(G/N)_{(L)}}$. It can be shown that the primary unitary representations of G associated with the orbit of (L) correspond one-to-one in a natural way to the primary σ-representations of $(G/N)_{(L)}$.

At first sight the fact that we may need to study projective representations of $(G/N)_{(L)}$ would seem to indicate that one cannot use the generalized semi-direct product theory to find the irreducible representations of complicated groups by "induction." At the very first step we could be led from the domain of ordinary representations to that of projective representations. However, it turns out that the analogy between ordinary and projective representations is sufficiently far-reaching to encompass the semi-direct product theory and the just indicated generalization.

Of course, the primary σ-representations of G associated with the orbit of (L) in $\hat{N}^{\sigma}$ (the equivalence classes of irreducible σ-representations of N) will correspond not to the primary σ-representations of $(G/N)_{(L)}$ but to the primary τ-representations for some multiplier τ of $(G/N)_{(L)}$ not necessarily of the form $\sigma(x,y)g(xy)/(g(x)g(y))$. Thus we can use the generalized theory inductively, provided that we enlarge the problem to

ISBN 0-8053-6702-0/0-8053-6703-9, pbk

that of finding all projective representations for every multiplier. It must be admitted of course that the problem is not an easy one. It includes the problem of finding the multiplier group M_G, which is a special case of a general problem in the cohomology theory of groups. Moreover, changing from one multiplier to an essentially different multiplier can change the structure of the representation theory quite as much as changing to another group.

As an illustration of this last remark consider the following example. Let G be a separable locally compact commutative group and let $\hat{G}$ be its dual. Let U and V be the regular representations of G and $\hat{G}$, respectively, and let us use the Fourier transform to realize V in $\mathcal{L}^2(G)$. $\mathcal{H}(U)=\mathcal{H}(V)=\mathcal{L}^2(G,\mu)$, where μ is Haar measure on G, $(U_x f)(y)=f(yx), V_\chi f(y)=\chi(y)f(y)$. Then a simple calculation shows that $U_x V_\chi=\chi(x)V_\chi U_x$ for all x and χ. Thus if $A=G\times\hat{G}$ and we define $W_{x,\chi}=U_x V_\chi$, we have

$$\begin{aligned} W_{(x_1,\chi_1)(x_2,\chi_2)} &= W_{x_1x_2,\chi_1\chi_2} \\ &= U_{x_1x_2}V_{\chi_1\chi_2} \\ &= U_{x_1}\chi_1(x_2)V_{\chi_1}U_{x_2}V_{\chi_2} \\ &= \chi_1(x_2)W_{x_1,\chi_1}W_{x_2,\chi_2} \end{aligned}$$

It follows that W is a σ-representation of A, where $\sigma(x_1,\chi_1;x_2,\chi_2)=\chi_1(x_2)$. Moreover, it is easy to see that W is irreducible. It follows that A, in spite of being commutative, admits irreducible σ-representations that are infinite-dimensional. Perhaps more surprising is the fact that for this particular σ every irreducible σ-representation of A is equivalent to W. To see this, note that for every σ-representation W of A we may write $W_{x,\chi}=W_{x,0}W_{0,\chi}=U_x V_\chi$, where U and V are ordinary representations of G and $\hat{G}$, respectively. Moreover, if U and V are arbitrary unitary representations of G and $\hat{G}$, respectively, then $W_{x,\chi}=U_x V_\chi$ defines a σ-representation of A if and only if $U_x V_\chi=\chi(x)V_\chi U_x$ for all x and χ. Let P be the projection valued measure on G defining V. Then $U_x V_\chi=\chi(x)V_\chi U_x$ if and only if P is a system of imprimitivy for U based on G, where the action of G on G is group multiplication. Since this action is transitive and the subgroup leaving a pointy fixed contains only the identity, we see that there is a unique irreducible solution.

Actually we can find multipliers for commutative groups yielding primary projective representations that are not of Type I. For further details and proofs concerning the above mentioned extension of the semi-direct product theory, see G. W. Mackey, "Unitary representations of group extensions I," *Acta Math*, vol. 99 (1958), 265–311.

ISBN 0-8053-6702-0/0-8053-6703-9, pbk

Notes and References (For Sections 4 Through 11)

The extension of the Frobenius theory from finite groups to compact Lie groups was begun by Schur in 1924 and brought to a considerable degree of completeness by Hermann Weyl in the next few years. Schur's basic idea was to use an integration process on the group manifold due to Hurwitz as a substitute for summation over the group. Weyl combined this idea with results of E. Cartan on representations of Lie algebras to obtain the irreducible unitary representations of the classical semi-simple compact Lie groups and also (in collaboration with F. Peter) proved the existence of "sufficiently many" irreducible unitary representations for every compact Lie group. Weyl was also the first to recognize the group theoretical character of Fourier analysis and to point out that the classical theory of expansions in surface harmonics is a matter of decomposing the unitary representation of the rotation group defined by its action on the two sphere.

The existence of Haar measure was proved by A. Haar in 1933 and this opened the door to extending the Frobenius theory from finite groups and compact Lie groups to arbitrary locally compact groups. Haar himself pointed out that, using his invariant measure as a substitute for the Hurwitz integral, one could immediately extend the general theory (including the Peter Weyl theorem) to arbitrary compact groups. (Haar only considered the separable case.)

The extension to locally compact groups that are not compact did not come at once. The difficulties are great enough so that one had to forget Frobenius for awhile and deal first with the commutative case where all irreducible unitary representations are one dimensional. The Pontrjagin–van Kampen duality theory was published in 1934 and 1935 and a few years later A. Weil and others noticed that one can define a "Fourier transform" for functions on any locally compact commutative group and prove a generalization of the Plancherel formula. The theory of locally compact groups that are either compact or commutative, which was made possible by the discovery of Haar measure and Pontrjagin–van Kampen duality, was exposed at length in two (rather different) influential books that have become classics: *Topological Groups* by L. Pontrjagin published in 1939 (slightly earlier in Russian) and *L'integration dans les groupes topologiques et ses applications à l'analyse* by A. Weil published in 1940.

In the theory presented in the books of Pontrjagin and Weil, nothing is said about the analog for locally compact commutative groups of the theorem on the essentially unique decomposition of a finite-dimensional representation of a finite group as a direct sum of irreducible representations. Consider, however, the finite-dimensional unitary representations of the additive group R of all real numbers. Every such representation $t \rightarrow U_t$

ISBN 0-8053-6702-0/0-8053-6703-9, pbk

may be written uniquely in the form $U_t = e^{itH}$, where H is a self-adjoint operator. Moreover, the diagonalization theorem for self-adjoint operators together with the theorem that two such operators are unitarily equivalent if and only if they have the same eigenvalues with the same multiplicity, is completely equivalent to a Schur–Frobenius-type decomposition theorem for *finite*-dimensional unitary representations of R. Now as shown by M. H. Stone in a famous paper published in 1930 the one-to-one correspondence between finite-dimensional unitary representations of R and self-adjoint operators in a finite-dimensional Hilbert space may be carried over to the infinite case by making use of Hilbert's spectral theorem of 1906 as extended to unbounded operators by von Neumann and by Stone himself. The unitary equivalence (spectral multiplicity) theory for bounded self-adjoint operators with possibly continuous spectra was worked out in the 1907 thesis of Hilbert's student Hellinger, improved by Hahn in 1911, and extended to the unbounded case by Stone in his celebrated treatise of 1932 "Linear transformations in Hilbert space and their applications to analysis." Putting these pieces together, one sees that the workers in spectral theory from Hilbert on were essentially developing a complete analog of the decomposition theory for group representations with R replacing the finite group and infinite dimensionality permitted. While the Pontrjagin–van Kampen duality theorem made it possible to generalize the theory for R to a completely parallel theory valid for all separable locally compact commutative groups, this fact seems to have escaped notice until 1944. In that year, M. I. Neumark, W. Ambrose, and R. Godement published independent papers proving the analog of the spectral theorem for arbitrary locally compact commutative groups. The fact that spectral multiplicity theory could also be carried over to the general (separable) case was not pointed out until the 1950s.

The rest of the material in Sections 4 through 11 is developed in detail in the author's Chicago lecture notes recently published in book form as mentioned in the Preface. For further historical remarks and references to the original literature, the reader is referred to that book as well as to the author's 1961 American Mathematical Society Colloquim lectures "Infinite dimensional group representations" published in Vol. 69 (1963) of the *Bulletin* of the society. The appendix to the book containing the Chicago notes contains a long survey of developments in the theory that took place between 1955 and 1975. We call attention here especially to subsections a, b, d, e, g, and h of Section 3 of that appendix as well as to the recent work of Howe and Bernstein on the Type I-ness of p-adic semi-simple Lie groups (end of subsection m).

In addition to the books of Dixmier, Gaal, and G. Warner cited in the Preface to the 1976 edition of the Chicago notes, the following accounts of the general theory of unitary group representations should also be men-

ISBN 0-8053-6702-0/0-8053-6703-9, pbk

tioned:

(1) A. Borel, "Representations de groupes localement compacts," *Springer Lecture Notes*, vol. 276 (1972), 98.

(2) R. Lipsman, "Group representations," *Springer Lecture Notes*, vol. 388 (1974) 166.

(3) A. A. Kirillov, *Elementary theory of representations*. Springer-Verlag, New York (1976) (Russian edition, 1972).

(4) A. O. Barut and R. Raczka *Theory of Group Representations and Applications*, PWN-Polish Scientific Publishers, Warsaw, 1977.

ISBN 0-8053-6702-0/0-8053-6703-9, pbk

12. Simple groups

Quite apart from the difficulty of finding the various multiplier groups and the possibility of encountering nonregular embeddings and normal subgroups having non-Type I projective representations, the whole inductive program comes to a halt when we reach a $(G/N)_L$ that is noncomutative and *simple* in the sense that it has no proper closed normal subgroups. In order to find the irreducible unitary representations of simple groups, various special methods have to be used and the problem remains unsolved for many interesting and important examples. In subsequent sections we shall give an account of some of these methods and the results that they have produced. However, before looking at their representations, let us attempt to get some idea of the extent and variety of the simple separable locally compact groups themselves. Since the connected component of the identity of a topological group is always a closed normal subgroup, it follows that any simple group is either connected or else *totally disconnected* in the sense that it has no connected subsets containing more than one point. Of course the discrete groups are totally disconnected, but it is also possible for a group to be nondiscrete and still be totally disconnected. The simple separable locally compact groups fall naturally into three classes: those that are connected, those that are totally disconnected but not discrete, and those that are discrete. It will be convenient to discuss these classes separately. The first and third may be further subdivided into those that are and that are not compact, but every compact totally disconnected group that is not discrete is a so called "projective limit" of finite groups and cannot be simple. Of course the compact discrete groups are just the finite ones. Thus we have the following five classes: the finite simple groups; the countable discrete simple groups; the totally discon-

George W. Mackey, Unitary Group Representations in Physics, Probability, and Number Theory

ISBN 0-8053-6702-0/0-8053-6703-9, pbk

nected nondiscrete simple groups; the compact connected simple groups; and the noncompact connected simple groups.

It turns out that the connected ones are the easiest to study systematically. Indeed, they have been completely classified and their *finite-dimensional* representations (unitary or not) have been completely determined. In particular, one knows completely all the irreducible unitary representations of every *compact* connected simple group. The fact that makes this systematic study possible is that connected simple locally compact groups are necessarily connected Lie groups. This follows from the work of Gleason, Montgomery, and Zippin on the famous fifth problem of Hilbert. A *connected Lie group* is by definition a topological group that is also a connected C^∞ manifold in such a way that the group operations are differentiable functions. We shall not bother to give a formal definition but will content ourselves with explaining the more important consequences. One of these is the possibility of introducing the so called *Lie algebra*. Let us define a *one-parameter subgroup* of a group G to be a continuous homomorphism of the additive group of the real line into G. Let $\mathfrak{L}_G$ denote the set of all one-parameter subgroups of G. If ϕ is in $\mathfrak{L}_G$ and f is a real-valued function on G, then for each x in G we may consider the function $t \rightarrow f(x\phi(t))$ and ask whether it is differentiable at $t=0$. If so, we say that f is differentiable at x in the direction ϕ and denote the result by the symbol $f_\phi(x)$. If $f_\phi(x)$ exists for all x, we say that f is differentiable in the direction ϕ and denote the function $x \rightarrow f_\phi(x)$ by the symbol f_ϕ. If $((f_{\phi_1})_{\phi_2} \cdots)_{\phi_n}$ exists for all $\phi_1, \ldots, \phi_n$ in $\mathfrak{L}_G$, let us say that f is infinitely differentiable. Let $\mathfrak{F}$ be the set of all infinitely differentiable functions on G. It is clear that $\mathfrak{F}$ is an algebra over the real numbers and that for each ϕ in $\mathfrak{L}_G$ the mapping $f \rightarrow f_\phi$ is a linear map of $\mathfrak{F}$ into $\mathfrak{F}$. Let us call it D_ϕ, that is, $D_\phi f = f_\phi$. So far everything we have said applies to arbitrary topological groups, though of course $\mathfrak{L}_G$ could reduce to a one point set. However, if we make the hypothesis that G is a connected Lie group, one can prove the following propositions:

(1) $\phi = \phi'$ if and only if $D_\phi = D_{\phi'}$.

(2) If ϕ and ϕ' are in $\mathfrak{L}_G$, then there exists θ in $\mathfrak{L}_G$ such that $D_\phi + D_{\phi'} = D_\theta$.

(3) If ϕ is in $\mathfrak{L}_G$ and t is a real number, then there exists θ in $\mathfrak{L}_G$ such that $tD_\phi = D_\theta$.

(4) If ϕ and ϕ' are in $\mathfrak{L}_G$, then there exists θ in $\mathfrak{L}_G$ such that $D_\phi D_{\phi'} - D_{\phi'} D_\phi = D_\theta$.

It follows from (1) that the θ's in (2), (3), and (4) are unique. Thus we may define $\phi + \phi'$ to be the unique θ such that $D_\phi + D_{\phi'} = D_\theta$ and $t\phi$ as the unique θ such that $tD_\phi = D_\theta$. Similarly, we may define $[\phi, \phi']$ to be the

ISBN 0-8053-6702-0/0-8053-6703-9, pbk

unique θ such that $D_\phi D_{\phi'} - D_{\phi'} D_\phi = D_\theta$. It is obvious that $\mathfrak{L}_G$ becomes a real vector space under the first two operations and it can be proved that:

(5) the dimension of $\mathfrak{L}_G$ is finite and equal to the dimension of G as a C^∞ manifold.

The multiplication $\phi,\theta \rightarrow [\phi,\theta]$ is distributive with respect to vector space operations but it is not associative. Instead, as one easily verifies, it satisfies the "Jacobi identity"

$$[[\phi,\phi'],\phi''] + [[\phi',\phi''],\phi] + [[\phi'',\phi],\phi'] = 0 \qquad (*)$$

as well as the identity

$$[\phi,\theta] = -[\theta,\phi] \qquad (**)$$

Quite generally one defines a *Lie algebra* over a field F to be a vector space over F equipped with a distributive multiplication satisfying (*) and (**). The set $\mathfrak{L}_G$ equipped with the indicated operations is then a Lie algebra over the real numbers (real Lie algebra) called the *Lie algebra of G*. One may think of $\mathfrak{L}_G$ as a sort of "infinitesmal" version of G. Its importance lies in the fact that it may be studied by purely algebraic means and that it almost determines G. As far as the latter is concerned we have the following propositions:

(6) If G_1 and G_2 are Lie groups that are both connected and *simply connected*, then G_1 and G_2 are isomorphic as topological groups if and only if $\mathfrak{L}_{G_1}$ and $\mathfrak{L}_{G_2}$ are isomorphic as Lie algebras.

(7) If G is any connected Lie group, then there exists a simply connected Lie group G' and a discrete subgroup H of the center of G' such that $\mathfrak{L}_G$ and $\mathfrak{L}_{G'}$ are isomorphic and G is isomorphic to G'/H.

(8) Every finite-dimensional real Lie algebra is the $\mathfrak{L}_G$ for some connected, simply connected Lie group G.

The groups G and G' of (7) are in a certain obvious sense "locally isomorphic" since H is discrete. Thus two connected Lie groups with isomorphic Lie algebras are locally isomorphic to the same simply connected Lie group and hence to each other. Thus it follows from (6), (7), and (8) that the isomorphism classes of finite-dimensional real Lie algebras correspond one-to-one in a natural way to the isomorphism classes of connected, simply connected Lie groups and also to the local isomorphism classes of connected Lie groups. In order to get a clearer grasp of the meaning of local isomorphism, it is useful to look at the commutative case. A connected Lie group G is commutative if and only if $[\phi,\theta]=0$ for all θ, ϕ in $\mathfrak{L}_G$. Thus the Lie algebras of commutative Lie groups are just those in

ISBN 0-8053-6702-0/0-8053-6703-9, pbk

which the bracket operation is trivial. Hence they correspond one-to-one to the finite-dimensional real vector spaces. It follows that there is just one connected simply connected commutative Lie group of every positive integral dimension n and it is clearly the additive group of a real n-dimensional vector space. The connected nonsimply connected ones are obtained from these by factoring out discrete subgroups. The simplest example is obtained by letting $n=1$ and taking H to be the subgroup of all integers.

Let $\mathfrak{L}$ be a Lie algebra. A subspace $\mathcal{I}$ of $\mathfrak{L}$ will be called an *ideal* if $[\phi,\theta]$ is in $\mathcal{I}$ whenever ϕ is in $\mathcal{I}$ and θ is in $\mathfrak{L}$. Every ideal $\mathcal{I}$ is a Lie algebra in its own right and the quotient $\mathfrak{L}/\mathcal{I}$ can be made into a Lie algebra in a natural way. If $\mathfrak{L}$ has no proper ideals, we shall say that $\mathfrak{L}$ is *simple*. It can be shown that $\mathfrak{L}_G$ is simple whenever G is simple. Conversely, if $\mathfrak{L}_G$ is simple, then the center Z of G is discrete and G/Z is simple. In particular, two simple connected Lie groups are isomorphic if and only if their Lie algebras are isomorphic. Thus we have a natural one-to-one correspondence between isomorphism classes of simple simply connected Lie groups and isomorphism classes of simple finite-dimensional real Lie algebras. In particular, it follows that the problem of classifying the connected simple locally compact groups is completely equivalent to the purely algebraic problem of classifying the simple finite-dimensional real Lie algebras.

This algebraic problem was solved by Eli Cartan in two papers published in 1894 and 1914, respectively. The first paper (Cartan's Ph.D. thesis) dealt with the auxiliary problem of finding all simple Lie algebras over the complex numbers. Given any real Lie algebra $\mathfrak{L}$ one may construct from it a complex Lie algebra $\mathfrak{L}_C$ called its *complexification* as follows: The underlying real vector space of $\mathfrak{L}_C$ is the direct sum of $\mathfrak{L}$ with itself so that the elements of $\mathfrak{L}_C$ are ordered pairs of elements in $\mathfrak{L}$. We make $\mathfrak{L}_C$ into a complex vector space by defining $i(\phi,\theta)=(-\theta,\phi)$. Then every element in $\mathfrak{L}_C$ can be written uniquely in the form $\phi+i\theta$, where $\phi=(\phi,0)$, $\theta=(\theta,0)$, and the set of all $(\phi,0)$ is identified with $\mathfrak{L}$. Finally, we define $[\phi+i\theta,\phi'+i\theta']$ as $([\phi,\phi']-[\theta,\theta'])+i([\phi,\theta']+[\theta,\phi'])$. One can prove that whenever $\mathfrak{L}$ is simple, then either $\mathfrak{L}_C$ is simple or else $\mathfrak{L}_C$ is the direct sum of two isomorphic simple complex Lie algebras. Thus every isomorphism class of real simple Lie algebras is canonically associated with an isomorphism class of complex simple Lie algebras. Since complex Lie algebras are easier to study than real ones, it is natural to break the problem up into two parts as follows. First find all simple complex Lie algebras. Then for each simple complex Lie algebra find all its possible "real forms." Cartan solved the first problem in his thesis and the second twenty years later. Actually the solution of the first problem has direct bearing on the classification of simple connected Lie groups. This is because each simple complex Lie algebra has to within isomorphism one and only one real form that is *compact* (i.e., the Lie algebra of a compact group). Thus to classify the complex simple Lie algebras is to classify the

ISBN 0-8053-6702-0/0-8053-6703-9, pbk

compact connected simple groups. Moreover, this is not all. Every complex simple Lie algebra becomes a real Lie algebra if we just ignore the complex scalar multiplication. This real Lie algebra is always simple and the simple Lie algebras so obtained are just those in which there exists a linear transformation J such that $J^2 = -1$ and $[J\phi, \theta] = J[\phi, \theta]$ for all ϕ and θ. The corresponding simple Lie groups are the so called *complex* simple Lie groups—a connected Lie group being called complex if the manifold of the group admits a complex analytic structure compatible with the other structures. Thus a classification of the simple complex Lie algebras also carries with it a classification of the complex simple Lie groups (and these are never compact). Given a compact simple real Lie group G we may form $\mathcal{L}_G$ and then the complexification $\mathcal{L}_{G,C}$. $\mathcal{L}_{G,C}$ will be a simple complex Lie algebra and the real Lie algebra we get by dropping the complex scalars will be that of a certain complex simple Lie group $\overline{G}$. If we let $(\phi, \theta)^* = (\phi, -\theta)$ in $\mathcal{L}_{G,C}$, we will have defined an involutory automorphism of $\mathcal{L}_{G,C}$ (as a real Lie algebra) whose set of fixed elements is $\mathcal{L}_G$. Correspondingly, we shall find that $\overline{G}$ admits an involutory automorphism such that G is the subgroup of all x with $x^* = x$.

Now let us describe the classification of complex simple Lie algebras established by Cartan and the associated classification of compact simple Lie groups and of complex simple Lie groups. Let X be a finite-dimensional complex vector space and let $\mathcal{L}(X)$ be the vector space of all linear transformations of X into X. $\mathcal{L}(X)$ becomes a Lie algebra if we set $[T_1, T_2] = T_1T_2 - T_2T_1$. The subspace of all elements of trace zero is an ideal $\mathcal{L}_0$ that is *simple* as a Lie algebra. The isomorphism class of this simple Lie algebra depends only on the dimension n of X. It is the trivial zero algebra when $n = 1$, but otherwise is nontrivial. The simple Lie algebra resulting when the dimension is $n+1$ is customarily denoted by the symbol A_n. Now let F be a nonsingular bilinear form on X such that $F(\phi,\theta) \equiv F(\theta,\phi)$ or $F(\phi,\theta) \equiv -F(\theta,\phi)$. (We say accordingly that F is symmetric or antisymmetric.) For each T in $\mathcal{L}(X)$ let T^F be defined by the condition $F(T\phi,\theta) \equiv F(\phi, T^F\theta)$. Then $(T_1T_2)^F = T_2^F T_1^F$ and the set of all T in $\mathcal{L}_0(X)$ with $T^F = -T$ is a Lie algebra. Let us call it $\mathcal{L}_0^F(X)$. It is not hard to show that the isomorphism class of $\mathcal{L}_0^F(X)$ depends only on the dimension of X and upon whether F is symmetric or antisymmetric. If F is symmetric, $\mathcal{L}_0^F(X)$ is denoted by D_n whenever X has dimension $2n$ and by B_n whenever X has dimension $2n+1$. If F is antisymmetric, then X necessarily has dimension $2n$ and then $\mathcal{L}_0^F(X)$ is denoted by C_n. The algebras B_n and C_n are all simple and D_n is simple for $n = 3,4,5,\ldots$. A_1, B_1, and C_1 are mutually isomorphic, B_2 is isomorphic to C_2 and A_3 to D_3. However, there are no other isomorphism. Thus the following four infinite sequences

$$A_1, A_2, A_3, \ldots \qquad C_3, C_4, \ldots$$
$$B_2, B_3, \ldots \qquad D_4, D_5, \ldots$$

ISBN 0-8053-6702-0/0-8053-6703-9, pbk

are mutually nonisomorphic simple Lie algebras. In addition to these four infinite classes of complex simple Lie algebras there are five "exceptional" complex simple Lie algebras. Their dimensions are 14, 52, 78, 133, and 248, and they are usually denoted by the symbols G_2, F_4, E_6, E_7, and E_8, respectively. There are no others. Every complex simple Lie algebra is isomorphic to one of the four infinite series or one of the five exceptional algebras. We shall not attempt to give the rather complicated description of the exceptional algebras.

For the four infinite series A_n, B_n, C_n, and D_n it is quite easy to describe the associated complex and compact simple groups. Let $SL(n+1,C)$ denote the group of all $n+1$ by $n+1$ complex matrices of determinant one. Its center Z is a finite group and the quotient by the center is simple. It is a complex simple Lie group whose Lie algebra is A_n. Now let G_F^X denote the group of all linear transformations in X that have determinant one and leave F fixed. Its center is always finite and its quotient by the center is a complex simple Lie group whose Lie algebra is $\mathfrak{L}_0^F(X)$. When X has dimension n and F is symmetric, G_F^X is denoted by the symbol $\mathrm{SO}(n,C)$. When F is antisymmetric and X has dimension $2n$, G_F^X is denoted by $\mathrm{Sp}(n,C)$. Thus B_n, C_n, and D_n are, respectively, the Lie algebras of $\mathrm{SO}(2n+1,C)/Z$, $\mathrm{Sp}(n,C)/Z$, and $\mathrm{SO}(2n,C)/Z$, where Z in each case denotes the finite center of the group in question.

To describe the corresponding compact groups we choose a Hilbert space inner product in X and do this in such a way that $(T^*)^F=(T^F)^*$ for all T, where T^* is the usual Hilbert space adjoint. Then $\mathrm{SL}(n,C)$, $\mathrm{SO}(n,C)$, and $\mathrm{Sp}(n,C)$ are all invariant under * and we may consider the subgroup defined by the condition $T=T^{*-1}$, that is, the subgroup of unitary elements of each of these groups. These subgroups are compact and their quotients by their centers are simple. The complexifications of their Lie algebras are the simple Lie algebras A_n, B_n, C_n, and D_n. Thus the compact simple group associated with A_n is just $\mathrm{SU}(n+1)/Z$, that is, the quotient by its center of the group of all unitary operators of determinant one on an $(n+1)$-dimensional Hilbert space; that associated with B_n is $\mathrm{SO}(2n+1)/Z$, the quotient by its center of the group of all orthogonal operators of determinant one on a $(2n+1)$-dimensional real Euclidean space; that associated with D_n is $\mathrm{SO}(2n)/Z$, the quotient by its center of the group of all orthogonal operators of determinant one on a $2n$-dimensional real Euclidean space; and that associated with C_n is $\mathrm{Sp}(n)/Z$, the quotient by its center of the group of all unitary operators on a $2n$-dimensional complex Hilbert space that leave an alternating form invariant.

We are now ready to discuss Cartan's solution to the second part of the problem, that of finding *all* simple real Lie algebras having a given complexification. Since it is only the complex simple Lie groups whose Lie algebras have nonsimple complexifications, we need only consider the problem of realizing simple complex Lie algebras in the form $\mathfrak{L}_C$. Now

ISBN 0-8053-6702-0/0-8053-6703-9, pbk

every element in an $\mathfrak{L}_C$ may be written uniquely in the form $\phi+i\theta$, where ϕ and θ are in $\mathfrak{L}$. Let $\sigma(\phi+i\theta)=\phi-i\theta$. Then σ is a conjugate linear mapping of $\mathfrak{L}_C$ onto itself such that σ^2 is the identity and $\sigma([\phi,\theta])=[\sigma(\phi),\sigma(\theta)]$. In other words, it is an involutory automorphism of $\mathfrak{L}_C$ as a real Lie algebra such that $\sigma(i\phi)=-i\sigma(\phi)$. Moreover, $\mathfrak{L}$ itself is the subspace on which σ is the identity. Hence every real form of a complex simple Lie algebra $\mathfrak{L}'$ is the set of all ϕ with $\sigma(\phi)=\phi$ for some involutory automorphism σ of the underlying real Lie algebra of $\mathfrak{L}'$. Actually one can prove that if σ is any automorphism of $\mathfrak{L}'$ as a real Lie algebra such that $\sigma i=-i\sigma$, $\sigma^2=I$, then the set of all ϕ with $\sigma(\phi)=\phi$ is a real Lie algebra whose complexification is $\mathfrak{L}'$. Thus the problem reduces to that of finding the possible σ's for every simple complex Lie algebra $\mathfrak{L}'$. There will usually be an infinite number of these but always only a small finite number that are nonconjugate and give nonisomorphic real forms. For the members of the four series A_n, B_n, C_n, and D_n it is not difficult to give a complete description of all possibilities.

The complex simple Lie algebra A_n may be realized as the Lie algebra of all $n+1$ by $n+1$ complex matrices of trace zero. An obvious involution σ having the required properties is σ_0, where $\sigma_0((a_{ij}))=(\bar{a}_{ij})$. Another is σ_1, where $\sigma_1((a_{ij}))=(-\bar{a}_{ji})$. If $\bar{I}$ is any matrix such that $\bar{I}^2=I$, then $\sigma_{\bar{I}}$, will be still another, where $\sigma_{\bar{I}}((a_{ij}))=\bar{I}^{-1}\sigma_0((a_{ij}))\bar{I}$. Let σ_2^q denote the particular $\sigma_{\bar{I}}$ in which $\bar{I}$ is a diagonal matrix with q 1's and $n+1-q$ (-1)'s on the diagonal. When $n+1$ is odd, every real form comes from one of these involutions and we have $\frac{1}{2}n+1+2=\frac{1}{2}n+3$ real forms of A_n of which one, that associated with σ_1, is the compact real form. The corresponding simple groups may be described as follows: That associated with σ_0 is the quotient by its center of the group $\mathrm{SL}(n+1,R)$ of all $n+1$ by $n+1$ real matrices with determinant one. That associated with σ_2^q is the quotient by its center of the group $\mathrm{SU}(n+1-q,q)$ of all $n+1$ by $n+1$ matrices of determinant one that are "unitary" with respect to an indefinite Hermitean inner product with q plus terms and $n+1-q$ negative ones. When $n+1$ is even, there is one further possibility. Let W be the matrix $\begin{pmatrix} 0 & I \\ -I & 0 \end{pmatrix}$, where I is the $\frac{1}{2}(n+1)$ by $\frac{1}{2}(n+1)$ identity matrix. Then $W^2=-1$ and σ_3 will have the required properties where $\sigma_3((a_{ij}))=W^{-1}\sigma_0((a_{ij}))W$. The corresponding simple Lie group is the quotient by its center of the group $\mathrm{SU}^*(n+1)$ consisting of the group of all members of $\mathrm{SL}(n+1,C)$ that commute with the transformation $(z_1,z_2,z_3,\ldots,z_{2k})\rightarrow(\bar{z}_{k+1},\bar{z}_{k+2},\ldots,\bar{z}_{2k},-\bar{z}_1,\ldots,-\bar{z}_k)$, where $2k=n+1$.

The simple Lie algebra B_n is the subalgebra of A_{2n} on which $\sigma_0\sigma_1^{-1}$ is the identity. It is carried into itself by σ_0, σ_1, and σ_2^q and these involutions define all possible real forms. That defined by $\sigma_0=\sigma_1$ is the compact one. The only noncompact real forms are those defined by σ_2^q with $q=1,2,\ldots,2n$. The simple Lie group associated with σ_2^q and B_n is the quotient

ISBN 0-8053-6702-0/0-8053-6703-9, pbk

by its center of $SO_0(q, 2n+1-q)$, the connected component of the identity of the group $SO(q, 2n+1-q)$ consisting of all $2n+1$ by $2n+1$ real matrices of determinant one that are "orthogonal" with respect to a quadratic form with q 1's and $2n+1-q$ (-1)'s.

The simple Lie algebra C_n is the subalgebra of A_{2n-1} on which $\sigma_3\sigma_0^{-1}$ is the identity. It is carried into itself by σ_0, σ_1, and σ_3 and the σ_2^q, and these involutions define all possible real forms. That defined by σ_1 is the compact one. The only noncompact real forms are those defined by $\sigma_0=\sigma_3$ and those defined by the σ_2^q for $q=1,2,\ldots,2n-1$. The simple Lie group associated with $\sigma_0=\sigma_3$ and C_n is the quotient by its center of $Sp(n,R)$, the analog of $Sp(n,C)$ with the real numbers replacing the complex ones. The simple Lie group associated with σ_2^q and C_n is the quotient by its center of $Sp(q,2n-q)$, the subgroup of all members of $Sp(n,C)$ that leave invariant an indefinite Hermitean form with q plus terms and $2n-q$ minus ones.

Finally, the simple Lie algebra D_n is the subalgebra of A_{2n-1} on which $\sigma_0\sigma_1^{-1}$ is the identity. It is carried into itself by $\sigma_0=\sigma_1, \sigma_3$ and the σ_2^q, and these involutions define all possible real forms. That defined by $\sigma_0=\sigma_1$ is the compact one as above. Thus the only noncompact real forms are those defined by σ_3 and the σ_2^q. The simple Lie group associated with σ_2^q and D_n is just that associated with σ_2^q and B_n except that $SO_0(q,2n+1-q)$ is replaced by $SO_0(q,2n-q)$. The simple Lie group associated with D_n and σ_3 is the quotient by its center of $SO^*(2n)$, the group of all matrices in $SO(2n,C)$ that leave invariant a certain skew Hermitean form.

In addition to the A_n, B_n, C_n, and D_n, each of the five exceptional simple complex Lie algebras has certain noncompact real forms, each of which corresponds to a certain simple noncompact connected group. We shall content ourselves with stating the numbers involved. The 14-dimensional G_2 has just one noncompact real form and the 52-dimensional F_4 has two. The Lie algebras E_6, E_7, and E_8 of dimension 78, 133, and 248 have, respectively, four, three, and two noncompact real forms. In all there are $10+1+2+4+3+2=22$ connected simple locally compact groups that have exceptional Lie algebras. Five of these are compact and five complex.

So much for the simple locally compact groups that are connected. We have three other classes to consider—the finite simple groups, the infinite discrete simple groups, and the nondiscrete totally disconnected simple groups. For none of these classes do we have a complete analysis of all possibilities. Indeed, such an analysis has been only seriously attempted for the finite simple groups. On the other hand, we have a rich collection of examples in each case. Let F be any separable locally compact field, and let $SL(n,F)$ denote the group of all n by n matrices over F of determinant one. Then $SL(n,F)$ is separable and locally compact in the obvious topology and its quotient by its center is simple. If p is a prime and F is the field of p-adic numbers or any finite extension of the field of p-adic numbers, then F will be totally disconnected and nondiscrete and so will

ISBN 0-8053-6702-0/0-8053-6703-9, pbk

the quotient of SL(n,F) by its center. If k is a nonnegative integer, let F_0 be the field of all rational functions of k variables with rational coefficients and let F be any algebraic extension of F_0. Then F will be a countable field to which we may give the discrete topology, and the quotient by its center of SL(n,F) will be a simple countable discrete group. Finally, let q be any positive integer of the form p^f, where p is a prime, and let F be the unique finite field with q elements. Then the quotient of SL(n,F) by its center will be a finite simple group. Thus we have an infinite number of examples in each of our three categories. Moreover, these examples are far from being exhaustive. There are other examples of countable and totally disconnected fields F for which we can form SL(n,F). More important, we can form analogs over a general separable locally compact field F of the various other series of connected simple Lie groups. For example, we can consider quadratic forms over F and construct analogs of SO(n,C), alternating forms and construct analogs of Sp(n,C), and so forth.

More surprisingly, one can construct analogs over a general field F of the exceptional Lie groups. In a fundamental paper published in the Tôhoku Journal in 1955, C. Chevalley gave a general method for constructing simple groups based on the following ideas: If G is a Lie group, every inner automorphism θ of G defines an automorphism θ' of the Lie algebra of G and thus we have a natural homomorphism of G into the group of all automorphisms of $\mathfrak{L}_G$. If G is simple, this homomorphism is an isomorphism and G is isomorphic to a certain group of automorphisms of $\mathfrak{L}_G$. Now given any locally compact field F we may construct Lie algebras over F and consider their automorphism groups. Moreover, given any simple Lie algebra over the reals, it is possible to choose a basis $\phi_1,\ldots,\phi_n$ in the corresponding vector space so that $[\phi_i,\phi_j]=\Sigma n_k^{ij}\phi_k$, where the n_k^{ij} are *integers*. Thus we may construct an analogous Lie algebra over any field F by setting $[\Sigma a_i\phi_i,\Sigma b_j\phi_j]=\Sigma a_i b_j n_k^{ij}\phi_k$, where the a_i and b_j are in F and n_k^{ij} is taken mod p if F has characteristic p. For further details and subsequent improvements and extensions of Chevalley's work we refer the reader to survey articles by J. Tits in the 1962 Stockholm Congress Proceedings and R. W. Carter in the Proceedings of the London Mathematical Society for 1965.

In addition to the finite simple groups obtainable from Lie algebras, there are two other *known* infinite series and six exceptional examples. One of the infinite series is rather trivial. For each prime p the cyclic group Z_p of order p is obviously simple. The other is the series $\mathcal{A}_1,\mathcal{A}_2,\ldots$ of alternating groups. For every positive integer n one defines the symmetric group S_n to be the group of all permutations of a set of n objects. Every permutation may be written as a product of transpositions (permutations that transpose two objects and leave everything else fixed) and while the representation is far from unique, the parity of the number of transpositions is unique. Thus we may speak unambiguously of even and odd

ISBN 0-8053-6702-0/0-8053-6703-9, pbk

permutations. The subset $\mathcal{A}_n$ of all even permutations is a normal subgroup of S_n of index 2. It is called the *alternating* group of n objects. It can be shown that $\mathcal{A}_n$ is simple whenever n is greater than four.

Notes and References

The work of Cartan on the classification of complex simple Lie algebras in his thesis of 1894 was a corrected and completed version of work done by Killing a few years earlier.

For a detailed account of the structure theory of semi-simple Lie groups, the reader is referred to the beautiful book of S. Helgason, *Differential Geometry and Symmetric Spaces*, Academic, New York (1962).

The situation with respect to the existence of "sporadic" simple finite groups has changed considerably since Section 12 was written in the spring of 1967. The number of known such groups had recently changed from 5 to 6 as a result of a discovery of a new example by Janko. Before that, it had been 5 for over half a century. Janko's discovery turned out to be just a beginning and the published version of W. Feit's address to the 1970 international Congress in Nice reported a threefold increase to 18. At the present writing (April 1978) I am told by an expert that 24 sporadic simple groups are definitely known to exist and that two more possibilities are under investigation.

ISBN 0-8053-6702-0/0-8053-6703-9, pbk

13. The irreducible unitary representations of compact simple groups

A compact simple group is either finite or connected. We have seen that the connected simple groups are all known and we shall devote most of this section to a description of the equally complete results that are known concerning their irreducible unitary representations. Much less is known in the finite case and we shall confine ourselves to a serious study of one subcase and some brief indications about others.

Every irreducible unitary representation of a compact group is necessarily finite-dimensional and every finite-dimensional irreducible representation of a compact group can be made unitary in essentially only one way. Thus the study of the irreducible unitary representations of a compact group is equivalent to the study of its irreducible representations in an abstract finite-dimensional complex vector space. It turns out that the latter study can be carried out just as well for noncompact groups as for compact ones so long as the group is simple and connected. Since the nonunitary finite-dimensional representations of noncompact simple connected groups are of considerable interest, we shall treat our problem in this greater generality. We shall begin our discussion in an even more general context. Let G be any connected Lie group and let V be any finite-dimensional representation of G. For each ϕ in the Lie algebra $\mathfrak{L}_G$ of G, the mapping $t \to V_{\phi(t)}$ is a representation of the additive group of the real line. By an easily proved and well-known theorem there exists a unique linear operator A_ϕ^V such that $V_{\phi(t)} = e^{tA_\phi^V}$. In fact,

$$A_\phi^V = \left(\frac{d}{dt}\left(V_{\phi(t)}\right) \right)_{t=0}$$

George W. Mackey, Unitary Group Representations in Physics, Probability, and Number Theory

ISBN 0-8053-6702-0/0-8053-6703-9, pbk

and it easily follows from this that the mapping $\phi \rightarrow A_\phi^V$ is linear and has the property that $A_{[\phi,\theta]}^V = A_\phi^V A_\theta^V - A_\theta^V A_\phi^V$ for all ϕ and θ in $\mathfrak{L}_G$. In general, a mapping $\phi \rightarrow A_\phi$ of $\mathfrak{L}_G$ into the linear transformations of some vector space X is said to be a *representation* of $\mathfrak{L}_G$ if it is linear and $A_{[\phi,\theta]} = A_\phi A_\theta - A_\theta A_\phi$ for all ϕ and θ in $\mathfrak{L}_G$. Thus every finite-dimensional representation V of G is canonically associated with a representation A^V of the Lie algebra $\mathfrak{L}_G$ of G. V is uniquely determined by A^V and when G is simply connected, every finite-dimensional representation of $\mathfrak{L}_G$ is of the form A^V. V is irreducible if and only if this is the case for A^V and so the problem of finding all (equivalence classes of) irreducible finite-dimensional representations of a simply connected Lie group G is completely reduced to the purely algebraic problem of finding all (equivalence classes of) irreducible finite-dimensional representations of its Lie algebra. For a connected Lie group G that is not simply connected, there is always an associated simply connected group G' whose Lie algebra is isomorphic with that of G and a discrete normal subgroup D of G' such that G'/D is isomorphic to G. The irreducible representations of G correspond one-to-one in a natural way to the irreducible representations of G' that are trivial on D.

Now let A be a representation of the Lie algebra $\mathfrak{L}_G$ of the connected Lie group G. Let us form the complexification $\mathfrak{L}_{G,C}$ of $\mathfrak{L}_G$ and for each (ϕ,θ) in $\mathfrak{L}_{G,C}$ let $A_{(\phi,\theta)}^C = A_\phi + iA_\theta$. Then A^C is a representation of $\mathfrak{L}_{G,C}$ that is also complex linear, that is, A^C is a representation of $\mathfrak{L}_{G,C}$ as a *complex* Lie algebra. Conversely, if B is any representation of $\mathfrak{L}_{G,C}$ as a complex Lie algebra, then $B_{(\phi,\theta)} = B_{(\phi,0)} + iB_{(\theta,0)}$ so $B = A^C$, where A is the restriction of B to the real subalgebra of all (ϕ,θ) with $\theta = 0$. Thus the study of the representations of a real Lie algebra is equivalent to the study of the representations of its complexification (as a complex Lie algebra). Now if $\mathfrak{L}_G$ is simple, $\mathfrak{L}_{G,C}$ will be either simple or else the direct sum of two isomorphic complex simple Lie algebras. Moreover, the second alternative will occur if and only if G is a complex analytic Lie group and then $\mathfrak{L}_G$ will be a complex Lie algebra and $\mathfrak{L}_{G,C}$ will be isomorphic to $\mathfrak{L}_G \oplus \mathfrak{L}_G$. In either event the study of the finite-dimensional irreducible representations of $\mathfrak{L}_{G,C}$ is equivalent to the study of the representations of a complex simple Lie algebra. The only difference between the two alternatives is that in the first the irreducible representations of $\mathfrak{L}_G$ correspond one-to-one to the irreducible representations of $\mathfrak{L}_{G,C}$ and in the second they correspond one-to-one to the pairs of irreducible representations of $\mathfrak{L}_G$ regarded as a complex Lie algebra.

These Lie algebra results have the following consequences for the associated Lie groups. Given the simple connected Lie group G, let G_C denote the simply connected Lie group whose Lie algebra is $\mathfrak{L}_{G,C}$. Then G_C will be a complex analytic Lie group whose center D is discrete and such that G_C/D is either simple or the direct product of two isomorphic simple groups. Moreover, G_C will contain a closed subgroup isomorphic to G', the

ISBN 0-8053-6702-0/0-8053-6703-9, pbk

simply connected covering group of G. Finally, the representations of G' corresponding to the representations of $\mathfrak{L}_{G,C}$ as a complex Lie algebra are just those whose matrix elements are complex analytic as functions on G_C and the restrictions of these to G' are just the finite-dimensional representations of G'. In other words, every finite-dimensional representation of G' is (uniquely) extendible to a representation of G_C whose matrix elements are complex analytic; and the problem of finding the irreducible finite-dimensional representations of G' (and hence of G) is just that of finding G_C and then finding the complex analytic irreducible representations of G_C.

It turns out that the simply connected simple complex Lie groups are sufficiently similar to one another so that the facts about their irreducible complex analytic representations may be stated in a uniform way for all of them at once. However, the statements are probably easier to understand in a familiar concrete instance that happens to be typical. Consider then the group $\mathrm{SL}(n,C)$ of all n by n complex matrices of determinant one. It is the simply connected covering group of the simple group obtained by factoring out its finite center. Let us call this group G and let D denote the subgroup of all diagonal matrices. D is commutative and is maximal commutative in the sense that no properly larger subgroup is commutative. Let N denote the subgroup of all matrices that are zero above the diagonal and one on the diagonal. Then the set ND of all products nd with n in N and d in D is a closed subgroup that is the semi-direct product of N and D. Moreover, N is the so called "commutator subgroup" of ND. Quite generally if A is a topological group, one can consider the closure A' of the group generated by elements of the form $xyx^{-1}y^{-1}$. This is clearly the intersection of all closed normal subgroups whose quotients are commutative. It is called the commutator subgroup. If $A' \neq A$, then one can form $A'' = (A')' \subseteq A', A'''$ and so forth. If $A^{(n)}$ reduces to the identity for some finite integer n, the group A is said to be *solvable*. It turns out that ND is solvable and that every solvable subgroup of G is a subgroup of some conjugate of ND. Thus ND is a maximal solvable subgroup and every maximal solvable subgroup is conjugate to ND. In other words, ND is characterized to within conjugacy by being a maximal solvable subgroup. It turns out that every complex Lie group with a simple Lie algebra has the property that its maximal solvable subgroups are mutually conjugate. They are called its *Borel subgroups*. They are all like ND in being semi-direct products of their commutator subgroups and a commutative group. These commutative groups are also mutually conjugate and are called the *Cartan subgroups* of the underlying complex Lie group. Each Cartan subgroup is isomorphic to a direct product of a finite number of replicas of the multiplicative group of the nonzero complex numbers and the number involved is called the *rank* of the group G.

Now let V be any finite-dimensional irreducible representation of G. The Cartan subgroup D is a direct product of a compact group and a

ISBN 0-8053-6702-0/0-8053-6703-9, pbk

vector group and the restriction of V to D is the unique complex analytic extension to D of its restriction to the compact factor. It follows easily from this that the restriction of V to D is a direct sum of one-dimensional representations of D. Each such is of course uniquely determined by a complex analytic homomorphism of D into the nonzero complex numbers. The particular ones that occur are called the weights of the representation V (with respect to the Cartan subgroup D). It is easy to see that the weights must determine V to within equivalence. Indeed, the matrices in G that are conjugate to a member of D form a dense set. Hence the character of V is uniquely determined by its values on D. But the restriction to D of the character of V is just the sum of all its weights—each being repeated according to its multiplicity. If we bring N into the picture, we can select one particular weight—the so called "highest weight" with respect to the Borel subgroup ND—that turns out to characterize V completely. Indeed it can be proved that the restriction of V to ND has just one one-dimensional invariant subspace. If ϕ is a nonzero vector in this subspace, then $V_{nd}\phi = \chi(d)\phi$, where χ is a weight of V with respect to D. This weight is called the *highest weight* and uniquely determines V. Indeed V can be explicitly described once χ is given as something very like an induced representation. Given χ, let χ' denote the homomorphism $nd \rightarrow \chi(d)$ of ND into the multiplicative group of all nonzero complex numbers and let $\mathcal{H}_\chi$ denote the vector space of all complex analytic functions f on G that satisfy the identity $f(yx) = \chi(y^{-1})f(x)$ for all y in ND and all x in G. Then $\mathcal{H}_\chi$ is clearly invariant under right translation and can be shown to be finite-dimensional. Let $(W_x^\chi f)(y) = f(yx)$. Then W^χ is a finite-dimensional representation of G whose space is $\mathcal{H}_\chi$. It can be shown to be equivalent to V. We note that the definition of W^χ differs from that of an induced representation in only two respects: (1) The requirement of square summability is replaced by that of complex analyticity; (2) neither χ nor W^χ is unitary.

Since every irreducible finite-dimensional complex analytic representation of G is uniquely of the form W^χ, where χ is a complex analytic homomorphism of D into the nonzero complex numbers, the problem of finding the most general such representation of G reduces to the following:

(1) Find the most general complex analytic homomorphism of D into the multiplicative group of the nonzero complex numbers.

(2) Determine which of these can occur as highest weights.

The answer to the first problem is almost obvious. D consists of all matrices

$$\begin{pmatrix} d_1 & & 0 \\ & \ddots & \\ 0 & & d_n \end{pmatrix} \quad \text{with} \quad d_1 d_2 \cdots d_n = 1$$

ISBN 0-8053-6702-0/0-8053-6703-9, pbk

and the general complex analytic homomorphism is defined by an n-tuple of integers $(k_1,k_2,\ldots,k_n)$,

$$\chi_{(k_1,\ldots,k_n)}\begin{bmatrix} d_1 & & 0 \\ & \ddots & \\ 0 & & d_n \end{bmatrix} = d_1^{k_1}\cdots d_n^{k_n}$$

Of course two n-tuples define the same homomorphism if they differ by an n-tuple of the form $(k,k,\ldots,k)$. The answer to the second problem is not obvious but may be stated very simply. $\chi_{(k_1,\ldots,k_n)}$ is a highest weight if and only if $k_1 \geqslant k_2 \geqslant \cdots k_n$.

Now let $m_j = k_j - k_{j+1}$ and let χ_j be that $\chi_{(k_1,\ldots,k_n)}$ for which $m_j = 1$ and the other m_i are zero. Then

$$\chi_{(k_1,\ldots,k_n)} = \chi_1^{m_1}\cdots\chi_{n-1}^{m_{n-1}}$$

and this representation is unique. Thus we may also describe the highest weights as the characters $\chi_1^{m_1}\cdots\chi_{n-1}^{m_{n-1}}$ where all the m_j are nonnegative.

Summing up, we see that the irreducible complex analytic representations of $G=\mathrm{SL}(n,C)$ are just the representations $W^{\chi_1^{m_1}\cdots\chi_{n-1}^{m_{n-1}}}$, where $m_1,\ldots,m_{n-1}$ vary independently over the nonnegative integers. $\mathrm{SL}(n,C)$ is the complexification of the compact group $\mathrm{SU}(n)$ and the restriction of $W^{\chi_1^{m_1}\cdots\chi_{n-1}^{m_{n-1}}}$ to $\mathrm{SU}(n)$ is irreducible and can be made unitary by an essentially unique choice of inner product. The irreducible unitary representations of $\mathrm{SU}(n)$ so obtained are inequivalent and include all irreducible unitary representations of $\mathrm{SU}(n)$.

In applications one is interested not only in the irreducible representations themselves but also in their tensor products with one another and their restrictions to subgroups. When the irreducible representations are finite-dimensional, as they are here, such questions can be settled by straightforward computation once the characters of the representations are known. In the case at hand the definition of W^{χ} is such that it is not obvious how to compute its dimension when χ is given. On the other hand, this dimension is clearly equal to $\mathrm{Trace}(W_e^{\chi})$. Thus it is of considerable interest to have a formula giving the character of W^{χ} in terms of χ. Since the conjugates of the elements of D are dense and the characters are continuous, it will suffice to have a formula for the restriction of this character to D. This will simply be the character of the restriction of W^{χ} to D and this restriction as we have already observed is a direct sum of one-dimensional representations. Hence the restriction to D of the character of W^{χ} will be a finite sum of complex analytic characters of D. However, this sum is easily described as a quotient of two other sums. To define these we first introduce a certain finite group of automorphisms of D called the *Weyl group*. In general let M' denote the group of all elements

ISBN 0-8053-6702-0/0-8053-6703-9, pbk

x of the compact group G such that the inner automorphism $y \to xyx^{-1}$ of G_C carries D into D and let M denote the subgroup of all y in M' that commute with every element of D. Then M'/M acts on D as a group of automorphisms. It is always finite and is called the Weyl group. When $G_C = \mathrm{SL}(n, C)$, $G = \mathrm{SU}(n)$ the Weyl group is the group of all automorphisms of D of the form

$$\begin{bmatrix} d_1 & & 0 \\ & \cdot \quad \cdot & \\ 0 & & d_n \end{bmatrix} \to \begin{bmatrix} d_{\sigma(1)} & & 0 \\ & \cdot \quad \cdot & \\ 0 & & d_{\sigma(n)} \end{bmatrix}$$

where σ is any permutation of the integers $1, \ldots, n$. We also introduce the so called adjoint representation of G_C. Each element of G_C defines an inner automorphism of the Lie algebra of G_C and hence a linear transformation of the underlying vector space. This map from G_C to linear transformations is a representation of G_C to linear transformations is a representation of G_C called the adjoint representation. In the case at hand the Lie algebra of G_C may be identified with the space of all n by n complex matrices of trace zero with $[A, B] = AB - BA$ and the adjoint representation is that which takes T into the linear transformation $A \to T^{-1}AT$. The adjoint representation restricted to D assigns to the diagonal matrix with diagonal elements d_i the linear transformation

$$(a_{ij}) \to \begin{bmatrix} 1/d_1 & & 0 \\ & \cdot \quad \cdot & \\ 0 & & 1/d_n \end{bmatrix} (a_{ij}) \begin{bmatrix} d_1 & & 0 \\ & \cdot \quad \cdot & \\ 0 & & d_n \end{bmatrix}$$
$$= (b_{ij}) \quad \text{where} \quad b_{ij} = a_{ij} d_j / d_i.$$

Thus for each i,j the set of all matrices which are zero except on the ith row and jth column form a one dimensional invariant subspace on which the adjoint representation reduces to the character $\chi_{(k_1, \ldots, k_n)}$ where $k_j = 1$, $k_i = -1$ and the others are zero. The characters of this special form then are the weights of the adjoint representation and are usually referred to as the *roots* of the Lie algebra in question. Let $\chi^0{}_{ij}$ denote the root defined by $d_1, \ldots, d_n \to d_j / d_i$. Thus $\chi^0{}_{ij} = 1/\chi^0{}_{ji}$, and if j is less than i we have

$$\chi^0{}_{ij} = \chi^0{}_{j+1,j} \chi^0{}_{j+2,j+1} \cdots \chi^0{}_{i,i-1}$$

Thus the roots of the form $\chi^0{}_{k+1,k}$ form a *base* for the set of all roots in the sense that every root is uniquely a product of integral powers of members of the base and that these powers are either all positive or all negative. We may speak accordingly of the *positive roots* and the *negative roots* with respect to a given base. Let θ' denote the product of all positive

ISBN 0-8053-6702-0/0-8053-6703-9, pbk

roots. One verifies without difficulty that $\theta' = \chi_1^2\chi_2^2 \cdots \chi_{n-1}^2$ so that $\theta' = \theta^2$, where $\theta = \chi_1\chi_2 \cdots \chi_{n-1}$. For each σ in the Weyl group let $\mathrm{sgn}(\sigma) = \pm 1$ according as σ is an even or odd permutation. Then the character of the representation W^χ is given by the following formula first found by Hermann Weyl:

$$\underset{x \text{ in } D}{\mathrm{Trace}(W_x^\chi)} = \frac{\Sigma \mathrm{sgn}(\sigma)\theta(\sigma(x))\chi(\sigma(x))}{\Sigma \mathrm{sgn}(\sigma)\theta(\sigma(x))}$$

where sum is over all σ in Weyl group.

There is an ambiguity in our definition of θ since the system of roots admits more than one base. The one we have chosen is the unique one with the following property: For each positive root χ let $\mathcal{L}_\chi$ be the subalgebra of the Lie algebra in which the adjoint representation reduces to this weight. Then the linear span of the $\mathcal{L}_\chi$ is the Lie algebra corresponding to the commutator subgroup of the Borel subgroup with respect to which the W^χ are defined.

As an example let us consider SU(2), which has the rotation group in three dimensions as a quotient. Here D is the set of all matrices of the form $\begin{pmatrix} d & 0 \\ 0 & 1/d \end{pmatrix}$, $n-1=1$, and $\chi_1\begin{pmatrix} d & 0 \\ 0 & 1/d \end{pmatrix} = d$. The possible weights are the characters χ_1^m, where $m = 0, 1, 2, \ldots$ and $\theta = \chi_1$. The Weyl group is of order 2, and if σ is its nontrivial member, $\sigma(X_1^p) = \chi_1^{-p}$. The Weyl formula now tells us that the restriction to D of the character of the irreducible representation whose highest weight is χ_1^m is

$$\frac{(\chi_1\chi_1^m - \chi_1^{-1}\chi_1^{-m})}{(\chi_1 - \chi_1^{-1})} = \chi_1^m + \chi_1^{m-2} + \chi_1^{m-4} + \cdots + \chi_1^{-m}$$

In particular, it is $m+1$ dimensional. Thus SL$(2, C)$ has to within equivalence just one complex analytic representation of every possible finite dimension and SU(2) has to within equivalence just one irreducible unitary representation of every nonnegative integral dimension. It is customary to denote the $(m+1)$-dimensional representation of SU(2) by the symbol $\mathfrak{D}_{m/2}$, where $m/2 = j$ takes the values $0, \frac{1}{2}, 1, \frac{3}{2}, 2, \ldots$. We verify at once that $\mathfrak{D}_j$ is trivial on the center of SU(2) if and only if j is an integer. Thus the rotation group has just one irreducible unitary representation for every odd dimension. Note that

$$(\chi_1^{2j} + \chi_1^{2j-2} + \cdots + \chi_1^{-2j})(\chi_1^{2j'} + \cdots + \chi_1^{-2j'})$$
$$= (\chi_1^{2(j+j')} + \cdots + \chi_1^{-2(j+j')}) + (\chi_1^{2(j+j'-1)} + \cdots + \chi_1^{-2(j+j'-1)}) + \cdots$$
$$+ (\chi_1^{2(j-j')} + \cdots + \chi_1^{-2(j-j')})$$

as one verifies by routine algebraic manipulation. Since $\mathfrak{D}_j \otimes \mathfrak{D}_{j'}$ is completely determined by its restriction to D, we deduce the important

ISBN 0-8053-6702-0/0-8053-6703-9, pbk

Clebsch–Gordon formula for decomposing the tensor product of two irreducible unitary representations of SU(2).

$$\mathfrak{D}_j \otimes \mathfrak{D}_{j'} = \mathfrak{D}_{|j-j'|} \oplus \mathfrak{D}_{|j-j'|+1} \oplus \cdots \oplus \mathfrak{D}_{j+j'}$$

As a further example, let us consider SU(3). Here D is the group of all 3-by-3 complex diagonal matrices d with diagonal elements d_1, d_2, d_3 such that $d_1 d_2 d_3 = 1$. $n-1=2$ and the possible highest weights are $\chi_1^{m_1}\chi_2^{m_2}$, where m_1 and m_2 are nonnegative integers. $\chi_1(d) = d_1$, $\chi_2(d) = d_1 d_2$, and $\theta = \chi_1\chi_2$. The Weyl group is the permutation group on three objects and the transforms of χ_1 and χ_2 are as follows:

	e	(12)(3)	(13)(2)	(23)(1)	(123)	(132)
χ_1	χ_1	$\chi_2\chi_1^{-1}$	χ_2^{-1}	χ_1	$\chi_2\chi_1^{-1}$	χ_2^{-1}
χ_2	χ_2	χ_2	χ_1^{-1}	$\chi_1\chi_2^{-1}$	χ_1^{-1}	$\chi_1\chi_2^{-1}$

Thus

$$\Sigma \operatorname{sgn}(\sigma)\theta(\sigma)\chi_1^{m_1}(\sigma)\chi_2^{m_2}(\sigma)$$
$$= \chi_1^{m_1+1}\chi_2^{m_2+1} + \chi_1^{-(m_1+m_2+2)}\chi_2^{m_1+1} + \chi_1^{m_2+1}\chi_2^{-(m_1+m_2+2)}$$
$$- \chi_1^{-m_1-1}\chi_2^{m_1+m_2+2} - \chi_1^{-m_2-1}\chi_2^{-m_1-1} - \chi_1^{m_1+m_2+2}\chi_2^{-m_2-1}$$

and

$$\Sigma \operatorname{sgn}(\sigma)\,\theta(\sigma) = \chi_1\chi_2 + \chi_1^{-2}\chi_2 + \chi_1\chi_2^{-2} - \chi_1^{-1}\chi_2^{2} - \chi_1^{-1}\chi_2^{-1} - \chi_1^{2}\chi_2^{-1}$$

Hence the character of the representation whose highest weight is $\chi_1^{m_1}\chi_2^{m_2}$ is given by the quotient of the two above expressions. With some trouble one can express this quotient as a sum of terms of the form $\chi_1^{j_1}\chi_2^{j_2}$. By counting these one obtains the dimension of the representation whose highest weight is $\chi_1^{m_1}\chi_2^{m_2}$. However, there is an easier way of getting the dimension that works for all n. It is based on a formula for obtaining the restriction to SU$(n-1)$ of any irreducible representation of SU(n). We note of course that we may identify SU$(n-1)$ with the subgroup of SU(n) consisting of all matrices whose first row is $1,0,0,\ldots,0$. Thus if L is any irreducible representation of SU(n) whose highest weight is $\chi_1^{m_1}\cdots\chi_{n-1}^{m_{n-1}}$, we may restrict to SU$(n-1)$ and ask for the highest weights of the irreducible representations of SU$(n-1)$ into which it decomposes. The answer is given by the following so called "branching law." Let $f_1,\ldots,f_n$ be positive integers such that $f_i - f_{i+1} = m_i$. Then there will be an irreducible constituent of our restriction for every set $f_1', f_2', \ldots, f_{n-1}'$ of integers that satisfy the inequalities

$$f_1 \geqslant f_1' \geqslant \cdots \geqslant f_{n-1}' \geqslant f_n$$

ISBN 0-8053-6702-0/0-8053-6703-9, pbk

and the highest weight of the representation associated with $f'_1,\ldots,f'_{n-1}$ will be

$$\chi_1^{f'_1-f'_2}\chi_2^{f'_2-f'_3}\cdots\chi_{n-2}^{f'_{n-2}-f'_{n-1}}$$

To see how this works let us consider the representations of SU(3) whose highest weights are χ_1, χ_2, and $\chi_1\chi_2$, respectively. For the first we may take $f_1=1$, $f_2=0$, and $f_3=0$. Then $1\geqslant f'_1\geqslant 0\geqslant f'_2\geqslant 0$ has just two solutions, 1,0 and 0,0, and these define the representations of SU(2) whose highest weights are χ_1^0 and χ_1^1, that is, $\mathfrak{D}_0$ and $\mathfrak{D}_{1/2}$. The sum of their dimensions is 3. Thus W^{χ_1} is three dimensional. For the second we may take $f_1=1$, $f_2=1$, and $f_3=0$. Our basic inequality here is $1\geqslant f'_1\geqslant 1\geqslant f'_2\geqslant 0$ and again has two solutions, 1,1 and 1,0. These define representations of SU(2) with highest weights χ_1^0 and χ_1^1 just as before. Thus W^{χ_1} and W^{χ_2} are both three dimensional. For the third we may take $f_1=2$, $f_2=1$, and $f_3=0$ so that our inequality is $2\geqslant f'_1\geqslant 1\geqslant f'_2\geqslant 0$. This has four solutions 2,1, 2,0, 1,1, and 1,0. The corresponding highest weights are χ_1, χ_1^2, χ_1^0, and χ_1 and these correspond to the representations $\mathfrak{D}_{1/2}$, $\mathfrak{D}_1$, $\mathfrak{D}_0$, and $\mathfrak{D}_{1/2}$ of SU(2). The sum of their dimensions is $2+3+1+2=8$. Thus $W^{\chi_1\chi_2}$ is an eight-dimensional irreducible representation of SU(3) whose restriction to SU(2) is $\mathfrak{D}_0+2\mathfrak{D}_{1/2}+\mathfrak{D}_1$. The fact that there exists an irreducible representation of SU(3) with the properties of $W^{\chi_1\chi_2}$ plays a fundamental role in the physics of elementary particles. One refers to the relevant part of the theory as the "eightfold way."

So much for the *connected* compact simple groups. The disconnected ones are necessarily finite and we must now discuss the irreducible representations of the finite simple groups. Unfortunately, there is relatively little that can be said. For only a few of the known finite simple groups do we know very much about the possible irreducible representations. The cyclic groups of prime order are commutative and their irreducible representations are one dimensional and obvious. Other than this we have only the classical results of Frobenius on the groups of even permutations—the so called alternating groups—and rather recent work of J. A. Green on GL(n,F), where F is a finite field. Green (*Trans. Am. Math. Soc.* **80** (1955), 402–449) gives a complete determination of the *characters* of the irreducible representations of GL(n,F) but does not describe the representations themselves. Earlier workers had studied GL(n,F) for low values of n. We shall not attempt to describe Green's rather complicated analysis.

The other special case in which one knows all the irreducible representations is important for later applications and for its connections with the representation theory of the compact simple groups. We shall conclude this section with a brief description of the principal facts. The applications actually concern the full symmetric group rather than its simple subgroup of index two, and the representation theory is more easily described for this larger group. Thus we shall describe the distinct irreducible representa-

ISBN 0-8053-6702-0/0-8053-6703-9, pbk

tions of the group S_n of all permutations of n objects. To begin with let us recall that for every finite group the number of distinct irreducible representations is equal to the number of conjugacy classes. In general, of course, there is no natural one-to-one correspondence between the two sets of objects and thus no natural parameterization of the one by the other. However, it is a curious and important fact that S_n is exceptional in this respect. Given any permutation σ of the integers $1,2,\ldots,n$, let G_σ be the cyclic subgroup of S_n generated by σ. Then the set $1,2,\ldots,n$ decomposes into orbits under the action of G_σ. Let $d_1,d_2,\ldots,d_r$ be the number of elements in these orbits chosen so that $d_1 \leqslant d_2 \leqslant \cdots \leqslant d_r$. Then $d_1+d_2+\cdots+d_r=n$ and we refer to any such sequence $d_1,\ldots,d_r$ as a *partition* of n. It is easy to see that every partition of n arises from some σ in S_n and that σ_1 and σ_2 are in the same conjugacy class if and only if their associated partitions are identical. Thus there must be just as many inequivalent irreducible representations of S_n as there are partitions of n. For example, the partitions of 5 are

$$1,1,1,1,1 \quad 1,1,1,2 \quad 1,1,3 \quad 1,4 \quad 1,2,2 \quad 2,3 \quad 5$$

so there must be seven inequivalent irreducible representations of S_5. We shall now show how to assign an irreducible representation V^p of S_n to each partition p of n in such a way that V^{p_1} and V^{p_2} are inequivalent whenever p_1 and p_2 are distinct partitions. It will follow that the V^p must constitute all the irreducible representations of S_n to within equivalence. Given the partition $p=d_1 \leqslant d_2 \leqslant \cdots \leqslant d_r$, let I_j denote the set of all integers m with $d_1+d_2+\cdots+d_{j-1}<m \leqslant d_1+\cdots+d_j$, and let S_n^p denote the subgroup of S_n consisting of all permutations that map each I_j setwise onto itself. The partition p is completely described by the matrix of r rows and d_r columns in which the first d_j elements of the jth row are one for all j and all other elements are zero. The transpose of this matrix is also the matrix of a partition and we call this partition the dual p' of p. We may now also form the subgroup $S_n^{p'}$ and compute without difficulty that there is a unique $S_n^p : S_n^{p'}$ double coset. It follows from this and a general theorem about induced representations that if χ and χ' are one-dimensional characters of S_n^p and $S_n^{p'}$, then the induced representations U^χ and $U^{\chi'}$ have at most a one-dimensional family of intertwining operators and have a one-dimensional family F if and only if χ and χ' coincide on the intersection of S_n^p and $S_n^{p'}$. Since the intersection of S_n^p and $S_n^{p'}$ is the identity, we can conclude that U^χ and $U^{\chi'}$ have to within a scale factor just one nontrivial intertwining operator and hence just one irreducible constituent in common. Now S_n has two one-dimensional characters—the trivial one taking every element to 1 and another that is one on the even permutations and -1 on the odd ones. Let us call them χ_1 and χ_{-1}, respectively. Let $\chi_{1,p}$ denote the restriction of χ_1 to S_n^p and let $\chi_{-1,p'}$ denote the restriction of χ_{-1} to $S_n^{p'}$. It follows from the preceding that $U^{\chi_{1,p}}$ and $U^{\chi_{-1,p'}}$ have a

ISBN 0-8053-6702-0/0-8053-6703-9, pbk

unique irreducible constituent in common. Let us denote this irreducible representation of S_n by A^p. It can be shown that A^p and A^{p^*} are inequivalent whenever $p \neq p^*$. Thus every irreducible representation of S_n is equivalent to one and only one A^p. We thus have a natural one-to-one correspondence between the partitions of n and the equivalence classes of irreducible representations of S_n.

We shall close this section with some remarks on connections between the representations of S_n and those of more general groups. Let $\mathcal{H}$ be a Hilbert space and let $\mathcal{H}^n$ denote the n-fold tensor product of $\mathcal{H}$ with itself. For every permutation σ in S_n there is a unique unitary operator W_σ on $\mathcal{H}^n$ such that $W_\sigma(\phi_1 \times \phi_2 \times \cdots \times \phi_n) = \phi_{\sigma(1)} \times \cdots \times \phi_{\sigma(n)}$. Clearly $\sigma \to W_\sigma$ defines a unitary representation of S_n whose space is $\mathcal{H}^n$. Let us decompose W into primary parts. Let $\mathcal{H}(W) = \mathcal{H}_{p_1} \oplus \cdots \oplus \mathcal{H}_{p_r}$ be the corresponding decomposition of $\mathcal{H}(W) = \mathcal{H}^n$, where $\mathcal{H}_{p_j}$ is the space of that primary constituent whose associated irreducible representation is that defined by the partition p_j. The two extreme cases $p = n$ and its dual $p' = 1, \ldots, 1$ correspond to the two one-dimensional representations of S^n and the subspaces $\mathcal{H}_p$ and $\mathcal{H}_{p'}$ are called the symmetric and anti-symmetric nth powers of $\mathcal{H}$. More generally we have a "semi-symmetric" nth power $\mathcal{H}_{p_j}$ of $\mathcal{H}$ for each partition p_j of n. It is convenient and suggestive to indicate these semi-symmetric powers by writing the exponent as a sum corresponding to the partition in question. Thus we may write $\mathcal{H}^{0+3}$ for the symmetrized third power of $\mathcal{H}$ and $\mathcal{H}^{1+1+1}$ for the anti-symmetrized third power. In addition, we have the semi-symmetrized power $\mathcal{H}^{1+2}$ corresponding to the two-dimensional irreducible representation of S_3 defined by the partition $3 = 1 + 2$.

Now let L be a unitary representation of some group G. Then L^n is by definition the representation $L \otimes \cdots \otimes L$, n factors, whose space is $(\mathcal{H}(L))^n$ and which is uniquely defined by the stipulation that

$$L_x^n(\phi_1 \times \phi_2 \times \cdots \times \phi_n) = L_x(\phi_1) \times L_x(\phi_2) \times \cdots \times L_x(\phi_n)$$

Observe that L_x^n and W_σ commute for all x and σ. Hence each L_x^n is in the commuting algebra of W and it follows that $(\mathcal{H}(L))_p^n$ is an invariant subspace of L^n. In this way L^n falls naturally into a discrete sum of subrepresentations, one for each partition of n. It is convenient to use the notation $L^{d_1 + \cdots + d_n}$ for the subrepresentation defined by $\mathcal{H}_p$, where $p = d_1, \ldots, d_n$. Thus for each n we have not only the symmetric nth power L^{0+n} of our irreducible representation and the anti-symmetric nth power $L^{1+1+\cdots+1}$, we also have a semi-symmetric nth power $L^{d_1+d_2+\cdots+d_r}$ for every partition $d_1, d_2, \ldots, d_r$ of n.

Now let $G = \mathrm{SU}(n)$ and let M be the "natural" n-dimensional unitary representation of G; that is, the representation taking each element of $\mathrm{SU}(n)$ into itself. Since M_x^m and W_σ commute, we obtain a representation

ISBN 0-8053-6702-0/0-8053-6703-9, pbk

V of $G\times S_m$ by setting $V_{(x,\sigma)}=M_x^m W_\sigma$. It can be shown that the subspaces $(\mathcal{H}(M))_p^m$ are irreducible under V. Hence V restricted to $(\mathcal{H}(M))_p^m$ is of the form $B^p\times A^p$, where A^p is the irreducible representation of S_n associated with the partition p and B^p is some irreducible unitary representation of SU(n). In this way we may assign a well-defined irreducible unitary representation of SU(n) to each pair (m,p) consisting of a positive integer m and a partition p of m. Since $m=d_1+d_2+\cdots+d_r$, we have really assigned an irreducible unitary representation of SU(n) to each finite sequence $d_1\leqslant d_2\leqslant\cdots\leqslant d_r$.

Actually the space $\mathcal{H}_p^m$ can reduce to zero when $\mathcal{H}$ is finite-dimensional and in fact does so when and only when $r>\dim(\mathcal{H}_p)$. Thus unless we permit this zero-dimensional representation, our statement is only strictly true when $r\leqslant n$. Finally, then, we obtain an honest irreducible representation of SU(n) for each finite positive sequence $d_1\leqslant d_2\leqslant\cdots\leqslant d_r$ of positive integers with $r\leqslant n$. It can be proved that $d_1\leqslant d_2\leqslant\cdots d_r$ and $d_1'\leqslant d_2'\leqslant\cdots\leqslant d_s'$ define equivalent representations of SU(n) if and only if $d_s'-d_{s-1}'=d_r-d_{r-1}$, $d_{s-1}'-d_{s-2}'=d_{r-1}-d_{r-2}$, and so forth, and that every irreducible unitary representation is equivalent to one of them. Thus we have an independent description of the irreducible unitary representations of SU(n). The connection between the two descriptions is quite simple. The highest weight of the representation defined by $d_1\leqslant d_2\leqslant\cdots\leqslant d_r$ is just $\chi_1^{d_r-d_{r-1}}\chi_2^{d_{r-1}-d_{r-2}}\cdots$.

The irreducible unitary representations of the other simple compact groups can be described in a strictly analogous way except that we cannot always start with the natural representation. Hermann Weyl's book the *Classical Groups* carries out the details for the compact (and complex) groups whose complexified Lie algebras belong to one of the four series A_n, B_n, C_n, and D_n. A full discussion for $U(n)$, GL(n,C), and S_n will be found in Chapter V of Weyl's *Group Theory and Quantum Mechanics*. The description of S_n in terms of induced representations is a slight reformulation of the description given in the last cited reference to Weyl. A detailed account from the induced representation point of view will be found in the Oxford lecture notes of A. J. Coleman.

Notes and References

As already stated in the notes to Sections 4 through 11, the theory of the unitary representations of the compact simple Lie groups was mainly worked out by Hermann Weyl in three papers published in the 1920s. A detailed account for the special case of the unitary groups will be found in the fifth chapter of Weyl's book *The Theory of Groups and Quantum Mechanics*.

The determination of the irreducible unitary representations of the finite simple groups has now advanced far beyond where it was left by Green in

ISBN 0-8053-6702-0/0-8053-6703-9, pbk

1955. In his talk at the Stone retirement conference in the spring of 1968 ("Eisenstein series over finite fields," in *Functional Analysis and Related Fields*, Felix Browder (ed.), Springer-Verlag, New York (1970). Harish Chandra showed that the representation theory of the finite Chevalley groups can be developed in a manner parallel to the unitary representation theory of the noncompact semi-simple Lie groups (Cf. Section 14). In particular, one has an analog of the "discrete series" and theorems allowing one to construct nondiscrete series members by inducing from proper subgroups. The representations one induces from the proper subgroups are built in turn from discrete series members so that the principal problem becomes that of finding the discrete series. The latter problem has been solved in very recent work of G. Lusztig and P. Deligne. "Representations of reductive groups over finite fields," *Ann. Math.*, vol. 103 (1976), 103–161. In all of this work, heavy use is made of the theory of algebraic groups over finite fields. Further details including the course of the development and the many contributions of other authors may be obtained by consulting the following references:

(1) The published lecture notes of a seminar on algebraic groups and related finite groups held at the Institute for Advanced Study during the academic year 1968–69. (*Springer Lecture Notes*, vol. 131 (1970)). Among other things, these notes contain the conjecture of MacDonald whose proof is one of the main results of the paper of Lusztig and Deligne cited above.

(2) The papers by T. Springer and P. Géradin in "Harmonic analysis on homogeneous species," vol. 26 of the American Mathematical Society Symposium on Pure Mathematics (1973).

(3) The Bourbaki seminar talk given by T. Springer in February 1973 (Expose 429).

(4) The article by Lusztig in the Proceedings of the 1974 International Congress in Vancouver.

ISBN 0-8053-6702-0/0-8053-6703-9, pbk

14. The irreducible unitary representations of noncompact simple groups

As remarked earlier the noncompact simple separable locally compact groups fall into three classes: the connected ones, the countable discrete ones, and the nondiscrete totally disconnected ones. As in the compact case, the connected groups are much easier to discuss than the discrete ones. Indeed, the situation is much more extreme. It follows from Thoma's theorem that a simple countably infinite discrete group can never be of Type I, and this makes it extremely unlikely that anyone will ever effect a complete determination of all the irreducible unitary representations of such groups. The totally disconnected groups form an intermediate class. Their systematic study has just begun and our information is rather meager. On the other hand, such information as we do have is encouraging. The group of all 2-by-2 matrices of determinant one over any nondiscrete separable locally compact field has been proved by Kirillov to be a Type I group and one can hope that most if not all of the totally disconnected analogs of the simple Lie groups will be proved to be of Type I.

In any event we shall begin with and devote most of our attention to the study of the connected case; that is, to the unitary representations of the noncompact connected simple Lie groups. These groups may be further subdivided into those that are complex and those that are not—the latter being referred to as the real groups. The complex groups have a simpler representation theory and the fact that they do may be attributed to the following circumstance. Let D be a Cartan subgroup of the complex group G and let $\tilde{D}$ denote the set of all elements that are conjugate to some element of D. Then $G-\tilde{D}$ is of measure zero with respect to Haar

George W. Mackey, Unitary Group Representations in Physics, Probability, and Number Theory

ISBN 0-8053-6702-0/0-8053-6703-9, pbk

measure. In other words, almost every element of G is conjugate to an element of D. The Weyl group acting on D divides its elements into equivalence classes each of which contains only a finite number of elements and two elements of D belong to the same G conjugacy class if and only if they are equivalent under the Weyl group. Thus the orbits of D under the Weyl group serve to parametrize almost all of the conjugacy classes of G. Because the equivalence classes of irreducible unitary representations of G are in a sense "dual" to the conjugacy classes of G, one might hope to find that "almost all" of the equivalence classes of irreducible unitary representations of G are parametrized by the orbits of the Weyl group in the dual of D. This turns out to be the case. Let us choose a nilpotent subgroup N of G such that ND is a semi-direct product of N and D and is a Borel subgroup of G. For each character χ of D let χ' denote its natural lifting to ND and form the induced representation $U^{\chi'}$. It can be proved that $U^{\chi'}$ is irreducible whenever all of the transforms of χ under the Weyl group are distinct and then $U^{\chi_1'}$ and $U^{\chi_2'}$ are equivalent if and only if χ_1 and χ_2 belong to the same Weyl orbit. The irreducible unitary representations of G of the form $U^{\chi'}$ constitute "almost all" of the irreducible unitary representations of G in the following sense. If G is any Type I group, then the regular representation may be written in the form $\int_{\hat{G}} n(L)\,d\mu(L)$, where the class of the measure μ is uniquely determined and $n(L)$ is the multiplicity function and is determined almost everywhere mod μ. The uniquely determined measure class of μ is called the *Plancherel measure class*. (When G is commutative, it is just the Haar measure class in $\hat{G}$.) The irreducible unitary representations of G that are not of the form $U^{\chi'}$ are of measure zero with respect to the Plancherel measure class in $\hat{G}$. It follows in particular that we may decompose the regular representation into irreducibles of the form $U^{\chi'}$.

As an example let us consider the simple groups whose Lie algebras belong to the series $A_1, A_2, \ldots$, that is, the groups $\mathrm{SL}(n,C)/Z_n$, where Z_n is the center of $\mathrm{SL}(n,C)$. It turns out that the same rules apply to $\mathrm{SL}(n,C)$ as to $\mathrm{SL}(n,C)/Z_n$, and it is more convenient to discuss the former group. The group of all matrices in $\mathrm{SL}(n,C)$ that vanish except on the main diagonal is a Cartan subgroup and may be taken as D. N may be correspondingly chosen to be the group of all matrices of the form

$$\begin{bmatrix} 1 & 0 & \cdots & 0 \\ - & \ddots & & \vdots \\ - & - & \ddots & 0 \\ - & - & & 1 \end{bmatrix}$$

that is, all matrices that take the value one on the main diagonal and are zero above this diagonal. ND is the group of all matrices of determinant one that are zero above the main diagonal. As is well known, a matrix is

ISBN 0-8053-6702-0/0-8053-6703-9, pbk

similar to one in diagonal form whenever the characteristic polynomial has no multiple roots, and it is clear that the set of all such matrices is a dense set in SL(n, C) whose complement has measure zero. Thus almost every element in SL(n, C) is conjugate to a member of D as asserted. The Weyl group is just the set of all automorphisms of D of the form

$$\begin{bmatrix} d_1 & & 0 \\ & \ddots & \\ 0 & & d_n \end{bmatrix} \to \begin{bmatrix} d_{\sigma(1)} & & 0 \\ & \ddots & \\ 0 & & d_{\sigma(n)} \end{bmatrix}$$

where σ is an arbitrary permutation of the integers $1, 2, \ldots, n$. The most general member of $\hat{D}$ is of the form

$$\begin{bmatrix} d_1 & & 0 \\ & \ddots & \\ 0 & & d_n \end{bmatrix} \to \left(\frac{d_1}{|d_1|}\right)^{k_1} e^{i\log|d_1|r_1} \cdots \left(\frac{d_n}{|d_n|}\right)^{k_n} e^{i\log|d_n|r_n}$$

and is determined by an n-tuple of integers $(k_1, \ldots, k_n)$ and an n-tuple of real numbers $(r_1, \ldots, r_n)$. It follows from the general theory that $U^{\chi'}$ is irreducible whenever the pairs $(k_1, r_1), \ldots, (k_n, r_n)$ are distinct from one another. However, in this particular case it can be proved that *all* the $U^{\chi'}$ are irreducible. The representations $U^{\chi'}$ are said to belong to the *principal series* of irreducible unitary representations of SL(n, C). It is interesting to compare the members of the principal series with the finite-dimensional (nonunitary) representations of SL(n, C). If we replace the real numbers $r_1, \ldots, r_n$ by the pure imaginary numbers $-ik_1, \ldots, -ik_n$ and insist that the integers k_j be positive and nonincreasing, our member of $\hat{D}$ becomes the (nonunitary) complex homomorphism

$$\begin{bmatrix} d_1 & & 0 \\ & \ddots & \\ 0 & & d_n \end{bmatrix} \to d_1^{k_1} \cdots d_n^{k_n}$$

These homomorphisms lifted to ND were the ones used to define finite-dimensional representations of SL(n, C) by a modification of the inducing process.

Actually there is another modification of the inducing process that can be used to construct further irreducible *unitary* representations of SL(n, C) starting with suitable nonunitary complex homomorphisms of D. Let H be a closed subgroup of the separable locally compact group G and let L be a finite-dimensional representation of H that need not be unitary. Let $V(L)$ denote the space of L and let F_L denote the space of all continuous

ISBN 0-8053-6702-0/0-8053-6703-9, pbk

functions g from G to $V(L)$ that satisfy the identity $g(hx)=L_h g(x)$ for all h in H and all x in G and are zero outside of a set whose image in G/H is compact. Then F_L is invariant under translation and we obtain a (non-unitary) representation A^L of G whose space is F_L by defining $(A_x^L f)(y)=f(yx)$. Now let $\bar{L}$ be the representation of H whose space is the dual $V(L)^*$ of $V(L)$ and is defined by the identity $\bar{L}_x=L^*_{x^{-1}}$. Then if g is in $F_{\bar{L}}$ and f is in F_L, the function $x\to g(x)(f(x))$ is constant on the right H cosets. Suppose that G/H admits an invariant measure μ. Let $\int g(x)(f(x))\,d\mu(x)=f\!:\!g$. Then $f,g\to f\!:\!g$ is a bilinear form on $F_L\times F_{\bar{L}}$, which is invariant under translation. Suppose that there exists an anti-linear operator T from F_L to $F_{\bar{L}}$ such that $TA_x^L=A_x^{\bar{L}}T$ for all x. Let $(f,g)_T=f\!:\!Tg$ and suppose that T can be chosen so that $(f,f)_T>0$ whenever $f\neq 0$. Then T will convert F_L into a pre-Hilbert space. Each A_x^L will have a unique continuous extension B_x^L to the completion of this pre-Hilbert space and $x\to B_x^L$ will be a unitary representation of G. It can be shown that T may exist even when L is not unitary and, that if it does, the resulting unitary representation B^L is independent (to within equivalence) of the particular T chosen. Moreover, when L is unitary, B^L is equivalent to U^L as previously defined so that we may use the notation U^L even when L is not unitary without risk of ambiguity. In other words, U^L is a well-defined unitary representation of G for some nonunitary representations L of H. We remark finally that the restriction that G/H admit an invariant measure may be removed by using Radon–Nikodym derivatives much as this was done earlier for unitary L.

Returning to the problem at hand, we note that we may replace the real numbers $r_1,\ldots,r_n$ by arbitrary complex numbers and obtain a one-dimensional nonunitary representation of D and hence of ND. Let $\chi_{z_1\cdots z_n}^{k_1\cdots k_n}$ denote this representation. Gelfand and Naimark have shown that $U^{\chi_{z_1\cdots z_n}^{k_1\cdots k_n}}$ exists as a unitary representation of $\mathrm{SL}(n,C)$ for some choices of $k_1,\ldots,k_n$, and $z_1,\ldots,z_n$ in which the z_j are not all real. The irreducible representations they obtain in this way are said to belong to the *supplementary* series of irreducible unitary representations of $\mathrm{SL}(n,C)$. One defines the supplementary series in an analogous way for the other complex almost simple Lie groups. In the special case in which $n=2$, it turns out that $U^{\chi_{z_1 z_2}^{k_1 k_2}}=U^{\chi_{0,z_2-z_1}^{0,k_2-k_1}}$ exists only when z_2-z_1 is real or is of the form $-ia$, where $0<a<1$. These irreducible unitary representations of $\mathrm{SL}(2,C)$ parametrized by a are mutually inequivalent and are not equivalent to any member of the principal series. Moreover, it can be proved that every irreducible unitary representation of $\mathrm{SL}(2,C)$ (except the identity) is equivalent either to a member of the principal series or to a member of the supplementary series. Thus we have a complete determination of all irreducible unitary representations of $\mathrm{SL}(2,C)$. Since the quotient of $\mathrm{SL}(2,C)$ by its center is isomorphic to the homogeneous Lorentz group, we have a description of all possible unitary irreducible representations of the homogeneous Lorentz group. This determination was effected by Gelfand and Naimark in 1947.

ISBN 0-8053-6702-0/0-8053-6703-9, pbk

It is natural to conjecture that for $n>2$ as well every irreducible unitary representation of $\mathrm{SL}(n,C)$ is equivalent to $U^{\chi'}$ for some (not necessarily unitary) character χ of D. However, this conjecture is false, and whenever $n \geqslant 3$ certain further series of irreducible unitary representations of $\mathrm{SL}(n,C)$ may be defined. Let X be an n-dimensional complex vector space and let $X_1, X_2, \ldots, X_r$ be linearly independent subspaces whose linear union is X. Then $X = X_1 + \cdots + X_r$. Let N' denote the ring of all linear transformations A such that $A(X_j) \subseteq X_1 + \cdots + X_{j-1}$, $A(X_1) = 0$. Then every element of N' is nilpotent. Hence $I + A$ is nonsingular with inverse $I - A + A^2 - \cdots$ for each A in N' and the set of all such $I + A$ is a subgroup of the group of all linear transformations with determinant one. Let us call this subgroup N. Let D denote the group of all linear transformations T with determinant one such that $T(X_i) = X_i$ for all i. Then ND is a semi-direct product of N and D with N the normal subgroup, and every homomorphism χ of D into the multiplicative group of nonzero complex numbers defines a corresponding homomorphism of ND. Now every member of D is defined by a sequence $T_1, \ldots, T_r$ of linear transformations where T_j is the restriction of T to X_j. Moreover,

$$T \rightarrow \left(\frac{\det T_1}{|\det T_1|} \right)^{k_1} e^{i \log|\det T_1| z_1} \cdots \left(\frac{\det T_r}{|\det T_r|} \right)^{k_r} e^{i \log|\det T_r| z_r}$$

defines such a homomorphism for each choice of $k_1, \ldots, k_r$ and $z_1, \ldots, z_r$ where the k_j are integers and the z_j complex numbers. When all the z_j are real, this homomorphism defines a one-dimensional unitary representation of ND and the corresponding induced representation of $\mathrm{SL}(n,C)$ is a member of the so-called *degenerate series* provided that at least one of the X_j is more than one dimensional. When all the X_j are one dimensional, N and D are as defined previously and we recover the principal series. When all X_j are not one dimensional, we may get some unitary irreducible representations by inducing from nonunitary homomorphisms of D, and these are said to belong to the *supplementary degenerate series*. It is now natural to conjecture that all irreducible unitary representations of $\mathrm{SL}(n,C)$ are induced by unitary or nonunitary characters of the subgroup ND defined by some decomposition $X = X_1 + X_2 + \cdots + X_r$, but no satisfactory proof of this conjecture has yet been given. In 1950 Gelfand and Naimark published a book in which they described certain irreducible unitary representations of $\mathrm{SL}(n,C)$ and the other classical complex groups and asserted that it could be proved that every irreducible unitary representation of any of these groups was equivalent to one on their lists. Later attempts were made by Naimark and by Berezin to prove this assertion. However, the attempt of Naimark was in three parts and only the first two were published. Moreover, gaps were discovered in Berezin's argument that were never satisfactorily filled. Then Kunze and Stein found members of the supplementary series and the degenerate supplementary series that

ISBN 0-8053-6702-0/0-8053-6703-9, pbk

had been overlooked by Gelfand and Naimark, thus confirming the incompleteness of Berezin's attempted proof. At this writing $SL(2,C)/Z$ is the only complex simple group for which it can be proved that we have completely determined *all* irreducible unitary representations.

It is interesting and suggestive to compare the unitary irreducible representations of $SL(2,C)$ with those of a certain semi-direct product. SU(2) is a maximal compact subgroup of $SL(2,C)$ and the homogeneous space $SL(2,C)/SU(2)$ is a three-dimensional space with an invariant Riemannian metric. Thus $SL(2,C)$ may be regarded as the connected component of e in the group of all isometries of a curved or non-Euclidean version of ordinary three space. Actually the parallel is a little closer if we use $SL(2,C)/Z_2$ and $SU(2)/Z_2$ instead. Here Z_2 is the two element group consisting of $\begin{pmatrix} 1 & 0 \\ 0 & 1 \end{pmatrix}$ and $\begin{pmatrix} -1 & 0 \\ 0 & -1 \end{pmatrix}$ and $SU(2)/Z_2$ is isomorphic to the rotation group in three dimensions. "Locally" our homogeneous space is just like ordinary flat Euclidean space whose isometry group has two connected components. The one containing the identity is the semi-direct product of the translations and rotations $N \circledS R$. We wish to compare the irreducible unitary representations of $N \circledS R$ with those of $SL(2,C)/Z_2$. We recall that the orbits of R in $\hat{N}$ are just the spheres with center at e and that a cross section for the orbits is a ray from the identity to ∞. If we think of N as E^3, then the set of all $(a,0,0)$ with $a \geqslant 0$ is such a cross section. When $a=0$, the corresponding representations are just the (finite-dimensional) irreducible representations of R lifted up to $N \circledS R$. When $a>0$, the subgroup R_a is the commutative group of all rotations about the "a axis" and we have one irreducible unitary representation of $N \circledS R$ for each integer n. It is induced by the character $(x,y,z;\theta) \to e^{iax}e^{in\theta}$ of $N \circledS R_a$. It turns out that the finite-dimensional irreducible representations of $N \circledS R_a$ are of measure zero with respect to the Plancherel measure class, so that almost all of the unitary irreducible representations are induced representations parametrized by a positive real number a and an integer n. On the other hand, as we have just seen, almost all the irreducible unitary representations of $SL(2,C)$ are of the form U^{χ}, where χ is a character of the group of all matrices $\begin{pmatrix} d & 0 \\ x & 1/d \end{pmatrix}$ that reduces to the identity on those of the form $\begin{pmatrix} -1 & 0 \\ x & -1 \end{pmatrix}$. The most general such is

$$\begin{pmatrix} d & 0 \\ x & 1/d \end{pmatrix} \to \left(\frac{d}{|d|} \right)^{2n} e^{ia \log|d|}$$

where n is an integer and a is a real number. However, (n,a) and $(-n,-a)$ describe the same representation. Thus if we restrict a to nonnegative values, we get every representation of the indicated form just once. Those with $a=0$ are also of measure zero. Hence almost all irreducible unitary

ISBN 0-8053-6702-0/0-8053-6703-9, pbk

representations are parametrized by a positive real number a and an integer n just as with the Euclidean group $N Ⓢ R$.

Actually there is more to this parallel than a mere coincidence of parametrization. The group T of all $\begin{pmatrix} q & 0 \\ x & 1/q \end{pmatrix}$ with $q>0$ is a subgroup that acts freely and transitively on the homogeneous space; that is, for each pair of points p_1 and p_2 there is one and only one member of T that takes p_1 into p_2. As such it is analogous to the translation subgroup N of the Euclidean group. The group R_a was defined as the subgroup of all h in R which leave fixed the character $(x,y,z) \rightarrow e^{iax}$. However, since every element of R normalizes N, it could equally well have been defined as the subgroup of all elements of R that normalize N and leave the given character fixed. Its analog in $\mathrm{SL}(2,C)/Z_2$ would then be the group of all elements in SU(2) that normalize T and leave a given character fixed. This is easily seen to be the group D_0 of all matrices $\begin{pmatrix} e^{it} & 0 \\ 0 & e^{-it} \end{pmatrix}$ and $T Ⓢ D_0$ is just the group of all

$$\begin{pmatrix} qe^{it} & 0 \\ x & e^{-it}/q \end{pmatrix} = \begin{pmatrix} d & 0 \\ x & 1/d \end{pmatrix}$$

used in inducing the principal series for $\mathrm{SL}(2,C)$. Thus the representation of $\mathrm{SL}(2,C)/Z_2$ with parameter (a,n) is induced from a subgroup constructed in a manner strictly analogous to the one from which the corresponding representation of $N Ⓢ R$ is induced.

Now let G be any connected simple Lie group and let K be its maximal compact subgroup. We may think of G/K as an n-dimensional "curved space" that can be approximated locally by the tangent space V_p at the point p. The subgroup K_p leaving p fixed will be isomorphic to K and every element k of K_p will define a linear transformation A_k of V_p. In this way we obtain a homomorphism $k \rightarrow A_k$ of K_p into the group of all automorphisms of the additive group of V_p. Using this we may form a semi-direct product of the vector group V_p and the compact group K_p. This semi-direct product will be related to G just as the Euclidean group is to $\mathrm{SL}(2,C)/Z_2$, and it would be interesting to see to what extent one has a parallel between the representation theory of these groups generalizing the one discussed above. The relationship between G and $V_p Ⓢ K_p$ has been discussed from a rather different point of view by Robert Hermann in "The Gell-Mann Formula for Representations of Semi-Simple Groups," *Commun. Math. Phys.* **2** (1966), 155–164.

When we pass from complex groups to real ones, the problem becomes much more difficult and results are even less complete. The fundamental difficulty seems to be that in the real case there are always a number of mutually nonconjugate Cartan subgroups and no one that can play the central rôle played by the diagonal matrices in $\mathrm{SL}(n,C)$. Consider for

ISBN 0-8053-6702-0/0-8053-6703-9, pbk

example SL(2, R), the group of all 2-by-2 real matrices of determinant one. This has two quite distinct Cartan subgroups. One is the group of all matrices of the form $\begin{pmatrix} d & 0 \\ 0 & 1/d \end{pmatrix}$ and the other is the compact group of all matrices of the form $\begin{pmatrix} \cos\theta & \sin\theta \\ -\sin\theta & \cos\theta \end{pmatrix}$. The existence of these two Cartan subgroups is a reflection of the fact that a real matrix $\begin{pmatrix} a & b \\ c & d \end{pmatrix}$ is diagonalizable if and only if the roots of its characteristic equation are not only distinct but also real. An element whose characteristic equation has complex conjugate roots will be conjugate not to a diagonal matrix but to one of the form $\begin{pmatrix} \cos\theta & \sin\theta \\ -\sin\theta & \cos\theta \end{pmatrix}$, where $e^{\pm i\theta}$ are the two conjugate roots. For SL(n, R) there will be as many possibilities as there are nonnegative integers k with $2k \leqslant n$. For each such k there will be a Cartan subgroup containing conjugates of all matrices whose characteristic equations have k pairs of conjugate complex roots. When $n=4$, k can be 0, 1, or 2 and the corresponding Cartan subgroups consist respectively of all diagonal matrices

$$\begin{bmatrix} d_1 & & 0 & \\ & d_2 & & \\ 0 & & d_3 & \\ & & & d_4 \end{bmatrix}$$

with $d_1 d_2 d_3 d_4 = 1$, of all matrices of the form

$$\begin{bmatrix} d_1 & & & \\ & d_2 & & \\ & & d_3\cos\theta & d_3\sin\theta \\ & & -d_3\sin\theta & d_3\cos\theta \end{bmatrix}$$

with $d_1 d_2 d_3^2 = 1$ and of all matrices of the form

$$\begin{bmatrix} d_1\cos\theta_1 & d_1\sin\theta_1 & & \\ -d_1\sin\theta_1 & d_1\cos\theta_1 & & \\ & & d_2\cos\theta_2 & d_2\sin\theta_2 \\ & & -d_2\sin\theta_2 & d_2\cos\theta_2 \end{bmatrix}$$

with $d_1^2 d_2^2 = 1$. By the same heuristic reasoning used for the complex groups, one expects to find a family of irreducible unitary representations for each Cartan subgroup and to need the members of all families in order to get almost all representations with respect to the Plancherel measure class. This turns out to be a difficult program and one does not even know how to define the appropriate family in all cases.

ISBN 0-8053-6702-0/0-8053-6703-9, pbk

In describing what is known, let us begin with SL(2, R), the simplest of the real groups and one of the only two whose irreducible representations have been completely determined. This determination was carried out by the mathematical physicist V. Bargmann in a paper that is now regarded as one of the classics of the subject. The quotient of SL(2, R) by its center is isomorphic to SO(2, 1), the homogeneous Lorentz group in two space dimensions, and his paper is "Irreducible unitary representations of the Lorentz group," *Ann. Math.* **48** (1947), 568–640.

In part, the results are quite analogous to those for SL(2, C). For each unitary character χ of the group of all diagonal matrices $\begin{pmatrix} d & 0 \\ 0 & 1/d \end{pmatrix}$ one has a corresponding lifted character χ' of the group of all $\begin{pmatrix} d & 0 \\ a & 1/d \end{pmatrix}$ and with two exceptions the induced representations $U^{\chi'}$ are all irreducible and are distinct unless $\chi_1 = \chi_2^{-1}$. However, we are dealing with the multiplicative group of the reals rather than the complexes and the $U^{\chi'}$ are parametrized by a nonnegative real number and ± 1 rather than by a nonnegative real and an integer. The representations $U^{\chi'}$ are said to constitute the *principal series*, but no longer include almost all irreducible unitary representations of our group. There are further irreducible representations to be obtained by replacing χ by certain nonunitary characters. As with SL(2, C) the appropriate ones are described by the real numbers in a finite interval and the corresponding irreducible unitary representations are collectively of measure zero with respect to the Plancherel measure. They are said to form the *supplementary series*. It follows from what has been said that there must be further irreducible unitary representations of SL(2, R) and that collectively they must be of positive Plancherel measure. Moreover, we may expect them to be parametrized by the characters of the compact group K of all $\begin{pmatrix} \cos\theta & \sin\theta \\ -\sin\theta & \cos\theta \end{pmatrix}$, that is, the integers. It turns out that one can assign in a natural way a unitary irreducible representation of SL(2, R) to each χ in $\hat{K}$ except the identity. These irreducible representations are defined by a modification of the inducing process in which one restricts attention to those functions in the space of the ordinary induced representation U^{χ} that are in a certain sense analytic. In order to say what this sense is, it is useful to take the vector bundle point of view and consider the members of $\mathcal{H}(U^{\chi})$ as cross sections of an appropriate bundle. To this end let H denote the upper half of the complex plane and let $G = \mathrm{SL}(2, R)$ act on H in such a way that the transform of z by $\begin{pmatrix} a & b \\ c & d \end{pmatrix}$ is $(az+b)/(cz+d)$. As is well known G acts transitively on H and preserves its structure as a complex analytic manifold. We may identify H with G/G_{z_0}, where G_{z_0} is the subgroup carrying any fixed element of H into itself. If we take this element to be i, we find that G_{z_0} is the group of all $\begin{pmatrix} a & b \\ c & d \end{pmatrix}$ for which $(ai+b)/(ci+d) = i$, that is, $ai+b = di-c$ or $a = d, b = -c$. Together with

ISBN 0-8053-6702-0/0-8053-6703-9, pbk

$ad-bc=1$, this implies $a^2+b^2=1$. Hence $a=\cos\theta$, $b=\sin\theta$, $c=-\sin\theta$, and $d=\cos\theta$. In other words, $G_i=K=$group of all $\begin{pmatrix} \cos\theta & \sin\theta \\ -\sin\theta & \cos\theta \end{pmatrix}$ and we may identify H with G/K as indicated. Since a complex-valued function on G that satisfies the identity $f(yx)=f(x)$ for all y in K and all x in G may be identified with a function on H, it is clear what we mean when we say that a function on the space of U^I is holomorphic or complex analytic. Unfortunately, no square integrable function on G/K is holomorphic except the function identically zero, and the meaning of complex analytic is not immediately clear for members of $\mathcal{H}(U^\chi)$, where χ is not the identity character. But consider the one-dimensional "complex vector bundle" on H whose elements are the pairs (z,v), where z is in H and v is a tangent vector to H at z. The "dual" bundle, called the cotangent bundle, consists of all pairs (z,w), where w is a linear functional on the tangent space to H at z. The action of G on H induces an action of G on each of these bundles and the individual operators are bundle automorphisms. Since the subgroup G_i leaving i fixed is compact, we may make the tangent space at i and its dual into Hilbert spaces whose norm is G_i invariant. Thus we may assign each tangent space and its dual a Hilbert space structure in such a way that the operators of G are Hilbert bundle automorphisms. Taking the space of square integrable cross sections, we obtain two unitary representations of G induced by representations of the compact group G_i. The representations in question are just those defined in the tangent space to H at i and its dual be the natural action of G_i on these spaces. In general, a diffeomorphism leaving a point fixed will define an automorphism of the tangent space at that point—and hence of the dual space. We verify at once that the representation of G_i associated with the tangent bundle is

$$\begin{pmatrix} \cos\theta & \sin\theta \\ -\sin\theta & \cos\theta \end{pmatrix} \to e^{2i\theta}$$

and that the one associated with the dual bundle is

$$\begin{pmatrix} \cos\theta & \sin\theta \\ -\sin\theta & \cos\theta \end{pmatrix} \to e^{-2i\theta}$$

To obtain the representation of G induced by

$$\begin{pmatrix} \cos\theta & \sin\theta \\ -\sin\theta & \cos\theta \end{pmatrix} \to e^{2in\theta}$$

where n is a positive integer we have only to use the nth-order contravariant tensors instead of tangent vectors. In n is a negative integer, we use nth-order covariant tensors.

ISBN 0-8053-6702-0/0-8053-6703-9, pbk

The advantage of realizing the representations of G induced by these characters of G_i by cross sections of vector and tensor bundles is that it makes sense to say that such cross sections are holomorphic or complex analytic. Indeed a contravariant vector field defines a differential operator and we may say that it is holomorphic if this differential operator takes holomorphic functions into holomorphic functions. In a similar fashion we define holomorphy in the other cases. We now look at the subspace of our Hilbert space of square integrable cross sections consisting of holomorphic cross sections. This can reduce to zero but if it does not, it is clearly an invariant subspace that defines a subrepresentation of the representation of G associated with the given bundle. It turns out that these subrepresentations are always irreducible and nontrivial if and only if n is negative.

When $n = -1$, every holomorphic cross section is defined by a "differential" $f(z)\,dz$, where f is a complex-valued function that is holomorphic in the upper half-plane. As calculation shows, the transform of this cross section by $\begin{pmatrix} a & b \\ c & d \end{pmatrix}$ is

$$g(z)\,dz$$

where

$$g(z) = f\left(\frac{(az+b)}{(cz+d)}\right)(cz+d)^{-2}$$

Moreover, the square integral of the cross section $f(z)\,dz$ is

$$\int(|f(x+iy)|^2)y^2\,dx\,dy$$

Thus in this case our representation may be realized in the space of all holomorphic functions f on the upper half-plane for which

$$\int(|f(x+iy)|^2)y^2\,dx\,dy < \infty$$

in such a way that the unitary operator corresponding to $\begin{pmatrix} a & b \\ c & d \end{pmatrix}$ takes f into the function

$$z \to f\left(\frac{(az+b)}{(cz+d)}\right)(cz+d)^{-2}$$

When $n = -k$, every cross section is a function f multiplied by $(dz)^k$ and our representation may be realized in the space of all holomorphic functions f in the upper half-plane for which

$$\int(|f(x+iy)|^2)y^{2k}\,dx\,dy < \infty$$

ISBN 0-8053-6702-0/0-8053-6703-9, pbk

in such a manner that the unitary operator corresponding to $\begin{pmatrix} a & b \\ c & d \end{pmatrix}$ takes f into the function

$$z \to f\left(\frac{(az+b)}{(cz+d)}\right)(cz+d)^{-2k}$$

Thus for each of the characters

$$\begin{pmatrix} \cos\theta & \sin\theta \\ -\sin\theta & \cos\theta \end{pmatrix} \to e^{2ik\theta}, \qquad k=1,2,\ldots$$

we have defined in quite concrete explicit fashion a unitary irreducible representation of SL$(2,R)$ that acts in a space of holomorphic functions and is an irreducible subrepresentation of the representation of SL$(2,R)$ induced by the character in question. Let us call this representation V^{2k}. We note that the concrete formulas defining V^{2k} make sense for k half integral as well as integral and we may regard V^j as defined for all positive integral j. The difference between odd and even values is that the vector bundle interpretation is less easy to come by for odd j. A mildly tedious calculation shows that V^j restricted to the subgroup G_i is the direct sum of all characters

$$\begin{pmatrix} \cos\theta & \sin\theta \\ -\sin\theta & \cos\theta \end{pmatrix} \to e^{i(j+2m)\theta}, \qquad m=0,1,2,\ldots$$

Now consider the representation $g \to (V^j_{g^{-1}})^*$, where * denotes the Banach space adjoint of $V^j_{g^{-1}}$. This representation restricted to G_i will be the direct sum of the complex conjugates of the characters $e^{i(j+2m)\theta}$ and hence will be the direct sum of all characters $e^{-i(j+2m)\theta}$ for $m=0,1,2,\ldots$. It is natural to denote this representation by the symbol V^{-j}. Clearly V^{-j} is not equivalent to any V^k with k positive.

The irreducible representations V^j, j an integer $\neq 0$, constitute the celebrated "discrete series" of Bargmann. Bargmann's notation is $V^s = D^+_{j/2}$ for $j=1,2,3,\ldots$; $V^s = D^-_{-j/2}$ for $j=-1,-2,-3,\ldots$. Each one with $|j| \neq 1$ occurs as a *discrete* component of the regular representation, and correspondingly the "matrix elements" $(V^j_g\phi \cdot \theta)$ are all square integrable functions on SL$(2,R)$. One speaks of the V^j as square integrable representations. Finally, Bargmann has proved that every irreducible unitary representation of SL$(2,R)$ is either a member of the principal series, the supplementary series, or the discrete series.

It is interesting and significant that the members of the discrete series may be described in a rather different manner that relates them quite closely to members of the principal and supplementary series. Let χ_r be the one-dimensional (nonunitary) representation of the subgroup of all $\begin{pmatrix} d & 0 \\ a & 1/d \end{pmatrix}$ that takes $\begin{pmatrix} d & 0 \\ a & 1/d \end{pmatrix}$ into $e^{r\log|d|}$, where r is a real number,

ISBN 0-8053-6702-0/0-8053-6703-9, pbk

and let us try to define U^{χ_r} as a unitary representation as explained above. The attempt succeeds for $-2<r<0$ and this yields the supplementary series (or more exactly, that part of it which is the identity on the center of SL$(2,R)$). For other values of r it seems to fail, but something rather special happens when r is an even integer. The space in which we try to introduce our invariant Hilbert space norm turns out to contain two invariant subspaces. When the even integer r is negative, these two invariant subspaces are disjoint and one of them admits an invariant Hilbert space norm. The corresponding unitary representation of SL$(2,R)$ is irreducible and can be shown to be equivalent to the member of the discrete series labeled V^{-r} above. When the even integer r is zero or positive, the two invariant subspaces intersect in a finite-dimensional subspace whose dimension is $r+1$. Consider the quotient spaces of our two invariant subspaces by this finite-dimensional invariant subspace. One of these admits an invariant Hilbert space norm and the corresponding unitary representation of SL$(2,R)$ can be shown to be irreducible. This irreducible unitary representation can be shown to be equivalent to the member of the discrete series labeled $V^{-(r+2)}$ above. Thus as r varies over the even integers we get all members of the discrete series that are the identity on the center of SL$(2,R)$. We get the others by replacing χ_r by its product with the character

$$\begin{pmatrix} d & 0 \\ a & 1/d \end{pmatrix} \rightarrow \operatorname{sgn}(d)$$

and going through an analogous construction. It is interesting to notice that the finite-dimensional invariant subspaces of the U^{χ_r} for r integral are just the finite-dimensional (nonunitary) irreducible representations of SL$(2,R)$. Notice also that describing the members of the discrete series from the U^{χ_r} leads to a parametrization of the series by all integers rather than by the nonzero integers as above. It is also significant that we get a natural one-to-one correspondence between the finite-dimensional irreducible representations and certain members of the discrete series.

Next let us look at the group $G=\mathrm{SL}(n,R)$, where $n>2$. In this case the problem of determining all equivalence classes of irreducible unitary representations is still unsolved. However, Gelfand and Graev have determined almost all irreducible unitary representations in the sense of the Plancherel measure class. Moreover, as suggested above, they find a "series" of irreducible unitary representations associated with each conjugacy class of Cartan subgroups of G and parametrized by the unitary characters of a member of this conjugacy class. It will be convenient to describe these series as an illustration of a general principle discovered somewhat later by Harish–Chandra.

Let G be any connected Lie group with finite center whose Lie algebra is simple or a direct sum of simple Lie algebras. Let K be a maximal compact

ISBN 0-8053-6702-0/0-8053-6703-9, pbk

subgroup. By a fundamental theorem of Iwasawa there exists a closed nilpotent subgroup N and a commutative subgroup D isomorphic to a finite-dimensional vector space such that every element of G is uniquely of the form ndk, where n is in N, d is in D, and k is in K, and ND is a semi-direct product of D and the normal subgroup N. When $G=\mathrm{SL}(n,R)$, we may take N as the group of all matrices that are one on the diagonal and zero above it. K is the group of all orthogonal matrices and D is the group of all diagonal matrices with postive elements. In the general case it can be shown that any Cartan subgroup is conjugate to the direct product of a closed vector subgroup D_0 of D and a closed commutative subgroup K_0 of K. Let G_0 be the group of all elements that commute with all elements of D_0. It is possible to write G_0 in the form $D_0 \times G_0'$ where G_0' modulo the center of G is a finite extension of a group G_0'' containing the connected component K_{00} of K_0 as a Cartan subgroup. Moreover, $D_0 \times G_0'$ normalizes a nilpotent subgroup N_0 and the choices can be made so that $N_0 \subseteq N$. Let S_0 be the semi-direct product of N_0 with $D_0 \times G_0'$. If χ is a unitary character of D_0 and L is an irreducible unitary representation of G_0', we obtain a unitary representation of G by inducing the representation $ndg \rightarrow \chi(d)L_g$ from S_0 to G. Harish–Chandra conjectures that $U^{\chi L}$ is irreducible for almost all choices of χ and L. If this is so and if we can find a family of irreducible unitary representations of G_0' parametrized by the characters of K_{00}, then as L varies over these, $U^{\chi L}$ will vary over a family parametrized by the characters of $D_0 \times K_{00}$ as required.

Now the irreducible unitary representations of G_0' are in general easily obtainable from those of G_0'' and this is a group containing the compact Cartan subgroup K_{00}. To this extent and in this sense Harish–Chandra reduced the problem of associating a series of representations to each Cartan subgroup to the case in which this subgroup is compact. In the special case in which $G=\mathrm{SL}(n,R)$, we may take D_0 to be the set of all diagonal matrices with entries $d_{11}, d_{22}, \ldots, d_{nn}$ such that $d_{11}=d_{22}, d_{33}=d_{44}, \ldots, d_{2t-1,2t-1}=d_{2t,2t}$ for some $t \leqslant \frac{1}{2}n$ and the corresponding K to be the set of all matrices of the form

$$\begin{pmatrix} \cos\theta_1 & \sin\theta_1 & & & & & & & \\ -\sin\theta_1 & \cos\theta_1 & & & & & & & \\ & & \cos\theta_2 & \sin\theta_2 & & & & & \\ & & -\sin\theta_2 & \cos\theta_2 & & & & & \\ & & & & \ddots & & & & \\ & & & & & \cos\theta_t & \sin\theta_t & & \\ & & & & & -\sin\theta_t & \cos\theta_t & & \\ & & & & & & & u_{2t+1} & \\ & & & & & & & & \ddots \\ & & & & & & & & & u_n \end{pmatrix}$$

ISBN 0-8053-6702-0/0-8053-6703-9, pbk

where all elements not on the diagonal or in one of the 2-by-2 blocks through the diagonal are zero and $u_j = \pm 1$ for all j. G_0 is then the group of all matrices in G having the same block form as above but with arbitrary members of $GL(2,R)$ in the 2-by-2 blocks and arbitrary nonzero real numbers in place of the u_j. We may clearly write $G_0 = G_0' \times D_0$, where G_0' differs from G_0 in that the elements of $GL(2,R)$ are replaced by elements in $SL'(2,R)$ and the u_j again become ± 1. ($SL'(2,R)$ is the group of all 2-by-2 real matrices with determinant ± 1.) In this case G_0'' is the direct product of t replicas of $SL(2,R)$ and K_{00} is the direct product of the subgroups of all $\begin{pmatrix} \cos\theta & \sin\theta \\ -\sin\theta & \cos\theta \end{pmatrix}$. From our study of $SL(2,R)$ we know a family of irreducible unitary representations of this group parametrized by the characters of K_{00}. Any such character is described by a t-tuple of integers $(k_1,\ldots,k_t)$ and takes the member of K_{00} described by $\theta_1,\ldots,\theta_t$ into $e^{i(k_1\theta_1 + k_2\theta_2 + \cdots + k_t\theta_t)}$. We associate with it the tensor product $V^{k_1} \times V^{k_2} \times \cdots \times V^{k_t}$, where V^k is the kth member of the discrete series for $SL(2,R)$ if $k \neq 0$ and the identity if k is. We leave to the reader the simple task of passing from these representations of G_0'' to related irreducible representations of G_0' and hence of S_0. The corresponding induced representations are irreducible and are those found by Gelfand and Graev.

The case of $SL(n,R)$ is rather special in that the study of G_0'' reduces to that of $SL(2,R)$, a problem already solved by Bargmann. In general, things do not work out so easily. Consider for example the group $SU(p,q)$ of all complex matrices of determinant one leaving fixed an indefinite Hermitean form with p plus signs and q minus signs. In this case there is a conjugacy class of Cartan subgroups for each $r = 0,1,2,\ldots,\min(p,q)$ and the G_0'' for r is $SU(p-r,q-r)$. It turns out that $SU(p',q')$ modulo its maximal compact subgroup has the structure of a complex analytic manifold and that one can define a discrete series of representations of it associated with the characters of a compact Cartan subgroup by generalizing the Bargmann construction. Indeed Harish–Chandra has shown how to do this whenever G_0'' modulo its maximal compact subgroup has a complex analytic structure. Unfortunately, there are many cases in which G_0'' does not have this property and Harish–Chandra's construction yields nothing. What to do is still an unsolved problem. The simplest case is that of $SO(4,2)$ and this has been studied in detail by Dixmier (*Bull. Soc. Math. de France*, **89** (1961)) using Lie algebra methods. He actually succeeds in determining *all* irreducible unitary representations and verifies the existence of the appropriate discrete series. However, he does not define it in such a way as to suggest the solution of the general problem. On the other hand, Harish–Chandra has quite recently been able to find the characters of the irreducible unitary representations in the general case.

As we have said before, one is by no means finished when one knows all the irreducible unitary representations of the groups in which one is interested. There is also the problem of finding the tensor product of pairs

ISBN 0-8053-6702-0/0-8053-6703-9, pbk

of irreducible representations and that of finding how the irreducible representations decompose when restricted to various subspaces. Now we have seen the very great extent to which irreducible unitary representations may be thrown into the form U^L, where L is an (often one-dimensional) irreducible unitary representation of a subgroup. Thus it is of considerable interest that there exist certain general theorems that give us information both about the tensor products of induced representations and about their restrictions to subgroups. We shall close this section by stating these theorems and indicating some applications.

Let L be a unitary representation of the closed subgroup H_1 of the separable locally compact group G and let H_2 be a second closed subgroup. We wish to consider the structure of the representation obtained by forming U^L and then restricting to H_2. Recall that U^L admits a canonical system of imprimitivity P^L based on G/H_1. Of course P^L will also be a system of imprimitivity for the restriction of U^L to H_2, but this system of imprimitivity will be transitive only if the action of H_2 on G/H_1 is transitive. In general, it will not be. The H_2 orbit of the point H_1x in G/H_1 will be the set of all right H_1 cosets contained in the double coset H_1xH_2 and it follows immediately that the orbits of G/H_1 under the action of H_2 correspond one-to-one to the $H_1:H_2$ double cosets. Let us consider the important special case in which there are only countably many $H_1:H_2$ double cosets. Then G/H_1 breaks down into countably many H_2 invariant subsets and each of these defines a subspace of $\mathcal{H}(U^L)$ that is invariant under the U_h^L with h in H_2. Then U^L restricted to H_2 has a natural direct sum decomposition with one summand for each $H_1:H_2$ double coset that is not of measure zero. On each of these components P^L restricts to be a transitive system of imprimitivity and we conclude that the component itself is induced by a unitary representation of some subgroup of H_2. The subgroup of course will be that leaving H_1x fixed where $H_1x\,H_2$ is the double coset in question. But $H_1xh=H_1x$ if and only if $H_1xhx^{-1}=H_1$, that is, xhx^{-1} belongs to H_1, that is, h belongs to $x^{-1}H_1x$. Thus the subgroup is $H_2\cap x^{-1}H_1x$ and one verifies without difficulty that the representation is the restriction to $H_2\cap x^{-1}H_1x$ of the representation $h\rightarrow L_{xhx^{-1}}$ of $x^{-1}H_1x$. We have sketched a proof of the following theorem.

Let H_1 and H_2 be closed subgroups of the separable locally compact group G such that there are only countably many $H_1:H_2$ double cosets H_1xH_2. Let L be a unitary representation of H_1 and for each x in G let L^x denote the representation of $x^{-1}H_1x$ that takes h into $L_{xhx^{-1}}$. Then the representation of H_2 induced by the restriction of L^x to $H_2\cap x^{-1}H_1x$ depends to within equivalence only on the double coset to which x belongs and may be denoted by the symbol L^d, where $d=H_1xH_2$. The restriction of U^L to H_2 is equivalent to the direct sum of all the L^d over all double cosets d which are not of Haar measure zero.

ISBN 0-8053-6702-0/0-8053-6703-9, pbk

It is not difficult to weaken the restriction that there are only countably many double cosets. Let us say that H_1 and H_2 are *regularly related* if there exists a Borel cross section for the $H_1:H_2$ double cosets. When this is the case, the space $D(H_1,H_2)$ of all $H_1:H_2$ double cosets is standard in the Borel structure that it inherits from G. Moreover, the Haar measure class in G defines a measure class C in $D(H_1,H_2)$—the unique one such that A is a null set if and only if the inverse image of A in G is of Haar measure zero. The theorem quoted in the previous paragraph remains true for subgroups that are only regularly related provided that we replace our direct sum by a direct integral with respect to the measure class C.

By way of application let us sketch a proof of the theorem that members of the principal series for $\mathrm{SL}(n,C)$ are irreducible. We let H_1 be the subgroup of all matrices that are zero above the main diagonal and let L be the representation

$$\begin{pmatrix} d_1 & & & \\ & d_2 & & \\ & & \ddots & \\ & & & d_n \end{pmatrix} \to \chi_1(d_1)\cdots\chi_n(d_n)$$

Then the representations U^L are by definition members of the principal series. Let H_2 be the subgroup of $\mathrm{SL}(n,C)$ consisting of all matrices whose top row is $1,0,\ldots,0$. Then an easy calculation shows that almost every element is in the double coset containing

$$\begin{pmatrix} & & & -1 \\ & & \cdot^{\cdot^{\cdot}} & \\ & 1 & & \\ 1 & & & \end{pmatrix}$$

Thus there is just one double coset whose Haar measure is not zero and the restriction of U^L to H_2 is induced by a one-dimensional representation of

$$H_2 \cap \begin{pmatrix} & & & -1 \\ & & \cdot^{\cdot^{\cdot}} & \\ & 1 & & \\ 1 & & & \end{pmatrix} H_1 \begin{pmatrix} & & & -1 \\ & & \cdot^{\cdot^{\cdot}} & \\ & 1 & & \\ 1 & & & \end{pmatrix}^{-1}$$

and this equals the group of all matrices in H_2 which vanish below the

ISBN 0-8053-6702-0/0-8053-6703-9, pbk

main diagonal. Now H_2 is a semi-direct product of $SL(n-1, C)$ and the commutative group of all matrices of the form

$$\begin{pmatrix} 1 & & & \\ a_2 & 1 & 0 & \\ \vdots & 0 & \ddots & \\ a_n & 0 & & 1 \end{pmatrix}$$

and we compute that U^L restricted to H_2 reduces to the result of lifting to H_2 a member of the principal series for $SL(n-1, C)$. Using induction we see that it will suffice to show that members of the principal series of $SL(2, C)$ are irreducible. To prove this we take $H_1 = H_2$ and apply our theorem again. In this case H_1 is a semi-direct product of two commutative groups and our theory of semi-direct products allows us to conclude that the result is irreducible. We leave the details to the reader.

The theorem alluded to about tensor products is actually a corollary of the subgroup theorem. Let L and M be irreducible unitary representations of the closed subgroups H_1 and H_2 of G. Then $U^L \otimes U^M$ is the restriction to the "diagonal" $\tilde{G}$ of the representation $U^L \times U^M$ of $G \times G$. It is easy to see that $U^L \times U^M$ may be put into the form $U^{L \times M}$ and hence is a representation of $G \times G$ induced by a representation of the subgroup $H_1 \times H_2$. Applying the subgroup theorem, we realize $U^L \otimes U^M$ as a direct integral over the $H_1 \times H_2 : \tilde{G}$ double cosets of certain induced representations of G. Working out what these are and noticing that there is in fact a natural one-to-one correspondence between $H_1 \times H_2 : \tilde{G}$ double cosets on the one hand and $H_1 : H_2$ double cosets on the other hand, we arrive at the following theorem:

> Let H_1 and H_2 be closed regularly related subgroups of the separable locally compact group G and let L and M be unitary representations of H_1 and H_2, respectively. For each pair (x,y) of elements of G, let L^x be the representation $h \to L_{xhx^{-1}}$ of $x^{-1}H_1x$, let M^y be the representation $h \to M_{yhy^{-1}}$ of $y^{-1}H_2y$, and let $A^{(x,y)}$ be the tensor product of their restrictions to $x^{-1}H_1x \cap y^{-1}H_2y$. Let $V^{(x,y)}$ be the representation of G induced by $A^{(x,y)}$. Then to within equivalence $V^{(x,y)}$ depends only on the $H_1 : H_2$ double coset to which xy^{-1} belongs and $U^L \otimes U^M$ is equivalent to the direct integral over these double cosets of the $V^{(x,y)}$.

In many interesting cases this result combined with suitable auxiliary considerations leads to a complete decomposition of $U^L \otimes U^M$ into its irreducible constituents.

ISBN 0-8053-6702-0/0-8053-6703-9, pbk

Notes and References

Work on the irreducible unitary representations of the noncompact semi-simple Lie groups began in 1947 with the long papers of Bargmann, Harish Chandra and Gelfand, and Neumark on SL(2, R) and SL(2, C). As mentioned in the text, the problem of finding *all* irreducible unitary representations proved rather intractable and emphasis soon shifted to finding enough to decompose the regular representation and prove a generalization of the Plancherel formula. Gelfand and Neumark completed this program for the classical complex groups in 1950, treating each of the four infinite classes separately. A few years later in 1954 Harish Chandra took care of the exceptional complex groups in an analysis that applied uniformly to all complex semi-simple Lie groups. The real groups proved to be much more intractable. Gelfand and Graev studied SL(n, R) and other special cases in the early 1950s, and around the same time Harish Chandra began a unified attack on the general case. Harish Chandra's program reducing the problem to finding the "discrete series" is briefly described in the text. By 1962 he had succeeded in finding the entire discrete series (more precisely, their characters) for an arbitrary semi-simple Lie group with a finite center, and announced the results in 1963 in the *Bulletin of the American Mathematical Society*. The announcement states the results in four theorems whose proofs turned out to be quite long and complicated. The proof of Theorem 1 alone was spread out over five papers whose combined length is about 210 pages. The proofs of Theorems 2, 3, and 4 occupy another three papers totaling 240 pages. All eight papers were published between 1964 and 1966. Two each appeared in the *American Journal of Mathematics*, the *Annals of Mathematics*, and *Acta Mathematica*. The other two appeared in the *Transactions of the American Mathematical Society* and the *Publications Mathematiques de le I.H.E.S.*

A very detailed summary of the main ideas in Harish Chandra's work on the discrete series will be found in the lectures given by V. S. Varadarajan at the American Mathematical Society symposium on harmonic analysis on homogeneous spaces held in the summer of 1972. These were published in 1973 in the volume already cited in the notes and references for Section 12. Later (*Springer Lecture Notes*, vol 576 (1977)) Varadarajan gave a complete exposition with proofs. Still another complete exposition will be found in the book of Warner mentioned in the notes and references for Sections 4 through 11.

While finding the discrete series was certainly the main step, there remained the problems of showing that the other series could be obtained by the inducing process as described in the text, of completing the decomposition of the regular representation, and of finding the Plancherel measure. Harish Chandra announced a solution to these problems in the

ISBN 0-8053-6702-0/0-8053-6703-9, pbk

course of delivering the Colloquium Lectures of the American Mathematical Society in the summer of 1969. The contents of these lectures were published in 1970 (in the society *Bulletin*) and the detailed proof of the Plancherel formula was completed in three long papers published in 1975 and 1976. They appeared, respectively, in the *Journal of Functional Analysis*, *Inventiones Mathematicae*, and the *Annals of Mathematics*.

The papers of Harish Chandra which appeared in 1969 and after are filled with terminology from the theory of automorphic forms such as "Eisenstein series," "Eisenstein integral," "cusp form," "constant term," and so forth. Moreover, Harish Chandra acknowledges that his work was heavily influenced by what he calls the "philosophy of cusp forms" which he "learned from R. P. Langlands." This reflects an extremely interesting development in which the applications of the theory of unitary group representations to the theory of automorphic forms discovered in the 1950s (see Section 30 of this book) led in the late 1960s to important new insights into the unitary representation theory of semi-simple groups as well as their finite and p-adic analogs. Let $G=\mathrm{SL}(2,R)$ and let Γ be a suitably restricted discrete subgroup of G. As explained in Section 30, the structure of the representation U^{I_Γ} of G induced by the identity representation of Γ is closely related to the theory of automorphic forms for the subgroup Γ. In particular, the decomposition of automorphic forms as sums of Eisenstein series and cusp forms (Section 29) is precisely reflected in the decomposition of U^{I_Γ} into continuously and discretely decomposable pieces—provided that one considers the nonanalytic automorphic forms introduced by Maass in 1949 along with the classical forms of Dedekind, Klein, Poincaré, and Hurwitz. The classical theory associates an Eisenstein series with each cusp of the fundamental region for the natural action of Γ on the upper half-plane, and the Maass analog of these Eisenstein series are canonically associated with intertwining operators from U^{I_N} to U^{I_Γ}, where N is the subgroup of $\mathrm{SL}(2,R)$ consisting of all matrices of the form $\begin{pmatrix} 1 & 0 \\ a & 1 \end{pmatrix}$. The closed linear span $\mathfrak{M}$ of the ranges of these intertwining operators is precisely the continuously decomposable part of U^{I_Γ} so that U^{I_Γ} restricted to $\mathfrak{M}^\perp$ decomposes discretely. The central point of the philosophy of cusp forms may now be stated very simply as follows: The intertwining operators for U^{I_N} and U^{I_Γ} that are defined by Eisenstein series and that remove the continuous part of U^{I_Γ} may be generalized in two directions: (a) One may replace $\mathrm{SL}(2,R)$ by an arbitrary semi-simple Lie group with a finite center. (b) One may replace U^{I_Γ} by the regular representation of G or by its subrepresentations of the form U^L, where L varies over the irreducible representations of a maximal compact subgroup K. The discrete part of U^L is just a direct sum of members of the discrete series for G each occurring with finite multiplicity and every discrete series member occurs in some U^L. Isolating the discrete part of the U^L thus amounts to isolating the discrete series for G.

ISBN 0-8053-6702-0/0-8053-6703-9, pbk

In the generalization from $\mathrm{SL}(2,R)$ to an arbitrary semi-simple Lie group with a finite center, the role of the subgroup N is played by the "nil radicals" of the "parabolic subgroups" of G. In the special case in which $G=\mathrm{SL}(n,R)$, the parabolic subgroups are just those that contain conjugates of the subgroup B of all matrices in $\mathrm{SL}(n,R)$ that are zero above the main diagonal. There are only a finite number of conjugacy classes of parabolic subgroups and each such has a maximal nilpotent normal subgroup called its nil radical. The notion of parabolic subgroup originated in the theory of algebraic groups as "globalized" by A. Borel in 1956 (*Ann. Math.* vol. 64, 20–82). However, in many contexts including the present one, it may be replaced by an equivalent notion based on the concept of a B-N pair introduced by J. Tits in 1962 (*Comptes Rendues, Acad. Sci. Paris*, Vol. 254, 2910–2912) Let $G=\mathrm{SL}(n,\mathfrak{F})$, where $\mathfrak{F}$ is any field, let B denote the subgroup of all matrices that vanish above the main diagonal, and let N denote the normalizer of the subgroup of all matrices that vanish except on the main diagonal. (This use of N should not be confused with that in our discussion of U^{I_N} above). Then B and N satisfy certain simple axioms that imply among other things that $H=B\cap N$ is normal in N and that the elements of the quotient group $W=N/H$ correspond one-to-one in a natural way to the $B:B$ double cosets. Since W is finite and every subgroup containing B is clearly a union of $B:B$ double cosets, there can be only a finite number of such subgroups. A pair of subgroups of any group which satisfies these axioms is called a B-N pair, or a Tits system, and the conjugates of the subgroups containing B are called the parabolic subgroups.

As explained in Section 30, the construction of Eisenstein series in the sense of Maass involves a nontrivial analytic continuation. Moreover, Selberg, in extending his "trace formula" (cf. Section 30) to the case in which G/Γ is not compact, had to use these analytically continued Eisenstein series to remove (in effect) the contribution of the continuous part of U^{I_N}. For semi-simple Lie groups other than $\mathrm{SL}(2,R)$ this analytic continuation is especially difficult and Langlands was the first to prove that it could be carried out under fairly general circumstances. From 1965 until 1976 Langland's proof was available only in privately circulated mimeographed form although a 17 page summary was published in 1966 (vol. IX of the Proceedings of the American Mathematical Society Symposia on Pure Mathematics). However, during the academic year 1966–67, Harish Chandra gave a course of lectures at the Institute for Advanced Study based on Langlands' unpublished typescript. In it, he presented most of the material in the first six of the seven sections in a somewhat altered and improved form. Notes for the course were written up by J.M.G. Mars and published in 1968 ("Automorphic forms on Semi-simple Lie groups," *Springer Lecture Notes*, Vol. 62). Langlands' original version is now available as Vol. 544 of *Springer Lecture Notes* under the title "On

ISBN 0-8053-6702-0/0-8053-6703-9, pbk

the functional equation satisfied by Eisenstein series." Harish Chandra tells us that he was led to the "philosophy of cusp forms" through his study of this (then unpublished) work of Langlands. However, he later came to realize that much of this philosophy is expressed in Gelfand's 1962 Stockholm Congress address and is based on joint work of Gelfand and Graev (1959) and Gelfand and Pyatetzki-Shapiro (1964). In this connection see the discussion at the end of Section 30 on the relationship between double coset intertwining operators and the horocycle method of Gelfand and Graev.

While Harish Chandra's determination of the discrete series by way of the characters of its members was sufficient for many purposes, including the proof of the Plancherel formula, there remained the problem of explicitly constructing the irreducible representations themselves. Considerable interest in this problem was stimulated by a specific conjecture of Langlands, which appeared at the end of a second paper published in the same Symposium Proceedings as the summary of his work on Eisenstein series. In 1968 a proof of a large part of Langlands' conjecture was announced by W. Schmid and the proof itself was published in 1971. Alternative but closely related treatments of the problem were published in 1970 and 1972 by M.S. Narasimhan and K. Okamoto, by R. Parthasarathy, and by R. Hotta. The "more degenerate" members of the discrete series caused difficulties that prevented any of these authors from constructing the entire discrete series. These difficulties were overcome quite recently in various ways by several mathematicians, in particular, by Schmid. Schmid's paper "L^2 cohomology and the discrete series" appears on pages 375–394 of vol. 103 (1976) of the *Annals of Mathematics*. It contains references to the related work mentioned above as well as to an independent formulation of Langlands' conjecture by Kostant. A brief description of the nature of the conjecture will be found in subsection 3*j* of the appendix to the author's published Chicago lecture notes.

A quite different approach to the construction of the discrete series will be found in a very recent paper of Atiyah and Schmid ("A geometric construction of the discrete series for semi simple Lie groups," *Inventiones Mathematicae*, vol. 42 (1977), 1–62). Unlike the authors cited above, they do not assume Harish Chandra's work on the characters of the discrete series, but present an independent proof of their existence. Moreover (in part by appealing to the Atiyah–Singer index theorem as well as to the recent L^2 index theorem of Atiyah), they succeed in presenting their argument in considerably less space than Harish Chandra required. A key device consists in exploiting the existence of discrete subgroups with compact homogeneous spaces. For a much more complete and detailed account of work on the explicit construction of the discrete series the reader is referred to a 1977 Bourbaki Seminar Report of Duflo.

While there are still very few noncompact semi-simple Lie groups for

ISBN 0-8053-6702-0/0-8053-6703-9, pbk

which the irreducible unitary representations have been completely classified, there are indications that this situation will change radically in the near future. The problem is largely that of analyzing the structure of certain induced representations that are known to decompose discretely; the work on intertwining operators for induced representations described in subsection 3f of the "appendix" cited above is developing into a powerful tool for this purpose. Moreover, Langlands has reduced the closely related problem of classifying the irreducible "admissible" representations to that of classifying those irreducible representations whose characters are tempered distributions; and a solution to the latter problem has been announced by Knapp and Zuckerman (*Proc. Nat. Acad. Sci., U.S.A.*, vol. 73 (1976), 2178–2180). Langlands' reduction is presented in mimeographed Institute for Advanced Study lecture notes (1973). A classification in different terms, but closely related to that of Langlands, Knapp, and Zuckerman, has been very recently announced by D. Vogan (*Proc. Nat. Acad. Sci., U.S.A.*, vol. 74 (1977), 2649–2656). The problem in passing to a classification of the irreducible unitaries consists in deciding which of the above are unitary.

Under the stimulus provided by the introduction of noncommutative adele groups into the group representational study of automorphic forms (cf. Notes and References to Section 30), there is now a rapidly growing literature devoted to the representation theory of the p-adic analogs of the semi-simple Lie groups. A fairly detailed summary up to the summer of 1975 will be found on pages 316 to 328 of the "appendix" already cited several times above. The statement on page 327 to the effect that G_A for SL(2) is not of Type I is incorrect. A much longer exposition with full proofs of much of the material has been published by I.N. Bernstein and A.V. Zelevinski (*Uspehi Mat. Nauk*, vol. 31 (1976)). An English translation has appeared in *Russian Mathematical Surveys*, vol. 31 (1976), 1–68.

The material in the text relating the representation theory of SL(2, C) to that of the Euclidean group turns out to have a considerable generalization. Let G be any connected semi-simple Lie group with a finite center and let K be its maximal compact subgroup. Let V be the tangent space to the manifold G/K at the point defined by K. Then each element of K defines an automorphism of V and one can form the corresponding semi-direct product $V \circledS K$. Although details have only been checked in certain cases, there seems to be a natural one-to-one correspondence between almost all (equivalence classes of) irreducible unitary representations of G, on the one hand, and almost all (equivalence classes of) irreducible unitary representations of $V \circledS K$ on the other. The nature of this correspondence can perhaps best be understood by viewing it as a result of composing three other correspondences as follows: (1) A much more elementary correspondence between conjugacy classes in G and $V \circledS K$, respectively. (2) The Gelfand–Harish Chandra principle described

ISBN 0-8053-6702-0/0-8053-6703-9, pbk

in the text that associates a family of equivalence classes of irreducible unitary representations of G to each conjugacy class of Cartan subgroups and parametrizes its members by Weyl orbits in the duals of these subgroups. (3) An analog of the Gelfand–Harish Chandra principle that holds for many groups other than semi-simple Lie groups including semi-direct products of the indicated form. The reader will find the beginning of a development of these ideas in two papers by the author: "On the analogy between semi-simple Lie groups and certain related semi-direct product groups," in *Lie Groups and Their Representations*, I. Gelfand (ed.), Hilgar (1975), 339–363, and "On the structure of the set of conjugate classes in certain locally compact groups," *Symposia Mathematica*, vol. XVI (1975).

ISBN 0-8053-6702-0/0-8053-6703-9, pbk

15. Ergodic theory

Let N be the additive group of the Euclidean plane and let H be the infinite cyclic group. Let β_θ be the automorphism of N defined by a rotation about the origin through the angle θ, and let G_θ be the semi-direct product of N and H formed by letting h be a generator of H and setting $\alpha_{h^k}(n)=\beta_\theta^k(n)$. If β_θ has finite order, that is, if θ is a rational multiple of π, then it is easy to see that the orbits of H in $\hat{N}$ admit an analytic cross section so that the theory of Section 10 gives us a complete account of the irreducible unitary representations of G_θ. On the other hand, when θ is an irrational multiple of π, it does not seem easy to find such a cross section, and one can in fact find infinitely many ergodic, H invariant measure classes in $\hat{N}$, and associated with each an infinite number of irreducible unitary representations of G_θ that are not of transitive type. Indeed, for each $r>0$, let ν_r be the measure such that $\nu_r(E)$ is the "arc length measure" of the intersection of E with the circle about 0 of radius r in $\hat{N}$. It is obvious that ν_r is invariant, and a simple argument with Fourier series shows that it is ergodic. More generally, let $G=N \circledS H$ be any semi-direct product of separable locally compact groups (N commutative), for which $\hat{N}$ admits a measure class C, invariant and strictly ergodic under the H action. Choose $\nu\in C$, form $L^2(\hat{N},\nu)$, and let B be the unitary representation of H defined by $B_hf(\chi)=f([\chi]h)\sqrt{\rho(\chi,h)}$ where ρ is the appropriate Radon–Nikodym derivative. Let P denote the associated projection valued measure, and let A be the unitary representation of N associated with P by the spectral theorem. Then, $n,h\to A_nB_h$ is a unitary representation of G that is easily seen to be irreducible. Indeed, any projection that commutes with all A_n commutes with all P_E, and hence is of the form P_{E_0}. But $B_hP_{E_0}B_{h^{-1}}=P_{[E_0]h^{-1}}$, so if P_{E_0} commutes with all B_h, then $P_{[E_0]h}=P_{E_0}$. Since ν is ergodic, this implies that $P_{E_0}=0$ or I. On the other hand, the measure

George W. Mackey, Unitary Group Representations in Physics, Probability, and Number Theory

ISBN 0-8053-6702-0/0-8053-6703-9, pbk

class of A is that of ν and is not essentially transitive. When ν *is* essentially transitive, the above construction still goes through and gives us the irreducible representation of G defined by the orbit of ν and the one-dimensional identity representation of H_{χ_0}. Thus the particular irreducible representations of G of the form $U^{\chi I}$, where I is the one-dimensional identity representation of H_χ and which correspond one-to-one to the orbits of H in $\hat{N}$, may be looked upon as a small subset of a much larger family of irreducible unitary representations of G that correspond one-to-one to the ergodic invariant measure classes of H in $\hat{N}$. Now attached to a given orbit, there are, in general, many irreducible representations of G other than $U^{\chi I}$. Indeed, there is one, $U^{\chi L}$, for every irreducible unitary representation L of H_χ. Accordingly, it is natural to wonder whether there are more irreducible representations of G attached to a given ergodic invariant measure class and whether they in any sense have the same structure as the set of all irreducible unitary representations of a group. We shall see below that the answer to the second question is yes and that the answer to the first question is usually yes also. It should already be clear that there is a sense in which an ergodic action of a group may be viewed as a generalization of a closed subgroup, and it turns out that the analogy between subgroups and ergodic actions thus suggested is a very far reaching one.

Before describing this analogy in greater detail, let us look a little more closely at the strictly ergodic actions themselves. At first sight, their very existence is a little surprising. Consider the simplest example—the one implicitly involved in the group G_θ studied above. Let S be the multiplicative group of all complex numbers of modulus one, and let θ_0 be an irrational multiple of π. Then the additive group Z of all integers acts via the mapping $(e^{i\theta})n = e^{i(\theta + n\theta_0)}$. Each orbit is countable and any union of orbits is invariant. Surely S must have proper invariant subsets. Strangely, though, one can prove that no collection of orbits can be a Borel set—or even a measurable set—unless it has measure zero or has a complement of measure zero.

This peculiar phenomenon seems to have been first observed in 1928. G. D. Birkhoff and Paul Smith happened upon it in their studies of dynamics, and four years later it was recognized by von Neumann that strict ergodicity was just the concept that workers in the foundations of statistical mechanics had been groping for ever since they had learned that the real line cannot act continuously and transitively on a space of dimension greater than one. It can act ergodically on spaces of arbitrarily high dimension. This discovery and the proofs of the "ergodic theorems" of von Neumann and Birkhoff aroused great interest in the mathematical world, and, by the end of 1932 the brand new subject of "ergodic theory" was underway and had begun to accumulate a sizable literature. In particular, there was a basic paper of von Neumann entitled "Zur Operatorenmetho-

ISBN 0-8053-6702-0/0-8053-6703-9, pbk

den in der Klassichen Mechanik," which contained, among others, the following results:

(a) A proof that every action of the real line (with a finite invariant measure) can be uniquely decomposed in a certain sense, as a "direct integral" of ergodic actions.

(b) A complete classification of all ergodic actions of the real line, having a so called "pure point spectrum" (see below for the definition of this concept).

(c) Examples of ergodic real line actions with no point spectrum

(d) A proof of the mean ergodic theorem.

Actually von Neumann's work seems to have been inspired, not by the paper of Birkhoff and Smith, but by a short note of Koopman published in 1931. Koopman noticed that one can get a unitary representation of the real line from any dynamical system by letting $U_t f(u) = f([u]t)$, where $(u,t) \rightarrow [u]t$ is the action of the real line on the phase space Ω of the system defined by the passing of time, and f is in $\mathfrak{L}^2(\Omega, \zeta)$, where ζ is the invariant "Liouville measure" in Ω. He proposed using Stone's theorem to associate $t \rightarrow U_t$ with a self-adjoint operator H, and to study the connection between the action and the spectral properties of H.

From our point of view, Koopman's suggestion amounts to the remark that one should study ergodic (or other) actions by studying the group representation: $(U_x f)(s) = f(sx)\sqrt{\rho_x(s)}$ in $\mathfrak{L}^2(S, \mu)$, where μ is a member of the invariant measure class and ρ_x is the appropriate Radon–Nikodym derivative. In particular, an action of the real line with a "pure point spectrum" is one whose induced unitary representation is a discrete direct sum of irreducibles—or, equivalently, is defined by a measure class concentrated in a countable set.

It is illuminating to see what Koopman's suggestion amounts to for *transitive* actions of general separable locally compact commutative groups. A transitive action of G is defined by a closed subgroup K of G, and the Koopman representation for the action is just the representation of G induced by the one-dimensional identity representation of K. We have already remarked that this representation is multiplicity free and is defined by a measure class in $K^{\perp}$, namely, the measure class of the Haar measure of $K^{\perp}$. It follows at once that the transitive actions of G are uniquely determined by their Koopman representations or "spectra" and one wonders to what extent this result is true for strictly ergodic actions. von Neumann's result for the real line is that an ergodic action with pure point spectrum is multiplicity free, that the irreducible representations that occur form a subgroup of $\hat{G}$, and that this subgroup determines the action

ISBN 0-8053-6702-0/0-8053-6703-9, pbk

up to an evident equivalence. Moreover, every countable subgroup Γ of $\hat{G}$ occurs and the action of G may be recovered from it as follows: Give Γ the discrete topology and let $\hat{\Gamma}$ be its compact dual. Then each $x \in G$ defines a member of $\hat{\Gamma}$ via $\chi \to \chi(x)$ and we obtain in this way a continuous homomorphism φ of G in $\hat{\Gamma}$ whose range is dense in $\hat{\Gamma}$. The ergodic action with spectrum Γ has $\hat{\Gamma}$ with Haar measure as its space and $\gamma x = \gamma \cdot \varphi(x)$ (where $\cdot$ means group multiplication). Strict ergodicity holds whenever Γ is not closed in $\hat{G}$. Although von Neumann considered only the real line, we have formulated his result in such a way that it makes sense for arbitrary separable locally compact commutative groups, and the result has now been proved to hold in this generality.

More generally, still let ϕ be any continuous homomorphism of the commutative separable locally compact group G into a commutative separable locally compact group $\tilde{G}$. Then there exists a unique dual homomorphism φ^* of $\hat{\tilde{G}}$ into $\hat{G}$ defined by the identity $\varphi^*(\chi)(x) = \chi(\varphi(x))$. Clearly $\varphi^{**} = \varphi$ when we identify $\hat{\hat{G}}$ with G. For each s in $\tilde{G}$ and each $x \in G$, let us set $sx = s \cdot \varphi(x)$, where "$\cdot$" denotes group multiplication. Then Haar measure in $\tilde{G}$ is invariant and it is easy to see that the action is ergodic if and only if $\varphi(G)$ is dense in $\tilde{G}$, and is strictly ergodic if and only if $\varphi^*(\hat{\tilde{G}})$ is not closed in $\hat{G}$. Now $\varphi(G)$ is dense in $\tilde{G}$ if and only if φ^* is one-to-one. Thus $\varphi \to \varphi^*$ gives us a one-to-one correspondence between the strictly ergodic actions defined by the dense imbeddings of quotient groups of G and the different ways of making nonclosed subgroups of $\hat{G}$ locally compact by giving them stronger topologies. The simplest way of doing the latter is to give a countable subgroup the discrete topology. The examples constructed in this way are just those with a pure point spectrum and also just those with a finite invariant measure. However, there are many others. It is not difficult to prove that if the subgroup Γ of $\hat{G}$ can be given a stronger topology under which it is separable and locally compact, then this can be done in only one way. Thus the ergodic actions of G defined by dense imbeddings correspond one-to-one to certain subgroups of $\hat{G}$. The spectrum of such an action is always multiplicity free and is defined by the Haar measure of the defining subgroup (relative to its locally compact topology). So far, the ergodic actions we have been able to construct have been uniquely determined by their spectra. However, as we shall see below, the suggested general conjecture is false, even for commutative groups. For noncommutative groups, it is false even for transitive actions of finite groups. It is possible for U^{I_1} and U^{I_2} to be equivalent when I_1 and I_2 are the one-dimensional identity representations of nonconjugate subgroups.

The general method given above for constructing strictly ergodic actions generalizes easily to noncommutative groups. Let φ be a continuous homomorphism of the separable locally compact group G onto a dense subgroup of some other separable locally compact group $\tilde{G}$, and let H be any closed subgroup of $\tilde{G}$. Let $S = \tilde{G}/H$ and let $[s]x = s\varphi(x)$ $(x \in G)$,

ISBN 0-8053-6703-9, pbk

where the second expression derives its sense from the fact that S is a $\tilde{G}$-space. This action of G will always be ergodic with respect to the unique G invariant measure class in S and will be strictly ergodic if and only if $\varphi(G) \neq \tilde{G}$. When S is compact, the spectrum of the action will be a discrete direct sum of irreducible representations of G, and when $\tilde{G}$ is compact, these irreducible representations will all be finite-dimensional. Conversely, one has the following partial generalization of the pure point spectrum theorem. Given an arbitrary ergodic action whose spectrum is a discrete direct sum of *finite-dimensional* irreducible representations, it may always be obtained from a system $\varphi, \tilde{G}, H$ as above with G *compact*.

A class of strictly ergodic actions very different from those defined above by dense imbeddings, may be defined quite easily for countable discrete groups. Let G be a countable discrete group (not necessarily commutative) and let M be any countable G-space (not necessarily transitive) in which every orbit has an infinite number of elements. (The most general such M is, of course, obtained by choosing a finite or countably infinite sequence $H_1, H_2, \ldots$ of subgroups of G and letting M be the disjoint union of the coset spaces G/H_i.) Let ν be a measure in a standard Borel space S such that $\nu(S)=1$, and let S^M denote the space of all functions from M to S. S^M may be regarded as an infinite direct product of replicas of the measure space S, ν and as such is itself a standard Borel space with a measure $\tilde{\nu}$. S^M becomes a G-space if we set $(fx)(m) = f([m]x)$ for each $f \in S^M$, $x \in G$, $m \in M$. It is clear that ν is an invariant measure, and it can be proved that it is ergodic. If no point of S has finite measure, then S, ν is isomorphic to the measure space A, μ, where μ is Haar measure in a nonfinite separable compact commutative group A. In this case, S^M itself is a separable compact group in a natural way, and $f \to fx$ is an automorphism. The ergodicity of the action of G in S^M follows more or less immediately from the following lemma: If the compact commutative group A is a G-space in such a way that $a \to [a]x$ is an automorphism of A for all x, then Haar measure is ergodic with respect to this action if and only if the dual action of G on $\hat{A}$ has no finite orbits except for the identity. The proof of the lemma is an elementary exercise in Fourier analysis on compact groups (cf. Halmos BAMS 1944 for the special case in which G is infinite cyclic). When S does have points of finite measure, it is always an equivalence class space for one that does not, and ergodicity can be proven from the above and the remark that a quotient action of an ergodic action is ergodic.

The spectrum of the action of G on $S^M, \tilde{\nu}$ is especially easy to describe in the special case in which $M = G$ and the action of G on M is group multiplication. It turns out to be the direct sum of the one-dimensional identity and of infinitely many replicas of the regular representation of G. Since it is very far from being multiplicity free, this spectrum cannot be the spectrum of an ergodic action defined by a dense imbedding, at least when

ISBN 0-8053-6702-0/0-8053-6703-9, pbk

G is commutative. Actually, there is a probabilistic sense in which the actions defined as above and those defined by dense imbeddings are at opposite extremes—but more of that in the next section, where we shall discuss the very interesting relationship between strict ergodicity and probability theory.

The definition of the action of G on $S^M, \tilde{\nu}$ may be generalized so as to apply when G and M are nondiscrete (actually only the discreteness of M is really needed in the discussion as given above), but this generalization is not immediate and, in fact, the special case in which $M = G =$ additive group of the real line is the "flow" associated with Wiener's model of Brownian motion. We shall postpone our discussion until the next section.

For a given countable discrete G and $M = G$, one can construct what appear to be different ergodic G spaces by forming S^M and $\tilde{\nu}$ with different choices of $S, \tilde{\nu}$. Of course, there is only one possibility in which there are no points of finite measure, but as many possibilities for finite and countable S, as there are ways of writing 1 as a finite or infinite sum of smaller positive numbers. All of these G-spaces have identical spectra and we can wonder whether they are actually isomorphic as G-spaces. For the special case in which G is the additive group of the integers, this was a well-known unsolved problem in ergodic theory for many years. It was finally solved in the negative in 1959 by Kolmogorov and Sinai. Using concepts from information theory, they assigned a nonnegative real number called the "entropy" to each ergodic action of the integers having a finite invariant measure. The entropy depends only on the isomorphism class of the action, and for the actions on S^M can be made to take on all positive values by suitably choosing S, ν. Thus there are at least continuum many nonisomorphic ergodic actions of the integers all having identical spectra. While no one seems to have worked out the details, it seems likely that one can prove an analogous result for general countable discrete groups by applying the entropy concept to its infinite cyclic subgroups. To return to the ergodic actions of the integers, it is of interest to note that the actions with pure point spectrum all have entropy zero though they may have different spectra and thus not be isomorphic. Of course, the entropy concept also does not differentiate between all the actions on spaces of the form S^M, ν, and it is an open question whether these actions are isomorphic when they have the same entropy.

One can construct ergodic actions different from either of the types constructed above by forming products. If C_j is an invariant measure class in the standard Borel G_j space S_j for $j = 1,2$, then $C_1 \times C_2$ is an invariant measure class in the $G_1 \times G_2$ space $S_1 \times S_2$, and it is easy to show that $C_1 \times C_2$ is ergodic if and only if C_1 and C_2 are. Suppose now that $G_1 = G_2 = G$ and that C_1 and C_2 are ergodic. In addition to the ergodic action of $G \times G$ on $S_1 \times S_2$, $C_1 \times C_2$, we may restrict to the "diagonal" subgroup of $G \times G$ consisting of all x,y with $x = y$ and obtain an action of

ISBN 0-8053-6702-0/0-8053-6703-9, pbk

G in $S_1 \times S_2$, $C_1 \times C_2$. This action need not be ergodic but often will be. Indeed, if C_1 and C_2 admit finite invariant measures, then it can be shown that the action of G on $S_1 \times S_2$, $C_1 \times C_2$ is ergodic if and only if no finite-dimensional irreducible component of the spectrum of the action of G on S_1, C_1 other than the identity is equivalent to any finite-dimensional irreducible component of the action of G on S_2, C_2. In particular, if G is a countable discrete group, we obtain an ergodic action by taking the product of $S^G, \tilde{\nu}$ with any other ergodic action. By choosing the second action as one with a pure point spectrum, we obtain an action whose spectrum is the direct sum of infinitely many replicas of the regular representation and of all finite-dimensional irreducible representations in the spectrum of the second action. This action is clearly inequivalent to any we have considered hitherto.

It should be clear by now that the possible strictly ergodic actions of a given group are much more varied than the possible transitive actions. The transitive actions correspond one-to-one to the conjugacy classes of closed subgroups, and in case G is the additive group of the integers there is just one of these for each positive integer. On the other hand, there is a strictly ergodic action of G for each infinite countable subgroup of the circle group, one for each positive entropy, and one for each combination of the two. In addition, there are many others—perhaps an unanalyzable multiplicity of them. Certainly no one has yet succeeded in giving what looks like an exhaustive list or made a plausible conjecture about what such a list should look like. One does not even know what spectra can occur. If one admits the ergodic actions as generalized subgroups, as one must in following through on the program begun in Section 10, even the additive group of the integers appears as a much more complicated object than one would have thought possible.

One can put some order into the very difficult if not impossible problem of classifying the strictly ergodic actions of a separable locally compact group G by taking quite seriously the analogy between ergodic actions and closed subgroups. For example, if H is a closed subgroup of G, then every closed subgroup K of H is also a closed subgroup of G. Moreover, two closed subgroups conjugate in H are also conjugate in G. Thus every transitive action of H is canonically associated with a transitive action of G. This suggests that we seek a canonical construction permitting us to pass from an arbitrary ergodic action of an arbitrary closed subgroup H of G to an ergodic action of the whole group G. The construction has been found and may be described as follows: Let H act ergodically on S, C. First form the auxiliary space $S \times G$, and make it into an $H \times G$ space by setting $(s,x)(h,y) = [s]h, y^{-1}xh$. The action of H above defines an equivalence relation in $S \times G$. Let $\widetilde{S \times G}^H$ denote the space of all equivalence classes, and let ϕ be the mapping taking s,x in $S \times G$ into its equivalence class. Since the action of G on $S \times G$ commutes with the action of H, the

ISBN 0-8053-6702-0/0-8053-6703-9, pbk

action of G maps each H equivalence class into another and we obtain a well defined action of G on $\widetilde{S\times G^H}$ by setting $\widetilde{\varphi(s,x)}y=\varphi(s,y^{-1}x)$. Using φ we transfer the Borel structure in $S\times G$ into $\widetilde{S\times G^H}$ in the usual way. If μ is a finite measure in S and ν is a finite measure in the Haar measure class, then $E\rightarrow\mu\times\nu(\varphi^{-1}(E))$ defines a measure in $\widetilde{S\times G^H}$, whose class $\tilde{C}$ is independent of the choices of μ and ν. This measure class in $\widetilde{S\times G^H}$ is ergodic and the action of $\widetilde{S\times G^H},\tilde{C}$ is the required ergodic action of G defined by the given ergodic action of H. This construction is closely related to the construction for passing from a unitary representation L of a closed subgroup to the induced representation U^L of the whole group. Indeed, when one takes the Hilbert G bundle point of view, the two can be regarded as different special cases of the same construction. We shall not give details here, but content ourselves with the remark that the spectrum of the induced ergodic action of G is the representation U^L of G, where L is the spectrum of the inducing ergodic action of H. In particular, we may find a strictly ergodic action of the real line whose spectrum is not multiplicity free by starting with such an action of the cyclic subgroup of all multiples of some $\lambda>0$.

That the analogy between ergodic actions and closed subgroups is quite far reaching may be seen by studying the following generalization of the above construction of $\widetilde{S\times G^H},\tilde{C}$ from S,C. In converting $S\times G$ into an $H\times G$ space, let us replace the definition $(s,x)(h,y)=[s]h,y^{-1}xh$ by $(s,x)(h,y)=[s]h,y^{-1}x\pi(s,h)$, where π is any Borel function from $S\times H$ to G satisfying the "cocycle" identity

$$\pi(s,h_1h_2)=\pi(s,h_1)\pi([s]h_1,h_2) \qquad (*)$$

As is easy to see, satisfaction of this identity is a necessary and sufficient condition that $S\times G$ actually be an $H\times G$ space under the indicated action. Moreover, H need no longer be a closed subgroup of G. We now proceed as above to obtain an ergodic action of G on the space of H orbits. However, we must be careful about one point. In the general case, the space of H orbits need not be a standard Borel space. In that case we must apply a generalization of the von Neumann decomposition theorem to decompose the H action into ergodic parts and consider the G action on these parts rather than on the orbits.

To see the meaning of this generalization, consider the special case in which the H action is transitive so that $S=H/K$ for some closed subgroup K of H. Let $s_0=K$ and consider π restricted to $s_0\times K$. The identity (*) becomes $\pi(s_0,h_1h_2)=\pi(s_0,h_1)(s_0,h_2)$ so that π becomes simply a homomorphism π^0 of K into G. It turns out that this homomorphism determines π in all of its essential features. π_1 and π_2 define homomorphisms of K into G that are conjugate via inner automorphism if and only if they are cohomologous cocycles in the sense that there exists a Borel function a from S to G

ISBN 0-8053-6702-0/0-8053-6703-9, pbk

such that $\pi_2(s,x)=a(s)\pi_1(s,x)a^{-1}(s,x)$ for all s,x. Moreover, when π_1 and π_2 are cohomologous, the G actions that they define are isomorphic in a natural way. Finally, it is easy to see that the G action defined by π is precisely the transitive action of G defined by the closure of $\pi^0(K)$. The condition that the H orbits form a standard Borel space is exactly that $\pi^0(K)$ be closed. This construction differs from our earlier one in replacing the natural imbedding of K in G by an arbitrary Borel (and hence continuous) homomorphism of K in G. Note again the analogy with induced representations—especially with the correspondence between transitive Hilbert G bundles and unitary representations of the subgroup K_{s_0}. One can, in fact, redo the above in terms of appropriately defined "group bundles."

The significance of the above construction for our analogy between ergodic actions and closed subgroups is that it suggests an analog for the notion of homomorphism of a closed subgroup into another group; namely, a cohomology class of cocycles in the sense of (*). For each such we have a definition of what it means for its range to be closed and a definition of the closure of this range as a certain ergodic action of the second group. We shall not give further details, but it is a short step from this to the notion of a homomorphism of the "virtual subgroup" of G_1 defined by some ergodic action to the "virtual subgroup" of G_2 defined by some ergodic action of G_2. Continuing in this way, we are led to a notion of "similarity" between ergodic actions of different groups that reduces to honest isomorphisms of the defining closed subgroups where the actions are transitive. Finally, we may define a virtual group as a similarity class of ergodic actions and prove that the possible ergodic actions of a fixed group correspond one-to-one to the possible systems consisting of a virtual group and an imbedding (in a sense that we have not defined) of this virtual group into the group. As suggested by the preceding, it turns out that most of the notions of group theory make sense for virtual groups. We have already seen how to define homomorphisms of virtual groups into groups. Taking the groups to be the unitary groups, we have a notion of unitary representation for virtual groups. For these, one can define equivalence, direct sums, and all the notions of group representation theory. In these terms we may give a partial answer to the question that motivated our account of ergodic theory. Given a semi-direct product $G=N \circledS H$ and an ergodic invariant measure class C in N, consider the virtual subgroup of H associated with this ergodic action. Then there is a natural one-to-one correspondence between the (equivalence classes of) primary unitary representations of this virtual group and the primary representations of G associated with C. This correspondence reduces to the one described in Section 10 when the action is transitive so that the virtual subgroup is real. We say that the answer is a partial one because it seems to be difficult if not impossible to classify the irreducible representations of

ISBN 0-8053-6702-0/0-8053-6703-9, pbk

virtual groups. In particular, virtual subgroups of the integers can have non-Type I primary representations, and hence infinite-dimensional irreducible representations.

There has been a tendency for ergodic theory to stay close to its historical origins and concern itself mainly with ergodic actions of the real line and the integers. We hope that the above discussion has convinced the reader that it should be developed in the same generality as the theory of unitary group representations and that the two subjects—properly regarded—are inseparably intertwined.

ISBN 0-8053-6702-0/0-8053-6703-9, pbk

16. Applications to probability theory

The notion of random variable or "chance variable" was for a long time a rather vague concept and there were grounds for wondering whether probability theory could be regarded as a genuine branch of mathematics subject to the same standards of rigor as algebra and analysis. However, in the 1920s and 1930s it gradually came to be recognized that one obtained a thoroughly satisfactory mathematical model for all probabilistic phenomena by considering "events" to be points in a measure space S,μ with $\mu(S)=1$, $\mu(E)$ being the probability of a given "event" being in the subset E of S. From this point of view a random variable is a measurable function on S and a stochastic process is a family of random variables indexed by the points of time or by a subset of the points of time. Given random variables $f_1,f_2,f_3,\ldots,f_r$, one obtains a measurable map T of S into the space of all r-tuples of function values by setting $T(s)=f_1(s),\ldots,f_r(s)$. Setting $\mu_T(F)=\mu(T^{-1}(F))$ one obtains a probability measure in the space of all r tuples of possible values. This is the so-called joint probability distribution of the random variables $f_1,f_2,f_3,\ldots,f_r$. The random variables are said to be independent if this measure is a product of measures in the factor spaces, and other probabilistic notions are defined analogously.

In line with our previous work and also with the ideas of probabilists such as Blackwell, we shall modify the above set up slightly by insisting that the space S of events be a standard Borel space and that random variables be Borel functions. This simplifies technical problems and causes no essential loss in generality. A stochastic process, defined on the whole real line or on a proper closed subgroup thereof, is said to be stationary if the joint distribution of the random variables $f_{x_1},\ldots,f_{x_n}$ is the same as that of $f_{x_1+t},\ldots,f_{x_n+t}$ for any $x_1\cdots x_n,t$ in the group in question. Now one is interested in sets of random variables indexed by parameters other than

George W. Mackey, Unitary Group Representations in Physics, Probability, and Number Theory

ISBN 0-8053-6702-0/0-8053-6703-9, pbk

time—in particular, the study of turbulence in hydrodynamics leads quite naturally to random variables indexed by the points of *space*. Indeed, in 1935 N. Wiener introduced the term *chaos* to describe a family of random variables indexed by the points of Euclidean n-space, and corresponding to the notion of stationary stochastic process, he introduced the notion of *homogeneous chaos*. We shall find it convenient to go a step further and replace E^n by an arbitrary transitive standard Borel G-space $\mathcal{M}$, having an invariant measure ν in its unique invariant measure class. In the special case in which $\mathcal{M}$ is a countable discrete space we may pass from a family of random variables indexed by the points of $\mathcal{M}$ to a family indexed by the finite subsets of $\mathcal{M}$, by assigning to each finite subset E of $\mathcal{M}$ the random variable $f_E=\Sigma_{m\in E}f_m$. This assignment of random variables to sets obviously has the following additivity property:

$$f_{E_1\cup E_2}=f_{E_1}+f_{E_2} \quad \text{whenever} \quad E_1\cap E_2=0 \qquad (*)$$

Conversely if $E\to f_E$ is an assignment of random variables to finite subsets of $\mathcal{M}$ such that (*) holds, then $f_E=\Sigma_{m\in E}f_{\{m\}}$ and we see that our assignment can be obtained as above from a family of random variables indexed by the points of $\mathcal{M}$. In the general case, the notion of a random variable indexed by the points of $\mathcal{M}$ is *not* equivalent to the notion of a random variable indexed by the Borel subsets of finite measure, and there are reasons for preferring the latter to the former. Generalizing Wiener's definition, we define a chaos on a standard measure space $\mathcal{M},\nu$ as follows: It is an assignment of a random variable f_E to each Borel subset E of $\mathcal{M}$ of finite measure in such a manner that (*) holds and $f_E=0$ whenever $\nu(E)=0$. If there exists a Borel function g in $\mathcal{M}\times S$ such that $f_E(s)=\int_E g(m,s)\,d\nu(m)$ for each E and almost all s, then $E\to f_E$ is derivable from a point assignment, namely, $m\to g(m,\cdot)$, where $g(m,\cdot)$ is the random variable $s\to g(m,s)$. When this is the case the chaos is said to be *differentiable*. Now suppose that $\mathcal{M}$ is a transitive Borel G-space and that ν is invariant under the G action. For each $x\in G$ and each finite family $E_1,E_2,\ldots,E_k$ of disjoint Borel subsets of finite measure, we may compare the joint distribution of the random variables $f_{E_1},\ldots,f_{E_k}$ with that of the random variables $f_{E_1x}\cdots f_{E_kx}$. If these two probability measures are always equal, we shall say that our chaos is *homogeneous*. A stationary stochastic process, of course, is just a homogeneous chaos in which $G=\mathcal{M}=$additive group of real line or $G=\mathcal{M}=$additive group of the integers.

Given a homogeneous chaos $E\to f_E$ defined on the G-space $\mathcal{M}$, for each $s\in S$ we have an additive set function $E\to f_E(s)$ on $\mathcal{M}$. Let us denote this additive set function by α_s and consider the mapping $s\to\alpha_s$. There is no real loss in generality in supposing that this mapping is one-to-one because if it is not we may identify points s_1 and s_2 in S whenever $\alpha_{s_1}=\alpha_{s_2}$ and replace S,μ by the corresponding quotient space $\tilde{S},\tilde{\mu}$. Points thus identified

ISBN 0-8053-6702-0/0-8053-6703-9 pbk

correspond to indistinguishable events as far as the random variables f_E are concerned. Now let $\mathfrak{F}_{\mathfrak{M}}$ denote the space of all additive set functions α on the Borel subsets of $\mathfrak{M}$ which have the property that $\nu(E)=0$ implies $\alpha(E)=0$. Make this into a Borel space by giving it the smallest Borel structure for which all maps $\alpha \to \alpha(E)$ are Borel functions, and make $\mathfrak{F}_{\mathfrak{M}}$ into a G-space via $(\alpha x)(E) = \alpha([E]x^{-1})$. Then $s \to \alpha_s$ maps μ into a measure in $\mathfrak{F}_{\mathfrak{M}}$ defined on at least some Borel sets and invariant under the given action. When $\mathfrak{M}$ is countable and discrete, $\mathfrak{F}_{\mathfrak{M}}$ is a standard Borel G-space and the transfer of μ to $\mathfrak{F}_{\mathfrak{M}}$ is invariant under the G action. Moreover, the image of S in $\mathfrak{F}_{\mathfrak{M}}$ has a complement of measure zero, so that we may replace S by $\mathfrak{F}_{\mathfrak{M}}$ without changing anything essential. Thus, in this case at least, the action of G on $\mathfrak{M}$ induces a measure preserving action of G in S related to it in such a way that

$$f_{[E]x^{-1}}(s) \equiv f_E([s]x)$$

When $\mathfrak{M}$ is not discrete things become more complicated. However, a corresponding result is true. One can replace S, μ by a standard Borel G-space and a G invariant probability distribution in such a way that

$$f_{[E]x^{-1}}(s) \equiv f_E([s]x)$$

Moreover, the pair S', μ' consisting of the G-space and the invariant measure μ' are uniquely determined.

This means that every homogeneous chaos defined on the G-space $\mathfrak{M}$ is canonically associated with a measure preserving action of G on a probability space S, μ. Moreover, given the $\mathfrak{M}, \nu$ and S, μ actions, the chaos is completely specified by giving the function $f_E(s)$ of Borel subsets of E and points of S. Consider the subspace $\mathfrak{L}_0$ of $\mathfrak{L}^2(\mathfrak{M}, \nu)$ consisting of all Borel functions that take on only a finite number of values and take each nonzero value on a set of finite measure. These functions are just those of the form $c_1\varphi_{E_1} + \cdots + c_k\varphi_{E_k}$, where the c_j are constants and the E_j are disjoint Borel sets of finite measure. A chaos defines a linear map of $\mathfrak{L}_0$ into the Borel functions on S, namely, $c_1\varphi_{E_1} + \cdots + c_k\varphi_{E_k} \to c_1 f_{E_1} + \cdots + c_k f_{E_k}$. Conversely, given any linear map T of $\mathfrak{L}_0$ into the Borel functions on S, we obtain a unique chaos by setting $f_E = T(\varphi_E)$. In both spaces we regard two functions as identical when they differ on sets of measure zero. Now $\mathfrak{L}_0$ is a dense subspace of $\mathfrak{L}^2(\mathfrak{M}, \nu)$ which is invariant under all the unitary translation operators U'_x, where $(U'_x g)(m) = g([m]x)$. Moreover, the set $\mathfrak{B}$ of all Borel functions on S is, of course, invariant under the translation operator V'_x, where $(V'_x f)(s) = f([s]x)$. A straightforward calculation shows that the chaos defined by T is homogeneous if and only if $TU'_x = V'_x T$ for all x; that is, if and only if T is an intertwining

ISBN 0-8053-6702-0/0-8053-6703-9, pbk

operator for the two group representations U' and V'. In other words, the most general homogeneous chaos (modulo certain identifications in terms of null sets) on a given transitive G-space $\mathfrak{M}$ is obtained by choosing a triple S,μ,T, where S,μ is a standard Borel G-space with an invariant probability measure μ, and T is an intertwining operator for the group representations U' and V' defined by the G-spaces $\mathfrak{M}$ and S.

Suggestive as it is, this correspondence does not connect homogeneous chaoses with the theory we have been developing because V' is not unitary. However, it is only a mild restriction to consider only those chaoses in which every random variable f_E has a "finite dispersion"; and to say, that f_E has a finite dispersion is just the probabilists' way of saying that $f_E \in \mathfrak{L}^2(S,\mu)$. If we let V be the restriction of V' to $\mathfrak{L}^2(S,\mu)$, we see that homogeneous chaoses with finite dispersion on $\mathfrak{M}$ are just the triplets S,μ,T, where S,μ is as before and T is intertwining operator for U' and V. If T happens to be a bounded operator in the $\mathfrak{L}^2(\mathfrak{M},\nu)$ norm, it will have a unique continuous extension to all of $\mathfrak{L}^2(\mathfrak{M},\nu)$ and will thus define an intertwining operator $\tilde{T}$ for the unitary representations U and V of G in the sense discussed in the early sections of these lectures. In particular, the unitary representation V of G must have a subrepresentation equivalent to a subrepresentation of U. While T *need* not be bounded, the chaoses for which it *is* bounded form an interesting and important class.

Consider the particular case in which our chaos is differentiable and let m_0 be a fixed reference point in $\mathfrak{M}$. Then because of the homogeneity, we have $f_{\{[m_0]x\}}(s) = f_{(m_0)}([s]x^{-1})$ and the complete assignment $m \to f_m$ is determined by the single function $f_{\{m_0\}} = f_0$ on S. Conversely, given f_0 the chaos is uniquely determined by setting $f_{\{m\}}(s) = f_0([s]x^{-1})$ whenever $m = m_0x$. Let H_{m_0} be the group of all $h \in G$ with $[m_0]h = m_0$. Clearly a particular Borel function f_0 on S will arise from a chaos in this way if and only if it has the following two properties:

(a) For almost every s_0 the "sample function" $x \to f_0([s]x)$ is almost everywhere equal to a function that is constant on the right H_{m_0} cosets.

(b) The sample functions $x \to f_0([s]x)$, regarded as functions on $\mathfrak{M} = G/H_{m_0}$, are integrable over subsets of $\mathfrak{M}$ of finite measure.

Thus, to determine a differentiable chaos, one has only to select a probability G-space S,μ and a Borel function f_0 on S. In the case in which each $f_{\{m\}}$ has finite dispersion, f_0 will be in $\mathfrak{L}^2(S,\mu)$. By subtracting the same constant $\int f_0\, d\mu$ from each f_m, we may suppose that our random variables have mean zero. In that case, the inner product $\int f_{m_1} \cdot \bar{f}_{m_2}\, d\mu$ has a definite probabilistic interpretation. It is called the *correlation coefficient* between f_{m_1} and f_{m_2} and is a partial measure of the extent to which they fail to be independent. This coefficient is zero when they are independent, but not

ISBN 0-8053-6702-0/0-8053-6703-9, pbk

conversely. Of course, $f_{m_1}=V_{x_1}f_0$, $f_{m_2}=V_{x_2}f_0$, where $m_1=m_0x_1^{-1}$, $m_2=m_0x_2^{-1}$, so

$$\int f_{m_1}\cdot\bar{f}_{m_2}\,d\mu=(V_{x_1}f_0\cdot V_{x_2}f_0)$$
$$=(V_{x_2-x_1}f_0\cdot f_0)$$

where $x\rightarrow V_x$ is, as above, the representation of G in $\mathcal{L}^2(S,\mu)$ defined by the action of G on S. Thus the set of all possible correlations is completely described by the single complex-valued function on G, $x\rightarrow(V_xf_0\cdot f_0)$. It is not hard to show that the intertwining operator T defined above cannot be continuous unless the function $x\rightarrow(V_xf_0\cdot f_0)$ defines a function $\mathfrak{M}=G/H_{m_0}$ that is in $\mathcal{L}^2(\mathfrak{M},\nu)$; that is, unless the correlation between f_{m_1} and f_{m_2} goes to zero as m_2 gets further from m_1. Thus the chaoses in which $\tilde{T}$ exists and that may be defined by honest interwining operators between unitary group representations are, roughly speaking, those in which distant random variables are not too closely correlated.

Functions of the form $x\rightarrow(V_xf_0\cdot f_0)$ play an important role in one approach to the general theory of unitary group representations. It is not hard to prove that if W^1 and W^2 are arbitrary unitary representations of the separable locally compact group G, and φ^1 and φ^2 are members of $\mathcal{H}(W^1)$ and $\mathcal{H}(W^2)$, respectively, then $(W_x^1(\varphi^1)\cdot\varphi^1)\equiv(W_x^2(\varphi^2)\cdot\varphi^2)$ if and only if there exists a unitary map A setting up an equivalence between the subrepresentation of W^1 generated by φ^1 and the subrepresentation of W^2 generated by φ^2 and carrying φ^1 into φ^2. One calls a vector φ in the space of a representation W a *cyclic vector* if the smallest invariant subspace containing φ is the whole space. If a cyclic vector exists, one says that the representation is cyclic. Thus there is a one-to-one correspondence between pairs consisting of a cyclic unitary representation and a cyclic vector for it on the one hand, and functions of the form $x\rightarrow(V_x(f_0)\cdot f_0)$ on the other. Functions of the indicated form may be characterized alternately as those continuous functions on the group such that $\sum_{i,j=1}^n f(x_ix_j^{-1})c_i\bar{c}_j$ is a positive-definite quadratic form in the c's for every finite subset $x_1\cdots x_n$ of elements of the group. Accordingly, they are called positive-definite functions.

It can be proved that a Type I unitary representation of a separable locally compact group G is cyclic if and only if no irreducible constituent occurs with multiplicity greater than its dimension. Thus when G is commutative the cyclic representations are just the multiplicity free ones. Given a positive definite function f on G, we may write it in the form $f(x)=(V_x(\varphi_0)\cdot\varphi_0)$ for some multiplicity free unitary representation V of G. Let ζ be a finite member of the measure class in $\hat{G}$ definig V and let $\tilde{\varphi}_0$ be

ISBN 0-8053-6702-0/0-8053-6703-9, pbk

the member of $\mathfrak{L}^2(\hat{G},\zeta)$ corresponding to φ_0. Then

$$\begin{aligned} f(x) &= (V_x(\varphi_0)\cdot\varphi_0) \\ &= \int \chi(x)|\tilde{\varphi}_0(\chi)|^2 d\zeta(\chi) \\ &= \int \chi(x)\, d\alpha(\chi) \end{aligned}$$

where α is the finite measure

$$E \to \int_E |\tilde{\varphi}_0(\chi)|^2 d\zeta(\chi)$$

It is not hard to show that α is independent of the choice of ζ and we have the generalized Bochner–Herglotz theorem setting up a natural one-to-one correspondence between positive-definite functions on G and finite measures on $\hat{G}$. When the positive-definite function is that determined by the correlation coefficients of a differentiable homogeneous chaos, the corresponding measure on $\hat{G}$ is called the spectral measure of the chaos. Note that it determines but is not determined by the spectrum of the associated action of G on S,μ. Any α that is absolutely continuous with respect to the measure class associated with the spectrum of the action may occur. Thus it is easy to arrange that $\int\chi(x)\,d\alpha(\chi)$ be *not* in $\mathfrak{L}^2(G)$.

Up to now we have said nothing about the ergodicity, or lack of it, of the action of G on the fundamental probability space S,μ. We now wish to make a number of remarks indicating that strict ergodicity is very much the rule in cases of genuine probablistic interest.

First of all consider a discrete stationary stochastic process so that $G=\mathfrak{N}$ = additive group of all integers. Let us apply the von Neumann theorem to decompose the G action on S,μ into ergodic parts and let f be the function on S describing the process. Each "sample function" $n\to f([s]n)$ will belong to an s in some ergodic part. Suppose that strict ergodicity did not exist. Then the action of G on the ergodic part to which s belongs would be essentially transitive. Indeed, omitting a set of s's of μ measure zero, we can conclude that every other s would lie on an ergodic part on which G acts transitively. Since these transitive actions have a finite invariant measure, they must be defined by a subgroup of G other than the identity. Thus, for almost all s, the "sample function" $n\to f([s]n)$ would be periodic. If, for example, we were dealing with coin tosses, this means that with probability one, the sequence of heads and tails actually found would repeat itself exactly every N throws for some N. In other words, a mathematical model for probability yielding the sort of chaotic behavior actually expected and observed, would be impossible if strict ergodicity did not exist.

ISBN 0-8053-6702-0/0-8053-6703-9, pbk

Our next remark is that the action of G on S,μ is ergodic and strictly so whenever $\mathfrak{M}$ is discrete and the f_m are mutually independent. The proof of this fact was indicated in some detail in Section 15. Indeed the class of examples described using $\mathfrak{M}, S$ etc. is exactly the class one obtains by constructing the actions associated with those homogeneous chaoses in which G is discrete and the random variables are independent. When $\mathfrak{M}$ is not discrete, the f_m cannot be independent for a differentiable chaos. The nondifferentiable case will be discussed below. Actually, one can go much further and show that a large number of the homogeneous chaoses that occur naturally lead to actions of G on S,μ that are strictly ergodic.

The real significance of ergodicity in a probabilistic context may perhaps be best appreciated from the following remark whose truth will be come more apparent as we proceed. When the action of G on S,μ is ergodic, then for almost all s, knowing the single sample function $x \rightarrow f([s]x)$ gives us sufficient information to reconstruct the entire chaos. If the action is not ergodic and we attempt to reconstruct the chaos from a single sample function $x \rightarrow f([s]x)$ as though it were, then the chaos we construct has as its associated action the action of G on the ergodic part to which s belongs. Suppose, for example, that we have two coins, one of which is fair and one of which gives heads two thirds of the time. Suppose that we choose one of these coins by tossing some other fair coin, and then make an infinite sequence of tosses with the coin chosen. If we fit this probabilistic situation into the above context, the action of G on S will fall into two ergodic parts, one corresponding to the fair coin and one to the unfair one. Reflection upon this example, which is typical, should convince the reader that for many purposes one can restrict attention to the case in which the action of G on S is ergodic.

Now let us examine some of the implications of ergodicity and for simplicity let us concentrate our attention on stationary discrete stochastic processes where $G = \mathfrak{M} =$ additive group of the integers. Much can be done in somewhat more general cases and we shall give some indications. However, the situation in the completely general case is still rather obscure. The celebrated ergodic theorem of G. D. Birkhoff—stated for the integers rather than for the real line—can be completely formulated in probabilistic terms. It says that given any stationary discrete stochastic process for which the random variables f_n all have a mean value, that is, for which $\int_S |f_n(s)|\,d\mu(s) < \infty$, then for almost all s the sample function $n \rightarrow f_n(s)$ has a mean value in the sense that

$$\frac{f_n(s) + f_{n+1}(s) + \cdots + f_{n+m}(s)}{m+1}$$

tends to a limit as $m \rightarrow \infty$. (This is the part that is difficult to prove.) It goes on to say that this temporal mean value is equal to the common mean value $\int f_n(s)\,d\mu(s)$ of all the random variables, whenever the action is

ISBN 0-8053-6702-0/0-8053-6703-9, pbk

ergodic. The ergodic theorem is just a rather general form of the "law of large numbers" in probability theory—the law that sample function means actually exist and are independent of the sample chosen. Of course, the ergodic theorem does not just apply to the basic function f_0, but applies to any $\mathfrak{L}^1$ function on S. If, for example, we let ψ be the characteristic function of a Borel subset A of the space of all k-tuples of numbers and apply the ergodic theorem to $\psi(f_{n_1}, f_{n_2}, \ldots, f_{n_k})$, we see that the probability assigned to A by the joint probability distribution of $f_{n_1} \cdots f_{n_k}$ may be computed from almost any sample function by computing a suitable "frequency." It is in this way that one can recover the whole process by examining almost any sample function. Note that in making the necessary calculations one does not need the whole sample function, but only its values on the integers $\leqslant n_0$ (or $\geqslant n_0$). Thus knowing the "complete past" of almost any sample function, one may recover the whole process and make intelligent "best guesses" about the future. In the famous (linear) prediction theory of Kolmogoroff and Wiener, one defines best guess in such a way that one does not need the whole process to determine it. One needs only the auto correlation function

$$\int f_n \bar{f}_m \, d\mu(x) = (U_{n-m} f_0 \cdot f_0)$$

and this may be computed directly from the past of almost any sample function by the formula

$$U_k f_0 \cdot f_0 = \lim_{M \to \infty} \frac{1}{M} \sum_{n=1}^{M} \overline{a_{-n}} \, a_{-n-k} \quad \text{where } \{a_n\} = \{f([s]n)\}$$

for some s is the sample function in question. In the Kolmogoroff–Wiener theory, there is a sharp distinction between the perfectly and imperfectly predictable cases, which may be expressed in terms of the measure α which determines $U_k f_0 \cdot f_0$ via $U_k f_0 \cdot f_0 = \int e^{ik\theta} d\alpha(\theta)$. We may write α uniquely in the form $\alpha_1 + \alpha_2$ where α_1 is singular with respect to Lebesgue measure and α_2 is absolutely continuous with Radon Nikodym derivative ρ. If $\int_{-\infty}^{\infty} \log \rho(t) \, dt = -\infty$, then one has "perfect prediction" in the sense that the complete past uniquely determines the complete future. In the opposite case the complete past only gives an assignment of probabilities to the possible futures and when $\alpha_1 = 0$ the theory gives a neat algorithm for computing from ρ those future values that minimize the mean square error of the prediction.

The dichotomy between the perfectly and imperfectly predictable cases is related to (but not identical with) our earlier one between cases in which $\tilde{T}$ does not and does exist. Recall our necessary condition that $\tilde{T}$ exist: $d\alpha / d\theta$ must exist and be in $\mathfrak{L}^2(\hat{G})$. Clearly when we have perfect predictability, because $\alpha_2 \equiv 0$, it is also true that $\tilde{T}$ does not exist. A typical case in

ISBN 0-8053-6702-0/0-8053-6703-9, pbk

which we have both perfect prediction and the nonexistence of $\tilde{T}$ is that in which our ergodic action of G has pure point spectrum. Let $\chi_1, \chi_2, \ldots$ be the elements of the countable subgroup of $\hat{G}$ that defines the action, and let φ_j be a member of $\mathcal{L}^2(S,\mu)$ such that $\varphi_j([s]x)=\chi_j(x)\varphi_j(s)$. Then $f_0(s)=\sum_{j=1}^{\infty} c_j\varphi_j(s)$. Hence a typical sample function $x\to f_0([s]x)$ may be written in the form

$$f_0([s]x)=\sum_{j=1}^{\infty} c_j\varphi_j(s)\chi_j(x)$$
$$=\sum_{j=1}^{\infty} c_j(s)\chi_j(x)$$

The sense in which this expression for the sample function converges is most straightforward when f_0 is such that $\sum_{j=1}^{\infty}|c_j|<\infty$. Then $\sum|c_j(s)|<\infty$ for all s and is, in fact, independent of s, so that $\sum_{j=1}^{\infty} c_j(s)\chi_j(x)$ converges uniformly to the sample function. Thus our sample functions are all "uniformly almost periodic functions" on G in the sense that they are uniform limits of finite linear combinations of characters. It follows from the general theory of almost periodic functions on a commutative locally compact group that an almost periodic function on the line or on the integers cannot be zero for $x\leqslant a$ without being identically zero. Thus perfect predictability must always hold when the associated ergodic action has pure point spectrum. It is interesting to notice that the formulas for computing the coefficients $c_j(s)$ from the individual sample function $x\to f_0([s]x)$, which we may derive from the ergodic theorem as above, coincide exactly with the formula provided by the theory of almost periodic functions. Indeed, in the special case of an ergodic action with pure point spectrum, we may regard the ergodic theorem as a consequence of the theory of almost periodic functions. On the other hand, as first pointed out by A. Weil, the theory of almost periodic functions may be reduced to the Peter Weyl theory of compact groups by using a dense imbedding of the group in a compact group. Given any dense imbedding of a group G in a separable compact group $\overline{G}$, every continuous function on $\overline{G}$ is almost periodic when restricted to G and, conversely, every almost periodic function on G is the restriction to G of a continuous function on some compact group $\overline{G}$ in which G is densely imbedded.

We have seen above that whenever the action of G on S has pure point spectrum, then the decomposition of $\mathcal{L}^2(S,\mu)$ under G is reflected in a corresponding decomposition of the space of sample functions. However, this case is not only very special but is also in a sense rather "unprobabilistic" since the past determines the future rather completely. It is natural to ask what sort of translation invariant decomposition is available for the sample functions of more general processes—in particular, those in which

ISBN 0-8053-6702-0/0-8053-6703-9, pbk

the random variables are independent. This is basically the question that Wiener discusses in his famous paper in the 1930 Acta entitled "Generalized Harmonic Analysis." However, he carries on the discussion with G the real line rather than the integers and does not phrase it as we have done. To see the connection, recall that, by the ergodic theorem, the auto correlation function

$$k \rightarrow \lim_{M\rightarrow\infty} \frac{1}{M} \sum_{n=1}^{M} \bar{a}_{-n} a_{-n-k}$$

exists for almost any sample function $n \rightarrow a_n = f([s]n)$. Moreover in the real line case this function is necessarily continuous. Wiener starts with a function whose autocorrelation function is assumed to exist and be continuous. It follows from what we have said above that this autocorrelation function is of the form $t \rightarrow \int_{\hat{G}} \chi(t)\, d\alpha(\chi)$, where α is a finite measure in $\hat{G}$. Moreover, when the function we start with is a sample function for an ergodic stochastic process, α defines the measure class of the cyclic subrepresentation of the subrepresentation of V generated by f_0 in $\mathcal{L}^2(S,\mu)$. For Wiener, $\hat{G}$ is the additive group of the real line and emphasis is put upon the function $S(x) = \alpha([0,x])$, which he calls the "integrated periodogram" of the function being analyzed. One expects to find that the function being analyzed may be written in the form $t \rightarrow \int \chi(t) a(\chi)\, d\alpha(\chi)$, where $a(\chi)$ is a suitable Borel function on $\hat{G}$. Such formulas exist, however, only when $a(\chi)$ is replaced by something more general than a function—in modern language $a(\chi)\, d\alpha(\chi)$ is a distribution. From the point of view of these lectures, the harmonic analysis of sample functions is implicitly contained in the decomposition of the representations W of G in $\mathcal{L}^2(S,\mu)$ defined by the action of G on S. For example, for every finite division of $\hat{G}$ into disjoint Borel sets $\hat{G} = E_1 \cup E_2 \cup \cdots \cup E_r$, the projection valued measure P^V defines a corresponding decomposition of $\mathcal{L}^2(S,\mu)$ as a direct sum of orthogonal invariant subspaces. Writing $f_0 = f_0^1 + \cdots + f_0^r$, where $f_0^j = P_{E_j}^V(f_0)$, we have a corresponding decomposition of each sample function $x \rightarrow f_0^j([s]x) = f_0^1([s]x) + \cdots + f_0^r([s]x)$. The component $x \rightarrow f_0^j([s]x)$ is a function on G where spectral measure α_j is concentrated in E_j.

It is interesting to note that although Wiener had statistical ideas in mind, he did not assume that the functions he analyzed were sample functions of a stochastic process. Indeed, the modern foundations of probability theory did not exist at the time, and the ergodic theorem had yet to be proved. On the other hand, in his theory of Brownian motion developed in the early 1920s, Wiener had essentially constructed the first rigorous example of a *continuous* stationary stochastic process with so called "independent increments" (see below for definition). In his paper on generalized harmonic analysis, the sample functions of this process are used to obtain examples of functions to which his theory applies and for which the associated spectral measure is not discrete.

ISBN 0-8053-6702-0/0-8053-6703-9, pbk

Though we have placed our chief emphasis on the case in which G is the additive group of the integers, our remarks apply with only obvious simple changes to that in which G is the additive group of the real line. In so far as we deal with those actions of G on S that have a pure point spectrum, many of them apply equally well to arbitrary locally compact commutative groups and even to noncommutative ones as well. The role of the average

$$\lim_{N\to\infty} \frac{1}{N}(f(1)+\cdots+f(n))$$

is played by integration with respect to Haar measure in the compact group in which G is densely imbedded. The chief obstacle in the way of extending much, if not all, of the theory to quite general locally compact G is the lack of a suitably general definition of average of a function over a group and a correspondingly general ergodic theorem. Substantial progress in this direction has been made by Calderon (*Ann. Math.*, **58** (1953), 182–191), but work remains to be done. Of course, the natural orderings of the integers and the real line play an important role in certain questions—especially in prediction theory—and it is not easy to find a natural generalization of this part of the theory.

We have yet to fulfill the promise made in Section 15 that we would describe an analog of the second method of constructing strictly ergodic actions that would apply to nondiscrete $\mathfrak{M}$. When $\mathfrak{M}$ is discrete it is easy to see there exist homogeneous chaoses defined on $\mathfrak{M}$ with the property that the $f_{\{m\}}$ are all independent and that such a chaos is completely determined by specifying any single random variable $f_{\{m_0\}}$. This random variable can be arbitrary. The construction of the process from the random variable is classical and has essentially been given above. When we try to generalize to the nondiscrete case, we find that there cannot exist a *differentiable* chaos with nondiscrete $\mathfrak{M}$ in which the $f_{\{m\}}$ are all independent. In fact, the correlation function $\int f_{\{m_1\}}\cdot\overline{f_{\{m_2\}}}\,d\mu$ would be zero for $m_1 \neq m_2$ and this contradicts its continuity unless $\mathfrak{M}$ is discrete. On the other hand, one can look among the nondifferentiable chaoses and attempt to find an assignment $E \to f_E$ such that f_{E_1} and f_{E_2} are independent whenever $E_1 \cap E_2 = 0$. To show that such an assignment exists is nowhere nearly so straightfoward a business as it is in the discrete case, and it is one of Wiener's great merits to have done this. Wiener dealt with the special case in which $G=\mathfrak{M}=$additive group of the real line; but, as we shall see below, the general case is easily derived from this one using results about isomorphisms of measure spaces. The details of Wiener's proof appear in his famous paper "Differential Space" (*J. Math. Phys.*, **2** (1923), 131–174). Since any interval I on the real line may be written as a sum of N disjoint intervals I_j, where N can be arbitrarily large and where all I_j are mutually congruent, it follows that each f_I must be the sum of N independent identically distributed random variables where N can be as large as we

ISBN 0-8053-6702-0/0-8053-6703-9, pbk

please. On the other hand, the celebrated central limit theorem in probability theory tells us that such a sum has a distribution very close to a normal distribution whenever N is large—at least if mild restrictions are put on the individual summands. This suggests that there is a rather narrow range of possibilities for the distributions of the random variables f_E and that one of these is that f_E be normally distributed with mean zero and dispersion depending only on $\nu(E)$. The theorem, which is true and which Wiener essentially proved, is this: There is a unique way of assigning a random variable f_E to each Borel subset of the real line of finite measure in such a manner that:

(1) f_{E_1} and f_{E_2} are independent whenever $E_1 \cap E_2 = 0$,

(2) f_E is normally distributed with mean zero and $\int |f_E|^2 d\mu = \nu(E)$ (ν is Lesbegue measure on the line).

Now it is a fundamental theorem in measure theory that if ν_1 and ν_2 are atom free measures in standard Borel spaces $\mathfrak{M}_1$ and $\mathfrak{M}_2$ and $\nu_1(\mathfrak{M}_1) = \nu_2(\mathfrak{M}_2)$, then there exists a one-to-one Borel map θ of $\mathfrak{M}_1$ on $\mathfrak{M}_2$ such that $\nu_2(\theta(E)) = \nu_1(E)$. It follows immediately that the result of Wiener quoted above is still true when the real line and Haar measure are replaced by an arbitrary standard Borel space $\mathfrak{M}$ and an arbitrary nonatomic measure ν therein. Now let ν be an invariant measure in the standard Borel G-space $\mathfrak{M}$, and let $\mathfrak{M}$ be nondiscrete. Let $E \to f_E$ be the chaos in $\mathfrak{M}$ that results from applying Wiener's theorem as indicated above. Since the distribution of f_E depends only on $\nu(E)$, it follows at once that this chaos is homogeneous. The corresponding action of G on S can be shown to be ergodic. This construction of an ergodic action from independent random variables is the promised analogue of the discrete construction given in Section 15.

Let us study the spectrum of such an action. Consider the intertwining map T taking $c_1\varphi_{E_1} + \cdots + c_n\varphi_{E_n}$ into $c_1 f_{E_1} + \cdots + c_n f_{E_n}$. We compute at once that

$$\|c_1 f_{E_1} + \cdots + c_n f_{E_n}\|^2 = |c_1|^2 \nu(E_1) + \cdots + |c_n|^2 \nu(E_n)$$
$$= \|c_1\varphi_{E_1} + \cdots + c_n\varphi_{E_n}\|^2$$

Thus T is norm preserving so $\tilde{T}$ exists and is a unitary map of $\mathfrak{L}^2(\mathfrak{M}, \nu)$ onto the closed subspace of $\mathfrak{L}^2(S, \mu)$ generated by the f_E. It follows that the spectrum of the action contains a subrepresentation equivalent to the representation U of G defined by the action of G on $\mathfrak{M}, \nu$. To find the rest of the spectrum, we observe that any product of the functions f_E is in $\mathfrak{L}^2(S, \mu)$ and that the sums of all such products constitute a dense subspace of $\mathfrak{L}^2(S, \mu)$. Exploiting this observation it can be shown that the spectrum

ISBN 0-8053-6702-0/0-8053-6703-9, pbk

of the action is equivalent to the following representation of G:

$$I \oplus U \oplus U \textcircled{s} U \oplus U \textcircled{s} U \textcircled{s} U \oplus U \textcircled{s} U \textcircled{s} U \textcircled{s} U \oplus + \cdots$$

where $U \textcircled{s} U \textcircled{s} \cdots U$, k factors, is the so-called "symmetrized kth power" of U, and I is the one-dimensional identity. If A and B are two representations of the same group, then the tensor product $A \otimes B$ is the representation $x \rightarrow A_x \otimes B_x$ acting in the tensor product of the spaces $\mathcal{H}(A)$ and $\mathcal{H}(B)$. Now the group of all permutations on k objects acts in a natural way on the tensor product of k Hilbert spaces. Moreover, this action commutes with the operators of the representation $A \otimes A \cdots \otimes A$. Thus the subspace of all tensors that are left fixed by all permutations is invariant under all $A_x \otimes A_x \cdots \otimes A_x$. The subrepresentation it defines is called the symmetrized kth power of U. This relationship between the representation of U and the spectrum of the action of G on S turns out to be identical with the relationship in quantum field theory between the representation of the Euclidean group defined by the entire field and the representation of the Euclidean group associated with the associated "particles" or "quanta." We shall give further details in a later section.

A very simple proof of the existence of the Wiener chaos may be obtained by exploiting the duality of locally compact commutative groups. The generalized chaos for the special case in which $\mathcal{M} = G$ and G is commutative is completely defined by a unitary map of $\mathcal{L}^2(G)$ into $\mathcal{L}^2(S,\mu)$, where μ is a probability measure in a Borel space S. Now the Plancherel theorem gives us a natural unitary map of $\mathcal{L}^2(G)$ onto $\mathcal{L}^2(\hat{G})$. Combining these two unitary maps, we get a unitary map of $\mathcal{L}^2(\hat{G})$ into $\mathcal{L}^2(S,\mu)$ and it is easily verified that this new map defines a generalized Wiener chaos whenever the first one does. Taking G discrete and $\hat{G}$ compact and noticing that the discrete analog of a Wiener chaos is easily defined, we obtain the existence of the Wiener chaos on any space of finite total measure. To get the infinite case we have only to write $\mathcal{M}$ as a sum of disjoint sets of equal finite measure and take the product of the probability spaces for the Wiener chaoses of the factors.

In the special case considered by Wiener with $G = \mathcal{M}$ = additive group of the real line, one can obtain a stochastic process parametrized by the real numbers by setting $g_t = f_{[0,t]}$. Using the usual mapping from points of S to sample functions (and resolving certain technical difficulties), one obtains a measure in the space of sample functions. One of Wiener's main results is that almost all of these sample functions are continuous and that almost none are differentiable. Because of the first fact, one obtains a natural translation invariant measure in the space of continuous functions. This is the famous Wiener measure. The second fact is closely related to the fact that the Wiener chaos is not differentiable. The individual random

ISBN 0-8053-6702-0/0-8053-6703-9, pbk

variables g_t, of course, are not independent but differences $g_{t_1}-g_{t_2}, g_{t_3}-g_{t_4}, \ldots, g_{t_m}-g_{t_{m-1}}$ are whenever the intervals are disjoint. Accordingly, one speaks of a process with "independent increments."

We have consistently dealt with random variables that take on real or complex values. One can deal also with vector valued random variables and carry through much of the preceding theory. However, new questions arise that we shall not attempt to discuss here. We remark only that G can act on the vector space in which the random variables take their values as well as on $\mathfrak{M}$. When this happens the representation U of G arising out of the action of G on $\mathfrak{M}$ is no longer the representation of G induced by the identity representation of the isotropy subgroup H_{m_0} of G, but the representation U^L induced by some other representation L of H_{m_0}.

If we restrict ourselves to the "more truly probabilistic" cases in which the action of G on S,μ is ergodic and the correlations are small enough so that $\tilde{T}$ exists, we may describe the general homogeneous chaos with homogeneity group G as being defined by the following system:

(1) a pair of closed subgroups of G, one real and one virtual;

(2) a unitary representation L, of the real subgroup;

(3) an intertwining operator T for U^L and the representation V of G induced by the identity representation of the virtual subgroup.

In concluding this section we remind the reader that probability theory is by no means exclusively concerned with chaoses that are homogeneous and stochastic processes that are stationary. However, there is a device, which does not seem to have been studied systematically—if indeed it has been studied at all—for bringing group theoretical methods into the study of more general processes and chaoses. We illustrate the general idea with discrete stochastic processes. If $\cdots f_{-2}, f_{-1}, f_0, f_1, f_2, \ldots$ is any sequence of random variables on S,μ, we may use the mapping $s \rightarrow f_{-2}(s), f_{-1}(s), f_0(s), \ldots$ to transfer the action of the integers on themselves to an action of the integers on S. (We suppose as usual that the f_j separate the points of S.) The process is then defined by the G-space S, where G is the additive group of the integers, the function f_0 and the measure μ. However, unless the process is stationary, μ will not be invariant and need not even be quasi-invariant. Now let

$$\mu_n(E)=\mu(E_n) \quad \text{and} \quad \mu_\infty(E)=\sum \frac{1}{2^{|n|}}\mu_n(E)$$

Then $\mu_\infty(E)$ will always be quasi-invariant and the invariant measure class to which it belongs will not depend upon the choice of the convergence factors $1/2^{|n|}$. Thus an arbitrary discrete stochastic process is canonically associated with an invariant measure class in a standard Borel G-space and hence with a unitary representation of G that we may call the spectrum of

ISBN 0-8053-6702-0/0-8053-6703-9, pbk

the process. More generally, let G be any separable locally compact group and let μ be any measure in the standard Borel G-space S. Let C_1 be the measure class of μ, let C_2 be the measure class of Haar measure in G, and let $C_3 = C_1 \times C_2$ in $G \times S$. The mapping $\theta: s,x \rightarrow sx, x^{-1}$ is an involutory Borel automorphism of $G \times S$ that maps C_3 into some other measure class C_3^{θ}. Let C be the common measure class of all measures $\nu + \nu^1$, where $\nu \in C_3$ and $\nu^1 \in C_3^{\theta}$. Then C is invariant under θ and its image $\tilde{C}$ in S is quasi-invariant. This quasi-invariant measure reduces when G is the integers to the measure class of μ_∞ defined above. Using this construction it is clear that we can discuss ergodicity and define a spectrum for any continuous stochastic process or chaos.

Notes and References (For Sections 15 and 16)

For an account of later developments in the theory of irregular semi-direct products and group extensions, the reader is referred to subsection 3c of the appendix to the author's Chicago lecture notes. The unpublished paper of Ramsay cited there has now appeared (*Acta Math.*, vol. 137 (1976), 17–48).

A considerable generalization of the theory of ergodic actions with pure point spectrum was worked out by R. J. Zimmer in his 1975 Harvard thesis (*Illinois J. Math.*, vol. 20 (1976), 373–409 and 555–558). While this generalization was inspired by Furstenberg's remarkable structure theory for distal dynamical systems (*Am. J. Math.*, vol. 85 (1963), 477–515), to which it is closely related, it could easily have been motivated by the virtual group concept. One has only to remember that it makes sense to induce unitary representations from a "smaller" to a "larger" virtual subgroup and study the case in which the identity representation of the smaller induces a representation of the larger that has finite-dimensional discrete components. The resulting theory includes and "explains" the older notion of ergodic action with quasi-discrete spectrum introduced by Halmos and von Neumann and studied by Abramov. More recently, Zimmer has used the virtual group notion to suggest and explain a number of concepts in ergodic theory—in particular, a notion of "amenability" for ergodic actions (*J. Fnl. Anal.*, vol. 27 (1978), 350–372). In her 1976 Harvard thesis, C. Series has used the virtual group notion in a systematic study of the ergodic actions of product groups (Pacific Journal of Mathematics vol. 70 (1977), 519–547).

As indicated in the Preface, a much fuller treatment of the material in Sections 15 and 16 will be found in the author's 1972 lectures at Texas Christian University, Fort Worth. In particular, Section 11 of these Fort Worth lectures contains a brief description of Ornstein's remarkable partial converse of the Kolmogorov–Sinai entropy theorem; in Section 5 there is a detailed explanation of how one can use group duality to recover a

ISBN 0-8053-6702-0/0-8053-6703-9, pbk

stationary stochastic process from the past of almost any sample function. In Section 8 there is a brief description of the startling theorem of Dye implying that two virtual subgroups of the integers are isomorphic whenever the associated ergodic actions have finite invariant measures. One of the most interesting developments inspired by Dye's work has been the work of Krieger on the case in which there is no finite invariant measure but only an invariant measure class (*Z. Wahrscheinlichkeits theorie und verw. Geb.*, vol. 11 (1969), 83–119). Krieger found that Dye's isomorphism theorem no longer holds and succeeded in giving an extremely interesting analysis of the various possible isomorphism classes. This analysis turned out to be closely related to recent work on the structure of Type III von Neumann algebras as explained in subsection 3e of the "appendix" cited above. In the course of their work on the latter problem, Takesaki and Connes found that the virtual group point of view can be used to provide insight into some of the invariants that occur. See, for example, the address given by Connes at the 1974 International congress in Vancouver and II. 6 of a paper by Connes and Takesaki entitled "The flow of weights on factors of Type III" (*Tohoku Math. J.*, vol. 29 (1977), 473–575).

A connection between ergodic theory and the theory of unitary group representations that is not mentioned in Sections 15 and 16 began with the same 1952 paper of Gelfand and Fomin which first related unitary group representations to automorphic forms (cf. Section 30). The main point of that paper is that one can use the unitary representation theory of SL(2, R) to give a new proof of the ergodicity of the "geodesic flow" on a surface of constant negative curvature. Generalizing this result has led to a considerable literature summarized in subsection 3k of the Chicago book appendix. In very recent work Zimmer has shown how to use analogous methods to prove that strictly ergodic actions remain strictly ergodic when restricted to certain subgroups (*Ann. Math.*, vol. 106 (1977), 573–588).

To avoid conflict with another use of the term "strictly ergodic," the author now uses the term "properly ergodic." Moreover, it has become common to use the term "random field" for what is called a "chaos" in the text.

The use of Pontryagin duality to prove the existence of the Wiener chaos is due to S. Kakutani.

17. The foundations of quantum mechanics

For many years it has been a fundamental goal of physics to explain all material phenomena in terms of the motions of a large but finite number of "fundamental particles." Until 1900, at least, attempts to achieve this goal were based upon the hypothesis that the state of the universe at any instant of time was completely determined by the positions and velocities of these particles. Any question that one could ask about the material content of the universe was supposed to be calculable from these positions and velocities. Moreover, the values of these positions and velocities at any future time were supposed to be obtainable from those at a given instant by integrating a system of second-order differential equations. It was the business of the physicist to find the exact form of these equations, to deduce the properties of their solutions, and to find rules for passing from the positions and velocities of the fundamental particles to the phenomena that are actually observed in everyday life and measured in the laboratory.

This program had a certain limited success, but as experimental techniques progressed and physicists began to be able to deal directly with extremely small pieces of matter, it became clear that it would have to be modified in a fairly fundamental way. This necessity presented itself in the form of a series of experimental results that could not be reconciled with the hypotheses of the program. However, after a suitable modification had been found, cogent a priori reasons for the impossibility of the original program were discovered and these can be made to suggest the nature of the necessary modification. The difficulty seems to lie in the ultimate meaninglessness of the statement that a given fundamental particle has a definite position and a definite velocity at a definite time. If one thinks carefully about how the necessary measurements are to be made and

George W. Mackey, Unitary Group Representations in Physics, Probability, and Number Theory

remembers that one is dealing directly with the smallest particles that exist, one finds that position and velocity measurements intefere in such a way as to frustrate all attempts to assign a definite experimental meaning to statements of the indicated form. If one decides on this account that such statements may be devoid of meaning and that it is futile to attempt to base a theory of the universe on meaningless concepts, one is forced to seek a substitute in which the state of the universe at a given instant of time is described otherwise than by giving the positions and velocities of the fundamental particles.

Quantum mechanics is a substitute that in many respects is extremely adequate. It has had an impressive string of successes and is essentially universally accepted as the framework within which physical phenomena are to be explained. However, there are serious difficulties in attempting to reconcile it with the theory of relativity, and these difficulties are keenly felt when one attempts to go beyond chemistry and study the structure and properties of the atomic nucleus. Perhaps it will ultimately have to be modified in some fundamental way. However, we shall not concern ourselves with this possibility to any serious extent. In this section we shall describe quantum mechanics as though the theory of relativity did not exist. In a later one, we shall study what happens when attempts are made to reconcile the two theories.

Quantum mechanics should not be confused with its precursor, the so-called "old quantum theory." The latter was developed between 1900 and 1925 and consisted in combining classical mechanics with arbitrary "quantum rules" in a mysterious and logically incoherent manner in order to obtain ad hoc explanations of particular experiments. Quantum mechanics itself developed in the three or four years following 1925 after fundamental theoretical discoveries of Schrödinger and Heisenberg had provided some extremely suggestive clues as to what the new theory should be like. So many physicists were involved and so much of the communication between them was by word of mouth that it is almost impossible to assign an author to the finished theory or to trace the course of its development from the discoveries of Schrödinger and Heisenberg. On the other hand, it is more or less indisputable that the mathematically rigorous form of the theory, in which the spectral theorem plays a key role, is due to the mathematician J. von Neumann.

The very name quantum mechanics is somewhat misleading. The discretization of observable values that it suggests does not play a role in the conceptual foundations of the subject as it did in the old quantum theory. It is indeed true that certain observables are permitted to take on only a discrete range of values. However, from the point of view of the foundations this is an incidental consequence of the axioms of the subject and in no way a part of them.

The relationship between the notion of state in classical mechanics (i.e., a set of positions and velocities for the fundamental particles) and its

ISBN 0-8053-6702-0/0-8053-6703-9, pbk

quantum mechanical substitute can be made especially transparent by first generalizing the notion of state in classical mechanics. In doing this let us put aside our grandiose scheme of setting up a model for *all* the particles of the universe and speak instead of finite systems of particles that need be neither fundamental nor all–inclusive. We suppose only that we may discuss their motion without knowing either such internal structure as they may have or the positions and velocities of other particles. Let us choose a rectangular coordinate system in space and let Ω denote the $6N$-dimensional space of all possible position coordinates and velocity coordinates of our N particles. Classically a state is a point ω in Ω. We now observe that Ω is a Borel subset of Euclidean $6N$ space and as such is a standard Borel space, and we define a *statistical state* to be a *probability measure* in Ω, that is, a measure α such that $\alpha(\Omega)=1$. One has a statistical state to deal with when one does not know the ω for a given system but knows that it is more likely to be in some parts of Ω than others. $\alpha(E)$ is interpreted as the probability that the "real state" ω is in the Borel set E. Of course a "real state" can always be regarded as a special kind of statistical state. The real state ω_0 may be identified with the probability measure α such that $\alpha(E)=1$ if ω_0 is in E and $\alpha(E)=0$ if ω_0 is not in E. We note that the differential equations of motion whose integration allows us to compute ω at time t when ω at time $t_0<t$ is known also allows us to compute the statistical state α at time t when α at time $t_0<t$ is known. Indeed let U_t be the mapping of Ω into Ω such that $U_t(\omega)$ is the state of our system t time units after it was ω. Let α be the statistical state at a definite time and let β be the statistical state t time units later. Then $\omega \in E$ at the second time instant if and only if ω at the first instant is such that $U_t(\omega) \in E$. Thus $\beta(E)=\alpha(U_t^{-1}(E))$. In other words, β is the measure $E \to \alpha(U_t^{-1}(E))$. Clearly the theory dealing with changes in time of statistical states includes that dealing with changes in time of real states.

By a *dynamical variable* or *observable* we shall always mean a real-valued Borel function on Ω. For example, the y coordinate of the jth particle is an observable and the most general observable is a function of all the position and velocity coordinates. A given statistical state α assigns a definite probability measure on the real line to each observable f. $f(\omega)$ is in the Borel subset I of the real line if and only if $\omega \in f^{-1}(I)$, and in the statistical state α this happens with the probability $\alpha(f^{-1}(I))$. $I \to \alpha(f^{-1}(I))$ then is a probability measure on the real line that tells us what we know about possible values of f when we are in the statistical state α. We observe that a statistical state is uniquely determined by the real line probability measures it assigns to the various observables.

We now make the crucial remark that the notions of observable and statistical state in no way depend upon the fact that Ω is a point set. It is not necessary to refer to the points of Ω but only to the partially ordered set $\mathfrak{L}_\Omega$ of all Borel subsets of Ω and its "ortho-complementation" operation $E \to \Omega - E$. This is important because $\mathfrak{L}_\Omega$ has a perfect analog in quantum

ISBN 0-8053-6702-0/0-8053-6703-9, pbk

mechanics whereas the closest analog there is to a point of Ω is something of a radically different character. Before describing the quantum analog of $\mathcal{L}_\Omega$, let us study the notion of ortho-complemented set in general and verify that we need only the ortho-complemented partially ordered set structure of $\mathcal{L}_\Omega$ in order to define observables and statistical states.

Let $\mathcal{L}$ be any partially ordered set and let $a \to a'$ be a mapping of $\mathcal{L}$ into $\mathcal{L}$. We shall call this mapping an *ortho-complementation* if it has the following properties:

(i) $a_1 \leqslant a_2$ implies $a_2' \leqslant a_1'$ for all a_1, a_2 in $\mathcal{L}$.

(ii) $a'' = a$ for all a in $\mathcal{L}$.

(iii) If $a_1, a_2, \ldots$ are such that $a_i \leqslant a_j'$ for all $i \neq j$, then there exists a least member $\bar{a}$ of $\mathcal{L}$ such that $\bar{a} \geqslant a_i$ for all i. We denote it by $a_1 + a_2 + \cdots$.

(iv) $a + a' \geqslant b$ for all a and b.

(v) For all a and b with $a \leqslant b$ we have $b = a + (b' + a)'$.

Note that $a_i \leqslant a_j'$ if and only if $a_j \leqslant a_i'$. When this happens we say that a_i and a_j are orthogonal and write $a_i \perp a_j$. Note also that (iv) implies that $\mathcal{L}$ has a greatest element which we denote by I and that $a + a' = I$ for all a. I' is of course the unique least element of $\mathcal{L}$ and will be denoted by 0. In the particular case in which $\mathcal{L} = \mathcal{L}_\Omega$ and $E' = \Omega - E$, $E \leqslant F'$ means that E and F are sets with no points in common, that is, orthogonality reduces to disjointness. Moreover, $E_1 + E_2 + \cdots$ is just the set theoretic union of the E_j and (iv) and (v) are obvious trivialities.

Now a statistical state was defined to be a probability measure on Ω, that is, a function α from $\mathcal{L}_\Omega$ to $0 \leqslant x \leqslant 1$ such that:

(i) $\alpha(E_1 \cup E_2 \cup \cdots) = \alpha(E_1) + \alpha(E_2) + \cdots$ whenever $E_i \cap E_j = \varnothing$;

(ii) $\alpha(\Omega) = 1$.

But (i) may be rewritten as $\alpha(E_1 + E_2 + \cdots) = \alpha(E_1) + \alpha(E_2) + \cdots$ whenever $E_i \perp E_j$ and (ii) may be rewritten as $\alpha(I) = 1$. Thus the notion of probability measure on Ω is equivalent to the specialization to $\mathcal{L}_\Omega$ of the following general notion: A *probability measure* on the ortho-complemented partially ordered set $(\mathcal{L}, ')$ is a function α from $\mathcal{L}$ to $0 \leqslant x \leqslant 1$ such that:

(i) $\alpha(a_1 + a_2 + \cdots) = \alpha(a_1) + \alpha(a_2) + \cdots$ whenever $a_i \perp a_j$, $i \neq j$;

(ii) $\alpha(I) = 1$.

To see how to formulate the notion of observable using only the ortho-complemented partially ordered set $\mathcal{L}_\Omega$, let us note that if we wish to compute the probability distribution of the observable defined by the function f on Ω in a given statistical state α, we just pass to the set function

ISBN 0-8053-6702-0/0-8053-6703-9, pbk

$A \to f^{-1}(A)$ mapping Borel subsets of the real line into $\mathcal{L}_\Omega$ and then form the measure $A \to \alpha(f^{-1}(A))$. Note next that f is completely determined by the mapping $A \to f^{-1}(A)$. Thus an observable may be identified with a certain kind of $\mathcal{L}_\Omega$ valued function on the Borel subsets of the real line. To describe observables without mentioning the points of Ω we have only to characterize those $\mathcal{L}_\Omega$ valued set functions that are of the form $A \to f^{-1}(A)$. This is easy to do and here is the answer. A function $A \to E_A$ from the Borel subsets of the real line to $\mathcal{L}_\Omega$ is of the form $A \to f^{-1}(A)$ for some real-valued Borel function f if and only if it has the following properties:

(i) E_{A_1} and E_{A_2} are disjoint whenever A_1 and A_2 are disjoint.

(ii) If $A_1, A_2, \ldots$ are mutually disjoint, then $E_{A_1 \cup A_2 \cup \cdots} = E_{A_1} \cup E_{A_2} \cup \cdots$.

(iii) $E_\varnothing = \varnothing$, where $\varnothing$ is the empty set.

Translating disjointness into orthogonality and so on, we see that $A \to E_A$ is of the form $A \to f^{-1}(A)$ if and only if $A \to E_A$ is an $\mathcal{L}_\Omega$ valued measure on the real line in the sense of the following general definition. Let S be an analytic Borel space and let $(\mathcal{L}, ')$ be an ortho-complemented partially ordered set. Then an $\mathcal{L}$ valued measure on S is a mapping $A \to a_A$ from the Borel subsets of S to the members of $\mathcal{L}$ such that:

(i) a_{A_1} and a_{A_2} are orthogonal whenever A_1 and A_2 are disjoint.

(ii) If $A_1, A_2, \ldots$ are mutually disjoint, then $a_{A_1 \cup A_2 \cup \cdots} = a_{A_1} + a_{A_2} + \cdots$.

(iii) $a_\varnothing = 0$, where $\varnothing$ is the empty set.

Starting with the ortho-complemented partially ordered set $\mathcal{L}_\Omega$, we may then define the observables and statistical states of classical mechanics as follows. The (real-valued) observables are the $\mathcal{L}_\Omega$ valued measures on the real line. The statistical states are the probability measures on $\mathcal{L}_\Omega$. The probability measure on the real line associated with the observable $A \to a_A$ in the statistical state α is $A \to \alpha(a_A)$. From this point of view quantum mechanics differs from classical mechanics in a way that is very easy to state. In quantum mechanics $\mathcal{L}_\Omega$ is replaced by $\mathcal{L}_{\mathcal{H}}$, where $\mathcal{L}_{\mathcal{H}}$ is the ortho-complemented partially ordered set of all projection operators of some separable infinite-dimensional Hilbert space $\mathcal{H}$. The operation $a \to a'$ here is of course the operation of taking a closed subspace into its orthogonal complement. Everything else remains the same; statistical states are probability measures on $\mathcal{L}_{\mathcal{H}}$, observables are $\mathcal{L}_{\mathcal{H}}$ valued measures, etc. We have not said anything about how to formulate the time development of our system in a way that makes sense for a general $\mathcal{L}$, but this is an ommission that is easy to rectify. Confining ourselves to "reversible systems," we may suppose that each U_t on Ω is one-to-one and onto and define $U_{-t} = U_t^{-1}$. Then each U_t defines an automorphism $\overline{U}_t$ of $\mathcal{L}_\Omega$ and $t \to \overline{U}_t$ is a homomorphism of the additive group of the real line into

ISBN 0-8053-6702-0/0-8053-6703-9, pbk

the group of all automorphisms of $\mathcal{L}_\Omega$ and $\alpha(\overline{U}_t(E))$ is a Borel function of t for each E in $\mathcal{L}_\Omega$ and each statistical state α. Moreover, if α is the statistical state of the system at a certain time instant, then t time units later it is β, where $\beta(E)=(U_t^{-1}(E))$. This prescription makes sense for a general $\mathcal{L}$ and specializes for $\mathcal{L}_{\mathcal{H}}$ to the rule describing the time development of a system in quantum mechanics.

Before examining the detailed implications of the replacement $\mathcal{L}_\Omega \to \mathcal{L}_{\mathcal{H}}$, we make a number of remarks about its significance. Three questions suggest themselves at once.

(a) Why $\mathcal{L}_{\mathcal{H}}$ rather than some other ortho-complemented partially ordered set?

(b) Why any ortho-complemented partially ordered set at all?

(c) How does this replacement resolve the difficulty about the nonexistence of exact simultaneous position and momentum coordinates?

Our first remark relevant to question (c) is that the difficulties in assigning simultaneous position and velocity coordinates to very small particles do not arise in assigning simultaneous probability distributions to these observables. One arrives at a probability distribution by making the same measurement a great many times on different replicas of similarly prepared systems. One can measure different observables on different samples chosen from the same "statistical ensemble" and these measurements cannot interfere. This important point is discussed in detail in Section 1 of Chapter IV of von Neumann's book. It follows that we may always suppose that statistical states exist where we define a statistical state to be an assignment of a probability measure to every observable. Let $\mathcal{O}$ denote the set of all observables in our system and let $\mathcal{S}$ denote the set of all statistical states. For each O in $\mathcal{O}$ and each S in $\mathcal{S}$ and each Borel subset A of the real line, let $p(O,S,A)$ denote the probability that a measurement of O in the statistical state S will lead to a value in A. It is possible to write down a set of a priori plausible axioms about the behavior of the function p from which one can deduce the following theorem: There exists a (uniquely determined) ortho-complemented partially ordered set $(\mathcal{L},')$, a one-to-one mapping of $\mathcal{O}$ onto the $\mathcal{L}$ valued measures on the real line and a one-to-one mapping of $\mathcal{S}$ into a convex subset of the probability measures on $\mathcal{L}$ such that $p(O,S,A)=\alpha(a_A)$ whenever α is the probability measure corresponding to S and a is the $\mathcal{L}$ valued measure corresponding to O. This theorem is proved in Section 2 of Chapter II of the author's published lecture notes *Mathematical Foundations of Quantum Mechanics*, (Benjamin–Cummings, Reading, Mass. (1963)). It is our answer to question (b). To complete our answer to question (c) let us consider a pair of observables represented by the $\mathcal{L}$ valued measures $A \to a_A$ and $A \to b_A$. For each statistical state α we have a probability measure on the real line

ISBN 0-8053-6702-0/0-8053-6703-9, pbk

assigned to each of these observables $A \to \alpha(a_A)$ and $A \to \alpha(b_A)$. As α varies over the statistical states, we obtain a family of pairs of probability measures. When $\mathcal{L} = \mathcal{L}_\Omega$ one can always find statistical states for which each member of the pair is a point measure. However, for other choices of $\mathcal{L}$, for example, for $\mathcal{L} = \mathcal{L}_{\mathcal{H}}$, there are many pairs a, b with the property that any α that makes $A \to \alpha(a_A)$ approximate a point measure is such that $A \to \alpha(b_A)$ is rather highly dispersed. Changing $\mathcal{L}$ can indeed produce a model for observables and statistical states in which "classical states" (i.e., states in which every observable has a value) not only do not exist but are not even approximable by statistical states.

Question (a) has the least satisfying answer. The most compelling reason for replacing $\mathcal{L}_\Omega$ by $\mathcal{L}_{\mathcal{H}}$ is that "it works," that is, it leads to the model of the universe that physicists now favor and that has been successful in explaining a wide range of phenomena. It would be satisfying if one could prove that the $\mathcal{L} = \mathcal{L}_{\mathcal{H}}$ is the only hypothesis compatible with some set of a priori plausible axioms, but such a proof is not presently available. On the other hand, one can convince oneself without great difficulty that the choice $\mathcal{L} = \mathcal{L}_{\mathcal{H}}$ is by no means as arbitrary as it may appear at first sight. If one makes additional assumptions about $\mathcal{L}$ inspired less by logical necessity than by the faith that nature is relatively regular and rejects "pathology" in choosing its mathematical models, one can produce a partially heuristic argument suggesting that $\mathcal{L}$ is most probably a direct sum or direct integral of subsets each of which is either $\mathcal{L}_{\mathcal{H}}$ or something very much like it. This has been done on pages 72–74 of the author's published lecture notes.

Now let us put alternatives our of our minds and examine the implications of the choice $\mathcal{L} = \mathcal{L}_{\mathcal{H}}$. First, however, a word about probability measures on $\mathcal{L}$ and the statistical states that they represent. Given any two such α and β and any real number t with $0 < t < 1$, we may form $t\alpha + (1-t)\beta$ and observe at once that it is a probability measure on $\mathcal{L}$. We denote the corresponding statistical state by $t\alpha + (1-t)\beta$. We may think of this state as arising if we know that we are in the statistical state α with the probability t and in the statistical state β with the probability $(1-t)$. Accordingly, we think of $t\alpha + (1-t)\beta$ as representing "less knowledge of the system" than either α or β. We shall call a statistical state *pure* if it cannot be written in the form $t\alpha + (1-t)\beta$, where $\alpha \neq \beta$. When $\mathcal{L} = \mathcal{L}_\Omega$, the pure states are easily seen to be just the classical states, that is, the statistical states assigning the probability one to a single point of Ω. In this case every observable has a definite value in a pure state. The situation is quite otherwise when $\mathcal{L} = \mathcal{L}_{\mathcal{H}}$. Even for the statistical states that are pure there exist observables that have highly dispersed probability distributions in this state. The celebrated "proof" of von Neumann that "hidden variables" do not exist in quantum mechanics is just this (easily proved) observation.

ISBN 0-8053-6702-0/0-8053-6703-9, pbk

Assuming that $\mathcal{L}=\mathcal{L}_{\mathcal{H}}$, let us see what we can say about the structure of observables and statistical states. First of all let ϕ be any unit vector in $\mathcal{H}$ and for each projection P let $\alpha_\phi(P)=(P\phi,\phi)$. It is obvious that α_ϕ is a probability measure on $\mathcal{L}_{\mathcal{H}}$ and so each ϕ defines a statistical state. We obtain further statistical states by taking convex combinations of the α_ϕ; $\gamma_1\alpha_{\phi_1}+\gamma_2\alpha_{\phi_2}+\cdots$ with $\gamma_j \geqslant 0$ and $\gamma_1+\gamma_2+\cdots=1$ is a probability measure on $\mathcal{L}_{\mathcal{H}}$ for any choice of the ϕ_j. These, however, are all there are. Every probability measure on $\mathcal{L}_{\mathcal{H}}$ is of the indicated form. This conjecture of the author was proved by A. M. Gleason. His difficult and ingenious proof appears in *J. Math. and Phys.*, vol. 6 (1957), 885–894. It is not hard to show that one can always choose the ϕ_j to be orthogonal. One can even choose them to form a complete orthonormal set by throwing in a few zero γ_j's. But then we have

$$\sum \gamma_j\alpha_{\phi_j}(P)=\sum \gamma_j(P\phi_j,\phi_j)=\mathrm{Tr}(PA)$$

where A is the unique bounded linear operator such that $A\phi_j=\gamma_j\phi_j$. Thus the statistical states correspond one-to-one to the self-adjoint operators A such that $\mathrm{Tr}(A)=1$ and A has all nonnegative proper values. It follows almost immediately that the state represented by A is pure if and only if A has a one-dimensional range. If ϕ is any unit vector in this range, then $\mathrm{Tr}(AP)=(P\phi,\phi)$, so the pure states are just those of the form α_ϕ, where ϕ varies over the unit vectors of H and ϕ and $e^{ia}\phi$ define the same state for all real a. Just as one can do most of classical mechanics without mentioning the statistical states, so can we do most of quantum mechanics without mentioning any states but the pure ones. Since all quantum mechanical states are statistical, we shall omit this adjective from now on. The statistical states are important in classical statistical mechanics and similarly the impure states are important in quantum statistical mechanics. The operator A describing a (generally impure) state is called a *von Neumann density operator*. A von Neumann density matrix is of course just the matrix of a von Neumann density operator with respect to some basis.

So much for the states. What about the observables? These correspond one-to-one to the $\mathcal{L}_{\mathcal{H}}$ valued measures on the real line and one verifies at once that an $\mathcal{L}_{\mathcal{H}}$ valued measure is just what we have previously called a projection valued measure. It is true that the definition of projection valued measure includes the requirement that $P_{E\cap F}=P_EP_F=P_FP_E$ and the specialization of $\mathcal{L}$ valued measure to $\mathcal{L}_{\mathcal{H}}$ only requires the special case of this in which $E\cap F=0$ so that $P_{E\cap F}=0$. However, it is easy to see that the more general statement follows from the special one and the additivity requirement. Now by the spectral theorem we know that there is a natural one-to-one correspondence between projection valued measures and self-adjoint operators. Combining these two correspondences we obtain a natural one-to-one correspondence between observables and self-adjoint operators in the Hilbert space $\mathcal{H}$. Taking this correspondence in conjunc-

ISBN 0-8053-6702-0/0-8053-6703-9, pbk

tion with our earlier description of pure states we obtain von Neumann's famous unification and summary of the various statistical statements of quantum mechanics. *The observables in a quantum mechanical system correspond one-to-one to the self-adjoint operators in a separable infinite-dimensional Hilbert space $\mathcal{H}$. The pure states correspond to the unit vectors in H in such a way that two unit vectors correspond to the same pure state if and only if one is a scalar multiple of the other. Let A be a self-adjoint operator corresponding to a certain observable and let ϕ be a unit vector describing a certain state. Then the results of measuring the observable corresponding to A in the state corresponding to ϕ are described by the probability measure $E \to (P_E^A(\phi), \phi)$, where $E \to P_E^A$ is the projection valued measure associated with A by the spectral theorem.*

It is illuminating to look at the special case in which T has a so-called pure point spectrum; that is, in which H has an orthonormal basis $\phi_1, \phi_2, \ldots$ such that $T(\phi_j) = \lambda_j \phi_j$. Then, assuming in addition that T is multiplicity free, the above general statement implies the following. In the pure state described by the unit vector ϕ, let $c_j = (\phi, \phi_j)$ so that $\phi = \Sigma c_j \phi_j$ and $\Sigma |c_j|^2 = 1$. Then the probability that the observable described by T has the value λ_j is $|c_j|^2$ and the probability that this observable has a value other than one of the λ_j is zero.

Since we must immediately go back to the projection valued measure in order to compute probabilities, one may well wonder why we bother to introduce the operators at all. We shall now exhibit several ways in which the operators themselves are useful. Let T be a self-adjoint operator such that the corresponding observable has an expected value in the pure state defined by ϕ. Then this expected value is $\int x\, d(P_x(\phi), \phi)$. But by the spectral theorem this integral is just $(T(\phi), \phi)$. Hence we can compute the expected value of a given observable in a given state directly from the operator and the state vector—without the intervention of the projection valued measure. Now it is obvious that an observable is bounded if and only if the corresponding operator is a bounded operator. Let T_1 and T_2 be bounded self-adjoint operators and let us ask whether the sum of the corresponding observables exists. By definition this sum will exist if and only if there exists a third bounded self-adjoint operator T_3 such that

$$(T_3(\phi), \phi) = (T_1(\phi), \phi) + (T_2(\phi), \phi)$$

for all ϕ and then T_3 will be the operator corresponding to the sum. T_3 clearly does exist and is just the operator sum $T_1 + T_2$ of T_1 and T_2. In quantum mechanics sums of bounded observables do exist. They are much more easily dealt with if we describe observables by operators rather than by projection valued measures.

Let g be a real-valued Borel function defined in the spectrum of the self-adjoint operator T. Let θ_T denote the corresponding observable. Then

ISBN 0-8053-6702-0/0-8053-6703-9, pbk

we may form the observable $g(\theta_T)$ and the self-adjoint operator $g(T)$ and ask how they are related. It is gratifying to find that a simple analysis of the definitions concerned shows them to be equal. Thus the algebra of self-adjoint operators is consistent with the algebra of observables. In particular, we see that $(T^k(\phi),\phi)=\int x^k(dP_x(\phi),\phi)$ for all k. Thus we may find all the "moments" of $E\to(P_E(\phi),\phi)$ just knowing ϕ and the iterates of T.

It follows from an axiom that the time evolution of our system is defined by a homomorphism $t\to\beta_t$ of the real line into the group of all automorphisms of $\mathcal{L}_{\mathcal{H}}$ such that $\alpha(\beta_t(P))$ is a Borel function of t for each α and P. Now an easy adaptation of the proof of the fundamental theorem of projective geometry shows that for each t $\exists$ a unitary or anti-unitary transformation U_t of $\mathcal{H}$ on $\mathcal{H}$ such that $U_t(\mathfrak{M})=\beta_t(\mathfrak{M})$ for each closed subspace $\mathfrak{M}$ of $\mathcal{H}$. U_t is unique up to multiplication by a complex constant of unit modulus. Since $\beta_{t_1+t_2}=\beta_{t_1}\beta_{t_2}$ it follows that $U_{t_1+t_2}$ and $U_{t_1}U_{t_2}$ define the same automorphism of $\mathcal{L}_{\mathcal{H}}$ and hence that there exists a complex constant of unit modulus $c(t_1,t_2)$ such that $U_{t_1+t_2}=U_{t_1}U_{t_2}c(t_1,t_2)$. In particular, $U_t=U_{t/2+t/2}=(U_{t/2})^2c(t/2,t/2)$. Since the square of an anti-unitary operator is always unitary, it follows that every U_t is unitary. Let $P_{\mathfrak{M}}$ be the projection on $\mathfrak{M}$. Then $P_{U_t(\mathfrak{M})}=U_t^{-1}P_{\mathfrak{M}}U_t$. Hence

$$\alpha_\phi(U_t(\mathfrak{M}))=(P_{\mathfrak{M}}U_t(\phi),U_t(\phi))=\alpha_{U_t(\phi)}(\mathfrak{M})$$

Thus if ϕ is a unit vector representing a pure state of our system at any given time, then t time units later the state is represented by the unit vector $U_t(\phi)$. In other words, the purity of states is preserved by the passage of time and the action of the U_t on the unit vectors gives directly the action of time on the pure states. From the fact that $\alpha(\beta_t(\mathfrak{M}))$ is a Borel function in t, we conclude that $t\to(U_t(\phi),\phi)$ is a Borel function. Hence $(U_t(\phi+\psi),\phi+\psi)$, $(U_t(\phi),\phi)$, and $(U_t(\psi),\psi)$ are all Borel functions of t. Hence $(U_t(\phi),\psi)$ is a Borel function of t for all ϕ and ψ. In other words, $t\to U_t$ is a projective unitary representation of the additive group of the real line. It can be shown that the multiplier group for the additive group of the real line is trivial. In other words, the arbitrary constants in the U_t may be chosen so that $c(t_1,t_2)\equiv 1$. Thus we may suppose that $t\to U_t$ is actually an ordinary unitary representation. This representation is not quite unique. If we multiply U_t by a complex number $a(t)$ such that $t\to a(t)$ is a Borel function, then $t\to a(t)U_t$ will be an ordinary unitary representation if and only if $|a(t)|=1$ and $a(t_1+t_2)=a(t_1)a(t_2)$. This happens, as we know, if and only if $a(t)=e^{ilt}$ for some fixed real number 1. In other words, the unitary representation $t\to U_t$ giving the dynamics of a quantum mechanical system is uniquely determined up to multiplication by a one-dimensional representation $t\to e^{ilt}$. We shall refer to U as *the* dynamical group of the system in spite of the mild nonuniqueness involved in its definition.

ISBN 0-8053-6702-0/0-8053-6703-9, pbk

By Stone's theorem we may write U_t in the form e^{-itH}, where H is a self-adjoint operator. H is called the *Hamiltonian operator* or simply the *Hamiltonian* of the system.

It is uniquely determined by U, but the ambiguity in U produces a small corresponding ambiguity in H. H is uniquely determined by the system up to an additive constant multiple of the identity.

Let ϕ be a unit vector representing a particular pure state of our system. By its *trajectory* we shall mean the mapping $t \rightarrow \phi_t = U_t(\phi)$ assigning a unique vector describing what this pure state was or becomes at all past and future times. Writing $\phi_t = e^{-iHt}(\phi)$ we see that

$$\frac{d}{dt}[\phi_t] = -iHe^{-iHt}(\phi) = -iH(\phi_t);$$

a "formal" equation to which various vigorous meanings may be attached depending upon what topology is to be used in defining

$$\lim_{\Delta t \rightarrow 0} \frac{\phi_{t+\Delta t} - \phi_t}{\Delta t}$$

The "differential equation" $d\phi/dt = -iH(\phi)$ is the abstract form of what is known as *Schrödinger's equation*. In many specific instances one has a canonical concrete realization of the underlying Hilbert space as a space of functions and H is a "known" differential operator. ϕ_t then becomes a function of several variables $t, q_1, \ldots, q_\nu$ and $d\phi/dt = -iH(\phi)$ takes the form

$$\frac{\partial \phi}{\partial t}(q_1, \ldots, q_\nu, t) = -iH(\phi(q_1, \ldots q_\nu, t))$$

where applying H to ϕ means computing a sum of functions of $q_1, \ldots, q_\nu$ multiplied by partial derivatives with respect to the q_j. We shall give examples later. The system of ordinary differential equations that, in classical mechanics, tells us how the coordinates and velocities change in time is replaced in quantum mechanics by a first-order partial differential equation that tells us how the complex valued "state function" changes in time.

Notice that H not only defines the dynamics of our system but plays another role as well. As a self-adjoint operator it defines an observable. Clearly the (essentially) unique observable that defines the dynamics must be of special importance. It is a constant multiple of the quantum analog of the classical mechanical observable known as the (total) energy of the system. We carry over the name and call this observable (or rather a suitable constant multiple of it) the *energy observable* or simply the *energy* of the system. We shall also speak of the *energy operator*. In fact, the time

ISBN 0-8053-6702-0/0-8053-6703-9, pbk

has come to cease making the rather pedantic distinction between observables and self-adjoint operators on the one hand and between pure states and unit vectors on the other. Whenever it suits us we shall speak of the "observable A" when A is in fact a self-adjoint operator—and similarly for states. One may wonder why H itself is not called the energy rather than some (as yet undefined) constant multiple of it. The answer lies in the fact that energy cannot be defined in classical mechanics until one has chosen a "unit of mass" in addition to units of time and distance. In quantum mechanics one can define energy to be H without ever introducing the notion of mass. In a sense quantum mechanics provides a "natural" unit of mass. However, this unit differs from all the arbitrary units in use at the time quantum mechanics was discovered. Thus the "natural" energy notion of quantum mechanics differs by various constant multiples from all the "semi-artificial" classical energy notions. The conversion factor that the natural quantum mechanical energy must be multiplied by to get the quantum analog of the conventional energy is denoted by $\hbar$. It's numerical value depends of course upon our choice of fundamental units and can be made equal to one by choosing these properly. It turns out, as we shall see in more detail later, that $\hbar = h/2\pi$, where h is a fundamental constant introduced by Planck in 1900 at the very beginning of the old quantum theory.

Given an observable A we shall say that it is an integral of the motion if its expected value in every state is constant in time. This would be the case if and only if $(AU_t(\phi), U_t(\phi)) = (A(\phi),\phi)$ for all ϕ; that is, if and only if $(U_{t^{-1}}AU_t\phi,\phi) = (A(\phi),\phi)$ for all ϕ. But this is so if and only if $U_t^{-1}AU_t = A$, that is, if and only if A commutes with U_t for all t. Using the spectral theorem and the fact that $U_t = e^{-itH}$ and H have the same projection valued measure, we conclude that A is an integral of the motion if and only if H and A commute in the sense that every projection P_E^A in the projection valued measure for A commutes with every projection P_F^H in the projection valued measure for H. In particular, we see that the energy H is always an integral of the motion. This is the quantum mechanical form of the "law of conservation of energy."

Stone's theorem of course applies not only to the dynamical group $t \to U_t$, but to any one-parameter group of automorphisms of our system. Each such is associated with an essentially unique one-parameter unitary group and hence with a self-adjoint operator that is uniquely determined up to an additive constant. Thus every one-parameter group of automorphisms is associated with an observable (modulo an additive constant) and conversely. The one-parameter group of automorphisms will commute with all U_t and hence preserve the dynamics if and only if the generating self-adjoint operator commutes with H; that is, is an integral of the motion. Thus we have a natural one-to-one correspondence—modulo additive constants—between one-parameter groups of automorphisms that

ISBN 0-8053-6702-0/0-8053-6703-9, pbk

are also automorphisms with respect to the dynamics and observables which are integrals of the motion. In other words, every one-parameter symmetry group of the whole system has associated with it an essentially unique observable that is an integral of the motion.

The most obvious and important examples of one-parameter symmetry groups of quantum mechanical systems come from the one-parameter groups of symmetries of physical space. One observes that physical phenomena seem to be the same in one point of space as in another, and that in fact physical laws take the same form when expressed in any two rectangular coordinate systems—even if one is obtained by rotating the other. One expresses this observation in quantum mechanics by an axiom that we will give more fully below but that includes postulating the existence of a fundamental homomorphism of the group $\mathfrak{E}$ of all rigid motions of space into the group of all automorphisms of $\mathcal{L}_{\mathcal{H}}$. $\mathfrak{E}$ is by definition the group generated by the group T of all translations and the group R_{s_0} of all rotations about any one fixed point s_0. $\mathfrak{E}$ is then a semi-direct product of T and R_{s_0}. We do not include the spatial reflections in R_{s_0}. As in the case of the dynamical group, each automorphism of $\mathcal{L}_{\mathcal{H}}$ is described by a unitary operator unique up to a multiplication constant. Anti-unitary operators do not occur because every element of $\mathfrak{E}$ is the square of some other element. (This would not be the case if we included spatial reflections and our analysis would then be more complicated.) Thus our fundamental homomorphism is described by a fundamental projective unitary representation $\alpha \to V_\alpha$ of the Euclidean group $\mathfrak{E}$ in our Hilbert space $\mathcal{H}$. Moreover (for an isolated system), we postulate that $V_\alpha U_t V_\alpha^{-1} = U_t$ as the analytic statement of the fact that the laws of motion are unchanged when we translate and rotate the system. Each one-parameter subgroup of $\mathfrak{E}$, that is, each homomorphism γ of the additive group of the real line into $\mathfrak{E}$, defines a projective unitary representation of the real line $x \to V_{\gamma(x)}$ and hence by Stone's theorem an observable. Since $V_\alpha U_t V_\alpha^{-1} = U_t$, this observable will always be an integral. Now the subgroup T of $\mathfrak{E}$ is isomorphic to the additive group if a three-dimensional vector space over the real numbers and every one-dimensional subspace is describable in physical terms as the group of all translations in some fixed direction. If we introduce a unit of distance and choose as a basis vector a unit vector τ_0 in this subgroup, then $x \to x\tau_0$ is a one-parameter subgroup. The corresponding observable is called the *total linear momentum* in the given direction. If we have chosen a fixed rectangular coordinate system, then we have three distinguished directions one for each coordinate axis and can speak of the x, y, and z *components* of the total linear momentum. The total linear momentum in any direction is easily seen to be a linear combination with constant coefficients of the x, y, and z components just defined. At this level linear momentum is defined only up to an additive constant. Later we shall see how to choose this constant in a canonical fashion for a wide

ISBN 0-8053-6702-0/0-8053-6703-9, pbk

range of physical systems. The fact that total linear momentum in a given direction is an integral of the motion is the quantum mechanical version of the "law of conservation of linear momentum." Just as with energy the classical definition of linear momentum depends upon a choice of a "unit of mass." Thus if we are to have concordance with classical definitions, we must define the linear momentum observables to be certain constant multiples of the natural one. The conversion factor is the same as for energy—the number we have called $\hbar$.

We get another family of one-parameter subgroups of $\mathfrak{S}$ by choosing an s_0 and a line through s_0 and considering the subgroup of R_{s_0} leaving all points on this line fixed. In this case there is no need to choose a unit of length since the radian is a "natural" angle measure and serves to parametrize rotations by real numbers. Proceeding as before, we are led to associate an observable with every line is space that is called the *angular momentum* about that line. The fact that the angular momentum observables are integrals of the motion is the quantum mechanical version of "the law of conservation of angular momentum."

In classical particle mechanics the total linear momentum in the x direction is defined to be the sum $\Sigma_{j=1}^{N} m_j \dot{x}_j$, where m_j is the so called mass of the jth particle and $\dot{x}_j$ is the x velocity component of the jth particle. We shall see in the sequel that momentum and velocity are similarly related in quantum mechanics. However, we have yet to introduce the notion of velocity in quantum mechanics and now proceed to do so. Let $U_t = e^{-iHt}$ be the dynamical group, let A be any self-adjoint operator, and let ϕ be a unit vector. Then if $U_t(\phi)$ is in the domain of A for all t in some neighborhood of 0, we have the following formula for the time dependence of the expected value of A in the trajectory of ϕ:

$$t \rightarrow (AU_t(\phi), U_t\phi) = (Ae^{-iHt}(\phi), e^{-iHt}\phi)$$

The derivative with respect to t at $t=0$ is then

$$-i(AH(\phi),\phi) + i(HA(\phi),\phi) = i((HA - AH)(\phi),\phi)$$

Now when H and A are both bounded $i(HA - AH)$ is itself a bounded self-adjoint operator. Clearly it defines an observable whose expected value in every state is the time derivative of the expected value of A in that state. It is natural to call the observable $i(HA - AH)$ the time derivative of the observable A. When A and H are not bounded, there are difficulties about defining $i(HA - AH)$. However, whenever the domains of A and H are so related that $i(HA - AH)$ may be defined as a self-adjoint operator in an unambiguous fashion, it is natural to call the corresponding observable the time derivative of that of A. We shall define velocity observables by applying this definition to the operators defining the coordinates of our

ISBN 0-8053-6702-0/0-8053-6703-9, pbk

particle. Notice that the time derivative of A is zero if and only if H and A commute so that an integral of the motion is just an observable whose time derivative is zero.

Notes and References

The treatment of the basic principles of quantum mechanics given here is a minor variant of that which appears in the author's book *Mathematical Foundations of Quantum Mechanics*, Benjamin-Cummings, Reading, Mass. (1963). The axioms listed in that book are implicit in a paper by the author ("Quantum mechanics and Hilbert space," *American Mathematical Monthly*, vol. 64 (1957), 45–57) and were discovered while preparing lectures for a course given at Harvard. The emphasis on the lattice of closed subspaces goes back to a paper by Garrett Birkhoff and J. von Neumann ("The logic of quantum mechanics," *Ann. Math.*, vol. 37 (1936), 823–843). While the author did not understand that paper until he had worked out his own ideas on the subject, he owes much to the direct influence of Garrett Birkhoff who taught him to think in lattice theoretic terms.

For a much more thorough development of the ideas and viewpoints of this section, the reader is referred to the first volume of *The Geometry of Quantum Theory* by V. S. Varadarajan. This book, published in 1968 by van Nostrand, New York, has been re-issued under the imprint of Springer-Verlag.

For a discussion of the axioms mentioned above in the context of a general survey of the axiomatics of quantum mechanics and an account of later developments by other authors, the reader is referred to A. Wightman's treatment of Hilbert's sixth problem in volume 27 (1976) of the American Mathematical Society's *Symposia on Pure Mathematics*. This book contains reports on the current status of the Hilbert problems written by a variety of authors. Wightman's lengthy and informative article entitled "The mathematical treatment of problems in physics" also contains an interesting account of recent work on the "hidden variables problem" with emphasis on the role played by Gleason's theorem.

Foundations of Quantum Mechanics by J. M. Jauch, Addison-Wesley, Reading, Mass. (1968), is a full-length textbook for physicists that emphasizes the point of view of this section and the next.

ISBN 0-8053-6702-0/0-8053-6703-9, pbk

18. The quantum mechanics of one-particle systems

In order to go beyond the abstract framework of the preceding section and deal with specific physical problems, we must have a method for deciding which self-adjoint operators correspond to the observables of interest. If we know the Hamiltonian H and which operators correspond to the coordinates of the particles under consideration, we may find the operators corresponding to the velocity components by the rule for computing time derivatives. In this section we shall discuss what can be said about these operators in the rather idealized but important special case in which there is a single particle isolated from the rest of the universe. One speaks of a single *free* particle.

Let S denote physical space and let $\mathfrak{E}$ denote the Euclidean group of all rigid motions of space so that S is an $\mathfrak{E}$ space. We may think of the position of our "particle" as an "S valued observable" and describe it by a projection valued measure $E \to P_E$ defined on S. P_E is then the self-adjoint operator corresponding to the real-valued observable that is one when the particle is in E and zero when it is not. Given P we may immediately write down the projection valued measure on the line corresponding to any "generalized coordinate"; that is, to any real-valued Borel function g in S. It is $I \to P_{g^{-1}(I)}$. Now let $f_1(s)$, $f_2(s)$, and $f_3(s)$ be the x, y, and z coordinates, respectively, of the point s in S with respect to some rectangular coordinate system. Let X, Y, and Z be the self-adjoint operators corresponding to these observables. We may construct P from X, Y, and Z as follows. Let $P^X(P^Y, P^Z)$ denote the projection valued measure in the line defining $X(Y,Z)$. Then for each triple E_1, E_2, E_3 of Borel subsets of the real line

$$P_{(E_1 \times E_2 \times E_3)} = P^X_{E_1} P^Y_{E_2} P^Z_{E_3}$$

George W. Mackey, Unitary Group Representations in Physics, Probability, and Number Theory

ISBN 0-8053-6702-0/0-8053-6703-9, pbk

and this uniquely determines P. Indeed if X, Y, and Z are arbitrary mutually commuting self-adjoint operators there will exist a unique projection valued measure P on Euclidean three-space such that

$$P_{E_1 \times E_2 \times E_3} = P^X_{E_1} P^Y_{E_2} P^Z_{E_3}$$

Thus we do not need the motion of an "S valued observable" to justify the introduction of P. On the other hand, there is no real difficulty in extending our earlier analysis and defining M valued observables, where M is any Borel space.

For all that we have said so far P could be any projection valued measure on S. However, possibilities for it become sharply limited if we invoke Euclidean invariance. Let $\alpha \rightarrow V_\alpha$ denote the fundamental projective representation of $\mathcal{E}$ described above. Then $V_\alpha^{-1} P_E V_\alpha$ must be the projection corresponding to $P_{[E]\alpha}$ by analysis of what one means by invariance. Let us assume for the moment that V is an ordinary representation of E. Then to say that $V_\alpha^{-1} P_E V_\alpha = P_{[E]\alpha}$ is just to say that P is a system of imprimitivity for V based on S. Now S is a transitive E-space. Hence the imprimitivity theorem applies and tells us that the pair P, V is determined to within equivalence by a unitary representation L of the subgroup R_{s_0} leaving a point s_0 of S fixed. Specifically P, V is equivalent to the pair P^L, V^L, where U^L is the unitary representation of E induced by the unitary representation L of R_{s_0} and P^L is the associated system of imprimitivity.

The simplest possibility is that in which L is the one-dimensional identity representation. In that case $\mathcal{H}(U^L) = \mathcal{L}^2(S, \mu)$, where μ is Lebesque measure, $(U_\alpha^L \psi)(s) = \psi(s\alpha)$ and $(P_E^L \psi)(s) = \phi_E(s)\psi(s)$, where ϕ_E is the characteristic function of E. Thus we may realize our Hilbert space as $\mathcal{L}^2(S, \mu)$ in such a fashion that $(V_\alpha \psi)(s) = \psi(s\alpha)$ and $(P_E f)(s) = \phi_E(s)\psi(s)$. It follows at once that for any Borel function f on S the corresponding observable is multiplication by f. In particular, if we introduce coordinates so that $\psi(s) \sim \psi(x,y,z)$, then the x coordinate corresponds to the operator $\psi(x,y,z) \rightarrow x\psi(x,y,z)$ and similarly for the other coordinates. Also, the formula $(P_E(\psi),\psi)$ for the probability of finding the particle in the set E when the system is in the state ψ becomes

$$\int \phi_E \psi(s) \bar{\psi}(s)\, d\mu(s) = \int_E |\psi(s)|^2\, d\mu(s)$$

Thus the absolute value squared of the state function ψ has an immediate interpretation as a "probability density." The reader should not jump to the conclusion that only $|\psi|$ has physical significance. We may write $\psi = |\psi| e^{iA}$ and we shall see below that A determines the rate of change of $|\psi|$ with time.

ISBN 0-8053-6702-0/0-8053-6703-9, pbk

Let α_a^1 denote the member of $\mathcal{E}$ that corresponds to translation by a units in the x direction. Then $(V_{\alpha_a^1}\psi)(x,y,z)=\psi(x+a,y,z)$. Thus

$$\frac{d}{da}\bigg|_{a=0} V_{\alpha_a^1}(\psi)(x,y,z)=\frac{\partial\psi}{\partial x}(x,y,z)$$

Thus $V_{\alpha_a^1}=e^{-iaT}$, where $T(\psi)=i\partial\psi/\partial x$. Thus, applied to differentiable functions, the operator corresponding to the linear momentum in the x direction is $\psi\to i\hbar\partial\psi/\partial x$. Similarly, the y and z components of the linear momentum are $\psi\to i\hbar\partial\psi/\partial y$ and $\psi\to i\hbar\partial\psi/\partial z$ and the three components of the angular momentum are

$$\psi\to i\hbar\left(y\frac{\partial}{\partial z}-z\frac{\partial}{\partial y}\right)\psi,\quad i\hbar\left(z\frac{\partial}{\partial x}-x\frac{\partial}{\partial z}\right)\psi,\quad\text{and}\quad i\hbar\left(x\frac{\partial}{\partial y}-y\frac{\partial}{\partial x}\right)\psi$$

The preceding conclusions follow from the hypothesis that L is the one-dimensional identity representation of R_{s_0}. Let us see how they are affected if we consider some other choice for L. Every element α of $\mathcal{E}$ may be written uniquely in the form $\alpha_R\alpha_T$, where α_T is a translation and α_R is in R_{s_0}. Hence the formula $g(\xi x)=L_\xi g(x)$ in the definition of $\mathcal{H}(U^L)$ tells us that g is uniquely determined by its values on the subgroup of translations. Since $\alpha\to s_0\alpha$ maps the translations one-to-one onto S, we may identify members of $\mathcal{H}(U^L)$ with functions from S to $\mathcal{H}(L)$. Indeed one verifies without difficulty that we may realize our Hilbert space as $L^2(S,\mu,\mathcal{H}(L))$ in such a fashion that $U_\alpha^L\psi(s)=L_{\alpha_R^{-1}}\psi(s\alpha)$ and $P_E(\psi)(s)=\phi_E(s)\psi_E(s)$. As far as the position and linear momentum observables are concerned, things are just as before except that ψ is now a *vector valued* function taking values in $\mathcal{H}(L)$ and the probability density for position is $s\to\|\psi(s)\|^2$. The coordinate and linear momentum operators are still given by the same formulas. However, in computing angular momenta there is a difference. A member α_R of R_{s_0} enters the formula for U_α^L in two places and each angular momentum operator is correspondingly a sum of two terms. The first term is just as in the case of trivial L except that one applies the differential operators to vector valued functions rather than to scalar valued ones. The other term is of the form $\psi\to\psi'$, where $\psi'(s)=A\psi(s)$ and A is a self-adjoint operator in $\mathcal{H}(L)$. If $r\to(\alpha_R)_r$ is the relevant homomorphism of the real line into R_{s_0}, then A is the self-adjoint operator corresponding by Stone's theorem to $r\to L_{(\alpha_R)_r}$. To see what A is like, it suffices to consider the case in which L is irreducible because in more general cases we need only take the direct sum of the A's corresponding to the irreducible constituents. Now, as we have remarked much earlier, the group R_{s_0} has (to within equivalence) just one irreducible unitary representation of every odd finite dimension. It is customary to refer to the $(2j+1)$-dimensional one as D_j. The subgroup of all rotations about a fixed axis is isomorphic to

ISBN 0-8053-6702-0/0-8053-6703-9, pbk

the additive group of all real numbers modulo the subgroup of integral multiples of 2π. It is not hard to verify that the restriction of D_j to this subgroup is the direct sum of the characters $\theta \to e^{ik\theta}$ with $k=0, \pm 1, \pm 2, \ldots, \pm j$. It follows easily that the corresponding A has a $0, \pm \hbar, \pm 2\hbar, \ldots, \pm j\hbar$ as its eigenvalues. Thus when L is the irreducible representation D_j and $j \neq 0$, thc total angular momentum about an axis is in a natural way the sum of two observables. One of these is a differential operator that is just the same as when $j=0$ except for acting on vector valued functions. The other commutes with all the position and linear momentum observables and takes on just the $2j+1$ values $0, \pm \hbar, \pm 2\hbar, \ldots, \pm j\hbar$. One calls the first term the *orbital angular momentum* and the second the *spin angular momentum*. Physicists like to think of this second term as due to the particle's "spinning" on its axis. Hence the word spin. The maximum value that the spin angular momentum can take on divided by $\hbar$ is called the *spin* of the particle. For those particles in which L is irreducible we see that the coordinate and momentum observables are determined to within equivalence by specifying the spin. We are tempted to state further that the spin must be a positive integer but we have not yet explored the possibility that V might be only a projective representation.

In order to see what can be said about the possible multipliers for $\mathcal{E}$, we invoke a theorem about the possible multipliers for semi-direct products for which we shall have further need later on. Let $G = N \circledS H$ be a semi-direct product of the separable locally compact groups N and H. Then every multiplier for G is a trivial multiplier multiplied by a multiplier υ of the following form:

$$\upsilon(n_1, h_1, n_2, h_2) = \sigma(n_1, h_1(n_2))\omega(h_1, h_2)g(n_2, h_1)$$

where n_1 and n_2 are in N, h_1 and h_2 are in H. σ is a multiplier for N, ω is a multiplier for H, g is a complex valued Borel function in $N \times H$, and g and σ satisfy the following two identities

$$\sigma(h(n_1), h(n_2)) = \frac{\sigma(n_1, n_2)g(n_1 n_2, h)}{g(n_1, h)g(n_2, h)} \tag{*}$$

$$g(n, h_1 h_2) = g(h_2(n), h_1)g(n, h_2) \tag{**}$$

Moreover, for every choice of σ, ω, and g in which σ and g satisfy (*) and (**), we obtain a multiplier υ by setting

$$\upsilon(n_1, h_1, n_2, h_2) = \sigma(n_1, h_1(n_2))\omega(h_1, h_2)g(n_2, h_1)$$

The proof of this theorem is quite easy and the reader might like to attempt it as an exercise. It appears in pages 303 and 304 of the author's paper in *Acta Math.*, vol. 99 (1958), 265–311. Note that (*) says that σ and

ISBN 0-8053-6702-0/0-8053-6703-9, pbk

its transform by h are "similar" (i.e., one is a trivial multiplier times the other) and that $n \to g(n,h)$ defines the trivial multiplier. In the special case in which G is the direct product of N and H so that $h(n) \equiv n$, (*) and (**) reduce to

$$g(n_1 n_2, h) \equiv g(n,h) g(n_2,h) \tag{*}'$$

$$g(n, h_1 h_2) = g(n,h_1) g(n,h_2) \tag{**}'$$

In other words, there must exist a homomorphism $n \to \chi_n$ of N into the group of all one-dimensional characters of H such that $g(n,h) \equiv \chi_n(h)$. Thus the most general multiplier for $N \times H$ is similar to the product of a multiplier for N a multiplier for H and a function of the form $n_1, h_1, n_2, h_2 \to \chi_{n_1}(h_2)$. Since the additive group of the real line has only trivial multipliers, it follows that every multiplier v for a two-dimensional real vector space must be similar to one such that $v(x_1, y_1, x_2, y_2) = e^{iax_1y_2}$, where a is a fixed real number. Such a multiplier cannot be trivial (when $a \neq 0$) because it is not symmetric. Proceeding by induction, it can be shown that every multiplier in the additive group of a finite-dimensional real vector space must be similar to one of the form $\phi_1, \phi_2 \to e^{iA(\phi_1, \phi_2)}$, where $A(\phi_1, \phi_2)$ is a bilinear form such that $A(\phi_1, \phi_2) = -A(\phi_2, \phi_1)$.

Now let us apply this remark and our theorem to the special case in which $G = \mathcal{E}$, $H = R_{s_0}$, and N is the group of translations in three space. Using the fact that a bilinear form on an odd-dimensional space must be singular, it follows that the multiplier σ cannot satisfy (*) unless it is trivial. Thus we may choose v so that $\sigma \equiv 1$ and then (*) and (**) become

$$g(n_1 n_2, h) = g(n_1,h) g(n_2,h)$$

$$g(n, h_1 h_2) \equiv g(h_2(n), h_1) g(n,h_2)$$

The first identity tells us that there exists a map $h \to \chi_h$ of H into $\hat{N}$ such that $g(n,h) = \chi_h(n)$. The second tells us that $\chi_{h_1h_2} \equiv ([\chi_{h_1}]h_2)\chi_{h_2}$. Now $\chi \in \hat{N}$ and the latter is isomorphic to the additive group of a three-dimensional vector space. If we write N additively, our identity becomes $\chi_{h_1h_2} = [\chi_{h_1}]h_2 + \chi_{h_2}$. Let us integrate the vector valued function $h \to \chi_h$ with respect to Haar measure in H and let χ_0 be the result. Integrating the identity with h_2 fixed, we get $\chi_0 = [\chi_0]h_2 + \chi_{h_2}$. Returning to multiplicative notation we see that $\chi_h \equiv \chi_0 [\chi_0^{-1}]h$ for some fixed χ_0. Let $a(n,h) = \chi_0(n)$. Then

$$\frac{a((n_1,h_1)(n_2,h_2))}{a(n_1,h_1)a(n_2,h_2)} = \frac{\chi_0(n_1)\chi_0(h_1(n_2))}{\chi_0(n_1)\chi_0(n_2)} = (([\chi_0]h_1)\chi_0^{-1})(n_2)$$

On the other hand

$$g(n_2,h_1) = \chi_{h_1}(n_2) = \chi_0(h_1)\chi_0^{-1}h_1(n_2)$$

ISBN 0-8053-6702-0/0-8053-6703-9, pbk

Thus

$$g(n_2,h_1)=\frac{a((n_1,h_1)(n_2,h_2))}{a(n_1,h_1)a(n_2,h_2)}$$

In other words $n_1,h_1;n_2,h_2 \to g(n_2,h_1)$ is a trivial multiplier and we may adjust the constants in the V_α so that $g(n,h)\equiv 1$. We have thus shown that we may always choose the constants so that $v(n_1,h_1;n_2,h_2)=w(h_1,h_2)$, where w is a multiplier for R_{s_0}. As we have already remarked there *is* a non-trivial multiplier for R_{s_0} and, up to multiplication by trivial multipliers, only one. Let w_0 be a representative of this unique similarity class of nontrivial multipliers for R_{s_0} and let v_0 be the corresponding multiplier for E, $v_0(n_1,h_2;n_2,h_2)=w_0(h_1,h_2)$. Whenever the constants in the V_α *cannot* be chosen so that $\alpha \to V_\alpha$ is an ordinary representation of E, they can be chosen so that it is a v_0 representation. We have then just one other case to consider.

Suppose now that V is a v_0 representation. There is no difficulty in reformulating the imprimitivity theorem so that it applies to projective representations. In fact, the only thing that needs any modification at all is the definition of induced representation. If H is a closed subgroup of a separable locally compact group G and σ is any multiplier for G, then the restriction of σ to H is a multiplier for H and we may speak of the σ representations of H. If L is any such, we define the σ representation U^L of G *induced* by L just as in the case in which $\sigma=1$, except for the following slight changes:

(i) The identity $f(\xi x)=L_\xi f(x)$ must be replaced by $f(\xi x)=\sigma(\xi,x)L_\xi f(x)$.

(ii) The identity $(U_x^L f)(y)=f(yx)$ must be replaced by $(U_x^L f)(y)=f(yx)/\sigma(x,y)$.

Applying the imprimitivity theory reformulated for projective representations and noticing that v_0 restricted to R_{s_0} is just w_0, we see that there exists an w_0 representation L of R_{s_0} so that the pair P,V is equivalent to the pair P^L,U^L, where U^L is the v_0 representation of $\mathfrak{E}$ induced by L and P^L is the associated system of imprimitivity. The rest of the argument proceeds essentially as before with exactly the same conclusions about the position observables and the linear momentum observables. The angular momentum observables are also the same as before except that we have new possibilities for the operators A and hence for the spin. It turns out that to within equivalence R_{s_0} has exactly one irreducible w_0 representation of each nonzero even dimension. The one of dimension $2k$ is denoted by the symbol D_j, where $j=k-\frac{1}{2}$. Thus D_j is defined whenever j is an integer or half of an integer. It is a $(2j+1)$-dimensional representation of R_{s_0} whenever j is an integer and it is a $2j+1$ w_0 representation of R_{s_0} whenever

ISBN 0-8053-6702-0/0-8053-6703-9, pbk

j is a half integer. In any case j is the spin of the particle when $V=U^{D_j}$. Thus the integral spin possibilities we encountered in discussing the case in which V was an ordinary representation must be supplemented by a series of half integral spins $\frac{1}{2}, \frac{3}{2}, \frac{5}{2}$, and $\frac{7}{2}$. It turns out that the basic particles in atomic structure—the electron, the proton, and the neutron—are all particles of spin $\frac{1}{2}$.

No matter what V is, we see that we may realize our Hilbert space as $\mathcal{L}^2(S, M, \mathcal{H}_0)$, where $\mathcal{H}_0$ is some other Hilbert space, in such a fashion that the coordinate and momentum observables are

$$f \to xf, \quad f \to yf, \quad f \to zf, \quad f \to i\hbar\frac{\partial f}{\partial x}, \quad f \to i\hbar\frac{\partial f}{\partial y}, \quad f \to i\hbar\frac{\partial f}{\partial z}$$

Denoting these operators by Q_1, Q_2, Q_3, P_1, P_2, P_3, respectively, we see that they satisfy the "commutation relations":

$$Q_k Q_j - Q_j Q_k = P_k P_j - P_j P_k = 0, \qquad Q_j P_k - P_k Q_j = \frac{\hbar}{i} I \delta_k^j$$

In the original "matrix mechanics" of Heisenberg these commutation relations were assumed as part of the discoverer's informal axiom system—the motivation being the fact that such relations are satisfied by the Poisson brackets in classical mechanics. It is interesting to see that they are in fact a consequence of Euclidean invariance.

Let us look a little more closely at the relationship between the Heisenberg commutation relations and Euclidean invariance. Rather than deal with the commutation relations as written down above, it is more convenient to deal with them in "integrated form." Let $A_t^j = e^{itQ_j}$ and $B_s^j = e^{isP_j}$. Then $(A_t^1 f)(x,y,z) = e^{itx} f(x,y,z)$ with corresponding expressions for A^2 and A^3 and $(B_s^1 f)(x,y,z) = f(x - \hbar s, y, z)$ with corresponding expressions for B^2 and B^3. Now the commutation relations for the A^j and B^k are

$$\begin{aligned} A_t^k A_s^j - A_s^j A_t^k &= B_t^j B_s^k - B_s^k B_t^j = 0 \\ A_t^j B_s^j &= B_s^j A_t^j e^{its\hbar} \\ A_t^j B_s^k &= B_s^k A_t^j \quad \text{if } j \neq k \end{aligned} \tag{**}$$

for all t and s. Moreover, on a formal level at least, these commutation relations are completely equivalent to those of Heisenberg. That is, no matter what operators Q_1, Q_2, Q_3, P_1, P_2, P_3 we choose, these operators will satisfy the Heisenberg relations if and only if the A^j and B^k satisfy (**).

Now let us define $A_{t_1,t_2,t_3} = A_{t_1}^1 A_{t_2}^2 A_{t_3}^3$ and $B_{s_1,s_2,s_3} = B_{s_1}^1 B_{s_2}^2 B_{s_3}^3$. Then to say that the A^j and B^k satisfy (**) is exactly the same thing as to say that:

(1) $t_1, t_2, t_3 \to A_{t_1,t_2,t_3}$ is a unitary representation of the additive group of all triples of real numbers.

ISBN 0-8053-6702-0/0-8053-6703-9, pbk

(2) $s_1,s_2,s_3 \to B_{s_1,s_2,s_3}$ is a unitary representation of the additive group of all triples of real numbers.

(3) $A_{t_1,t_2,t_3}B_{s_1,s_2,s_3} = B_{s_1,s_2,s_3}A_{t_1,t_2,t_3}e^{i\hbar(t_1s_1+t_2s_2+t_3s_3)}$.

Now every character of the additive group N of all triples of real numbers is of the form $s_1,s_2,s_3 \to e^{i\hbar(t_1s_1+t_2s_2+t_3s_3)}$ for a unique triple t_1,t_2,t_3. Thus we may identify N with its own dual and when we do so in the indicated way we may replace (3) by the equation

$$A_\chi B_n = B_n A_\chi \chi(n) \qquad \dagger$$

In other words, to find the most general sextuple of unitary representations of the real line satisfying (**) is the same as to find the most general pair consisting of a unitary representation B of N and a unitary representation A of $\hat{N}$ satisfying †. Now using the spectral theorem for locally compact commutative groups, we may replace the unitary representation A of $\hat{N}$ by a projection valued measure P^A on N and a simple computation shows that † is equivalent to

$$P_E^A B_n = B_n P_{[E]n}^A \qquad \dagger\dagger$$

in other words, to the condition that $E \to P_E^A$ should be a system of imprimitivity for B based on N. It follows that we may use the imprimitivity theorem to determine all solutions of (**), that is, to the Heisenberg commutation relations in integrated form. Now in the case at hand, the group is N and so is the space on which it acts, that is, N acts on itself via group multiplication. The subgroup leaving a point fixed is thus the subgroup containing only the identity. Since the identity subgroup has a unique irreducible solution, so do the Heisenberg commutation relations. In other words, the solutions arising from the pairs P, V are to within equivalence the only possible solutions. These solutions of course differ from one another only through the dimension of the Hilbert space $\mathcal{H}_0$ in $L^2(S, M, \mathcal{H}_0)$. The dimension of $\mathcal{H}_0$ is the number of times the unique irreducible solution is repeated as a direct summand.

Let us observe that the projection valued measure P^A coming from Q_1, Q_2, and Q_3 via A^1, A^2 and A^3 may be regarded as a projection valued measure on space and then is identical with the one assigning Q_1, Q_2, and Q_3 as self-adjoint operators corresponding to the x, y, and z coordinates. Thus, via the spectral theorem, the Heisenberg commutation relations in integrated form are just exactly what we get from our basic identity connecting P and V if we restrict V to the translation subgroup and assume that this restriction is an ordinary representation. In other words, it is almost correct to say that to assume the Heisenberg commutation relations is just to assume translational invariance. It is not quite correct because from translational invariance alone we cannot conclude that V

ISBN 0-8053-6702-0/0-8053-6703-9, pbk

can be made into an ordinary representation of the translation group. This additional fact about V is a consequence of Euclidean invariance. It is also a consequence of the assumption that the momenta in various directions commute. Of course, it is much more satisfactory to assume Euclidean invariance than to assume the Heisenberg commutation relations. It is not only that the assumption of Euclidean invariance has considerably more a priori plausibility. In addition, we can deduce more from it: the form of the angular momentum operators and the concept of spin.

We digress momentarily to point out some facts about the relationship of the uniqueness of the irreducible solutions of the Heisenberg commutation relations to more general group theoretical theorems. Once we have thrown them into the form † it is clear that we have a problem that still makes sense when N is replaced by an arbitrary separable locally compact group. The transition to †† can be made in the general case and once made the problem makes sense even if N is not commutative. It is the problem of finding the most general pair consisting of a unitary representation B of N and a projection valued measure P on N such that $P_E B_n = B_n P_{[E]n}$ for all $n \in N$ and all Borel subsets E of N. The fact that this problem has a unique irreducible solution is just the imprimitivity theorem in the special case in which the subgroup is the identity. Historically, the imprimitivity theorem was first proved in this special case, and this special case was motivated by the observation that the theorem on the uniqueness of the Heisenberg commutation relations may be generalized as just described (see: On a theorem of Stone and von Neumann, *Duke Math. Journal*, vol. 16 (1949), 313–325).

Before turning our attention to dynamics, let us observe that much of the preceding analysis is independent of the hypothesis that space be Euclidean. If we suppose only that space is homogeneous in the sense that it admits a transitive group of distance preserving transformations, we can give an analysis of position and momentum observables that is as complete as that given above. If $\mathfrak{G}'$ is the given transitive group, then the role of the rotation group will be played by the subgroup R'_{s_0} leaving a point of s_0 fixed and we may not only classify particles by the representations of R'_{s_0} that appear but introduce a notion of spin. In the special case in which space is the surface of a sphere in Euclidean four space, it even turns out that R'_{s_0} is isomorphic to the ordinary rotation group. When the trivial representation of R'_{s_0} is involved so that we have a particle of "zero spin" we may take our Hilbert space to be $\mathfrak{L}^2(S,\mu)$ in such a way that the operator defining the observable corresponding to any "global" coordinate, that is, to any real-valued Borel function on S, is multiplication by that function. In general there will be no analog to the *linear* momentum observables. Every one-parameter subgroup of $\mathfrak{G}'$ will define an observable that will be expressible in the spin zero case as a first-order differential operator. However, there will be no distinction between linear

ISBN 0-8053-6702-0/0-8053-6703-9, pbk

momentum and angular momentum. When space is not flat, one has no close analog to the P's and Q's and one is forced to express the Heisenberg commutation relations in group theoretical terms.

We have now discussed the implications of Euclidean invariance for the position and momentum observables fairly exhaustively but have had nothing to say about the dynamical group $t \to V_t$ and hence about how the system changes with time. We shall begin with the case in which our particle has spin zero in order to discuss some of the main points without being bothered by too many complications. We shall also suppose, as in most of the above, that space is Euclidean. Having chosen an origin and a coordinate system, we may identify space with the set E_3 of all triples x,y,z of real numbers in such a way that our Hilbert space is $\mathfrak{L}^2(E_3,\mu)$, where μ is Lebesque measure and P and V are as described above. In order to exploit the fact that all U_α commute with all V_t, it will be convenient to transfer P and V via the Fourier transform to $\mathfrak{L}^2(\hat{E}_3,\nu)$. Of course, we may identify $\hat{E}_3$ with the set of all triples of real numbers because every triple a,b,c defines a character $\chi_{a,b,c}$ via the formula $\chi_{a,b,c}(x,y,z)=e^{i(ax+by+cz)}$. By the theory of the Fourier transform we have a unitary map T of $\mathfrak{L}^2(E_3,\mu)$ into $\mathfrak{L}^2(\hat{E}_3,\nu)$ that takes each f in $\mathfrak{L}^2(E_3,\mu)\cap\mathfrak{L}^1(E_3,\mu)$ into $\hat{f}$, where

$$\hat{f}(a,b,c)\equiv\frac{1}{(2\pi)^{3/2}}\iiint f(x,y,z)e^{i(ax+by+cz)}\,dx\,dy\,dz$$

and inversely

$$f(x,y,z)=\frac{1}{(2\pi)^{3/2}}\iiint \hat{f}(a,b,c)e^{-i(ax+by+cz)}\,da\,db\,dc$$

Now an easy calculation shows that replacing $\hat{f}$ by $e^{-i(x_0a+y_0b+z_0c)}\hat{f}$ replaces f by $x,y,z\to f(x+x_0,y+y_0,z+z_0)$ and replacing $\hat{f}$ by $a,b,c\to\hat{f}(a+a_0,b+b_0,c+c_0)$ replaces f by $fe^{i(a_0x+b_0y+c_0z)}$. Moreover, replacing $\hat{f}$ by $a,b,c\to\hat{f}(\beta(a,b,c))$, where $\beta(a,b,c)=\beta_{11}a+\beta_{12}b+\beta_{13}c,\beta_{21}a+\beta_{22}b+\beta_{23}c,\beta_{31}a+\beta_{32}b+\beta_{33}c$ replaces f by $a,b,c\to f(\beta'(x,y,z))$, where $\beta'(xyz)=\beta_{11}x+\beta_{21}y+\beta_{31}z,\beta_{12}x+\beta_{22}y+\beta_{32}z,\beta_{13}x+\beta_{23}y+\beta_{33}z$. Let $\hat{U}_t=TU_tT^{-1}$. Since U_t commutes with all V_α, $\hat{U}_t$ must commute with all $TV_\alpha T^{-1}$ for all t and α. Now if α is translation by x_0,y_0,z_0, then $TV_\alpha T^{-1}$ is (as we have just seen) the operator of multiplying by $e^{-i(x_0a+y_0b+z_0c)}$. Since $\hat{U}_t$ commutes with all V_α, it commutes with all such multiplications and hence with all multiplications by characteristic functions of Borel subsets of E_3. Hence $\hat{U}_t$ must be multiplication by a complex-valued function on E_3. Call it $h_t(a,b,c)$. Since $\hat{U}_t$ is unitary $|(h_t(a,b,c))|\equiv 1$ and since $t\to\hat{U}_t$ is a homorphism $h_{t_1+t_2}(a,b,c)=h_{t_1}(a,b,c)h_{t_2}(a,b,c)$. In other words, $h_t(a,b,c)\equiv e^{ith'(a,b,c)}$, where h' is some real-valued Borel function on E_3. For any choice of h' the $\hat{U}_t$ that it defines will commute with V_α for all translations α but not

ISBN 0-8053-6702-0/0-8053-6703-9, pbk

necessarily for rotations. In order that $\hat{U}_t$ commute with V_α for all *rotations* α and all t, it is clearly necessary and sufficient that $h'(\beta'(a,b,c))= h'(a,b,c)$ almost everywhere on E_3 for every rotation $x,y,z \to \beta(x,y,z)$. In other words, it is necessary and sufficient that (mod sets of measure zero) $h'((a,b,c))=\rho((a^2+b^2+c^2))$, where ρ is some Borel function defined for all positive real numbers. Thus U_{t_2} must be of the form $T^{-1}\hat{U}_tT$, where $\hat{U}_t$ is multiplication by $e^{it\rho((a^2+b^2+c^2))}$ and ρ is some real-valued Borel function defined on the positive real axis. Conversely, given any such ρ, there is a unique corresponding dynamical group that commutes with all V_α.

Given the function ρ, let us see what conclusions we can draw about the behavior of our (spin zero) one-particle system. Recall first that $U_t= e^{-i(H/\hbar)t}$, where H is the energy operator. Since $\hat{U}_t$ is multiplication by $e^{it\rho(a^2+b^2+c^2)}$, it follows that THT^{-1} is multiplication by $-\hbar\rho(a^2+b^2+c^2)$. Now $Ti(\partial/\partial x)T^{-1}$ is multiplication by a and similarly for $i(\partial/\partial y)$ and $i(\partial/\partial z)$. Thus if

$$P_1=i\hbar\frac{\partial}{\partial x},\quad P_2=i\hbar\frac{\partial}{\partial y},\quad P_3=ih\frac{\partial}{\partial z}$$

are the operators corresponding to the x,y,z components of linear momentum, then multiplication by $a^2+b^2+c^2$ is

$$T\frac{(P_1^2 \div P_2^2+P_3^2)}{\hbar^2}T^{-1}$$

Hence

$$H=-\hbar\rho\left(\frac{(P_1^2+P_2^2+P_3^2)}{\hbar^2}\right)$$

where ρ applied to the self-adjoint operator $(P_1^2+P_2^2+P_3^2)/\hbar^2$ makes sense by virtue of the spectral theorem. Let us set $\rho_0(r)=-\hbar\rho(r/\hbar^2)$. Then $H=\rho_0(P_1^2+P_2^2+P_3^2)$. Our first conclusion then is that the energy of our particle is a function ρ_0 of the square $P_1^2+P_2^2+P_3^2$ of the total linear momentum P_1,P_2,P_3 Of course the operator corresponding to $P_1^2+P_2^2+P_3^2$ is

$$-\hbar^2\left(\frac{\partial^2}{\partial x^2}+\frac{\partial^2}{\partial y^2}+\frac{\partial^2}{\partial z^2}\right)=\hbar^2\nabla$$

where ∇ denotes the Laplacian

$$-\left(\frac{\partial^2}{\partial x^2}+\frac{\partial^2}{\partial y^2}+\frac{\partial^2}{\partial z^2}\right)$$

ISBN 0-8053-6702-0/0-8053-6703-9, pbk

Thus the dynamical operator is just $\rho(\nabla)$. In other words, Euclidean invariance implies that the dynamical operator must be some function ρ of the Laplacian, and when units are chosen so that $\hbar=1$ this is just the function relating energy to total linear momentum. To know this function is to know the system completely.

For a given ρ let us now see whether velocity observables exist and if so what they are. We have already seen that (modulo questions of domain) the velocity component in the x direction must be given by $i(HQ_1 - Q_1H)/\hbar$, where Q_1 is multiplication by x. Now $T(H/\hbar)T^{-1}$ is multiplication by $-\rho(a^2+b^2+c^2)$ and TQ_1T^{-1} is $+i\partial/\partial a$. Thus

$$T\frac{(i(HQ_1-Q_1H))}{\hbar}T^{-1}$$

is

$$\hat{f}\to\rho(a^2+b^2+c^2)\frac{\partial\hat{f}}{\partial a}-\frac{\partial}{\partial a}\rho(a^2+b^2+c^2)\hat{f}=-\rho'(a^2+b^2+c^2)2a\hat{f}$$

and does not exist unless ρ is differentiable in a suitable sense. Adding the assumption that velocity observables are to exist, we conclude that ρ must be a differentiable function. Fourier transforming back to the original space we see that the velocity observables must take the form

$$iA\frac{\partial}{\partial x},\quad iA\frac{\partial}{\partial y},\quad iA\frac{\partial}{\partial z}$$

where A is $-2\rho'(\nabla)$ and ρ is as before. In particular, the velocity components, the momentum components, and the energy all mutually commute and we may discuss functional relations between them without restriction. Let us note that since the energy observable is determined by the physics only up to an additive constant, ρ is also known only up to an additive constant. However, this constant disappears when we form ρ'. Thus ρ' and the velocity observables are completely determined by the physics. The sum of the squares of the velocity components is

$$A^2\nabla=4(\rho'(\nabla))^2\nabla=\sigma(\nabla)=\sigma\left(\frac{P_1^2+P_2^2+P_3^2}{\hbar^2}\right)$$

where $\sigma(r)=4\rho'(r)^2r$. Let us suppose that we know at least that σ is differentiable and that the derivative σ' is everywhere positive. Then σ will be a monotone function mapping $(0\leqslant r<\infty)$onto $0\leqslant s<a$, where $0\leqslant a\leqslant\infty$ and σ will have an inverse ω mapping $0\leqslant s<a$ on $0\leqslant x<\infty$. Using ω we may express both the energy and the momentum components in terms of the velocity components. Indeed $\nabla=\omega(W_1^2+W_2^2+W_3^2)$, where W_1, W_2, W_3

ISBN 0-8053-6702-0/0-8053-6703-9, pbk

are the three velocity components. Thus $A = -2\rho'(\omega(W_1^2 + W_2^2 + W_3^2))$ and the three momentum components are

$$\frac{i}{2}\hbar\left[\rho'(\omega(W_1^2 + W_2^2 + W_3^2))\right]^{-1} W_1$$

$$\frac{i}{2}\hbar\left[\rho'(\omega(W_1^2 + W_2^2 + W_2^3))\right]^{-1} W_2$$

$$\frac{i}{2}\hbar\left[\rho'(\omega)(W_1^2 + W_2^2 + W_3^2)\right]^{-1} W_3$$

while the energy is $\rho(\omega(W_1^2 + W_2^2 + W_3^2))$. The operator $[\rho'(\omega(W_1^2 + W_2^2 + W_3^2))]^{-1}$ is what we must multiply each velocity component by to get the corresponding momentum component. Because of the commutativity we also think of this as a numerical relationship. In a state in which the velocity components are almost sure to be v_1, v_2, and v_3, respectively, then the corresponding momentum components are almost sure to be

$$\frac{\hbar v_1}{2\rho'(\omega(v_1^2 + v_2^2 + v_3^2))}, \quad \frac{\hbar v_2}{2\rho'(\omega(v_1^2 + v_2^2 + v_3^2))}, \quad \frac{\hbar v_3}{2\rho'(\omega(v_1^2 + v_2^2 + v_3^2))}$$

The function

$$v_1, v_2, v_3 \to \frac{\hbar}{2\rho'(\omega(v_1^2 + v_2^2 + v_3^2))}$$

plays the same role as mass in classical mechanics in that it is the number we multiply the velocity components by to get the momentum components. If we assume that it is a constant m as in classical mechanics, then we conclude at once that $\rho(r) = -(\hbar/2m)r$ and that the dynamical operator is $(\hbar i/2m)\nabla$. In this case the Schrödinger's equation takes the form

$$\frac{\partial \psi}{\partial t} = -\frac{i\hbar}{2m}\left(\frac{\partial^2 \psi}{\partial x^2} + \frac{\partial^2 \psi}{\partial y^2} + \frac{\partial^2 \psi}{\partial z^2}\right)$$

and this is the accepted free particle Schrödinger equation (in the so called nonrelativistic theory). We shall see below that $\rho(r) = -(\hbar/2m)r$ is a consequence of what is called Galilean relativity. The fact that mass varies with velocity when Galilean relativity does not hold is especially interesting in view of the fact that the revision of Galilean relativity proposed by Einstein also implies a change of mass with velocity. In this theory one speaks of the "rest mass" of a particle—this being defined as the mass when the velocity is zero. Clearly one can define rest mass for our general quantum mechanical particle in exactly the same way. Another consequence of the relativity theory of Einstein is the existence of a critical

ISBN 0-8053-6702-0/0-8053-6703-9, pbk

velocity c that can never be exceeded. This also exists for our general quantum mechanical particle whenever ρ is such that σ is a bounded function. In other words, the existence of a variation of mass with velocity and of a highest possible velocity are consequences not only of Einstein's theory but of a large number of other possible denials of Galilean relativity. In fact, it is not the positive assertion of Einstein's theory that has these consequences so much as it is the denial of the particular positive assertions of Galilean relativity.

Without committing ourselves to the exact form of ρ, we can discuss the extent to which the dynamics of a free particle can be looked upon as a form of wave propagation. Let $\psi \in \mathfrak{L}^2(E^3, \mu)$ represent the state of our particle at $t=0$ and let us form the function

$$t,x,y,z \rightarrow \tilde{\psi}(x,y,z,t) = U_t(\psi)(x,y,z)$$

representing its trajectory. Fourier transforming we may write

$$\psi(x,y,z) = \frac{1}{(2\pi)^{3/2}} \iiint \hat{\psi}(a,b,c) e^{-i(ax+by+cz)}\, da\, db\, dc$$

and then

$$\tilde{\psi}(x,y,z,t) = \frac{1}{(2\pi)^{3/2}} \iiint \hat{\psi}(a,b,c) e^{-i(ax+by+cz)} e^{i\rho(a^2+b^2+c^2)t}\, da\, db\, dc \qquad (\dagger)$$

Now for each fixed a,b,c the integrand is a certain constant multiple of the function

$$x,y,z,t \rightarrow e^{-i(a^2+b^2+c^2)^{1/2}\left[\dfrac{ax+by+cz}{(a^2+b^2+c^2)^{1/2}} - \dfrac{\rho(a^2+b^2+c^2)}{(a^2+b^2+c^2)^{1/2}}t\right]}$$

when $a>0$, $b=c=0$ this becomes

$$e^{-ia\left(x - \frac{\rho(a^2)}{a}t\right)}$$

For each fixed t this is a periodic function of x with period $2\pi/a$. Moreover, as t changes the only effect on the graph of the function is to move it to the right with velocity $\rho(a^2)/a$. One speaks of a *wave* with wave length $2\pi/a$, and wave velocity $\rho(a^2)/a$. When b and c are not equal to zero, one speaks in a strictly analogous fashion of a "plane wave" of wave length

$$\frac{2\pi}{(a^2+b^2+c^2)^{1/2}}$$

ISBN 0-8053-6702-0/0-8053-6703-9, pbk

propagated in the direction of the unit vector

$$\frac{a}{(a^2+b^2+c^2)^{1/2}},\quad \frac{b}{(a^2+b^2+c^2)^{1/2}},\quad \frac{c}{(a^2+b^2+c^2)^{1/2}}$$

until the "wave velocity"

$$\frac{\rho(a^2+b^2+c^2)}{(a^2+b^2+c^2)^{1/2}}$$

Equation (†) tells us that the general solution of Schrödinger's equation is a "continuous superposition" of such plane waves with a, b, and c varying over all possible values. Let us consider the special case in which our original state is such that the momentum components are certain to have values in the sphere $(p_1-p_1^0)^2+(p_2-p_2^0)^2+(p_3-p_3^0)^2<\epsilon^2$. It is easy to see that this will be the case if and only if $\hat{\psi}$ is zero outside of the sphere

$$\left(a-\frac{p_1^0}{\hbar}\right)^2+\left(b-\frac{p_2^0}{\hbar}\right)^2+\left(c-\frac{p_3^0}{\hbar}\right)^2<\frac{\epsilon^2}{\hbar^2}$$

This is because $\hat{\psi}$ plays the same role for momenta that ψ plays for coordinates. But in this case $\tilde{\psi}$ will be a superposition of plane waves all having their parameters a,b,c within $\epsilon/\hbar$ of $p_1^0/\hbar, p_2^0/\hbar, p_3^0/\hbar$. If $\epsilon/\hbar$ is very small, then all the plane waves in the superposition will be going very nearly in the direction p_1^0, p_2^0, p_3^0 and will have wave lengths very near to

$$\frac{2\pi\hbar}{\left((p_1^0)^2+(p_2^0)^2+(p_3^0)^2\right)^{1/2}}=\frac{h}{\left((p_1^0)^2+(p_2^0)^2+(p_3^0)^2\right)^{1/2}}$$

This is the sense in which a quantum mechanical particle of linear momentum p may be described by a "wave" of wave length h/p propagated in the direction of motion of the particle. It is noteworthy that the relationship between momentum and wave length is independent of ρ. Of course the formulas for computing the wave velocity and the so called "group velocity" will depend upon ρ. It is interesting but not surprising to find the group velocity for the waves associated with a particle of momentum p to be exactly the value of the velocity observable when the momentum observable is p.

So much for the case of particles of spin zero. For particles of spin j, a similar but more complicated analysis is possible that involves $(2j+1)$-independent Borel functions of r. We shall content ourselves here with a description of how one arrives at these $2j+1$ functions and not attempt a discussion of their physical significance. As in the spin zero case, each U_t

ISBN 0-8053-6702-0/0-8053-6703-9, pbk

lies in the commuting algebra of the underlying representation V of the Euclidean group, but now V is the induced representation U^{D_j}. Instead of elementary Fourier analysis it is now more convenient to survey the possibilities for U_t via a direct integral decomposition of U^{D_j}. The irreducible unitary representations of $\mathfrak{E}$ (as well as the irreducible projective representations) may be determined by applying the semi-direct product theory described in Section 10. The result is as follows: For each $r>0$ let $\chi_r(x,y,z)=e^{irz}$, where x,y,z is a general element in the translation subgroup N of E. The subgroup of R_{s_0} leaving χ_r fixed is the group $R^z_{s_0}$ of all rotations about the z axis. For each integer n let $f_n(\theta)=e^{in\theta}$ for all $\theta\in R^z_{s_0}$. Then $x,y,z,\theta\to e^{irz}f_n(\theta)$ is a one-dimensional unitary representation of $NR^z_{s_0}$ and the corresponding induced representation $U^{\chi_r f_n}$ is irreducible. These irreducible representations are all inequivalent and every irreducible unitary representation of E is equivalent to some $U^{\chi_r f_n}$ except for the finite-dimensional representations obtained by "lifting" the unitary irreducible representations of R_{s_0}. If we apply a suitable infinite-dimensional version of the Frobenius reciprocity theorem and the theorem on restricting induced representations to subgroups described in Section 14, we are led to the conclusion that

$$U^{D_j}=\int_0^\infty\left[U^{\chi_r f_{-j}}\oplus U^{\chi_r f_{-j+1}}\oplus\cdots\oplus U^{\chi_r f_j}\right]dr$$

and in particular that U^{D_j} is multiplicity free. Correspondingly, the unitary representation U of the real line is a direct integral of unitary representation W^r, where each W^r_t is in the commuting algebra of $U^{\chi_r f_{-j}}\oplus U^{\chi_r f_{-j+1}}\oplus\cdots\oplus U^{\chi_r f_j}$. Thus each W^r_t takes each $\mathcal{H}(U^{\chi_r f_k})$ with itself and is in the commuting algebra of the irreducible representation $U^{\chi_r f_k}$. Hence it is multiplication by a complex number of modulus unity. In other words, each W^r is completely defined by $2j+1$ one-dimensional representations of the additive group of the real line $t\to e^{i\lambda^r_k t}$ where $k=-j,-j+1,\ldots,j$. Our $2j+1$ functions of r are just the functions $r\to\lambda^r_k=\rho_k(r)$. Although the arguments are slightly different, exactly the same results hold when projective representations are involved and j is a half integer.

The possibilities for the functions ρ_k become much narrower if we invoke the hypothesis of "Galilean relativity." This hypothesis suggests itself when one begins to think about the meaning, if any, of "absolute motion." Since most of the objects of which we are aware are at rest relative to one another, one naively thinks that the majority are (in some unanalyzed sense) in a state of "absolute rest" while the minority that are moving relative to the majority are in a state of "absolute motion." Doubts are cast on this naive outlook when one studies astronomy and sees that it is more natural to think of the earth as in motion and the sun at rest than vice versa. At a more sophisticated level one finds that the laws of classical mechanics are unchanged when one replaces a given reference system by

ISBN 0-8053-6702-0/0-8053-6703-9, pbk

one that moves with uniform velocity with respect to it. Thus no experiment dealing with phenomena to which the laws of classical mechanics apply can be used to distinguish a particular reference system as in a state of absolute rest. This state of affairs suggests that absolute motion may be a meaningless concept and that it makes sense only to talk of the motion of one body relative to another. To the extent that this is so the concept of space also loses part of its meaning. What from the point of view of one observer is a single fixed point in space is from the point of view of another a point moving uniformly in a straight line. Thus there is no natural permanent one-to-one correspondence between the points of space for one observer and those of another who is in motion relative to the first. In other words, there is no such thing as "absolute" space; space as well as motion is relative to a particular observer. On the other hand, at each instant of time a given point of the space of one observer is at a definite point of the space of another and we do have a natural one-to-one correspondence between the set of all pairs s,t and the set of all pairs s',t, where s varies over the points of space for one observer, s' varies over the points of space for the other, and t varies over the instants of time. Thus we have a single "absolute" four-dimensional affine manifold of "events" that we shall call space–time and denote it by Ω. For each observer Ω will be uniquely factorable as the product of space and time; however, this factorization will vary from observer to observer.

In identifying time for two observers we are continuing to be naive—a close examination shows that this is unjustified and leads to the celebrated "special relativity" theory of Einstein. However, this theory is very difficult to reconcile with quantum mechanics and we shall ignore it for the time being. We shall pretend that there *is* an absolute time, as indeed there *could* be if nature had not decided otherwise, and develop the consequences of this intermediate relativity theory for quantum mechanics. When one speaks of nonrelativistic quantum mechanics the "nonrelativistic" is not to be taken too literally. It means that the theory is not relativistic in the more complete sense of Einstein, not that it is not relativistic at all.

In broad terms the hypothesis of Galilean relativity is the hypothesis that no physical experiment can distinguish a reference system from one moving uniformly with respect to it so that the laws of physics—in particular, those of quantum mechanics must be the same in all such reference systems. Let us try to put this hypothesis in more concrete and directly applicable terms. Let O_1 be an observer who has set up a rectangular coordinate system in his space S_1 and let O_2 be another observer who has done likewise in his space S_2. Let the origin s_0^2 in S_2 be moving in the coordinate system of O_1 with a velocity whose components are v_1, v_2, v_3. At $t=0$ let s_0^2 have coordinates in O_1's system equal to x_0, y_0, z_0. Then at any other time these coordinates will be x_0+v_1t, y_0+

ISBN 0-8053-6702-0/0-8053-6703-9, pbk

v_2t, z_0+v_3t. If O_2 has chosen his coordinate system so that its axes are parallel to those of O_1 at $t=0$ they will be so at any later time and if O_1 assigns coordinates x,y,z,t to any "event" O_2 will assign $x+x_0+v_1t, y+y_0+v_2t, z+z_0+v_3t, t$ to this same event. More generally, however, O_2 might have chosen a different time origin so that when his time coordinate is zero, that of O_1 is t_0 and he might choose a new system of axes obtained from the axes parallel to those of O_1 by the orthogonal matrix α. Then if O_1 assigns coordinates x,y,z,t to an event, O_2 will assign coordinates x',y',z',t' to this event, where

$$x',y',z' = \alpha(x+x_0+v_1t, y+y_0+v_2t, z+z_0+v_3t)$$
$$t' = t+t_0$$

Now for each $x_0,y_0,z_0,v_1,v_2,v_3,t_0,\alpha$ and each point x,y,z,t of space–time, let $\gamma_{x_0,y_0,z_0,v_1,v_2,v_3,t_0,\alpha}(x,y,z,t)$ be the space–time point whose O_1 coordinates are the O_2 coordinates of x,y,z,t. Then $\gamma_{x_0,y_0,z_0,v_1,v_2,v_3,t_0,\alpha_0}$ is a one-to-one transformation of space–time into itself and to say that O_1 and O_2 observe the same laws is to say that what O_1 observes is invariant under the transformation $\gamma_{x_0,y_0,z_0,v_1,v_2,v_3,t_0,\alpha}$. In other words, the hypothesis of Galilean relativity may be formulated as the hypothesis that the group G of all one-to-one transformations of space–time of the form

$$x,y,z,t \to \alpha(x+x_0+v_1t, y+y_0+v_2t, z+z_0+v_3t), t+t_0$$

is a group of "symmetries" of space–time and that the laws of physics are invariant under this group in just the way that they are invariant under the Euclidean subgroup $\mathcal{E}$; that is, the group of all transformation $x,y,z,t \to \alpha(x+x_0, y+y_0, z+z_0), t$. The group $\mathcal{G}$ is called the *Galilean group*. It is a ten-dimensional Lie group that is clearly generated by $\mathcal{E}$, the translations in time, and the three-dimensional vector group of all transformations of the form $x,y,z,t \to x+v_1t, y+v_2t, z+v_3t, t$.

Applied to the quantum mechanics of an isolated system, the hypothesis of Galilean relativity demands that the basic projective representation $\alpha \to V_\alpha$ of $\mathcal{E}$ be replaced by a projective representation $\gamma \to W_\gamma$ of the whole Galilean group $\mathcal{G}$ which of course must reduce to V when restricted to $\mathcal{E}$. Now consider the dynamical group $t \to U_t$. Since $U_tV_\alpha = V_\alpha U_t$ for all t and $\alpha, \alpha, t \to V_\alpha U_t$ is in fact a projective representation of the direct product of $\mathcal{E}$ and the subgroup $\mathcal{E} \times T$ of $\mathcal{G}$ consisting of all translations in time. How is this related to the restriction of W to $\mathcal{E} \times T$? It takes only a moments reflection on the meaning of time translation as a space–time symmetry to see that the restriction of W to $\mathcal{E} \times T$ must coincide with the representation $\alpha, t \to V_\alpha U_t$. Thus our basic "givens" U and V are both included in the single projective representation W of $\mathcal{G}$. Our one-particle system is defined by the projection valued measure P on space and the pair U, V. If it is to be Galilean invariant, we must not only obtain U and V by restricting W

ISBN 0-8053-6702-0/0-8053-6703-9, pbk

to T and $\mathcal{E}$, but also put conditions on P that will ensure that spatial observables will obey laws which are the same for different Galilean related observers. While it is possible to find such conditions on P, we shall not bother to do so as it is possible to deduce our restrictions on the ρ_i entirely from the condition that U and V be restrictions of a single projective representation W of $\mathcal{G}$. Moreover, the argument leading to the restrictions on the ρ_i is quite a long one and will be given in outline only.

The idea is quite simple. We note first that $\mathcal{G}$ may be written as a semi-direct product of a six-dimensional commutative group and the direct product of the rotation group and the time translation group. Using this fact, we find the most general projective multiplier for $\mathcal{G}$, and for each choice of this multiplier σ we find the most general irreducible σ representation of $\mathcal{G}$ using the semi-direct product theory developed earlier. Finally, we consider the restriction of these representations to $\mathcal{E} \times T$ and show that for each spin there is just a one-parameter family of W's having the right properties when so restricted. Given an admissible W the corresponding functions ρ_i can be "read off" by restricting W to T.

Here are a few of the details. Let D be the subgroup of all spatial translations. Let L be the subgroup of all space–time transformations of the form $x,y,z,t \to x+v_1t, y+v_2t, z+v_3t, t$. Let R be the group of all rotations about some origin and let T be as above. We verify at once that D and L commute with one another so that $DL=LD$ is isomorphic to $D \times L$ and is a normal abelian subgroup of $\mathcal{G}$. Moreover, $\mathcal{G}$ is a semi-direct product of DL and the subgroup RT that is in turn isomorphic to the direct product of R and T. The action of RT on DL is easily computed. Indeed, that of $\alpha \in R$ is obvious and $t_0 \in T$ takes $x_1, x_2, x_3, v_1, v_2, v_3$ into $x_1 + t_0v_1, x_2 + t_0v_2, x_3 + t_0v_3$.

Using this semi-direct product decomposition of $\mathcal{G}$, it is not difficult to determine the possible multipliers. For each real λ let

$$\sigma_\lambda(x_1,x_2,x_3,v_1,v_2,v_3,t_0,\alpha;\ x_1',x_2',x_3',v_1',v_2',v_3',t_0',\alpha')$$
$$= e^{\frac{1}{2}i\lambda((v_1x_1'+v_2x_2'+v_3x_3')-(v_1'x_1+v_2'x_2+v_3'x_3))}$$

and let

$$\omega'(x_1,x_2,x_3,v_1,v_2,v_3,t_0,\alpha;\ x_1',x_2',x_3',v_1',v_2',v_3',t_0',\alpha')=\omega(\alpha,\alpha')$$

where ω is a nontrivial multiplier for R. Then every multiplier σ for $\mathcal{G}$ that is trivial when restricted to R can be shown to be similar to σ_λ for some uniquely determined real number λ and every other σ can be shown to be similar to $\sigma_\lambda\omega'$ for some uniquely determined real number λ.

The theory of the σ_λ and $\sigma_\lambda\omega'$ representations of G depends sharply on whether $\lambda=0$ or $\lambda\neq0$ and the representations are much easier to describe in the latter case. When $\lambda=0$, there are four separate families of irreduc-

ISBN 0-8053-6702-0/0-8053-6703-9, pbk

ible representations each described by a multi-dimensional parameter. However, a case by case analysis of their restrictions to $\mathcal{E} \times T$ shows that *none* of them arises in our problem. Thus we may confine our attention to the simple case in which $\lambda \neq 0$. For $\lambda \neq 0$ the most general irreducible representation $W^{j,E}$ of $\mathcal{G}$ is determined by an integer j and a real number E and may be described as follows. Its space is the space of square summable functions from E^3 to $H(D^j)$ and

$$W^{j,E}_{x_1,x_2,x_3,0,0,0,e,0}(g)(v_1,v_2,v_3) = e^{i\lambda(x_1v_1+x_2v_2+x_3v_3)}g(v_1,v_2,v_3)$$

$$W^{j,E}_{0,0,0,\bar{v}_1,\bar{v}_2,\bar{v}_3,e,0}(g)(v_1,v_2,v_3) = g(v_1+\bar{v}_1, v_2+\bar{v}_2, v_3+\bar{v}_3)$$

$$W^{j,E}_{0,0,0,0,0,0,\alpha,t_0}(g)(v_1,v_2,v_3) = e^{-iEt_0}e^{-(i\lambda/2)(v_1^2+v_2^2+v_3^2)t_0}D^j_\alpha g(\alpha^{-1}(v_1,v_2,v_3))$$

Straightforward calculations now show the following. In order that $W^{j,E}$ when restricted to $\mathcal{E}$ should be the representation defining a particle of spin l, we must have $j=l$ and there is no restriction on E. No reducible σ_λ representation of G can have a suitable restriction to E. The Hamiltonian obtained by taking the infinitesimal generator of the restriction of $W^{j,E}$ to T is just the operator of multiplication by $-(\lambda\hbar^2/2)(v_1^2+v_2^2+v_3^2)+E$. We compute similarly that the three linear momentum components are defined by the operators of multiplication by $\hbar\lambda v_1$, $\hbar\lambda v_2$, and $\hbar\lambda v_3$, respectively. Thus the Hamiltonian H is

$$-\frac{P_1^2+P_2^2+P_3^2}{2\lambda}+E$$

where P_1, P_2, and P_3 are the momenta. Hence if we return to the realization of our system in which P_1, P_2, and P_3, respectively, are the operators

$$\frac{\hbar}{i}\frac{\partial}{\partial x}, \quad \frac{\hbar}{i}\frac{\partial}{\partial y}, \quad \text{and} \quad \frac{\hbar}{i}\frac{\partial}{\partial z}$$

we find that H becomes

$$\frac{\hbar^2}{2\lambda}\left(\frac{\partial^2}{\partial x^2}+\frac{\partial^2}{\partial y^2}+\frac{\partial^2}{\partial z^2}\right)+E$$

In other words, whenever the representation determining the spin is irreducible, Galilean invariance forces the energy operator to be a constant plus a constant times the Laplacian.

Since adding a constant to H has no physical significance, we see that for each possible integral spin there is just a one-parameter family of possibilities for H and hence for the dynamics. Because of the relationship

ISBN 0-8053-6702-0/0-8053-6703-9, pbk

$E = p^2/2m$ between energy E and total momentum p in classical mechanics, it is natural to call $-\lambda$ the *mass* of our particle. When we do so we see that there is just one possible way of assigning Galilean invariant position, energy, and momentum operators for every positive mass and every integral spin. At first sight it might appear that one has also the possibility of particles of negative mass. However, there is an anti-unitary map that carries H into $-H$ and leaves all other observables fixed, and there is no loss in generality in considering only negative values of λ. There is no difficulty in carrying over the discussion to the case in which σ_λ is replaced by $\sigma_\lambda \omega^1$ and getting also just one possibility of every positive mass and half integral spin as well.

We complete this account of single-particle systems by returning briefly to the axiomatic considerations of Section 17. Recall that our decision to assume that the logic of our system is isomorphic to $\mathcal{L}_{\mathcal{H}}$, where $\mathcal{H}$ is a complex Hilbert space, was somewhat arbitrary—more precisely, it was based upon the practical consideration that it was known to work. Such a priori arguments as we could muster made no distinction between the real, complex, and quaternionic cases. One might well wonder to what extent one of the other hypotheses might also work, and this suggests that we attempt to carry out the developments of this section replacing the hypothesis of a complex Hilbert space by one of the alternatives. To the extent that a similar analysis is possible, we may discover an alternative physics that may or may not be consistent with experiment or we may find a contradiction that rules the hypothesis out once and for all. Whatever happens, we should attain a deeper understanding of the meaning of having complex scalars in our Hilbert space.

Let us assume now that the logic of our system is $\mathcal{L}_{\mathcal{H}}$, where $\mathcal{H}$ is now a *real* Hilbert space. The developments of Section 17 concerning the association of self-adjoint operators to observables and unit vector to states go through essentially without change. One still has a natural one-to-one correspondence between projection valued measures on the line and observables and there is a spectral theorem for self-adjoint operators in a real Hilbert space that is identical in form with that for self-adjoint operators in a complex Hilbert space. Gleason's theorem also holds for real Hilbert spaces and the most general pure state is defined by a unit vector ϕ in such a fashion that the probability distribution of the observable A in the state ϕ is given by the measure $E \rightarrow (P_E^A(\phi), \phi)$, where $E \rightarrow P_E^A$ is the projection valued measure associated with A by the spectral theorem. The dynamics continues to be given by a one-parameter group $t \rightarrow V_t$, where each V_t is one-to-one linear isometric and onto. Moreover, there is an analog of Stone's theorem—easily deduced from that theorem—which says that $V_t = e^{-Kt}$, where K is a "skew adjoint" operator, $K^* = -K$. Here, however, we find a difference and it is an important one. We cannot write $K = iH$, where H is self-adjoint because we have no complex scalars. Hence we

ISBN 0-8053-6702-0/0-8053-6703-9, pbk

cannot associate an observable with the dynamics and we have no obvious analog of the energy. Similarly, we cannot associate observables with one-parameter groups of motions in space and hence have no obvious analog of the linear and angular momentum observables. One might be tempted to conclude at once that a real Hilbert space is out of the question, but this would be quite wrong. First of all every complex Hilbert space becomes a real one if we neglect the complex scalars and conversely every infinite- or finite- and even-dimensional real Hilbert space may be so obtained from a complex one in many different ways. It may well happen that when we put in more structure, by considering actual physical systems, a natural choice for the complex scalars will emerge. Moreover, a priori it could happen that quantum mechanics fails to admit exact analogs of energy and momentum observables. All we need to discuss physical events are position observables and a dynamical group.

With this in mind let us see how far we can go toward adapting the preceding arguments to the case in which $\mathcal{H}$ is a real Hilbert space. Just as before we are led to describe the position observables by a projection valued measure P on physical space S and the spatial symmetry of the system by a projective representation U of the Euclidean group $\mathcal{E}$ by "orthogonal" (=real unitary) operators in H. Moreover, we are led in the same way to the conclusion that P and U must satisfy the commutation relations

$$U_\alpha^{-1} P_E U_\alpha = P_{[E]\alpha}$$

Now there is a complete analog to the imprimitivity theorem that holds for unitary representations in a real Hilbert space and this theorem applied in the present instance tells us that the pair U, P is completely determined by a real unitary representation L of the rotation subgroup R of $\mathcal{E}$. Indeed U, P is equivalent in an obvious sense to the pair U^L, P^L, where U^L and P^L are defined in evident analogy with the real case. This means in particular that our $\mathcal{H}$ may be realized as $\mathcal{L}^2(S, \mathcal{H}(L))$ in such a fashion that $P_E(f)(s) = \phi_E(s) f(s)$, where ϕ_E is the characteristic function of E and $(U_\alpha f)(s) = L_{a^{-1}} f((s)\alpha)$. In analyzing the possiblities for the dynamical group $t \to V_t$, it is convenient to consider a related complex Hilbert space $\mathcal{H}^c$ obtained by taking the direct sum of two replicas of $\mathcal{H} = \mathcal{L}^2(S, \mathcal{H}(L))$ and converting this space into a complex Hilbert space by defining $i(\phi_1, \phi_2) = -\phi_2, \phi_1$. We call $\mathcal{H}^c$ the *complexification* of $\mathcal{H}$. Every real linear operator T in $\mathcal{H}$ becomes a complex linear operator T^c in $\mathcal{H}$ when we define $T^c(\phi_1, \phi_2) = T(\phi_1), T(\phi_2)$ and in this way we may convert the P_E, the U_α and the V_t into operators P_E^c, U_α^c, and V_t^c in $\mathcal{H}^c$. We may clearly identify $\mathcal{H}^c$ with $\mathcal{L}^2(S, \mathcal{H}(L)^c)$ and thus find that P^c, U^c, and V_c^c have just the properties demanded of P, U, and V in the complex case except that our operators must leave the real and imaginary parts of $\mathcal{H}^c$ invariant. [If

ISBN 0-8053-6702-0/0-8053-6703-9, pbk

A is any real linear operator in $\mathcal{H}^c$ that takes the real and imaginary parts of $\mathcal{H}^c$ into themselves, then $A(\phi_1,\phi_2)=A_1(\phi_1),A_2(\phi_2)$, where A_1 and A_2 are real linear operators in $\mathcal{H}$. The A defined by A_1 and A_2 will be complex linear if and only if $-A_1(\phi_2),A_2(\phi_1)=i(A_1(\phi_1),A_2(\phi_2))=-A_2(\phi_2),A_1(\phi_1)$, that is, if and only if $A_1=A_2$. Thus complex linearity and the stated invariance property are sufficient to ensure that A is of the form A_1^c for some real operator A_1 in $\mathcal{H}$.]

The fact that V^c must leave the real and imaginary parts of $\mathcal{H}^c$ invarinat can be used at once to show that $\mathcal{H}(L)$ cannot be one dimensional; that is, our Hilbert space $\mathcal{H}$ cannot be simply $\mathcal{L}^2(S)$ with real-valued functions in analogy to the complex spin zero case. Indeed, our earlier discussion shows in this case that $V_t=e^{-Kt}$, where $K=if(\nabla^2)$ for some real-valued function f. Since V_t must carry real-valued functions into real-valued function, so must K. But this is clearly impossible unless $f\equiv 0$. Thus unless our system is completely static, in the sense that all states remain fixed in time, $\mathcal{H}(L)$ must be at least two dimensional. Now suppose that $\mathcal{H}(L)$ is exactly two dimensional and that L is the two-dimensional identity representation of the rotation group. Then an obvious adaptation of the argument given in the complex spin zero case tells us that K is defined by a 2×2 matrix of functions of the Laplacian, the diagonal elements being real-valued and the other two elements being complex conjugates of one another. Specifically, let θ_1,θ_2 be a basis for $\mathcal{H}(L)$ so that with respect to this basis the elements of $\mathcal{H}$ may be described by pairs of real-valued functions on S and the members of $\mathcal{H}^c$ by pairs of complex-valued functions. Then

$$K(\psi_1,\psi_2)=i[f_{11}(\nabla^2)\psi_1+f_{12}(\nabla^2)\psi_2;f_{21}(\nabla^2)\psi_1+f_{22}(\nabla^2)\psi_2]$$

where f_{11} and f_{22} are real valued and $f_{21}=\overline{f_{12}}$. Clearly $K(\psi_1,\psi_2)$ will be real valued whenever ψ_1 and ψ_2 are if and only if f_{12} and f_{21} are pure imaginary and $f_{11}=f_{22}=0$. But then $f_{12}=ig$ and $f_{21}=-ig$, where g is real valued and $K(\psi_1,\psi_2)=-g(\nabla^2)\psi_2,g(\nabla^2)\psi_1$. In other words, if we convert $\mathcal{H}$ itself into a complex Hilbert space by viewing the two real functions defining an element as the real and imaginary parts of a single complex-valued function, then K becomes $ig(\nabla^2)$. But now the dynamics, and position and momentum observables are exactly as they were in the spin zero case with a complex Hilbert space for state space. However, we cannot say that we are actually in that case. How could we be? The logic of our system is not isomorphic to the partially ordered set of all closed subspaces of a complex Hilbert space, but to that of a real one. But where does this difference reveal itself? It reveals itself in the existence of further observables. Our Hilbert space $\mathcal{H}$ has now been equipped with complex scalars, but it is not only the complex linear self-adjoint operators that correspond to observables. Every real linear self-adjoint operator also corresponds to an observable. In other words, by starting with a real Hilbert space and assuming

ISBN 0-8053-6702-0/0-8053-6703-9, pbk

that L is the two-dimensional identity representation of the rotation group, we arrive at a system that is identical with that in the complex scalar zero spin case except for one thing. There are many more observables than those generated by the position and momentum observables. These only account for the observables defined by the complex linear self-adjoint operator. One has also an observable associated with every real self-adjoint operator. Actually, there are in a sense not so many of these. In fact, there are two real self-adjoint operators J_1 and J_2 that commute with all complex linear self-adjoint operators and have the following further properties:

(1) $J_1^2 = J_2^2 = I$ (the identity)

(2) $iJ_1 = -J_1 i$, $iJ_2 = -J_2 i$, $J_1 J_2 = -J_2 J_1$

(3) Every real self-adjoint operator is uniquely of the form $A_0 + A_1 J_1 + A_2 J_2$, where A_0, A_1, and A_2 are complex linear self-adjoint operators.

In other words, there are only two really new observables. The others are algebraic combinations of these and the old ones. Since the two basic new observables commute with the old ones, they commute in particular with the Hamiltonian and are constant in time.

We may think of the system generated by the complex linear observables and the J_i as that obtained from a spinless single-particle system with a complex Hilbert space as state space and an "independent" quantum system with trivial dynamics and a two-dimensional real Hilbert space of states by simple "composition." We shall not try to discuss the notion of composition of independent systems in the abstract, but content ourselves with the remark that such a notion exists and may be defined in a straightforward way as long as the logics involved are not bizarre. At any rate, we see that there is nothing mysterious about the "extra observables" forced on us by the assumption of a real Hilbert space as logic. Given any quantum system with a complex Hilbert space defining the logic, we may obtain another whose logic is defined by a real Hilbert space by simply composing the given one with a new independent system whose logic is the set of all subspaces of a real two-dimensional Hilbert space. The argument given above tells us that under certain auxilary hypotheses this is the only way in which a single-particle system can have a real Hilbert space as its logic. It seems likely that the same conclusion can be drawn with much weaker conditions on $\mathcal{L}$ and perhaps for all choices of $\mathcal{L}$. However, it would take us too far afield to go into the matter here. The main point we want to make is that the ambiguity about what $\mathcal{L}$ should be can be analyzed and to some extent removed by applying the group theoretical notions of the present section. It would be interesting but difficult to attempt to apply these group theoretical notions to an *abstract* $\mathcal{L}$. Perhaps one could then prove that $\mathcal{L}$ *must* be of the form $\mathcal{L}_{\mathcal{H}}$ or something similar.

ISBN 0-8053-6702-0/0-8053-6703-9, pbk

Notes and References

A part of the axiomatic treatment of the quantum mechanics of a single particle presented in this section was worked out by the author under the stimulus of a vague account of closely related work of Wightman. For further details, including the influence of a paper of E. Wigner and T. D. Newton, see pages 337 and 338 of the published version of the author's Chicago lecture notes.

The fact that in Galilean relativity a change in the mass of the particle corresponds to a change in the cohomology class of a multiplier for the Galilean group was observed and discussed by V. Bargmann in a paper published in 1954 (*Ann. Math.*, vol. 59, 1–46). A determination of the irreducible projective representations of the Galilean group done in the spirit of Wigner's classic 1939 paper on the Poincaré group appears in a paper of J. M. Lévy Leblond (*J. Math. Phys.*, vol 4 (1963), 776–788). Another determination using the theory explained in this book was given two years later by J. Voisin (*J. Math Phys.*, vol. 6 (1965), 1519–1529). An extended discussion of Galilean invariance written by Lévy Leblond will be found in the second volume of *Group Theory and Its Applications*, E. M. Loebl (ed.), Academic, New York (1971).

The second volume of Varadarajan's *The Geometry of Quantum Theory* was published by Van Nostrand in 1970 and like volume 1 was re-issued by Springer-Verlag. It is primarily devoted to a very detailed presentation of one-particle quantum mechanics from the point of view of this section and includes a complete proof of the imprimitivity theorem.

The application of one-particle axiomatics to study the implications of using a real instead of a complex Hilbert space was worked out by the author and appears in these notes for the first time. Much the same conclusions were reached (from a different starting point and using different methods) by Stueckelberg in a paper published in 1960 ("Quantum theory in real Hilbert space," *Helv. Phys. Acta.*, vol. 33, 727–752.) One could presumably study the implications of using a quaternionic Hilbert space in much the same way. It might be interesting to do this and compare the results with those found by G. Emch in a thesis written under the stimulation of the work of Stueckelberg ("Mecanique quantique quaternionienne et relativite resteinte I, II," *Helv. Phys. Acta*, vol. 36 (1963), 739–788).

ISBN 0-8053-6702-0/0-8053-6703-9, pbk

19. Particle interactions

Let the Euclidean group $\mathcal{E}$ act on space S as in Section 18. Let S^n denote the direct product of n replicas of S ($n=1,2,\ldots$) and let $\mathcal{E}^n$ denote the group theoretical product of n replicas of $\mathcal{E}$. Then S^n becomes an $\mathcal{E}^n$ space if we set

$$(s_1,s_2,\ldots,s_n)(\alpha_1\cdots\alpha_n)=s_1\alpha_1,s_2\alpha_2,\ldots,s_n\alpha_n$$

Moreover, this action is clearly transitive and if s_0 is an "origin" for S, then the subgroup leaving $s_0,s_0,\ldots,s_0$ in S^n fixed is just R^n, the direct product of n replicas of R, the subgroup of $\mathcal{E}$ leaving s_0 fixed. The points of S^n correspond one-to-one to the possible "configurations" of a system of n particles in space. For each Borel subset E of S^n, we have the observable that is one if the configuration is described by a point in E and zero if it is described by a point not in E. This observable will be defined by a projection P_E. If we assume that the P_E are mutually compatible and that certain simple limiting relations hold, we are led to the conclusion that $E\to P_E$ is a projection valued measure in S^n just as in the case $n=1$ considered in Section 18. Just as in that section, once we are given P we may associate a self-adjoint operator with every "configuration observable," that is, with every real-valued Borel function g on S^n. The self-adjoint operator associated with g is that whose associated projection valued measure on the line is $E\to P_{g^{-1}(E)}$. In its most general form an *n-particle quantum system* is determined by giving a projection valued measure P on S^n and a unitary representation $t\to V_t$ of the real line that determines how the configurations change in time. Without further hypotheses, P could be any projection valued measure on S^n. We shall

George W. Mackey, Unitary Group Representations in Physics, Probability, and Number Theory

ISBN 0-8053-6702-0/0-8053-6703-9, pbk

only consider those n-particle systems in which the particles are "independently Euclidean invariant" in the sense that each element of $\mathcal{E}^n$ induces a permutation of the observables that is defined by an automorphism of the whole system. In less metaphysical language, we assume, given a projective representation, $\alpha \to U_\alpha$ of $\mathcal{E}^n$ such that $U_\alpha P_E U_\alpha^{-1} = P_{[E]\alpha^{-1}}$, in other words, we make the assumptions of Section 18 with S replaced by S^n and $\mathcal{E}$ by $\mathcal{E}^n$. However, we do *not* assume that the V_t and the U_α commute—only that V_t and U_α commute whenever α belongs to the "diagonal" $\tilde{\mathcal{E}}$ of $\mathcal{E}^n$, that is, whenever $\alpha = \alpha_1, \alpha_2, \ldots, \alpha_n$, where $\alpha_1 = \alpha_2 = \cdots = \alpha_n$.

Applying the imprimitivity theorem just as in Section 18, we conclude the existence of a multiplier ω for $\mathcal{E}^n$ and an ω representation M of R^n such that the pair U, P is equivalent to the pair U^M, P^M. In particular, the essentially distinct pairs U, P correspond one-to-one to the essentially distinct pairs ω, M. Now it follows from the fact that $\mathcal{E}$ has no nontrivial one-dimensional representations and our general theorem about multipliers for direct products that ω is the direct product of $\omega^1, \cdots \omega^n$, where the ω^j are multipliers for $\mathcal{E}$, that is,

$$\omega(\alpha_1, \cdots \alpha_n, \alpha_1', \cdots \alpha_n') = \omega^1(\alpha_1, \alpha_1')\omega^2(\alpha_2, \alpha_2') \cdots \omega^n(\alpha_n, \alpha_n')$$

Moreover, we have already seen that there are to within similarity just two multipliers for $\mathcal{E}$—the trivial one and one other. The other is trivial in the translations and is obtained by "lifting" the essentially unique nontrivial multiplier in R to be a multiplier in $\mathcal{E}$. We now conclude that the most general ω representation M of R^n has the form $L^1 \times L^2 \times \cdots \times L^n$, where L^j is an ω^j representation of R. To see what this implies about U^M and P^M, we invoke a general theorem about the relationship between inducing and the formation of Kronecker products. Let ω^1 and ω^2 be multipliers for the separable locally compact groups G_1 and G_2, respectively. Let N^1 and N^2 be ω^1 and ω^2 representations of the closed subgroups H_1 and H_2 of G_1 and G_2, respectively. Then the $\omega^1 \times \omega^2$ representation $U^{N^1 \times N^2}$ of $G_1 \times G_2$ induced by the $\omega_1^1 \times \omega_2^2$ representation $N^1 \times N^2$ of $H_1 \times H_2$ is equivalent to $U^{N^1} \times U^{N^2}$. It follows from this theorem that the Hilbert space of the n-particle system may be realized in the form $\mathcal{H}^1 \otimes \mathcal{H}^2 \otimes \cdots \otimes \mathcal{H}^n$, where $\mathcal{H}^j$ is the Hilbert space of the single-particle system defined by the ω^j representation L^j of R in such a fashion that

$$U_{\alpha_1 \cdots \alpha_n} = U_{\alpha_1} \times U_{\alpha_2} \times \cdots \times U_{\alpha_n}$$

and

$$P_{E_1 \times E_2 \times \cdots \times E_n} = P_{E_1} \times P_{E_2} \times \cdots \times P_{E_n}$$

ISBN 0-8053-6702-0/0-8053-6703-9, pbk

where the E_j are Borel subsets of S. Of course, no such simple statement can be made about V_t in the general case. However, it can happen that there exist representations $V^1 \cdots V^n$ of the additive group if the real line in $\mathcal{H}^1, \cdots \mathcal{H}^n$, respectively, of such a character that $V_t = V_t^1 \times V_t^2 \times \cdots \times V_t^n$, and when this is the case each V^j is uniquely determined up to multiplication by e^{iat} and each V_t^j commutes with each U_α^j. If so, we shall say that our particles *move independently* or that there is *no interaction* between the particles. Clearly when this happens we may reduce our study of the whole system to the study of the individual particles of which it is composed. Our goal in this section is to study some of the possible dynamical groups V that are *not* products $V^1 \times V^2 \cdots V^n$ so that the particles do influence each other and we have a so-called "interaction" between them.

Before actually embarking on this study, let us put the result of applying the imprimitivity theorem to the pair U, P in another form. Since $\mathfrak{E}$ is a semi-direct product of the translation group D and the rotation group R, $\mathfrak{E}^n$ is a semi-direct product of D^n and R^n. Moreover, using $s_0, s_0, \ldots, s_0$ as a reference point, we may identify D^n with S^n and use the same arguments as in the one-particle case to show that we may realize our Hilbert space as $\mathcal{L}^2(S^n, \mathcal{H}(M))$ in such a fashion that P_E is multiplication by the characteristic function of E and $(U_\alpha f)(s) = M_{\alpha^{-1}} f(s\alpha)$. $\mathfrak{E}^n$ has many more one-parameter subgroups than $\mathfrak{E}$ and each of these defines a unique observable just as in the one-particle case. For example, the one-parameter subgroup

$$x_1, y_1, z_1, \ldots, x_n, y_n, z_n \rightarrow x_1, y_1, z_1, x_2, y_2, z_2, x_{j-1}, y_{j-1}, z_{j-1}, x_j + a, y_j, z_j, \ldots$$

has as infinitesimal generator the operator $(1/i)(\partial/\partial x_j)$ and for obvious reasons we call $\hbar$ times the corresponding observable the linear momentum of the jth particle in the x direction. Similarly, we may assign angular momenta to the various particles, and we verify without difficulty that the linear or angular momentum of the whole system in a given direction (respectively about a given axis) is the sum over the particles of the corresponding linear (angular) momenta of the individual particles.

We shall now restrict the possible interactions by invoking Galilean invariance. As in the one-particle case we assume given a fundamental projective unitary representation W of the Galilean group $\mathcal{G} = (D \times L) ⓢ R \times T$ whose restriction to T is the dynamical group and whose restriction W^0 to $\mathfrak{E} = D ⓢ R$ coincides with the restriction of U to the diagonal $\tilde{\mathfrak{E}}^n$ of $\mathfrak{E}^n$. Of course, we may speak of the "coincidence" of a representation of $\tilde{\mathfrak{E}}^n$ with one of $\mathfrak{E}$ only because there is a natural isomorphism $\alpha \rightarrow \alpha, \alpha \cdots \alpha$ between these two groups. We suppose, in addition, that velocity observables exist for all particles and all directions and that these have the

ISBN 0-8053-6702-0/0-8053-6703-9, pbk

"obvious" invariance properties with respect to the restriction of W to L. Specifically, if X_j is the self-adjoint operator corresponding to the x coordinate of the jth particle and H is the self-adjoint operator such that $V_t = e^{-iHt}$, then the x component of the velocity of the jth particle is defined by a self-adjoint operator X_j' such that $i(HX_j - X_jH) = X_j'$ and if l is the member of L that takes x,y,z,t into $x+v_1t, y+v_2t, z+v_3t, t$, then $W_lX_j'W_l^{-1} = X_j' - v_1I$. This last statement of course just says that if we view the system from the standpoint of an observer moving with uniform velocity relative to the given one, then the x component of the velocity observable of the jth particle will be altered by having the x component of this uniform velocity substracted from it. Naturally we make analogous assumptions about the y and z velocity components.

We supplement our assumptions about the existence and invariance properties of velocity observables with two further assumptions:

(1) If $i \neq j$, then the velocity observables for the ith particle commute with the position observables for the jth particle.

(2) The system consisting of the position observables and the restriction of W to $\mathcal{G}^0 = D \times L \circledS R$ is unitarily equivalent to what is when the particles move independently. In other words, the relationship between position coordinates in different coordinate systems—moving or not has nothing to do with the dynamical operator.

In our analysis of single-particle systems we neglected to finish things off by relating the projection valued measure defining the position coordinates to the extension of U from $D \circledS R$ to $\mathcal{G}$. We shall not give details here but content ourselves with the remark that one can show that the operators defining the x, y, and z coordinates coincide up to a multiplicative constant with the self-adjoint infinitesimal generators V_x, V_y, V_z of the one-parameter unitary groups obtained by restricting W to the corresponding one-parameter subgroups of the three-dimensional vector group L. Moreover, this constant is just $\hbar$ divided by the particle mass. Applying assumption (2) we see that V_x, V_y, and V_z for the system of n particles must be, respectively, the operators of multiplication by the functions

$$\frac{1}{\hbar}(m_1x_1 + \cdots m_nx_n), \quad \frac{1}{\hbar}(m_1y_1 + \cdots + m_ny_n), \quad \text{and} \quad \frac{1}{\hbar}(m_1z_1 + \cdots + m_nz_n)$$

respectively, where the m_j are the particle masses. Thus

$$V_x = (1/\hbar)(m_1X_1 + \cdots + m_nX_n)$$
$$V_y = (1/\hbar)(m_1Y_1 + \cdots + m_nY_n)$$
$$V_z = (1/\hbar)(m_1Z_1 + \cdots + m_nZ_n)$$

ISBN 0-8053-6702-0/0-8053-6703-9, pbk

Now in differential form our invariance condition on the velocity observables is just

$$X_j'V_x - V_xX_j' = Y_j'Y_y - Y_yY_j' = Z_j'V_z - V_zZ_j' = \frac{1}{i}$$

But by hypothesis (1) X_j' commutes with X_k whenever $j \neq k$. Hence in the last relations we may replace V_x, V_y, and V_z by $(m_j/\hbar)X_j$, $(m_j/\hbar)Y_j$, $(m_j/\hbar)Z_j$, respectively, and from this we conclude that

$$X_j'X_j - X_jX_j' = Y_j'Y_j - Y_jY_j' = Z_j'Z_j - Z_jZ_j' = \frac{\hbar}{m_j}\frac{1}{i}$$

for all j. But

$$\frac{1}{i}\frac{\hbar}{m_j}\frac{\partial}{\partial x_j}X_j - X_j\frac{\hbar}{im_j}\frac{\partial}{\partial x_j} = \frac{\hbar}{m_j}\frac{1}{i}$$

and similarly for Y_j and Z_j. Thus

$$X_j' - \frac{\hbar}{im_j}\frac{\partial}{\partial x_j}$$

commutes with all X_j, Y_j, and Z_j and the same is true of

$$Y_j' - \frac{\hbar}{im_j}\frac{\partial}{\partial y_j} \quad \text{and} \quad Z_j' - \frac{\hbar}{im_j}\frac{\partial}{\partial z_j}$$

Hence we arrive at the following important conclusion about the stıucture of the velocity observables. For each $j = 1, 2, \ldots, n$ there exist self-adjoint operators A_j'', B_j'', and C_j'' *that commute with all the position observables* such that

$$X_j' = \frac{\hbar}{im_j}\frac{\partial}{\partial x_j} + A_j'', \quad Y_j' = \frac{\hbar}{im_j}\frac{\partial}{\partial y_j} + B_j'', \quad Z_j' = \frac{\hbar}{im_j}\frac{\partial}{\partial z_j} + C_j''$$

Actually, it turns out to be convenient to write

$$A_j'' = \frac{\hbar}{m_j}A_j', \quad B_j'' = \frac{\hbar}{m_j}B_j', \quad C_j'' = \frac{\hbar}{m_j}C_j'$$

so that

$$X_j' = \frac{\hbar}{m_j}\left(\frac{1}{i}\frac{\partial}{\partial x_j} + A_j'\right), \quad Y_j' = \frac{\hbar}{m_j}\left(\frac{1}{i}\frac{\partial}{\partial y_j} + B_j'\right), \quad Z_j' = \frac{\hbar}{m_j}\left(\frac{1}{i}\frac{\partial}{\partial z_j} + C_j'\right)$$

ISBN 0-8053-6702-0/0-8053-6703-9, pbk

The consequences of this result are most easily explored in the special case in which all of the particles have spin zero so that each L^j is the one-dimensional identity representation. In this case, the X_j, Y_j, and Z_j, as we know, form a complete commuting family. Hence there exist real-valued Borel functions A_j, B_j, and C_j such that A_j' is multiplication by A_j, B_j' is multiplication by B_j, and C_j' is multiplication by C_j. Now it is easy to write down a differential operator H^0 such that

$$i(H^0X_j - X_jH^0) = \frac{\hbar}{m_j}\left(\frac{1}{i}\frac{\partial}{\partial x_j} + A_j'\right)$$

and similarly for Y_j and Z_j. Such an operator is

$$\frac{\hbar}{2}\sum_{j=1}^{n}\frac{1}{m_j}\left[\left(\frac{1}{i}\frac{\partial}{\partial x_j} + A_j'\right)^2 + \left(\frac{1}{i}\frac{\partial}{\partial y_j} + B_j'\right)^2 + \left(\frac{1}{i}\frac{\partial}{\partial z_j} + C_j'\right)^2\right]$$

as a straightforward calculation shows. Then H and H^0 are (formally) self-adjoint operators such that $H^0X_j - X_jH^0 = X_j' = HX_j - X_jH$ for all j and similarly for Y_j and Z_j. Hence $H - H^0$ must commute with all X_j, Y_j, and Z_j. But since the X_j, Y_j, and Z_j form a complete commuting set of operators, there must exist a real-valued function $\mathcal{V}$ on S^n such that $(H - H^0)\psi = \mathcal{V}\psi$. In other words, our hypotheses (including the spinlessness of the particles) have led us to the conclusion that the dynamical operator H must have the form

$$\frac{\hbar}{2}\sum_{j=1}^{n}\frac{1}{m_j}\left[\left(\frac{1}{i}\frac{\partial}{\partial x_j} + A_j'\right)^2 + \frac{1}{i}\left(\frac{\partial}{\partial y_j} + B_j'\right)^2 + \frac{1}{i}\left(\frac{\partial}{\partial z_j} + C_j'\right)^2\right] + \mathcal{V}'$$

where $\mathcal{V}'$ is the operator $\psi \to \mathcal{V}\psi$ and $\mathcal{V}$, the A_j, B_j and C_j are real-valued functions on S^n. Of course these functions cannot be entirely arbitrary. For example, Euclidean invariance tells us that each one is invariant under

$$x_1, y_1, z_1, \ldots, x_n, y_n, z_n \to (x_1 y_1 z_1)\alpha, (x_2, y_2, z_2)\alpha, \ldots, (x_n, y_n, z_n)\alpha$$

where α is any member of $\mathcal{E}$. However, we shall not investigate these restrictions just now. Instead, let us see what an n-particle quantum system looks like in the "classical limit." To this end let us compute the second time derivative of the expected value of the x coordinate of the jth particle, that is,

$$\frac{d^2}{dt^2}\left[(X_j V_t(\psi), V_t(\psi))\right]_{t=0}$$

ISBN 0-8053-6702-0/0-8053-6703-9, pbk

where $V_t = e^{-itH}$ and H are as previously described. Quite generally we have

$$\frac{d}{dt}(AV_t(\psi), V_t(\psi)) = \frac{d}{dt}(Ae^{-iHt}(\psi), e^{-iHt}(\psi))$$
$$= i((HA - AH)e^{-iHt}\psi, e^{-iHt}\psi)$$

so

$$\frac{d^2}{dt^2}\left[(Ae^{-iHt}(\psi), e^{-iHt}(\psi))\right]_{t=0}$$
$$= ([-H(HA-AH)+(HA-AH)H](\psi), \psi)$$

Thus we must compute the double commutator $-H(HA-AH)+(HA-AH)H$ with $A = X_j$, Y_j, and Z_j. Now we already know that

$$HX_j - X_jH = \frac{1}{i}X_j' = -\frac{\hbar}{m_j}\left(\frac{\partial}{\partial x_j} - \frac{1}{i}A_j'\right)$$

with similar expressions for $HY_j - Y_jH$ and $HZ_j - Z_jH$. It remains then to compute the commutators of these expressions with H. Both the computations and the result become simpler if we change notation so that the $3n$ coordinates $x_1, y_1, z_1, \ldots, x_n, y_n, z_n$ become $q_1, q_2, \ldots, q_{3n}$. Correspondingly, let $\mu_1, \mu_2, \mu_3 = m_1$, $\mu_4, \mu_5, \mu_6 = m_2$, etc. and let $A_1', B_1', C_1', \ldots, A_2', \ldots$ be denoted in order by $a_1' \cdots a_{3n}'$, where a_j' is multiplication by a_j. Then

$$H = \frac{\hbar}{2}\sum_{j=1}^{3n}\frac{1}{\mu_j}\left(\frac{1}{i}\frac{\partial}{\partial q_j} + a_j'\right)^2 + \mathcal{V}'$$

and if we denote $X_1, Y_1, Z_1, X_2, \ldots$ by $Q_1, Q_2, \ldots$, our problem is to compute $(HQ_k - Q_kH)H - H(HQ_k - Q_kH)$ knowing that

$$HQ_k - Q_kH = -\frac{i\hbar}{\mu_k}\left(\frac{1}{i}\frac{\partial}{\partial q_k} + a_k'\right)$$

The computation is straightforward and the result is

$$-\frac{\hbar}{\mu_k}\left(\frac{\partial\mathcal{V}}{\partial q_k}\right)' - \frac{\hbar^2}{2\mu_k}\sum_{j=1}^{3n}\frac{1}{\mu_j}\left[\left(\frac{\partial a_j}{\partial q_k} - \frac{\partial a_k}{\partial q_j}\right)'\left(\frac{1}{i}\frac{\partial}{\partial q_j} + a_j'\right)\right.$$
$$\left. + \left(\frac{1}{i}\frac{\partial}{\partial q_j} + a_j'\right)\left(\frac{\partial a_j}{\partial q_k} - \frac{\partial a_k}{\partial q_j}\right)'\right]$$

ISBN 0-8053-6702-0/0-8053-6703-9, pbk

The expected value at ψ of the observable defined by this operator is, by definition, the second derivative of the expected value of Q_k. On the other hand

$$\frac{\hbar}{i\mu_j}\frac{\partial}{\partial q_j}+\frac{\hbar a_j'}{\mu_j}$$

is the velocity operator $\dot{Q}_j$. Thus we have

$$\frac{\mu_k}{\hbar}\frac{d^2}{dt^2}(Q_k(\psi),\psi)=$$

$$-\left(\frac{\partial \mathcal{V}}{\partial q_k}\psi,\psi\right)-\sum_{j=1}^{3n}\tfrac{1}{2}\left\{\left[\left(\frac{a_j}{q_k}-\frac{a_k}{q_j}\right)'\dot{Q}_j+\dot{Q}_j\left(\frac{\partial a_j}{\partial q_k}-\frac{\partial a_k}{\partial q_j}\right)\right]\psi,\psi\right\}$$

Suppose now that we have a state in which the position and velocity coordinates are highly concentrated and hence are almost certain to be near their expected values. It follows from the above equation that these expected values will change with time so that the following system of ordinary differential equation is very approximately satisfied:

$$\frac{\mu_k}{\hbar}\frac{d^2q_k}{dt^2}=-\frac{\partial \mathcal{V}}{\partial q_k}+\hbar\sum_{j=1}^{3n}\left(\frac{\partial a_k}{\partial q_j}-\frac{\partial a_j}{\partial q_k}\right)\frac{dq_j}{dt}$$

Setting $\mathcal{V}^0=\hbar\mathcal{V}$, $a_k^0=\hbar a_k$, the $\hbar$ drops out and our system takes the form

$$\frac{\mu_k d^2 q_j}{dt^2}=-\frac{\partial \mathcal{V}^0}{\partial q_k}+\sum_{j=1}^{3n}\left(\frac{\partial a_k^0}{\partial q_j}-\frac{\partial a_j^0}{\partial q_k}\right)\frac{dq_j}{dt}$$

These are the equations of motions of a classical dynamical system that we call the classical limit of our quantum system. This system of $3n$ second-order differential equations is equivalent to a system of $6n$ first-order differential equations that have a special form known in classical mechanics as Hamiltonian. Indeed let

$$H(q_1\cdots q_{3n},p_1\cdots p_{3n})=\sum_{j=1}^{3n}\tfrac{1}{2}\frac{\left(p_j+a_j^0(q_1\cdots q_{3n})\right)^2}{\mu_j}+\mathcal{V}^0(q_1\cdots q_{3n})$$

$$=\sum_{j=1}^{3n}\frac{p_j^2}{2\mu_j}+\sum_{j=1}^{3n}\frac{p_j a_j^0}{\mu_j}+\mathcal{V}^0+\tfrac{1}{2}\sum_{j=1}^{3n}\frac{\left(a_j^0\right)^2}{\mu_j}$$

ISBN 0-8053-6702-0/0-8053-6703-9, pbk

(where we have omitted the arguments in $\mathcal{V}^0$ and a_j^0) and consider the system of $6n$ first-order differential equations.

$$\frac{dq_k}{dt}=\frac{\partial H}{\partial p_k}$$

$$\frac{dp_k}{dt}=-\frac{\partial H}{\partial q_k}$$

With H as indicated we compute that

$$\frac{\partial H}{\partial p_k}=\frac{p_k}{\mu_k}+\frac{a_k^0}{\mu_k}$$

$$-\frac{\partial H}{\partial q_k}=-\frac{\partial \mathcal{V}^0}{\partial q_k}-\sum_{j=1}^{3n}\frac{p_j}{\mu_j}\frac{\partial q_j^0}{\partial q_k}-\sum_{j=1}^{3n}\frac{a_j^0}{\mu_j}\left(\frac{\partial a_j^0}{\partial q_k}\right)=-\frac{\partial \mathcal{V}^0}{\partial q_k}-\sum_{j=1}^{3n}\frac{\partial a_j^0}{\partial q_k}\left(\frac{p_j+a_j^0}{\mu_j}\right)$$

so that our equations become

$$\frac{dq_k}{dt}=\frac{p_k}{\mu_k}+\frac{a_k^0}{\mu_k}$$

$$\frac{dp_k}{dt}=-\frac{\partial \mathcal{V}^0}{\partial q_k}-\sum_{j=1}^{3n}\frac{\partial a_j^0}{\partial q_k}\left(\frac{p_j+a_j^0}{\mu_j}\right)$$

If we differentiate the first equation and use the second to eliminate dp_k/dt, we get

$$\mu_k\frac{d^2q_k}{dt^2}=-\frac{\partial \mathcal{V}^0}{\partial q_k}-\sum_{j=1}^{3n}\frac{\partial a_j^0}{\partial q_k}\left(\frac{p_j+a_j^0}{\mu_j}\right)+\sum_{j=1}^{3n}\frac{\partial a_k^0}{\partial q_j}\frac{dq_j}{dt}$$

Using the first equation to eliminate p_j we get finally

$$\mu_k\frac{d^2q_k}{dt^2}=-\frac{\partial \mathcal{V}^0}{\partial q_k}+\sum_{j=1}^{3n}\left(\frac{\partial a_k^0}{\partial q_j}-\frac{\partial a_j^0}{\partial q_k}\right)\frac{dq_j}{dt}$$

These are just the equations of motion of our quantum system in the classical limit.

We thus arrive at the following result. If a quantum system of n *spinless* particles satisfies our hypotheses concerning Galilean invariance, etc., then it has a unique classical limit and this classical limit is a classical mechanical system whose equations of motion are in Hamiltonian form, the

ISBN 0-8053-6702-0/0-8053-6703-9, pbk

Hamiltonian function H having the special form

$$\sum_{j=1}^{3n} \frac{p_j^2}{2\mu_j} + \sum_{j=1}^{3n} p_j \frac{a_j^0}{\mu_j} + \mathcal{V}^0 + \tfrac{1}{2} \sum_{j=1}^{3n} \frac{\left(a_j^0\right)^2}{\mu_j} = \sum_{j=1}^{3n} \tfrac{1}{2} \frac{\left(p_j + a_j^0\right)^2}{\mu_j} + \mathcal{V}^0$$

Note that this is the most general possible function of the q's and p's that has the following properties:

(a) For each fixed choice of the q's it is a quadratic polynomial in the p's.

(b) The quadratic polynomial is independent of the q's and is a diagonal quadratic function with positive coefficients.

From the point of view of classical mechanics the equations of motion have a number of apparently arbitrary features: (1) they are *second-order* differential equations; (2) these equations may be thrown into Hamiltonian form; (3) the Hamiltonian function has the special properties listed above. These arbitrary features are quite fully explained by the above analysis together with the hypothesis that classical mechanics is only an approximation to the "more exact truth" that is quantum mechanics. We have shown that the only classical n-particle systems having "quantum refinements" that are Galilean invariant and satisfy certain other simple natural conditions are those with the above indicated "arbitrary" properties.

The function $\mathcal{V}_0$ in the equations of motion has a simple intrinsic meaning. A trivial calculation shows that the function H is constant in time for any Hamiltonian system. This function of the coordinates and momenta (or velocities) is called the total energy. Since

$$\frac{p_j + a_j^0}{\mu_j} = \frac{dq_j}{dt}$$

the total energy expressed in terms of the q_j and their time derivatives is

$$\sum_{j=1}^{3n} \frac{\mu_j}{2} \left(\frac{dq_j}{dt} \right)^2 + \mathcal{V}_0$$

Thus the total energy is just the sum of $\mathcal{V}_0$ and a function of the velocities alone that is zero when all velocities are zero. In particular, $\mathcal{V}_0$ is the function giving the total energy in any configuration when all velocities are zero. It is called the *potential energy*. The function of the velocities alone is called the *kinetic energy*. Since their sum is a constant, we may think of the

ISBN 0-8053-6702-0/0-8053-6703-9, pbk

motion of the system as resulting in a gradual conversion of potential energy into kinetic energy and back again. In the important special case in which the coefficients $\partial a_k/\partial q_j - \partial a_j/\partial q_k$ of the dq_j/dt are all zero so that the accelerations d^2q_k/dt^2 are *independent of the velocities*, then the interaction and the equations of motion are completely determined (at least in the classical case) by the masses μ_j and the potential energy function $\mathcal{V}_0$. Indeed we have the very simple equation

$$\frac{\mu_k d^2 q_k}{dt^2} = -\frac{\partial \mathcal{V}_0}{\partial q_k}$$

Going back to our old notation the right-hand side of the expression for $\mu_k d^2\psi_k/dt^2$ is called the x component of the *force* that acts on the kth particle due to the presence of the other particles. Similarly, one defines the y and z components of the *force* acting upon each particle. Of course the forces define and are defined by giving the equations of motion. When the $\partial a_k^0/\partial q_j - \partial a_j^0/\partial q_k$ are not all zero, one speaks of velocity dependent forces. Otherwise, one says that the forces are "derivable from a potential."

Notice that the $\partial a_k^0/\partial q_j - \partial a_j^0/\partial q_k$ can all be zero even when the a_j^0 are not. For example, one can take $a_j^0 = \partial w/\partial q_j$, where w is any sufficiently differentiable function on S^n. Indeed, since S^n is simply connected, this is the most general possibility. This remark shows that the a_j^0 themselves cannot have direct physical significance—at least in classical mechanics—and that a given system can be described by a wide range of Hamiltonians. Given any system of a_j^0's the Hamiltonian with a_j^0 replaced by $a_j^0 + \partial w/\partial q_j$ will describe exactly the same motion. Since the a_j^0 enter into the quantum mechanical H in a more complicated way, it would seem at first sight that a given classical system might have many quantum refinements—one for each choice of w. However, this is not so. Given w, consider the unitary transformation $U^w : \psi \rightarrow e^{-iw/\hbar}\psi$. This unitary transformation commutes with all the position observables but

$$\left(U^w \frac{\partial}{\partial q_j} U^{w^{-1}}\right)(\psi) = e^{-iw/\hbar} \frac{\partial}{\partial q_j} e^{+iw/\hbar}\psi$$

$$= e^{-iw/\hbar}\left(e^{+iw/\hbar}\frac{\partial \psi}{\partial q_j} + \frac{i}{\hbar}\frac{\partial w}{\partial q_j}\psi e^{+iw/\hbar}\right)$$

$$= \frac{\partial \psi}{\partial q_j} + \frac{i}{\hbar}\frac{\partial w}{\partial q_j}\psi$$

ISBN 0-8053-6702-0/0-8053-6703-9, pbk

Thus transformation by U^w leaves the position observables alone and takes the velocity observable

$$-\frac{\hbar}{\mu_j}\left(\frac{\partial}{\partial q_j}-\frac{1}{i}\frac{a_j^0}{\hbar}\right)$$

into

$$-\frac{\hbar}{\mu_j}\left(\frac{\partial}{\partial q_j}-\frac{1}{i}\frac{1}{\hbar}\frac{\partial w}{\partial q_j}-\frac{1}{i}\frac{a_j^0}{\hbar}\right)=-\frac{\hbar}{\mu_j}\left(\frac{\partial}{\partial q_j}-\frac{1}{i\hbar}\left(a_j^0+\frac{\partial w}{\partial q_j}\right)\right)$$

that is into what it was before with a_j^0 replaced by $a_j^0+\partial w/\partial q_j$. In other words, changing a_j^0 without changing $\partial a_j^0/\partial q_k-\partial a_k^0/\partial q_j$ makes no difference in quantum mechanics either—appearances to the contrary notwithstanding. Actually, there is another ambiguity in the classical Hamiltonian (and except for the obvious additive constant) this is the only other. We may divide the μ_j, $\mathcal{V}^0$, and the a_j^0 by the same positive constant c without changing the equations of motion. On the other hand, this same change in quantum mechanics produces a distinct differential equation that makes different physical predictions. The effect in fact is exactly the same as that of replacing $\hbar$ by $c\hbar$ and $\hbar$ quite clearly does not divide out of the Schrödinger equation. As a matter of fact, $\hbar$ was only introduced in order to make certain notions consistent with corresponding ones in classical mechanics. If we had refrained from doing this, the $\hbar$ would have reappeared at this point as the parameter c whose variation changes the Schrödinger equation without changing its classical limit. We would have argued as follows. Each value we assign to c gives us a different Schrödinger equation, but they all have the same classical limit. Thus, given the classical limit, we cannot know what the corresponding Schrödinger equation is until we have done some experiment that reveals "quantum effects" and allows us to compute c. Actually, only one such experiment need be done. Choosing a particular c amounts to choosing a unit of mass. On the other hand, in a quantum system there is a natural unit of mass—that which makes $c=1$. That this unit is the same for all systems follows from the fact that we may consider *all* particles as part of one big system. One quantum experiment will determine the ratio between the natural mass unit and the arbitrary one we have chosen. Once this ratio is known, the quantum system with a given classical limit may be written down as soon as the classical equations of motion are known and the masses are known in terms of the given arbitrary unit. Any one of the classical experiments determining Planck's constant may be taken as this

ISBN 0-8053-6702-0/0-8053-6703-9, pbk

unique quantum experiment. To write down the quantum operator H for any classical system, we find the masses μ_j with respect to some unit of mass and compute $\hbar$ for this mass unit. Then choose any H for the system using the specified μ_j and write it in the form

$$H=\sum_{j=1}^{3n}\frac{\left(p_j+a_j^0\right)^2}{2\mu_j}+\mathcal{V}_0$$

The quantum H is then just the operator

$$\frac{1}{\hbar}\left[\sum_{j=1}^{3n}\frac{\left(\frac{\hbar}{i}\frac{\partial}{\partial q_j}+a_j^{0'}\right)^2}{2\mu_j}\right]+\frac{\mathcal{V}_0'}{\hbar}$$

that is, it can be obtained from H by replacing each p_j by the operator $(\hbar/i)(\partial/\partial q_j)$, replacing each function of the q's by multiplication by that function, and dividing the result by $\hbar$. We shall call the quantum system that results from this process the *canonical quantization* of the classical system. We remind the reader that throughout this discussion we have been quite cavalier in ignoring questions about domains of operators and analytic conditions on the functions a_j^0 and $\mathcal{V}^0$. Setting up the indicated one-to-one correspondence with these matters taken into account is still an unsolved problem except in rather special cases.

It is interesting to note that the construction of the operator U^w would not have been possible if S and hence S^n were not simply connected. In a nonsimply connected space one can change the a_j^0 in a manner that has quantum mechanical significance but no classical mechanical significance. Thus one cannot write down the Schrödinger equation from the classical Hamiltonian without doing further quantum experiments analogous to determining $\hbar$. In the author's opinion this fact is the real explanation of the so called Bohm–Aharonov effect. Bohm and Aharonov produced an apparent paradox by doing an experiment in which changing the a_j^0 without changing the $\partial a_j^0/\partial q_k-\partial a_k^0/\partial q_j$ made genuine physical differences on the quantum mechanical level. This was considered a paradox because it was "well known" that only the $\partial a_j^0/\partial q_k-\partial a_k/\partial q_j$ had "physical significance." However, in their experiment they rendered space effectively multiply-connected by "removing" part of it. Thus they made it possible for a physical difference to appear at the quantum level even though none existed classically.

ISBN 0-8053-6702-0/0-8053-6703-9, pbk

Even with the restrictions (as yet not taken into account) on $\mathcal{V}_0$ and the a_j^0 produced by the demand that Galilean invariance hold, there is a hopelessly large manifold of possibilities for these functions. Fortunately, in much of physics it seems to suffice (at least at the present time) to confine attention to what are called "two-body forces." In the special case of spinless particles with the a_j^0 all zero, this means that we may write $\mathcal{V}_0$ in the form

$$\mathcal{V}_0(x_1,y_1,z_1,\ldots,x_n,y_n,z_n)=\sum_{\substack{i<j\\ i,j=1}}^{n}\mathcal{V}_0^{i,j}(x_i,y_i,z_i,x_j,y_j,z_j)$$

More generally, we say that the interaction is given by two-body forces if for each i and j with $i<j$ there is a self-adjoint operator $H^{i,j}$ in $\mathcal{H}_i\otimes\mathcal{H}_j$ such that $H=\Sigma_{i<j}H^{i,j}$, where $H^{i,j}=H^{i,j}\otimes I^{i,j}$ and $I^{i,j}$ is the identity on

$$\mathcal{H}_1\otimes\mathcal{H}_2\otimes\cdots\otimes\mathcal{H}_{i-1}\otimes\mathcal{H}_{i+1}\otimes\cdots\otimes\mathcal{H}_{j-1}\otimes\mathcal{H}_{j+1}\otimes\cdots\otimes\mathcal{H}_n$$

As long as only two-body forces are in operation, we know all possible interactions for n-body systems as soon as we know all possible interactions for two-particle systems. Accordingly, it is reasonable to simplify the study of the possible interactions in the higher spin case by considering only two-particle systems.

Before studying two-particle systems in the higher spin case, let us examine the restrictions on $\mathcal{V}_0$ and the a_j^0 in the spin zero two-particle case demanded by invariance considerations. First of all, it follows from translational invariance that

$$\mathcal{V}_0(x_1+a,y_1+b,z_1+c,x_2+a,y_2+b,z_2+c)\equiv\mathcal{V}_0(x_1y_1z_1,x_2y_2z_2)$$

Hence there exists a function $\mathcal{V}_{00}$ of *three* variables such that

$$\mathcal{V}_0(x_1y_1z_1,x_2y_2z_2)=\mathcal{V}_{00}(x_1-x_2,y_1-y_2,z_1-z_2)$$

Moreover, it follows from rotational invariance that for every rotation α about $0,0,0$ we have

$$\mathcal{V}_{00}(\alpha(x_1-x_2,y_1-y_2,z_1-z_2))=\mathcal{V}_{00}(x_1-x_2,y_1-y_2,z_1-z_2)$$

This means that $\mathcal{V}_{00}$ must be a constant on spheres about $0,0,0$ and hence that there must exist a real function ρ of *one* variable defined only for $r>0$

ISBN 0-8053-6702-0/0-8053-6703-9 pbk

such that

$$\mathcal{V}_{00}(x_1 - x_2, y_1 - y_2, z_1 - z_2) = \rho\left(\left[(x_1 - x_2)^2 + (y_1 - y_2)^2 + (z_1 - z_2)^2\right]^{1/2}\right)$$
$$= \mathcal{V}_0(x_1, y_1, z_1, x_2, y_2, z_2)$$

In other words, the velocity-independent part of the interaction between two spinless particles is completely described by giving a single function ρ of a positive real variable. Conversely, it is easy to see that for any choice of ρ and $a_j^0 = 0$ we obtain a Galilean invariant system.

Perhaps the first great triumph of mathematical physics was Newton's discovery that the motions of the planets could be explained by assuming velocity-independent two-body forces with $\rho(r) = Gm_1m_2/r$. Here G is a universal constant and m_1 and m_2 are the masses of the particles. Let us suppose that we have a set of n planets with masses $m_1, m_2, m_3, \ldots, m_n$. Let $\rho_{ij}(r)$ describe the interaction between the ith and jth. Then

$$\rho_{ij}(r) = \frac{Gm_im_j}{r} = \left(\sqrt{\frac{G}{r}}\, m_i\right)\left(\sqrt{\frac{G}{r}}\, m_j\right)$$

Thus it is possible to assign a function $\sigma_i(r) = \sqrt{G/r}\, m_i$ to each planet such that $\rho_{ij}(r) = \sigma_i(r)\sigma_j(r)$. If we replace each $\sigma_i(r)$ by $\sigma_i'(r) = -\sigma_i(r)$, we still have $\rho_{ij}(r) = \sigma_i'(r)\sigma_j'(r)$. Other than this it is easy to show (barring certain degenerate cases and assuming continuity of the σ_i) that the σ_i are uniquely determined by the ρ_{ij} when they exist. In fact, if $\sigma_i(r)\sigma_j(r) = \tau_i(r)\tau_j(r)$, then

$$\frac{\sigma_i(r)}{\tau_i(r)}\,\frac{\sigma_j(r)}{\tau_j(r)} \equiv 1$$

Hence

$$\frac{\sigma_i(r)}{\tau_i(r)} = w(r) = \frac{\sigma_j(r)}{\tau_j(r)}$$

for all i and j. Hence $w^2(r) = 1$. Hence $w(r) \equiv 1$ or $w(r) \equiv -1$. Now for most of the particles that one encounters in physics the functions ρ_{ij} behave for large r like those for planetary motion in the sense that $\lim_{r\to\infty}\rho_{ij}(r)r$ exists. Let us call this limit e_{ij}. In the planetary motion case $e_{ij} = Gm_im_j$. In the general case $e_{ij} = Gm_im_j - e_ie_j$, where the e_i and e_j are uniquely determined, up to a universal sign change, by the argument given above. The number e_j

is called the *electric charge* or simply the *charge* of the particle in question. It is well defined once one has chosen units of mass, length, and time and decided once and for all which charges are to be called positive and which charges are to be called negative. Notice that ρ_{ij} is always positive in planetary motion. This implies that if two planets are at rest, their subsequent motion will bring them closer together—one says that they *attract* one another. On the other hand $-e_ie_j/r$ is negative whenever e_i and e_j have the same sign. Thus the term $-e_ie_j/r$ produces attraction only when e_i and e_j have opposite sign. When e_i and e_j have the same sign (and $|e_ie_i|>Gm_1m_2$) the particles recede from one another when initially at rest and one says that they *repel* one another. We may write

$$\rho_{ij}(r)=\frac{Gm_im_j}{r}-\frac{e_ie_j}{r}+\rho_{ij}^0(r)$$

where $\lim_{r\to\infty} r\rho_{ij}^0(r)=0$. The term Gm_im_j/r is always present (but usually is negligibly small relative to $-e_ie_j/r$ when this term is not zero). The "forces" it produces are called *gravitational* forces. Those produced by the term $-e_ie_j/r$ are called *electrical* forces, *coulomb* forces, or *electrostatic* forces. It turns out that all charges in the universe are integer multiples (positive, negative, or zero) of a certain unique smallest charge and that there exist particles having this charge and its negative. Among these there are particles called *electrons*. Electrons all have the same mass and no other known particle has smaller mass. (The photon and the neutrino are believed to have mass zero but for this very reason are not full-fledged particles in the sense in which we have been using the word.) By convention the arbitrary sign in the definition of charge has been chosen so that electrons have negative charge. The number e^2/Gm^2 is independent of our choice of units and measures the relative strengths of the electrostatic and gravitational interactions between two particles of mass m and charge $\pm e$. When the particle is an electron this number turns out to be 4.16×10^{42}. Gravitational forces are certainly negligible relative to electrostatic ones in this case.

One can also use Galilean invariance to deduce various restrictions on the a_j^0. However, we shall not give details as the interesting velocity dependent two-body forces arise only when the particles have spin.

Next let us suppose that each particle may have a spin and that the spin of the kth particle is described by a projective representation L^k of R. We shall not at this point assume that L^k is irreducible—only that it is finite-dimensional. The Hilbert space of our system will now be $\mathcal{L}^2(S^n,\mathcal{H}(L^1\times L^2\ldots L^n))$ with the linear momentum and positive operators being just as before except for acting on vector valued functions. We may argue just as before about the velocity observables, reaching the conclusion that

$$X_j' = \frac{\hbar}{j}\left(\frac{1}{i}\frac{\partial}{\partial x_j} + A_j'\right)$$

$$Y_j' = \frac{\hbar}{j}\left(\frac{1}{i}\frac{\partial}{\partial y_j} + B_j'\right)$$

$$Z_j' = \frac{\hbar}{j}\left(\frac{1}{i}\frac{\partial}{\partial z_j} + C_j'\right)$$

where A_j', B_j', and C_j' commute with all the position operators. However we may no longer conclude that such operators are necessarily multiplications by real-valued functions on S^n. This conclusion can only be drawn when the position operators form a complete commuting family and this is so only when the $\mathcal{H}(L^j)$ are all one dimensional. Instead, we conclude that for each j there exist functions A_j, B_j, and C_j from S^n to the *self-adjoint operators* in $\mathcal{H}(L^1 \times L^2 \times \cdots \times L^n)$ such that A_j' takes the vector-valued function $s_1 \cdots s_n \rightarrow \psi(s_1 \cdots s_n)$ into

$$s_1 \cdots s_n \rightarrow A_j(s_1 \cdots s_n)(\psi(s_1 \cdots s_n))$$

and similarly for B_j' and C_j'.

Just as in the spinless case it is easy to write down a differential operator H^0 such that

$$i(H^0 X_j - X_j H^0) = \frac{\hbar}{j}\left(\frac{1}{i}\frac{\partial}{\partial x_j} + A_j'\right)$$

and similarly for Y_j and Z_j. Such an operator is

$$\frac{\hbar}{2}\sum_{j=1}^{n}\frac{1}{m_j}\left[\left(\frac{1}{i}\frac{\partial}{\partial x_j} + A_j'\right)^2 + \left(\frac{1}{i}\frac{\partial}{\partial y_j} + B_j'\right)^2 + \left(\frac{1}{i}\frac{\partial}{\partial z_j} + C_j'\right)^2\right]$$

Then H^0 and our dynamical operator H are (formally) self-adjoint operators such that $H^0 X_j - X_j H^0 = X_j' = H X_j - X_j H$ for all j and similarly for Y_j and Z_j. Thus $H - H^0$ must commute with all X_j, Y_j, and Z_j. Just as in the case of the velocity observables it follows from this that there exists a function $\mathcal{V}$ from S^n to the self-adjoint operators in $H(L^1 \times \cdots \times L^n)$ such that $H - H^0 = \mathcal{V}'$ takes ψ into ψ', where

$$\psi'(s_1, s_2, \ldots, s_n) = \mathcal{V}(s_1 \cdots s_n)(\psi(s_1 \cdots s_n))$$

ISBN 0-8053-6702-0/0-8053-6703-9, pbk

Finally, then, we get the same formula for H as in the spinless case, namely,

$$\frac{\hbar}{2}\sum_{j=1}^{n}\frac{1}{m_j}\left[\left(\frac{1}{i}\frac{\partial}{\partial x_j}+A_j'\right)^2+\left(\frac{1}{i}\frac{\partial}{\partial y_j}+B_j'\right)^2+\left(\frac{1}{i}\frac{\partial}{\partial z_j}+C_j'\right)^2\right]+\mathcal{V}'$$

However, now A_j', B_j', C_j', and $\mathcal{V}'$ are defined by self-adjoint operator-valued functions on S^n instead of by real-valued functions on S^n.

Let us look at the two-particle case and see what the restrictions placed on A_j', B_j', C_j', and $\mathcal{V}'$ by translational and rotational invariance look like now that we are permitting spin. Although more general cases are interesting and important, we shall consider here only the case in which the A_j', B_j', and C_j' are all zero so that there are no "velocity dependent forces." Then the formula for H reduces to

$$-\frac{\hbar}{2}\frac{1}{m_1}\left(\frac{\partial^2}{\partial x_1^2}+\frac{\partial^2}{\partial y_1^2}+\frac{\partial^2}{\partial z_1^2}\right)-\frac{\hbar}{2}\frac{1}{m_2}\left(\frac{\partial^2}{\partial x_2^2}+\frac{\partial^2}{\partial y_2^2}+\frac{\partial^2}{\partial z_2^2}\right)+\mathcal{V}'$$

and this is just the H for two noninteracting particles plus an "interaction operator" J, where

$$J\psi(x_1,y_1,z_1,\ x_2,y_2,z_2)=\mathcal{V}'(x_1,y_1,z_1,\ x_2,y_2,z_2)\psi(x_1,y_1,z_1,\ x_2,y_2,z_2)$$

Now a self-adjoint operator J is defined by an operator-valued function $\mathcal{V}'$ as just indicated if and only if it commutes with all projection operators P_E, where P_E is multiplication by the characteristic function of the Borel subset E of $S^2=S\times S$. But to commute with all P_E is the same as to commute with all $P_{E_1\times E_2}=P_{E_1\times S}P_{S\times E_2}=P^1_{E_1}P^2_{E_2}$. Moreover, $P^1_{E_1}$ may be identified with $P^{L_1}_{E_1}$ and $P^2_{E_2}$ with $P^{L_2}_{E_2}$, where P^{L_j} is the canonical system of imprimitivity associated with the induced representation U^{L_j}. In other words, given two one-particle systems with spins defined by the representations L_1 and L_2, the most general interaction operator defined by a function $\mathcal{V}'$ as above is the most general self-adjoint operator J in $\mathcal{H}(U^{L_1}\times U^{L_2})$ that commutes with all $P^{L_1}_{E_1}\times P^{L_2}_{E_2}$ or, equivalently, with all $P^{L_1\times L_2}_{E}$. Such a J will be translationally and rotationally invariant if and only if it also commutes with all $U^{L_1}_x\times U^{L_2}_x$, that is, all $(U^{L_1}\otimes U^{L_2})_x$. Thus to find the most general velocity independent interaction between our two one-particle systems is the same as to find the most general self-adjoint operator in $\mathcal{H}(U^{L^1\times L^2})$ that commutes with all $(P^1\times P^2)_E$ and all $U^{L^1\times L^2}_{x,x}$. Let $\tilde{\mathfrak{G}}$ be the subgroup of $\mathfrak{G}\times\mathfrak{G}$ consiting of all x,y with $x=y$. Then $P^1\times P^2$ is a system of imprimitivity for the representation $U^{L^1\times L^2}$ of $\tilde{\mathfrak{G}}$ based on the nontransitive $\tilde{\mathfrak{G}}$ space $S\times S$.

ISBN 0-8053-6702-0/0-8053-6703-9, pbk

Quite generally, let a group G act nontransitively on a standard Borel space M in such a fashion that the space $\tilde{M}$ of all G orbits is a standard Borel space, let U be a projective representation of G, and let P be a system of imprimitivity for U based on M. For each Borel subset F of $\tilde{M}$, let $\tilde{P}_F = P_{F'}$, where F' is the inverse image of F in M. Then the $\tilde{P}_F$ commute with all the U_x as well as with all of the P_E. Now the projection valued measure $\tilde{P}$ on $\tilde{M}$ defines a direct integral decomposition of $H(U) = \int_{\tilde{M}} H^m \, d\nu(m)$ and because of the commutativities just indicated the system U, P will decompose and for almost all m we will have a unitary projective representation ${}^mU$ in H^m and a system of imprimitivity ${}^mP$ for it based in M. Moreover, it is easy to see that ${}^mP$ is supported by the orbit m so that each ${}^mP$ is reducible to a *transitive* system of imprimitivity. Let J be a self-adjoint operator that commutes with all U_x and all P_E. Then it will commute with all P_F and hence J will also decompose as a direct integral $J = \int J^m \, d\nu(m)$. Clearly J will commute with all P_E and all U_x if and only if, for almost all m, J^m commutes with all ${}^mU_x$ and all ${}^mP_E$. Thus our problem is that of finding the most general self-adjoint operator valued function $m \rightarrow J^m$, where for each m, J^m commutes with all ${}^mU_x$ and all ${}^mP_E$. But for each m, ${}^mP$ is a transitive system of imprimitivity for ${}^mU$. Hence for each m there exists a closed subgroup H_m of G and a unitary projective representation ${}^mA$ of H_m such that the pair ${}^mU$, ${}^mP$ is equivalent to the pair $U^{{}^mA}, P^{{}^mA}$. Moreover, there is an isomorphism of the intersection of the commuting algebra of $U^{{}^mA}$ and $P^{{}^mA}$ with the commuting algebra of ${}^mA$. In other words, each J^m is defined by a self-adjoint operator in the commuting algebra of ${}^mA$ and any member of this latter commuting algebra will do. We may formulate our conclusions now as follows. Choose a Borel cross section B for the G orbits in M. For each $m \in \tilde{M}$ let $\bar{m}$ be the corresponding point in B and let H_m be the subgroup of G leaving m fixed. Let ${}^mA$ be the representation of H_m that induces ${}^mU$ and let ${}^m\mathfrak{g}$ denote the vector space of all self-adjoint operators in the commuting algebra of ${}^mA$. Then the possible operators J correspond one-to-one to the Borel function $m \rightarrow J^m$ where $J^m \in \mathfrak{g}^m$, and we identify two Borel functions if they are equal almost everywhere with respect to the measure class in $\tilde{M}$ defined by $\tilde{P}$.

Let us now apply this result to the case at hand. Here $\tilde{M} = S \times S$ and $G = \tilde{\mathfrak{E}}$. Thus s_1, s_2 and s_1', s_2' are in the same orbit if and only if they are the same distance apart. Hence $\tilde{M}$ may be identified with the space of all nonnegative real numbers. A convenient cross section for the $\tilde{\mathfrak{E}}$ orbits is the set of all pairs s_0, s, where s_0 is fixed and s varies over all points on a fixed ray through s_0. For $s \neq s_0$, the subgroup of $\tilde{\mathfrak{E}}$ leaving s_0, s fixed will be the group of all α, α, where α is a rotation about the fixed ray. Thus in our case H_m is independent of m and is isomorphic to the group R_0 of all rotations about the fixed ray. Finally, ${}^mA$ is the representation of H_m that

ISBN 0-8053-6702-0/0-8053-6703-9, pbk

induces the contribution to $U^{L_1 \times L_2} \uparrow \tilde{\mathfrak{S}}$ of the $\tilde{\mathfrak{S}}$ orbit m and this may be computed to be the restriction to R_0 of $\tilde{L}_1 \otimes \tilde{L}_2$, where $\tilde{L}_1$ and $\tilde{L}_2$ are equivalent to L_1 and L_2, respectively. To within equivalence, then, A^m is also independent of m and may be computed once L_1 and L_2 are known. If L_1 and L_2 are both finite-dimensional, then $L_1 \otimes L_2 \upharpoonright R_0$ is finite-dimensional and has a finite-dimensional commuting algebra. Hence the possible J's may be put into a natural one-to-one correspondence with the Borel functions from the positive real numbers to a certain finite-dimensional real vector space (functions equal almost everywhere being identified as usual). When L_1 and L_2 are both the identity, that is, when our particles are spinless, we see that A^m is the one-dimensional identity and that the most general J is described by a real-valued function of the distance between the particles. This is, of course, just the ρ discussed above.

Now consider the special case in which $L_1 \simeq L_2 \simeq D_{1/2}$. Then $L_1 \otimes L_2 \simeq D_{1/2} \otimes D_{1/2} = D_0 \oplus D_1$ and $D_0 \oplus D_1 \uparrow R_0$ is the direct sum of three independent characters of R_0—one of which is repeated twice. The commuting algebra then is the direct sum of two replicas of the complex numbers and one 2×2 matrix algebra and the space of all self-adjoint elements is six dimensional. In this case, then, the most general possible J is described by six independent functions of the distance between the particles. Other cases are handled similarly with higher spins leading to larger numbers of independent functions of the distance. Reversing one's steps, the possible interactions in terms of these arbitrary functions can be written in a quite explicit manner. We shall not give details, but shall content ourselves with the remark that the matrix valued function $\mathcal{V}'$ always turns out to be a sum of terms of the form:

$$\rho\big((x_1 - x_2)^2 + (y_1 - y_2)^2 + (z_1 - z_2)^2\big)\{P(x_1 - x_2, y_1 - y_2, z_1 - z_2)M\}$$

where M is an operator in the space $H(L^1 \oplus L^2)$ and P is a homogeneous polynomial. When P is of the first degree, one says that the particles interact through "vector forces," and when P is of degree k with $k > 1$, that they interact through "tensor forces of order k."

ISBN 0-8053-6702-0/0-8053-6703-9, pbk

20. Bound states and the structure of matter

Given that particles can interact, the next question that arises is as to what the effects of this interaction are on the changes with time of the particle observables. One important effect is that their mutual interactions may cause particles to "coalesce" and form aggregates of particles that behave in many respects as though they were particles themselves. Such aggregates are called *bound states* of the original particles even though this use of the word "state" is not entirely consistent with our earlier use of the term.

To introduce the notion of bound state let $\mathcal{H}$ be the Hilbert space of an n-particle quantum system and let W be the underlying projective representation of $\mathcal{G} = D \times L Ⓢ R \times T$ with multiplier σ. σ is a product of the multipliers for the several particles and the multiplier for the jth particle depends only on its mass m_j and upon whether its spin representation L^j is a direct sum of D^k with integral or half integral k. Thus σ itself will depend only on $m = m_1, \ldots, m_n$ and upon whether an odd or even number of the particles have half integral spins. Whenever $n \geqslant 2$, the σ representation W cannot be irreducible and we may consider its decomposition into irreducibles. We recall that these are parametrized by two parameters j and E, where E is a real number and j takes on either $0, 1, 2, \ldots$ or $\frac{1}{2}, \frac{3}{2}, \frac{5}{2}, \ldots$ depending on σ. In general, the decomposition of W will be partly discrete and partly continuous. If it is entirely continuous, that is, if W has no irreducible subrepresentation, then the system admits no bound states. We hasten to add that a system which admits no bound states may well have subsystems that do. However, these partial bound states will not reveal themselves as irreducible subrepresentations of the W for the whole system

George W. Mackey, Unitary Group Representations in Physics, Probability, and Number Theory

ISBN 0-8053-6702-0/0-8053-6703-9, pbk

and will not be discussed at this time. Given a system whose W does admit irreducible subrepresentations, let us write

$$\mathcal{H} = \mathcal{H}_{E_1 j_1} \oplus \mathcal{H}_{E_2 j_2} \oplus \mathcal{H}_{E_3 j_3} \oplus \cdots \oplus \mathcal{H}_c$$

where each $\mathcal{H}_{E_r j_r}$ is an irreducible invariant subspace and H_c has no irrreducible invariant subspaces. We note that this decomposition will be unique so long as no pair E_s, j_s is equal to any pair E_t, j_t for $s \neq t$; that is, so long as W restricted to $H \ominus H_c$ is multiplicity free. However, there is no reason to expect such uniqueness to hold in general. Let us now choose any r and look at the subspace $\mathcal{H}_{E_r, j_r}$. Since it is invariant under W, it will in particular be invariant under the dynamical group. Hence (barring influences from particles not in the system), if the state vector for the system is in $H_{E_r j_r}$ at time $t=0$ it will stay there forever. Thus we may look at W restricted to $H_{E_r j_r}$ as defining an independent quantum system. Since W is irreducible, this system may be interpreted as that of a free particle of mass $m = m_1 + m_2 + \cdots + m_n$ and spin j_r. The position observables of this free particle are easily seen to be just those of the "center of gravity" of the system as a whole, that is,

$$\frac{m_1 X_1 + \cdots + m_n X_n}{m_1 + \cdots + m_n}, \qquad \frac{m_1 Y_1 + \cdots + m_n Y_n}{m_1 + \cdots + m_n}, \qquad \frac{m_1 Z_1 + \cdots + m_n Z_n}{m_1 + \cdots + m_n}$$

The free particle that our system appears to be when its state vector is in $H_{E_r j_r}$ is said to be a *bound state* of the particles in the system. Somewhat more insight into the nature of bound states can be obtained by looking at the decomposition of W in a somewhat different way. The subgroup $D \times L$ of $\mathcal{G}$ has a unique irreducible σ representation A and W restricted to $D \times L$ must be simply a multiple of this. Hence we may write $\mathcal{H}(W) = \mathcal{H}(A) \otimes \mathcal{H}_0$, where $W \upharpoonright D \times L$ is just $A \otimes I$ and every operator commuting with all $A_{d,l}$ is of the form $I \otimes B$, where B is an operator in $\mathcal{H}_0$. It follows from our previous work that $\mathcal{H}(A)$ is $\mathcal{L}^2(S)$ in a canonical fashion and that the action of $W \upharpoonright R \times T$ is the tensor product of the free spinless particle action on $\mathcal{H}(A)$ and some other action on $\mathcal{H}_0$. Thus we may regard our system as having been obtained by composing two independent systems—one being the "center of gravity" of the particles moving as a free spinless particle of mass $m = m_1 + \cdots + m_n$. Obviously the other must be interpreted as describing the motion of all the particles relative to the center of gravity as well as any "internal" motions these particles have. This result is a clear analog of the well-known result in classical mechanics according to which the center of mass of a system of mutually interacting particles

ISBN 0-8053-6702-0/0-8053-6703-9, pbk

moves uniformly in a straight line in a manner completely independent of the motions of the particles relative to this center of mass.

Now let $\alpha, t \to e^{-itH_0}B_\alpha = C_{\alpha,t}$ denote the "$\mathcal{H}_0$ component" of $W \restriction R \times T$. Then $H_0 B_\alpha = B_\alpha H_0$ for all α. Moreover, the irreducible subspaces of $\mathcal{H}(W) = \mathcal{L}^2(S) \otimes \mathcal{H}_0$ under W are just the subspaces of the form $\mathcal{L}^2(S) \otimes \mathcal{H}_{00}$, where $\mathcal{H}_{00}$ is a subspace of $\mathcal{H}_0$ invariant and irreducible under C. Let $\mathcal{H}_{00}$ be such a subspace. Then in $\mathcal{H}_{00}$ we will have $\alpha \to B_\alpha \simeq D^j$ for some j and $e^{-itH_0} \equiv e^{-itE_0}$ for some real number E_0. Clearly the corresponding subrepresentation of W will be that with parameters j and E_0. Now each vector ϕ in $\mathcal{H}_0$ defines a state for the motion of the whole system relative to its center of mass. This state changes in time according to the formula $t \to e^{-tH_0}\phi$. But if $\phi \in H_{00}$, $H_0(\phi) = E_0\phi$, so $e^{-itH_0}(\phi) = e^{-itE_0}\phi$, and since ϕ and $e^{i\theta}\phi$ define the same physical state we see that the state defined by ϕ is *constant in time*. Hence every "internal observable" (i.e., every observable defined by an operator of the form $I \times F$, where F is a self-adjoint operator in $\mathcal{H}_0$) will be a constant in time for every state vector in $\mathcal{L}^2(S) \times \mathcal{H}_{00}$. In other words, in a bound state the only thing that happens in the course of time is a change in the center of gravity of the system. The internal variables are all "frozen." Of course, they may be frozen in many different ways and it must be remembered that they are not frozen at definite values—they just have probability distributions that do not change with time. The different ways in which they may be frozen correspond to the different bound states that we get by taking different irreducible subrepresentations of W. If we change E_k without changing j_k, we do not change the nature of the free particle that our system simulates since the effect is just to change E_k by a constant. However, there is a real physical significance to the differences $E_k - E_k'$. When our system interacts with another, transitions from one bound state to another become possible and the energy difference $E_k - E_k^1$ must be made up by a corresponding energy change elsewhere. When there is a unique bound state of lowest E_j, one calls it the *ground state* of the system of particles. When we have studied "electromagnetic radiation" from the point of view of quantum mechanics, we shall see that it is possible for systems bound together by electrical forces to change from one bound state to another of lower energy by emitting the energy difference $E - E'$ as "electromagnetic radiation" of "frequency" $(E - E')/2\pi\hbar$. These frequencies can be measured with great accuracy and one can determine the energy levels E_j experimentally. However, when one compares the experimental values with the theoretical ones for systems containing several electrons, one finds that the theory predicts far too many energy levels. Moreover, there is a simple rule for telling which energy levels do and do not occur. To state this rule in its

ISBN 0-8053-6702-0/0-8053-6703-9, pbk

general form, let the n particles in our n-particle system be such that certain pairs are identical in the sense of having equal masses and spins and equal interactions with all other known particles. Then particles identical to the same particle will be identical to one another and we may group the particles into k subsets $S_1,\cdots,S_k$ having $m_1,m_2,\ldots,m_k$ particles respectively in such a manner that all particles in any subset are identical to one another. Let $n_j = m_1 + m_2 + \cdots m_j$. Now the Hilbert space of our whole system is a tensor product $\mathcal{H}_1 \otimes \mathcal{H}_2 \cdots \mathcal{H}_{n_1} \otimes \mathcal{H}_{n_1+1} \cdots \mathcal{H}_n$ of Hilbert spaces for the separate particles and we have natural maps of each $\mathcal{H}_i$ onto any $\mathcal{H}_j$ such that i and j are in the same S_t. Let $\mathbb{S}_j$ be the subgroup of all permutations π of the integers $1,2,\ldots,n$ such that $\pi(l)=l$ whenever $l \notin S_j$. Then the group generated by $\mathbb{S}_1 \cdots \mathbb{S}_k$ is isomorphic to $\mathbb{S}_1 \times \mathbb{S}_2 \times \cdots \times \mathbb{S}_k$. Now each $\pi \in \mathbb{S}_j$ defines a unitary operator in $\mathcal{H}_{n_j+1} \otimes \mathcal{H}_{n_j+2} \otimes \cdots \otimes \mathcal{H}_{n_{j+1}-1}$, namely, the unique unitary operator that maps $\phi_{n_j+1} \otimes \phi_{n_j+2} \otimes \cdots \otimes \phi_{n_{j+1}-1}$ into $\phi_{\pi(n_j+1)} \otimes \phi_{\pi(n_j+2)} \otimes \cdots \otimes \phi_{\pi(n_{j+1}-1)}$. Thus each $\pi \in \mathbb{S}_1 \times \mathbb{S}_2 \times \cdots \times \mathbb{S}_k$ defines a unitary operator N_π in $\mathcal{H}(W) = \mathcal{H}_1 \times \mathcal{H}_2 \times \cdots \times \mathcal{H}_n$ and $\pi \to N_\pi$ is a representation of $\mathbb{S}_1 \times \cdots \times \mathbb{S}_k$. Because of the identity of the particles it follows that $N_\pi W_\beta = W_\beta N_\pi$ for all $\pi \in \mathbb{S}_1 \times \cdots \times \mathbb{S}_k$ and all $\beta \in \mathcal{G} = D \times L \circledS R \times T$. Now consider the unique decomposition of N into primary representations. This will decompose $\mathcal{H}(W)$ as a direct sum of subspaces each associated with an irreducible representation of $\mathbb{S}_1 \times \cdots \times \mathbb{S}_n$, and since N_π and W_β commute each of these subspaces will be invariant under W as well. It follows then that every irreducible subrepresentation of W must be contained in the subspace associated with some irreducible representation of $\mathbb{S}_1 \times \cdots \times \mathbb{S}_k$ and hence that every bound state of the system must be invariantly associated with an irreducible representation of $\mathbb{S}_1 \times \cdots \times \mathbb{S}_k$. Now each $\mathbb{S}_j$ has exactly two one-dimensional representations—the identity and the representation that is one on even permutations and minus one on odd ones. Let χ_k be the identity if the particles in the jth subset have integral spin and the other one-dimensional representation if they have half integral spin. Let $\chi = \chi_1 \times \chi_2 \times \cdots \times \chi_k$. Then χ will be a one-dimensional unitary representation of $\mathbb{S}_1 \times \mathbb{S}_2 \times \cdots \times \mathbb{S}_k$. The rule telling which energy levels do and do not occur may now be stated as follows. Those and only those levels occur for which the corresponding irreducible subrepresentation of W is associated with χ. The question now arises as to how we can make our theory consistent with the omissions predicted by this rule. A moments reflection shows that one way of doing this is simply to replace our Hilbert space by the subspace $\mathcal{H}^\chi$ on which the representation N of $\mathbb{S}_1 \times \cdots \times \mathbb{S}_k$ reduces to a multiple of χ. This subspace is invariant under W and hence admits a well-defined dynamical group. We

ISBN 0-8053-6702-0/0-8053-6703-9, pbk

may simply suppose that the states in $\mathcal{H}^{\chi}$ are the only ones that occur in nature. Because of the time invariance of $\mathcal{H}^{\chi}$, this supposition is not violated by the passage of time and it automatically places the desired limitation on the energy levels for the bound states that occur. However, there are still difficulties. $\mathcal{H}^{\chi}$ is not invariant under all self-adjoint operators in $\mathcal{H}(W)$. Hence there must be missing observables as well as missing states. We must inquire into the nature of these missing observables and satisfy ourselves that their absence can be explained in a fashion consistent with our basic assumptions—or at least with intelligible modifications of these basic assumptions. In doing this it will suffice to study the special case in which all of the particles are identical. Indeed it is clear that $\mathcal{H}^{\chi}=\mathcal{H}^{\chi_1}\otimes\mathcal{H}^{\chi_2}\otimes\cdots\otimes\mathcal{H}^{\chi_n}$ so that we have only violated our rule about tensor products *within* each group of identical particles. To find the Hilbert space of states for a system consisting of several groups of particles that are only identical among themselves, we form the tensor product of the Hilbert spaces for the separate groups, just as usual.

Suppose then that all of our particles are identical and for definiteness let us take the case in which the spins are described by D^{j_s} with half integral j. Then our rule takes the simple form. The Hilbert space of our system is not the n-fold tensor product $\mathcal{H}_1\otimes\mathcal{H}_1\otimes\cdots\otimes\mathcal{H}_1$, where $\mathcal{H}_1$ is the Hilbert space for a single particle. Instead, it is the subspace of all tensors that are antisymmetric, that is, that go into their negatives under any odd permutation. To see in more concrete form what this amounts to let $\phi_1,\phi_2,\ldots$ be any orthonormal basis for $\mathcal{H}_1$. Then we obtain an orthonormal basis for $\mathcal{H}_1\otimes\mathcal{H}_1\otimes\cdots\otimes\mathcal{H}_1$ by taking all tensor products $\phi_{i_1}\otimes\phi_{i_2}\otimes\cdots\otimes\phi_{i_n}$ and the effect of a permutation π is to take $\phi_{i_1}\otimes\cdots\otimes\phi_{i_n}$ into $\phi_{\pi(i_1)}\otimes\cdots\otimes\phi_{\pi(i_n)}$. If $n=2$, then for all i and j with $i\neq j$

$$\frac{\phi_i\otimes\phi_j-\phi_j\otimes\phi_i}{\sqrt{2}}$$

is carried into its negative by the unique odd permutation. Moreover, it is easy to show that the elements of the form

$$\frac{\phi_i\otimes\phi_j-\phi_j\otimes\phi_i}{\sqrt{2}}$$

form an orthonormal basis for the space of all antisymmetric members of $\mathcal{H}_1\otimes\mathcal{H}_1$. If $n=3$, we obtain a basis by taking all elements of the form

$$\frac{i}{\sqrt{6}}\left[\phi_i\otimes\phi_j\otimes\phi_k-\phi_j\otimes\phi_i\otimes\phi_k+\phi_j\otimes\phi_k\otimes\phi_i\right.$$
$$\left.-\phi_k\otimes\phi_j\otimes\phi_i-\phi_i\otimes\phi_k\otimes\phi_j+\phi_k\otimes\phi_i\otimes\phi_j\right]$$

ISBN 0-8053-6702-0/0-8053-6703-9, pbk

where i, j, and k are different. The generalization is obvious. The fact that we have no elements in the antisymmetric product in which two particles have states defined by the same one-particle state vector is called the *Pauli exclusion principle.* It applies to all particles with half integral spin and has far reaching consequences for the structure of matter.

Now let us look at the effect of the antisymmetric reduction on the position observables. Recall that $\mathcal{H}_1 = \mathcal{L}^2(S, \mathcal{H}(L))$, where L defines the spin, and that $\mathcal{H}_1 \times \mathcal{H}_1 \times \cdots \times \mathcal{H}_1$ may be realized as $\mathcal{L}^2(S^n, \mathcal{H}(L) \otimes \mathcal{H}(L) \otimes \cdots \otimes \mathcal{H}(L))$ in such a fashion that the position observables are multiplication by functions f on S^n. It is not hard to see that such a multiplication will leave the subspace of antisymmetric tensors invariant if and only if it is symmetric in the sense that

$$f(s_{\pi(1)}, s_{\pi(2)}, \ldots, s_{\pi(n)}) = f(s_1, s_2, \ldots, s_n)$$

for all permutations π. This means that $x_1, x_2, \ldots, x_n$ individually do not leave the relevant subspace invariant but that $x_1 + x_2 + \cdots + x_n$, $x_1^2 + x_2^2 + \cdots + x_n^2$, $x_1^3 + \cdots + x_n^3$, and so forth all do. In other words, while we cannot talk about the x coordinate of the jth particle, we can talk about the sum of these x coordinates of all the particles, the sum of their squares, the sum of their cubes, and any other symmetric functions of these n x coordinates. What can this mean? To see the answer recall the fact from the theory of equations that the coefficients of a polynomial equation are symmetric functions of its roots. For example, if x_1 and x_2 are the two roots of $x^2 + ax + b = 0$, then $b = x_1 x_2$, $a = -(x_1 + x_2)$. Thus when one knows all possible symmetric functions of n numbers, one can find an equation whose roots are the numbers and find the numbers by solving the equation. However, one cannot tell which number is which. If $x_1 + x_2 = 5$, $x_1 x_2 = 6$, one knows either that $x_1 = 2$ and $x_2 = 3$ or that $x_1 = 3$ and $x_2 = 2$. However, there is no way of telling which.

In our system of n identical particles—after the antisymmetric reduction we still have enough observables to talk about the x, y, and z coordinates and hence the positions of all of the particles provided that we do not try to assign a particular position to a particular particle. We must regard interchanging two identical particles as an operation devoid of physical meaning. An analogy may help to make this statement less mysterious. Imagine a vibrating string stretched between two supports as indicated in diagram (a):

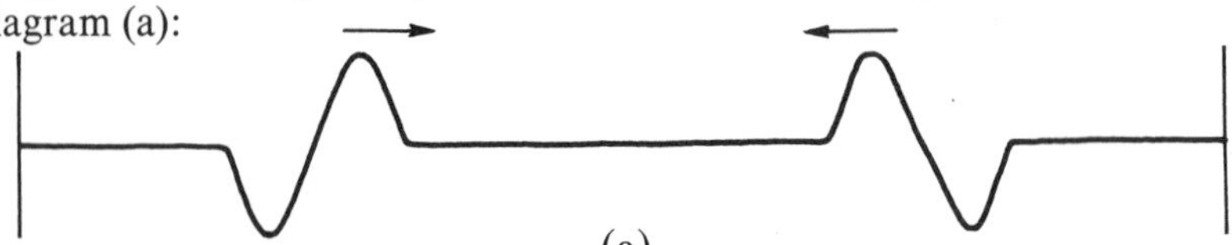

(a)

ISBN 0-8053-6702-0/0-8053-6703-9, pbk

where arrows indicate the directions of motion of the two disturbances. After a time the disturbances will pass one another and the string will be as indicated in diagram (b):

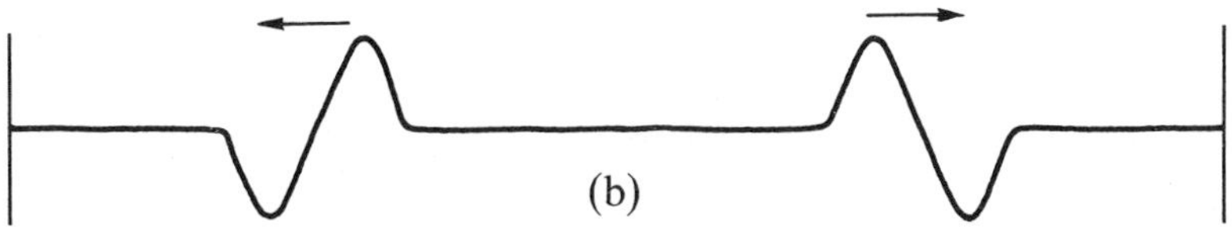
(b)

Later still they will be reflected at the supports, bounce back, and the string will be as in diagram (c):

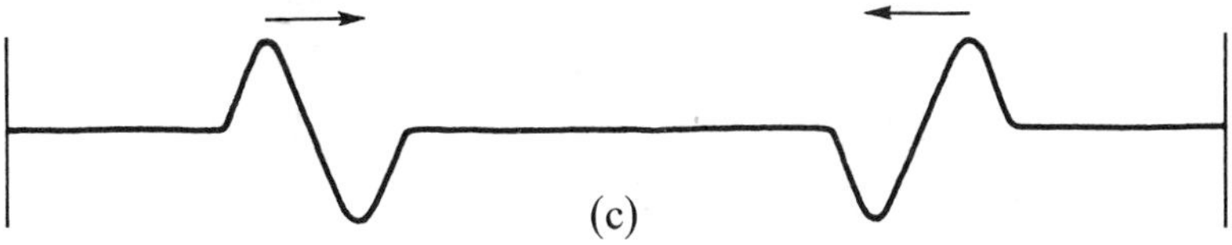
(c)

If we have concentrated our attention on the two disturbances regarded as individual entities, they seem to have changed places between the time of (a) and the time of (c). However, the system itself is in exactly the same state in diagram (c) as it was in diagram (a). It appears that quantum mechanical particles are even more wave like and less billiard ball like than we had thought. In any case, it turns out that this point of view can be consistently maintained and that one always can and must reduce the Hilbert space as indicated whenever identical particles are concerned. The argument is essentially the same for particles with integral spin. Again only symmetric position observables make sense and this is interpreted in the same way. On the other hand, the symmetrized tensor powers are somewhat larger than the antisymmetrized ones and one has no Pauli exclusion principle for particles with integral spin. The necessity of taking symmetric and antisymmetric subspaces has especially profound effects on quantum statistical mechanics. The modifications in the classical Maxwell Boltzman theory produced by the symmetric reduction were first worked out by Bose and Einstein, and one speaks of *Bose–Einstein statistics*. In the antisymmetric case, the names to be mentioned are those of Fermi and Dirac and one speaks of *Fermi–Dirac statistics*. The fact that the statistics are determined by the spin in the manner indicated is referred to as "the connection between spin and statistics." It has become customary to refer to particles requiring the antisymmetric reduction as *fermions* and to those requiring the symmetric one as *bosons*.

The fact that particle identity demands a reduction of the underlying Hilbert space is of course in contradiction to one of our basic assumptions —namely, that position observables exist for all the particles individually. A strict adherence to our deductive program would demand that we

ISBN 0-8053-6702-0/0-8053-6703-9, pbk

retrace our steps and try to carry out our analysis with configuration space S^n replaced by $\tilde{S}^n$, where $\tilde{S}^n$ is the space that S^n becomes when we identify the point $s_1, s_2, \ldots, s_n$ with $s_{\pi(1)} \cdots s_{\pi(n)}$ for all permutations n. One might well find other interaction possibilities in this way. However, so far as the author knows, such an analysis has never been seriously attempted and we shall not have time to take a stab at it here.

Bound states do not only manifest themselves as free particles. Suppose that we have two systems of particles M_1 and M_2 with M_1 having n_1 particles and M_2 having n_2 particles. Suppose that each system has bound states and that the particles in M_1 interact with the particles in M_2. Then in general the bound states themselves will appear to interact and the interaction may be computed from the interactions between the particles in M_1 and those in M_2 in a manner which may be outlined as follows. Let $\mathcal{H}^1$ and $\mathcal{H}^2$ be the Hilbert spaces of the two systems M_1 and M_2 and let

$$\mathcal{H}^1 = \mathcal{H}^1_1 \oplus \mathcal{H}^1_2 \oplus \cdots \oplus \mathcal{H}^1_c$$
$$\mathcal{H}^2 = \mathcal{H}^2_1 \oplus \mathcal{H}^2_2 \oplus \cdots \oplus \mathcal{H}^2_c$$

denote the decompositions of $\mathcal{H}^1$ and $\mathcal{H}^2$ that exhibit the bound states of the two systems. Then

$$\mathcal{H}^1 \otimes \mathcal{H}^2 = \sum_{i,j} \mathcal{H}^1_i \otimes \mathcal{H}^2_j \oplus \sum_j \mathcal{H}^1_c \otimes \mathcal{H}^2_j \oplus \sum_i \mathcal{H}^1_i \otimes \mathcal{H}^2_c \oplus \mathcal{H}^1_c \otimes \mathcal{H}^2_c$$

and the subspace $\mathcal{H}^1_i \otimes \mathcal{H}^2_j$ contains all states in which the first system is in the ith bound state and the second in the jth bound state. Let us for the moment make the simplifying but unrealistic assumption that no particle in M_1 is identical with any particle in M_2. Then $\mathcal{H}^1 \otimes \mathcal{H}^2$ will actually be the Hilbert space for the system $M_1 \cup M_2$. If no particle in M_1 interacts with any particle in M_2, then the dynamical group for the system will be the tensor product of the separate dynamical groups and the dynamical operator H will carry each $\mathcal{H}^i_1 \otimes \mathcal{H}^j_2$ into itself and define these two noninteracting free particles. In general, however, H will not be so simple. It will be described by a matrix of operators H_{i_1,j_1,i_2,j_2} (where any index may take the value c) and H_{i_1,j_1,i_2,j_2} is an operator from $\mathcal{H}^1_{i_1} \otimes \mathcal{H}^2_{j_1}$ to $\mathcal{H}^1_{i_2} \times \mathcal{H}^2_{j_2}$ such that $H^*_{i_1,j_1,i_2,j_2} = H_{i_2,j_2,i_1,j_1}$. The diagonal operator H_{i_1,j_1,i_1,j_1} will be self-adjoint and will describe the interaction between the i_1st bound state of the first system and the j_1st bound state of the second system—at least in states of the total system in which the operators connecting $\mathcal{H}^1_{i_1} \otimes \mathcal{H}^2_{j_1}$ with other subspaces have small values. On the other hand, as long as these operators are not zero, there will be a finite probability that

ISBN 0-8053-6702-0/0-8053-6703-9, pbk

one of our particles will make a transition into another bound state or even decompose. However, when the energy difference between the bound states we are interested in and all other bound states or states of decomposition is sufficiently large, our two bound states will behave as stable particles interacting via the dynamical operator H_{i_1,j_1,i_2,j_2}. When we have several bound states of nearly equal energy, we may lump together the corresponding Hilbert spaces $\mathcal{H}_{i_1}, \mathcal{H}_{i_2}, \mathcal{H}_{i_3}, \ldots \mathcal{H}_{i_k}$ into a direct sum $\mathcal{H}_{i_1} \oplus \mathcal{H}_{i_2} \oplus \cdots \oplus \mathcal{H}_{i_k}$ and regard our system as representing a stable particle that can exist in several forms.

Of course, things are really more complicated because in many of the most interesting cases M_1 does have particles that are identical with particles in M_2—both systems might contain electrons for example. When this happens

$$\mathcal{H}^1 \otimes \mathcal{H}^2 = \sum_{i,j} \mathcal{H}_i^1 \otimes \mathcal{H}_j^2 \oplus \sum_i \mathcal{H}_c^1 \otimes \mathcal{H}_i^2 \oplus \sum_i \mathcal{H}_i^1 \otimes \mathcal{H}_c^2 \oplus \mathcal{H}_c^2 \otimes \mathcal{H}_c^2$$

will not be the Hilbert space of our system. The Hilbert space $\mathcal{H}^\Delta$ of our system will be a certain subspace of $\mathcal{H}^1 \otimes \mathcal{H}^2$ that will not even contain $\mathcal{H}_i^1 \otimes \mathcal{H}_j^2$ and cannot be described in terms of $\mathcal{H}^1$ and $\mathcal{H}^2$ as above. How then shall we study the interaction between the "particles" defined by the bound states $\mathcal{H}_i^1$ and $\mathcal{H}_j^2$? First let us be explicit about the subspace $\mathcal{H}^\Delta$ of $\mathcal{H}^1 \otimes \mathcal{H}^2$ that is involved. $\mathcal{H}^1$ itself is a subspace of a larger space $\tilde{\mathcal{H}}^1$, where $\tilde{\mathcal{H}}^1$ is a tensor product of n_1 one-particle spaces and $\mathcal{H}^1$ is defined by a certain one-dimensional representation χ^1 of a certain subgroup G_1 of the permutation group on n_1 objects. Similarly, $\mathcal{H}^2$ is the subspace of $\tilde{\mathcal{H}}^2$ defined by a certain one-dimensional representation χ^2 by a certain subgroup G_2 of the permutation group on n_2 objects. Thus $\mathcal{H}^1 \otimes \mathcal{H}^2$ will be the subspace of $\tilde{\mathcal{H}}^1 \otimes \tilde{\mathcal{H}}^2$ defined by the representation $\chi^1\chi^2$ of the subgroup $G_1 \times G_2$ of the group $\mathcal{S}(n_1+n_2)$ of all permutations on n_1+n_2 objects. Let G_3 be the subgroup of *all* permutations in $S(n_1+n_2)$ that takes each particle in M_1 and M_2 into an identical particle. Then $G_1 \times G_2 \subseteq G_3$, and there exists a certain one-diemnsional representation χ^3 of G_3 that defines the subspace $\mathcal{H}^\Delta$ just as $\mathcal{H}^1$ and $\mathcal{H}^2$ are defined by χ^1 and χ^2. Moreover, χ^3 will agree on $G_1 \times G_2$ with $\chi^1\chi^2$. $\mathcal{H}^\Delta$ thus comes to us as a subspace of $\mathcal{H}^1 \otimes \mathcal{H}^2$, but since $\mathcal{H}_1 \otimes \mathcal{H}_2$ is defined by the restriction of χ^3 to a subgroup, it follows that $\mathcal{H}^\Delta \subseteq \mathcal{H}_1 \otimes \mathcal{H}_2$. Now let P^Δ be the projection of $\mathcal{H}_1 \otimes \mathcal{H}_2$ on $\mathcal{H}^\Delta$. Then P^Δ will map $\mathcal{H}_i^1 \otimes \mathcal{H}_j^2$ on a certain subspace of $\mathcal{H}^\Delta$ that we shall denote by $\mathcal{H}_{i,j}^\Delta$. The operator P^Δ may be factored uniquely into the product of a self-adjoint operator A in $\mathcal{H}_i^1 \times \mathcal{H}_j^2$ and a unitary operator V from the closure of the range of A to $\mathcal{H}_{ij}^\Delta$. Let H be the

ISBN 0-8053-6702-0/0-8053-6703-9, pbk

dynamical operator for the whole system of $n_1 + n_2$ particles and let $H_{ij,ij} = P_{ij}H$ restricted to H_{ij}^{Δ}, where P_{ij} is the projection on $\mathcal{H}_{ij}^{\Delta}$. Then $\mathcal{H}_{ij}^{\Delta}$ is the Hilbert space for that part of the total system in which the particles in M_1 and M_2 are bound into two composite particles as indicated. Moreover, $H_{ij,ij}$ is the dynamical operator for its motion in so far as this takes place with preservation of the bound states in question. To interpret vectors in $\mathcal{H}_{ij}^{\Delta}$ as describing composite particles, we must use V to transform them into vectors in $\mathcal{H}_i^1 \otimes \mathcal{H}_j^2$. This interpretation will be a little blurred when A is not constant, but this difficulty only becomes serious when the composite particles are relatively close together. The subject is actually a rather complicated one and we cannot go into more detail here. Our chief aim at this point is to explain how bound states of particles may act like particles themselves; even to the extent of having well-defined interactions—provided that we do not try to maintain this interpretation when the particles get too close together. Clearly the dynamical operator that we get from the above construction is rather different from what we would have if we did not have to deal with the symmetric and antisymmetric reductions. The "forces" resulting from the difference between the two dynamical operators are called "exchange forces" and have great importance in the quantum mechanical explanation of the formation of molecules.

At the beginning of this chapter we mentioned the fact that it had long been a goal of physics to "explain" all physical phenomena in terms of the motions of certain ultimate or fundamental particles. The discovery of quantum mechanics has brought this goal within sight—indeed, with certain qualifications to be explained below, it has actually been achieved as far as finding the basic physical laws is concerned. In the often quoted words of Dirac: "The underlying physical laws necessary for the mathematical theory of a large part of physics and the whole of chemistry are thus completely known and the difficulty is only that the exact application of these laws leads to equations much too complicated to be soluble."

Let me describe these laws. The basic idea is to regard all matter as made up of fundamental particles bound together through the bound-state mechanism. The program then is to find all fundamental particles and the interaction law for each pair of particles. Once this is done the problem of deciding what kind of matter can exist and what it properties are reduces to the purely mathematical one of finding the bound states for each subset of these particles and the ways in which these can interact.

There is a difficulty in deciding which particles are "fundamental" or "elementary" because of the fact that nonelementary ones can appear to be elementary especially when the "binding energy" is so high as to be hard to come by. In fact, physicists still are not sure what the "really"

ISBN 0-8053-6702-0/0-8053-6703-9, pbk

elementary particles are. On the other hand, large parts of physics and essentially all of chemistry can be explained on the assumption that the so-called "atomic nuclei" are elementary. It is now known that they are not. However, except for a few so called "radioactive" ones that decompose spontaneously, the energies required to decompose them are so great that one never encounters them during the course of ordinary phenomena and one may think of them as elementary. On this basis, the elementary particles and their interactions may be described as follows. There is an elementary particle of smallest mass called the *electron*. Its mass is 4.8×10^{-28} grams and it has spin $\frac{1}{2}$. Its charge is -4.8025×10^{-10} in the centimeter-gram-second system. There is another elementary particle called the *proton*. Its mass is 1836.5 times that of the electron and its charge is the negative of that of the proton. It also has spin $\frac{1}{2}$. The other elementary particles have charges that are positive integral multiples of the charge on the proton. Every positive integer from 1 to 92 occurs. These particles (including the proton) are called nuclei.

The masses of the nuclei are "almost" integer multiples of the mass of the proton. However, the integer to which this multiple is closest (called the mass number) is not the same as the "charge number." It is twice as big, or a little more, and there may be several nuclei with different masses for a given charge number. For example, there are four different nuclei with charge number 30. Their mass numbers are 64, 66, 68, and 70. Nuclei also have spins, but for most purposes one does not need to know exactly what these are. It suffices to know that nuclei with even mass numbers are bosons and that those with odd mass numbers are fermions. For definiteness one can assign spin 0 to the bosons (and this is right in most cases) and spin $\frac{1}{2}$ to nuclei fermions.

While the interactions between electrons and nuclei, electrons and electrons, and nuclei and nuclei are known to be more complicated, it is a very good approximation under all but the most extraordinary circumstances to assume that these interactions are the purely electrostatic ones determined by the appropriate charges. In other words, the interaction between an electron and a nucleus of charge number Z may be assumed to be the velocity and spin independent interaction defined by the function

$$x_1,y_1,z_1,x_2,y_2,z_2 \to \frac{(-e)(Ze)}{\left[(x_1-x_2)^2+(y_1-y_2)^2+(z_1-z_2)^2\right]^{1/2}}$$

For two electrons one replaces this function by

$$x_1,y_1,z_1,x_2,y_2,z_2 \to \frac{e^2}{\left[(x_1-x_2)^2+(y_1-y_2)^2+(z_1-z_2)^2\right]^{1/2}}$$

ISBN 0-8053-6702-0/0-8053-6703-9, pbk

and for two nuclei of charge number Z_1 and Z_2 one has an interaction determined by the function

$$x_1y_1z_1, x_2y_2z_2 \to \frac{Z_1Z_2e^2}{\left[(x_1-x_2)^2+(y_1-y_2)^2+(z_1-z_2)^2\right]^{1/2}}$$

Given any family of electrons and nuclei one supposes that their interaction is a sum of the above-described two-body interactions. All matter may be supposed to consist of nuclei and electrons grouped in various ways via the bound-state mechanism. One cannot form bound states out of nuclei alone or out of electrons alone (at least one cannot do so theoretically assuming only electrostatic interaction). The simplest bound states are those consisting of one nucleus and several electrons. When the number of electrons is equal to the charge number of the nucleus, the composite object that results is called an atom. One says that the atom is in an excited state or the ground state depending upon whether the bound state in question has the lowest possible energy of all bound states that can be formed with these particles. Given several atoms, there may or may not be a bound state for the system consisting of all the nuclei and all of the electrons in these atoms. If there is, one says that this bound state is a molecule. Again one can speak of ground states and excited states. The difference between the ground-state energy of a molecule and the sum of the ground-state energies of its constitient atoms is called the energy of formation of the molecule. To the extent that one can carry out the mathematics, one can predict from quantum mechanics and the above hypothesis which sets of atoms will combine into molecules and how much energy will be given off or absorbed when they do. Since it turns out to be possible to identify the atoms and molecules as defined above with the objects so named by chemists, we have in principle reduced chemistry to mathematics. If only the mathematics were more tractable we could dispense with chemists altogether. Actually, there is a flourishing intellectual discipline overlapping chemistry and physics called quantum chemistry. Its practitioners attempt to carry out the program of reducing chemistry to mathematics and much progress has been made. Not only can accurate detailed calculations be made for some of the smaller molecules, but also one can obtain a qualitative understanding of many chemical phenomena such as valence by studying the properties of interacting bound states mathematically. One can understand valence not only at the superficial level at which every atom has a single well-defined valence, but also at the deeper level where one element has several valences or molecules like benzene are formed in which the valence bonds are not fixed. In

ISBN 0-8053-6702-0/0-8053-6703-9, pbk

particular, once one understands the quantum mechanical origins of chemical binding, one sees that no picture as simple minded as that which used to appear in elementary chemistry books can even be more than approximately true.

We remark that the chemical properties of an atom depend entirely on the charge number of the nucleus and not at all on its mass number. Thus there are just 92 essentially different kinds of atoms. Atoms with the same charge number (usually called atomic number) and different mass number are called *isotopes*. Since one cannot separate isotopes by chemical means, the substances the chemist thinks of as containing all identical atoms actually contain a mixture of isomtopes. The so-called atomic weight of an atom with given atomic number is a weighted mean of the mass numbers of the various isotopes. One of the easier triumphs of quantum chemistry is an explanation of the periodic change in the chemical properties of an atom as one increases the charge number of the necleus from 1 to 92.

Now atoms and molecules are not the only possible bound states. Given almost any molecule there is a sequence of larger and larger bound states in which the nuclei and electrons that occur are just those in an integral number of copies of the molecule. This integer is unlimited in size and the nuclei in the bound states are arranged in the sort of pattern that can repeat itself indefinitely. When this bound state has sufficiently many nuclei in it, it becomes an object we can see and feel and we call it a crystal. Most of the objects that we call solids in everyday life are either crystals or mixtures of crystals. The properties of crystals are in principle deducible from our basic axioms just as are the properties of molecules—indeed there is no sharp distinction between a crystal and a molecule. On the other hand, in studying the properties of solid bodies one is usually interested in crystals large enough to be regarded as everyday objects and for such the number of nuclei must be enormously large. One milligram of a substance with a moderate atom weight such as 50 will contain 1.2×10^{20} nuclei. In studying systems of n particles in which n is so enormously large one cannot take account of what each individual particle is doing, but must correlate observable phenomena with statistical statements about what the $\gg 10^{20}$ particles are doing "on the average." A technique for doing this was first worked out in the context of classical mechanics and was responsible for some of the earliest successes in the program of explaining the properties of matter in terms of the motions of fundamental particles. This technique is known as statistical mechanics and has a straightforward adaptation to quantum mechanics. In either case, it provides an explanation of the meaning of heat and temperature and a deduction of the laws of thermodynamics. In addition, it leads to an

ISBN 0-8053-6702-0/0-8053-6703-9, pbk

algorithm for computing the thermodynamic properties of substances from the elementary particles composing them and their interactions. We shall describe it in Section 24. The branch of physics in which one attempts to explain the properties of solid material objects in terms of the quantum mechanical motions of the nuclei and electron is known, appropriately enough, as the theory of the solid state. One of its many successes is a simple explanation in terms of the Pauli exclusion principle, of the difference between substances which do and do not conduct electricity, and an understanding of the so called "semiconductors."

Of course, matter does not exist just in the solid state although practically all matter takes this form at sufficiently low temperatures. At higher temperatures matter can also exist in "liquid" and "gaseous form." A gas is simply a collection of molecules whose average kinetic energy is high compared to the energy of formation of a crystal. The various molecules behave as more or less free particles whose motions are slightly altered when they get too close to one another. In a liquid the kinetic energies are high enough to disrupt the rigid crystal structure but not enough to free the particles from their mutual influence. Liquids are much less well understood than either solids or gases and quantum mechanics has had as yet little influence on hydrodynamics. However, liquids are in principle as susceptible to quantum mechanical explanation as solids and gases. It turns out that classical statistical mechanics and quantum statistical mechanics make different predictions about the same phenomena, but that these differences decrease with increasing temperature ultimately tending to zero. In many particular cases, ordinary room temperatures are high enough for the classical limit to have been reached and it is for this reason that classical statistical mechanics had a certain success. Among the earliest triumphs of quantum mechanics were explanations of the various failures of classical statistical mechanics at low temperatures. Indeed, the old quantum theory, out of which quantum mechanics emerged, began with Planck's attempts to explain why the so called Rayleigh–Jeans law was valid at high temperatures but not low ones.

Our quotation from Dirac spoke of our now knowing the physical laws for "a large part of physics." Why not all of it? It would seem from our description that everything is included. What have we left out? First and foremost we have left out light and more generally electromagnetic radiation. In fact, we have even left out the forces giving rise to the phenomena of magnetism. Moreover, in accepting the nuclei as elementary particles we have thrown out the whole subject of nuclear physics, which discusses the nuclei as bound states of protons and certain unstable particles called neutrons and tries on this basis to understand the structure of these nuclei

ISBN 0-8053-6702-0/0-8053-6703-9, pbk

and the ways in which they may be transformed into one another. Finally, it turns out that the interactions we have assumed between charged particles are only approximations that become increasingly inaccurate as we consider particles with higher and higher kinetic energies. These omissions are all closely related to one another and to the discovery that Galilean relativity is only approximately correct. On the other hand, the more accurate special relativity theory of Einstein changes our picture of the physical world rather radically in certain important respects, and up till now it has not been possible to provide a consistent theory incorporating both special relativity and quantum mechanics. Many individual phenomena can be explained in whole or in part by combining the ideas of special relativity and those of quantum mechanics in an ad hoc piecemeal manner. However, there is no consistent general theory as there is for interacting particles under the assumption of Galilean relativity. To find such a theory is perhaps the great challenge of present-day theoretical physics. We shall discuss some of the difficulties and various approaches to them in Section 22.

Notes and References

In describing how the bound-state notion fits into our conceptual scheme, we have barely scratched the surface of an enormous and fascinating field. From the point of view of the practicing physicist or chemist, the key problem is that of computing the binding energies and geometrical properties of the bound states formed by particular sets of nuclei and electrons. This comes down in the end to finding the eigenvalues and eigenfunctions of rather complicated partial differential operators—a problem so difficult that little headway can be made without the use of high-speed computers. Even with their aid it is only in the last few years that it has been possible to make accurate computations for any but the simplest molecules. On the other hand, qualitative arguments involving perturbation theory and the exploitation of symmetry have yielded a considerable amount of insight. The exploitation of symmetry depends upon heavy use of the theory of group representations and, in fact, it is this use that physicists and chemists usually have in mind when they speak of applying group theory or group representation theory to quantum mechanics. We have omitted an account of this important topic here because of lack of time and because it is the finite-dimensional representation theory of finite and compact groups that is mainly involved. The fifth chapter of Hermann Weyl's classic *The Theory of Groups and Quantum*

ISBN 0-8053-6702-0/0-8053-6703-9, pbk

Mechanics contains an exposition of many of the key ideas, but is by no means easy to understand. Substantial simplifications can be made by making systematic use of the theory of induced representations and one of the many items in the author's agenda of future projects is to rewrite Weyl's exposition accordingly.

The special case in which there is only one nucleus is the one involved in explaining atomic spectra and has been particularly assiduously studied. Wigner, who is the principal original pioneer in this part of the subject, has explained it in detail in another well-known classic: *Group Theory and Its Application to the Quantum Mechanics of Atomic Spectra*, Academic, New York (1959) (original German edition, 1931). For more refined work on atomic spectra, G. Racah developed a rather complex extension of Wigner's original method in a series of articles published around 1940. For details about this and related material, the reader should consult *The Quantum Theory of Angular Momentum*, L. C. Biedenharn and H. Van Dam, Academic, New York (1965). In part a reprint collection, this book contains among other things copies of Racah's fundamental papers.

Although they are not written at a very sophisticated mathematical level, the author recommends the review article by Van Vleck and Sherman ("The Quantum Theory of Valence," *Rev. Mod. Phys.*, vol. 7 (1935), New York 168–228) and the book *Quantum Chemistry* by Walter, Eyring, and Kimball, Wiley, New York (1944), as good introductions to the theory of bound states containing several nuclei (i.e., molecules). *Valence*, by C. A. Coulson, Oxford, New York (1952), is more complete and mathematical.

The quantum theory of the solid state occupies a middle ground between the theory of molecular structure and statistical mechanics. For some brief remarks about the applications of unitary group representations in the spirit of this book and some references to the literature, see pages 354 through 356 of the published version of the author's Chicago lecture notes.

In spite of the high energy difficulties, many of the techniques for the group theoretical study of atoms and molecules can also be applied to the study of the structure of nucleii. This is true in particular of the methods of Wigner and Racah. For details, see the book of Biedenharn and Van Dam cited above and *Symmetry Groups in Nuclear and Particle Physics* by F. J. Dyson, Benjamin (1966).

ISBN 0-8053-6702-0/0-8053-6703-9, pbk

21. Collision theory and the S operator

Once one knows the dynamical operator for an n-particle system, one is prepared to discuss the "motions" of the system. We have begun this discussion by introducing the notion of bound state. If W is the underlying σ representation of the Galilean group, then the state vectors in each irreducible invariant subspace of $\mathcal{H}(W)$ may be regarded as the state vectors of a particle located at the center of mass of the system and in the absence of other particles this center of mass will move like a free particle. We have called this apparent particle a bound state of the system and have seen that many physical problems hinge on finding the bound states of systems of nuclei and electrons. However, W is never equal to the direct sum of its irreducible subrepresentations and state vectors in the continuous part of $\mathcal{H}(W)$ correspond to motions of the system in which the particles do not form a single bound state. Such motions occur typically when $n=2$ and one particle approaches the vicinity of another from a great distance. As it gets closer to the second particle its motion deviates more and more from that of a free particle. If the particles are incapable of forming a bound state or if their velocity is too great, the first particle will finally reach a point of closest approach and then will go off again with its free motion altered by its encounter with the "force" exerted by the second particle. One says that the first particle has been "scattered" by the second. Of course the second particle will also be set in motion and more generally we can consider two particles moving in such a way as to come close to one another and consider the extent to which their motions are altered by the encounter. These alterations will of course depend upon the momenta

George W. Mackey, Unitary Group Representations in Physics, Probability, and Number Theory

ISBN 0-8053-6702-0/0-8053-6703-9, pbk

of the two particles and upon how close together they would come, given their paths, if they did not interact. The resulting function of several variables can be measured directly in the laboratory, and can be computed —at least in the two-particle case—once the dynamical operator is known. Thus we have a way of checking hypotheses about particle interactions by experiment. One of the problems in attempting to understand the structure of nuclei is our incomplete knowledge of the dynamical group for a system consisting of two nucleons (nucleon is a collective word meaning neutron or proton). Our earlier statement according to which two protons interact (and repel one another) only through electrostatic forces is inaccurate not only at high velocities but also when the protons get too close together. There is an attractive interaction that falls off exponentially at large distances but counterbalances the electrostatic repulsion at short ones. In order to explore this, one performs collision experiments with nucleons moving at high enough velocities to get really close to one another, and the analysis of such experiments has become an important part of modern theoretical physics. The experiments themselves have led to the discovery of a vast array of new particles that may or may not be elementary and the program of studying the nonelectrostatic part of the interaction between nucleons has expanded to include a study of the interaction between these new particles as well. This study also involves the performing and analysis of scattering experiments.

Although most of these experiments are done under such conditions that the effects of special relativity cannot be ignored, it is still worthwhile to study scattering theory in the context of nonrelativistic quantum mechanics. Many of the concepts occur in the nonrelativistic case and may be discussed in a context free of the ambiguities and uncertainties of the relativistic theory.

We shall begin with a rather general discussion. Let $\mathcal{H}$ be the Hilbert space for an n-particle quantum system and let $t \to V_t = e^{-iHt}$ be the dynamical group. Let $t \to W_t = e^{-iH_0 t}$ be the dynamical group for the same system with the interaction removed, that is, for a system of n free particles with the same masses and spins as those we are considering. Let ϕ be a state vector for a state such that the corresponding trajectory $t \to V_t\phi$ is that for particles coming together from afar. Then as $t \to -\infty$, the trajectory $t \to V_t\phi$ becomes more and more like a free particle trajectory. In other words, there must exist a state vector ψ such that in some sense $V_t\phi - W_t\psi$ tends to zero as $t \to -\infty$. It turns out that in many cases of interest $V_t\phi - W_t\psi$ tends to zero in the Hilbert space norm and we shall only study such cases. Now if $V_t\phi - W_t\psi \to 0$ as $t \to 0$, it follows at once (since $\| W_t \| = 1$ for all t) that $W_t^{-1}V_t\phi - W_t^{-1}W_t\psi \to 0$ as $t \to 0$, that is, that $W_t^{-1}V_t\phi \to \psi$.

ISBN 0-8053-6702-0/0-8053-6703-9, pbk

Conversely, of course, if $W_t^{-1}V_t\phi$ has a limit ψ for some ϕ, then $V_t\phi - W_t\psi \to 0$. It is thus of considerable interest to study the subspace $\mathcal{M}^-$ of $\mathcal{H}$ consisting of all ϕ for which the limit in question exists and the operator S^- that takes ϕ into $\lim_{t\to-\infty} W_t^{-1}V_t\phi$. A straightforward argument shows that $\mathcal{M}^-$ is always a closed subspace and that S^- is a linear norm preserving map of $\mathcal{M}^-$ onto some other closed subspace $\mathcal{M}^{-\prime} = S^-(\mathcal{M}^-)$. Now let ϕ be a member of $\mathcal{M}^-$. Then for all a $W_t^{-1}V_tV_a\phi = W_t^{-1}V_{t+a}\phi = W_aW_{t+a}^{-1}V_{t+a}\phi$. Now $\lim_{t\to-\infty} W_t^{-1}V_t(\phi) = S^-(\phi)$ so $\lim_{t\to-\infty} W_aW_{t+a}^{-1}V_{t+a}(\phi) = W_aS^{-1}(\phi)$. Thus $V_a(\mathcal{M}^-) = \mathcal{M}^-$, that is, $\mathcal{M}^-$ is an invariant subspace of $\mathcal{H}(V)$ and $\mathcal{M}^{-\prime}$ is an invariant subspace of $\mathcal{H}(W)$. Moreover, since $S^-V_a(\phi) = W_aS^-(\phi)$, we see that S^- *is a unitary intertwining operator for* $V \restriction \mathcal{M}^-$ and $W \restriction \mathcal{M}^{-\prime}$. We deduce at once that $\mathcal{M}^-$ cannot be the whole of $\mathcal{H}$ unless V is equivalent to W or some subrepresentation of W. Thus $\mathcal{M}^-$ cannot be the whole of $\mathcal{H}$ if there are any bound states. On the other hand, whenever the interactions damp out sufficiently rapidly at infinite separation and there are no bound states (even of subsets of the particles), there is reason to believe (though rigorous proof may be lacking) that $\mathcal{M}^- = \mathcal{M}^{-\prime} = \mathcal{H}$. Whenever this happens, V and W are equivalent representations of the real line and S^- sets up the equivalence.

Now one can also study the behavior of the system as $t \to +\infty$ and give a completely parallel discussion. We define $\mathcal{M}^+$ to be the subspace of all ϕ for which $\lim_{t\to\infty} W_t^{-1}V_t\phi$ exists and define $S^+(\phi)$ as $\lim_{t\to\infty} W_t^{-1}V_t\phi$. Again S^+ is linear and norm preserving from $\mathcal{M}^+$ to $\mathcal{M}^{+\prime} = S^+(\mathcal{M}^+)$ and S^+ sets upon unitary equivalence between $V \restriction \mathcal{M}^+$ and $W \restriction (\mathcal{M}^+)'$. Consider now the important special case in which $(\mathcal{M}^-)' = (\mathcal{M}^+)' = \mathcal{H}$; that is, in which every free trajectory occurs both as the infinite past of some real trajectory and the infinite future of some real trajectory. Then if $t \to W_t\psi_1$ represents our trajectory in the infinite past, the actual trajectory will be $t \to V_t(S^-)^{-1}(\psi_1)$ and its infinite future will be described by the free particle trajectory $t \to W_t((S^+)(S^-)^{-1}(\psi_1))$ at least if $\mathcal{M}^+ \supseteq \mathcal{M}^-$. Let us add the assumption that $\mathcal{M}^+ = \mathcal{M}^-$. This means physically that any free system in the infinite past has a free infinite future, and conversely. Then the operator $\psi \to (S^+)(S^-)^{-1}(\psi) = S(\psi)$ is a unitary operator with the property that $t \to W_t(S(\psi))$ describes our system in the infinite future whenever $t \to W_t(\psi)$ describes it in the infinite past. It tells us how to transform the situation "before encounter" into "after encounter" without reference to what goes on "during encounter." It is called the scattering operator or the S operator of the system.

Since S^- sets up a unitary equivalence between W and the restriction of V to $\mathcal{M}^-$ and S^+ sets up a unitary equivalence between W and the

ISBN 0-8053-6702-0/0-8053-6703-9, pbk

restriction of V to $\mathfrak{M}^+$, and since $\mathfrak{M}^- = \mathfrak{M}^+$ by hypothesis, it follows that S sets up an equivalence between W and itself; that is, that S lies in the commuting algebra of W. If we assume Euclidean invariance, then the underlying representation $\alpha \rightarrow U_\alpha$ of the Euclidean group $\mathcal{E}$ will be such that $U_\alpha W_t = W_t U_\alpha$ and $U_\alpha V_t = V_t U_\alpha$ for all t and α. It follows that U_α commutes with S^+ and S^- for all α and hence with S. Thus S is actually in the commuting algebra of the representation $\alpha, t \rightarrow U_\alpha W_t$ of $\mathcal{E} \times T$. Finally, let us assume that our system is Galilean invariant. Then $\alpha \rightarrow U_\alpha W_t$ will be the restriction to $\mathcal{E} \times T$ of a certain projective representation $\tilde{W}$ of the Galilean group $D \times L Ⓢ R \times T = \mathcal{G}$ $(\mathcal{E} = D$ Ⓢ$R)$ and $\alpha \rightarrow U_\alpha V_t$ will be the restriction to $\mathcal{E} \times T$ of a certain projective representation $\tilde{V}$ of $\mathcal{G}$. Moreover, $\tilde{W}$ and $\tilde{V}$ restricted to $D \times L Ⓢ R$ will coincide. Now for each $l \in L$, $t \in T$, $tlt^{-1} \in D \times L$, so $\tilde{V}_{tlt^{-1}} = \tilde{W}_{tlt^{-1}}$. Hence $\tilde{V}_t \tilde{V}_l \tilde{V}_{t^{-1}} = \tilde{W}_t \tilde{W}_l \tilde{W}_{t^{-1}}$. Hence $V_t \tilde{V}_l \tilde{V}_t = W_t \tilde{W}_l \tilde{W}_t^{-1}$ and $\tilde{V}_l = \tilde{W}_l$. Hence $\tilde{W}_l$ commutes with $W_t^{-1} V_t$. Hence $\tilde{W}_l$ commutes with S^+, S^-, and S. Since S commutes with all $U_\alpha V_t$ with $\alpha \in D Ⓢ R$ and all $\tilde{W}_1$ with $l \in L$, it commutes with all $\tilde{W}_\beta$ with $\beta \in \mathcal{G}$.

To summarize: S is always in the commuting algebra of $t \rightarrow W_t$. If our system is Euclidean invariant, it is also in the commuting algebra of the underlying representation of the Euclidean group; and if our system is Galilean invariant, it is in the commuting algebra of the underlying projective representation of the Galilean group.

Before analyzing the implications of these invariance properties of S, let us look a little more closely at their physical significance. As a pedagogical device let us assume for the moment (contrary to fact) that $\mathcal{H}$ has an orthonormal basis each element of which is an eigenvector for all linear momentum operators and that such an eigenvector is completely described by the eigenvalues of all $3n$ linear momentum operators $P_1, P_2, \ldots, P_{3n}$. There will then be a certain countable subset Λ of the space of all $3n$-tuples of real numbers and for each member $p_1, \cdots, p_{3n}$ of Λ a unique member $\phi_{p_1, \cdots, p_{3n}}$ of our orthonormal basis such that $P_j(\phi_{p_1, \cdots, o_{3n}}) = p_j \phi_{p_1, \cdots, p_{3n}}$. Since the free particle energy is a function of the momenta, each $\phi_{p_1, \cdots, p_{3n}}$ will be carried into itself by all W_t and so will be a stationary state. If there is no interaction, the momenta will maintain their fixed values $p_1, \cdots, p_{3n}$ forever. However, as a result of the interaction, $\phi_{p_1, \cdots, p_{3n}}$ is carried into another free partic le state $S(\phi_{p_1, \cdots, p_{3n}})$ that need not be described by a basis vector but that will be described by a vector which can be expanded in terms of these basis vectors:

$$S(\phi_{p_1, p_2, \cdots, p_{3n}}) = \sum_{p'_1, \cdots, p'_{3n} \in \Lambda} s_{p_1, \cdots, p_{3n}, p'_1, \cdots, p'_{3n}} \phi_{p'_1, \cdots, p'_{3n}}$$

ISBN 0-8053-6702-0/0-8053-6703-9, pbk

Now by the general rules of quantum mechanics the probability that a measurement of the momenta of our particles will lead to a value in the subset A of n space when in the state $S(\phi_{p_1,\cdots,p_{3n}})$ is

$$\sum_{p'_1,\cdots,p'_{3n}\in A} |s_{p_1,\cdots,p_{3n},p'_1,\cdots,p'_{3n}}|^2$$

Thus if the momenta are $p_1,\cdots,p_{3n}$ before interaction, then the probability that they are $p'_1,\cdots,p'_{3n}$ after interaction is $|s_{p_1,\cdots,p_{3n},p'_1,\cdots,p'_{3n}}|^2$. Now the function

$$p_1,\cdots,p_{3n},p'_1,\cdots,p'_{3n}\rightarrow s_{p_1,\cdots,p_{3n},p'_1,\cdots,p'_{3n}}$$

is a matrix whose rows and columns are indexed by the members of Λ and this is the matrix of S with respect to the basis $\phi_{p_1,\cdots,p_{3n}}$. The elements of this matrix are called *probability amplitudes*. They are complex numbers whose squares are probabilities. Since the momentum observables do not have discrete spectra, the above remarks cannot be taken literally. However, there is a continuous analog of them that is true. If we assume spinless particles, then the linear momenta form a complete commuting set and we may realize our Hilbert space as $\mathcal{L}^2(E^{3n})$ in such a manner that

$$(P_j(\psi))(p_1,\cdots,p_{3n})=p_j\psi(p_1,\cdots,p_{3n})$$

that is, P_j is just multiplication by the jth coordinate. This realization is obtained from the usual one in which the coordinate observables are multiplication operators by means of a Fourier transform in $3n$ space. If the S operator were an *integral operator* in this realization, that is, if there existed a complex-valued function s on $E^{3n}\times E^{3n}$ such that

$$(S(\psi))(p'_1,\cdots,p'_{3n})=\int s(p_1,\cdots,p_{3n},p'_1,\cdots,p'_{3n})\psi(p_1,\cdots,p_{3n})$$

we could interpret $s(p_1\cdots p_{3n},p'_1\cdots p'_{3n})$ more or less as we did the matrix element in the above discussion. However, instead of probabilities we would have probability densities. Given a system of free particles with momenta $p_1\cdots p_{3n}$, after collision the momenta would be statistically distributed so that the probability of their being in a region A would be

$$\int_A |s(p_1\cdots p_{3n},p'_1\cdots p'_{3n}|^2 dp'_1\cdots dp'_{3n}$$

ISBN 0-8053-6702-0/0-8053-6703-9, pbk

Strictly speaking we cannot say that our free particles even have definite momenta—only that these momenta are certain to be in a region containing the point $p_1,\cdots,p_{3n}$ which we can make as small as we please. The above probability distribution is then interpreted to be the limit of the actual ones as the region gets smaller and smaller. However, S can be shown not to be an integral operator. Indeed, it can be shown to follow from conservation of momentum and energy that the probability distribution associated with the initial momenta $p_1,\cdots,p_{3n}$ must be concentrated in a set of Lebesgue measure zero, namely, the $(3n-4)$-dimensional subspace of E^{3n} corresponding to values of $p'_1,\cdots,p'_{3n}$ with

$$p'_1+p'_4+p'_7+\cdots+p'_{3n-2}=p_1+\cdots+p_{3n-2}$$

$$p'_2+p'_5+\cdots=p_2+\cdots$$

$$p'_3+p'_6+\cdots=p_3+p_6+\cdots$$

$$\sum\frac{(p'_i)^2}{\mu_i}=\sum\frac{(p_i)^2}{\mu_i}$$

To circumvent this difficulty, one first shows that S must commute with multiplication by the functions $p_1+p_4+p_7+\cdots$, $p_2+p_5+p_8+\cdots$, $p_3+p_6+p_9+\cdots$, and $\Sigma p_j^2/2\mu_j$, and so is a direct integral of operators acting in $\mathcal{L}^2(E^{3n-4})$. In the particular case in which $n=2$, at least these operators in $\mathcal{L}^2(E^2)$ are integral operators and one can read off the relevant probabilities from the "matrix coefficients." We omit the details.

Let us now consider the case of two spinless particles under the assumption of Galilean invariance. Given the momenta of the incident particles p_1,p_2,p_3,p_4,p_5,p_6, the probability distribution for the emergent momenta will be concentrated in the two-dimensional subset of E^6 consisting of all $p'_1,p'_2,p'_3,p'_4,p'_5,p'_6$ such that

$$p'_1+p'_4=p_1+p_4$$

$$p'_2+p'_5=p_2+p_5$$

$$p'_3+p'_6=p_3+p_6$$

$$\frac{p_1'^2}{2m_1}+\frac{p_2'^2}{2m_1}+\frac{p_3'^2}{2m_1}+\frac{p_4'^2}{2m_2}+\frac{p_5'^2}{2m_2}+\frac{p_6'^2}{2m_3}=\frac{p_1^2}{2m_1}+\frac{p_2^2}{2m_1}+\frac{p_3^2}{m_1}+\frac{p_4^2}{m_2}+\frac{p_5^2}{m_2}+\frac{p_6^2}{m_2}$$

ISBN 0-8053-6702-0/0-8053-6703-9, pbk

Clearly this two-dimensional subset of six-dimensional space will be a spherical surface in a three-dimensional hyperspace. The probability measure in this sphere will have a density that is the absolute square of a probability amplitude and we call this probability amplitude $s(p_1,p_2,\ldots,p_6,p_1',\ldots,p_6')$. It can be shown to follow from invariance properties that s depends essentially on only 2 variables instead of 12. We already know that s is zero when $p_1+p_4 \neq p_1'+p_4'+\cdots$, etc. In addition, it is found that s depends only on

$$\frac{p_1^2}{2m_1}+\frac{p_2^2}{2m_1}+\frac{p_3^2}{2m_1}+\frac{p_4^2}{2m_2}+\frac{p_5^2}{2m_2}+\frac{p_6^2}{2m_2}=E$$

and on

$$\begin{aligned}\Delta &= \left[(p_1-p_1')^2+(p_2-p_2')^2+(p_3-p_3')^2\right]^{1/2}\\ &= \left[(p_4-p_4')^2+(p_5-p_5')^2+(p_6-p_6')^2\right]^{1/2}\end{aligned}$$

provided our reference frame is one in which the center of mass of the two particles is "at rest," that is, provided we restrict ourselves to sextuples $p_1,p_2,p_3,p_4,p_5,p_6;\ p_1'\cdots p_6'$ satisfying $p_1+p_4=p_2+p_5=p_3+p_6=p_1'+p_4'+\cdots=0$. Here E is the constant total energy of the colliding particles and Δ is the so called *momentum transfer*. Thus the S operator and all its probability predictions for the case of two Galilean spinless particles is completely determined by a single complex-valued function F of the two variables E and Δ^2. If we separate the motion of the two particles into the motion of the center of gravity and the independent motion of the two particles with respect to the center of gravity, we see that we need consider only one particle since x_2, y_2, and z_2 are known assoon as $x_1-\bar{x}$, $y_1-\bar{y}$, and $z_1-\bar{z}$ and $\bar{x}$, $\bar{y}$, and $\bar{z}$ are both known. Of course

$$\bar{x}=\frac{m_1x_1+m_2x_2}{m_1+m_2}$$

and similarly for $\bar{y}$ and $\bar{z}$. Let $x_1-\bar{x}=x$, $y_1-\bar{y}=y$, $z_1-\bar{z}=z$. Let p,q,r be the corresponding momenta. Then the question becomes that of finding a probability distribution for the emerging particle momenta p',q',r' given p,q,r. The probability amplitude f turns out to depend only on $\cos\theta$, where θ is the angle between p,q,r and p',q',r' and on

$$E=\frac{p^2+q^2+r^2}{2m'}$$

ISBN 0-8053-6702-0/0-8053-6703-9, pbk

where m' is the "effective mass" of the single particle. Moreover, $\cos\theta = 1 - \Delta^2/4m'E$. Thus

$$F(E,\Delta^2)=f\left(E,1-\frac{\Delta^2}{4m'E}\right)$$

and we may choose either $\cos\theta$ or Δ^2 as our basic additional variable.

The function F of course depends upon the exact nature of the interaction between the particles, but there are certain important general statements that one can make about the F's coming from a broad class of interactions. When the interaction is velocity independent and the resulting potential vanishes sufficiently rapidly at infinity, one can show that for each $\Delta > 0$ the function $E \to f(E,\Delta^2)$ is the boundary value of a function that is analytic in the upper half-plane. Moreover, this function may be extended into the lower half-plane so as to be analytic (except for a finite number of poles on the negative real axis) on the whole complex plane minus the positive real axis. Using these facts and general properties of analytic functions one deduces the so called "dispersion relations" connecting the real part of $f(E,\Delta^2)$ with an integral over the imaginary part and the position and residues of the poles. These relations (or rather their relativistic analogs) turn out to be of considerable importance in analyzing scattering data in high-energy collisions.

Actually, one can also extend F so as to be defined for complex values of both E and Δ^2 in such a manner that F is analytic except on a finite number of hypersurfaces of real dimension two. The resulting analytic function of two complex variables has a relativistic analog that has proved to be of even greater importance in analyzing high-energy collisions. This technique was introduced in 1958 by S. Mandelstam and one often refers to "Mandelstam theory."

One can also describe the two-particle S operator in terms of an angle valued function of E and the angular momentum l called the phase shift. This description has a number of important and useful properties and leads in particular to the celebrated concept of "Regge pole." Moreover, it turns out to be possible to define it in an especially simple and elegant fashion using general properties of group representations.

In our two-particle scattering system let $\tilde{W}$ and $\tilde{V}$ as above be the underlying projective representations of the Galilean group $\mathcal{G}$, $\tilde{W}$ being the representation when there is no interaction. As we have seen, S lies in the commuting algebra of $\tilde{W}$. On the other hand, $\tilde{W} = \tilde{W}^1 \otimes \tilde{W}^2$, where $\tilde{W}^1$ and $\tilde{W}^2$ are the projective representation defining the two one-particle systems. If we continue to suppose that the particles are spinless, $\tilde{W}_1$ and

ISBN 0-8053-6702-0/0-8053-6703-9, pbk

$\tilde{W}_2$ will be irreducible, and to find W we must find the inner Kronecker product of two irreducible projective representations of $\mathcal{G}$. This may be done by using the fact that they are induced representations and using our general theorem about restricting induced representations to subgroups to obtain a similar general theorem about inner Kronecker products of induced representations. Here is how the deduction goes in outline. Let H_1 and H_2 be closed subgroups of the separable locally compact group G. Let L and M be unitary representations of H_1 and H_2, respectively, and consider the inner Kronecker product $U^L \otimes U^M$ of the induced representations U^L and U^M, respectively. Then $U^L \otimes U^M$ is by definition the restriction to the diagonal $\tilde{G}$ of $G \times G$ of the outer product $U^L \times U^M$ and this by an earlier theorem is $U^{L \times M}$, where $L \times M$ is of course a representation of the subgroup $H_1 \times H_2$. Thus $U^L \otimes U^M$ is the restriction to the subgroup $\tilde{G}$ of $G \times G$ of the representation of $G \times G$ induced by the representation $L \times M$ of the subgroup $H_1 \times H_2$ of $G \times G$. Applying our earlier result on restrictions of induced representations, we find that this restriction is a direct sum or direct integral over the $H_1 \times H_2 : \tilde{G}$ double cosets of certain induced representations of G. One shows easily that the $H_1 \times H_2 : \tilde{G}$ double cosets in $G \times G$ correspond one-to-one in a natural way to the $H_1 : H_2$ double cosets in G, and one obtains an expression for $U^L \otimes U^M$ as a direct integral or direct sum over the $H_1 : H_2$ double cosets of a certain induced representation of G that is entirely similar in form to our earlier result about the restriction of U^L to H_2. The representation of G associated with the double coset $H_1 \times H_2$ is the representation of G induced by the inner Kronecker product of the restrictions of L and M^x to $H_1 \cap x H_2 x^{-1}$. Using the resulting theorem, one can show in a straightforward way that $\tilde{W} = \tilde{W}^1 \otimes \tilde{W}^2$ is *multiplicity free* whenever $\tilde{W}^1$ and $\tilde{W}^2$ are the representations associated with *spinless* particles. Thus it is a direct integral of irreducible representation each occuring just once. Recall that the irreducible projective representation of $\mathcal{G}$ for a fixed σ are parametrized by a real number E and a discrete parameter l, where l takes on either the values $0, 1, 2, \ldots$ or the values $\frac{1}{2}, \frac{3}{2}, \frac{5}{2}, \ldots$ depending on σ. In this case, since the particles are spinless, l takes on the values $0, 1, 2, \ldots$. Let us denote the irreducible representation with parameter l and E by $A^{E,l}$. Then $W = W^1 \times W^2$ turns out to be $\int A^{E,l} d\mu(E, 1)$, where μ is the direct product of Lebesgue measure in the postive real axis with the "counting measure" on the nonnegative integers. Now any element T of the commuting algebra of $\tilde{W}$ must be a direct integral of operators in the Hilbert spaces $\mathcal{H}(A^{E,l})$, the component of $T^{E,l}$ in $\mathcal{H}(A^{E,l})$ being in the commuting algebra of $A^{E,l}$. Since $A^{E,l}$ is irreducible, $T^{E,l}$ is multiplication by a constant. When $T = S$, the scattering operator we know that each $S^{E,l}$ is unitary and hence is

ISBN 0-8053-6702-0/0-8053-6703-9, pbk

multiplication by a complex number of modulus one. We can write this in the form $e^{i2\delta(E,l)}$, where $2\delta(E,l)$ is an "angle," that is, a real number defined only up to multiples of 2π. $\delta(E,l)$ is called the *phase shift* for the *given energy* E and *angular momentum* l. If we know the phase shift for all $E>0$ and all $l=0,1,2,\ldots$, then we know S. Of course, we expect a formula allowing us to compute $f(E,\cos\theta)$ from $\delta(E,l)$, and such a formula is available. We shall simply write it down with no attempt at a derivation:

$$f(E,\cos\theta)=\sum_{l=0}^{\infty}\frac{2l+1}{(Em)^{1/2}}e^{i\delta(E,l)}\sin\delta(E,l)P_l(\cos\theta)$$

Here P_l is the lth Legendré polynomial. We may write this as

$$\sum_{l=0}^{\infty}(2l+1)a_l(E)P_l(\cos\theta)$$

where

$$a_l(E)=\frac{e^{i\delta(E,l)}\sin\delta(E,l)}{(Em)^{1/2}}$$

is called the *partial wave amplitude*. It often turns out that the series converges quite rapidly for points $E,\cos\theta$ of physical interest and that one needs to know $\delta(E,l)$ only for a few values of l—sometimes only one. If $l=0$ is the only important value of l, one speaks of "s wave" scattering. If $l=1$ is the important one, then it is a question of "p wave" scattering.

The partial wave amplitudes $E\to a_l(E)$ may be extended into the complex plane just as with $F(E,\Delta)$. Under the same hypotheses on the interaction one finds that $a_l(E)$ may be extended to be meromorphic on a two sheeted Riemann surface with branch point at the origin except for a cut on both sheets extending from $-\infty$ to $\mu<0$ on the x axis. The poles on one of these sheets lie on the positive real axis and correspond one-to-one to those *bound states* of the two-particle system that have spin l. Those on the other sheet may be anywhere but those near the positive real axis correspond one-to-one to "unstable bound states," that is, to trajectories that behave for a while as bound states but finally break-up into two particles moving apart. If the pole of $a_l(E)$ is at the point $E_0+i\lambda$, then it represents our unstable bound state of energy E_0 and spin l. The larger λ is the shorter the particles lifetime. When the lifetime is so short that the bound state decomposes before it can travel a measurable distance, it is

ISBN 0-8053-6702-0/0-8053-6703-9, pbk

called a *resonance*. Of course, there is no sharp distinction between resonances and unstable particles.

It is interesting that resonances and unstable particles not only have energies and spins but behave like stable particles in other respects. One can even discuss their interactions with one another.

An advance similar to that made by Mandelstam in showing that $F(E,\Delta^2)$ could be continued to be an analytic function of two complex variables was made by Regge in 1959. Regge showed that one could extend $a_l(E)$ in a natural way so as to be defined for all positive real l and that then $l \to a_l(E)$ could be regarded as a boundary value of an analytic function of two complex variables. The famous "Regge poles" are the poles of $l \to a_l(E)$ for fixed positive E. As we vary E these poles change continuously and we obtain a discrete family of functions $E \to \alpha_k(E)$, where $\alpha_k(E)$ is a pole of $l \to a_l(E)$ for each k. These are called Regge pole trajectories. Whenever $\alpha_k(E)$ passes through a positive integral value l_0, then the corresponding value of E is a pole of a_{l_0}. Thus the bound states of the system lie on Regge pole trajectories with several on each trajectory. In other words, the bound states may be grouped into families—all members of the same family lying on a single Regge trajectory. One can deal similarly with unstable bound states and resonances. Indeed, given any pole E_0 of $\alpha_{l_0}(E)$ in the complex E plane, we may obtain a nearby pole in the complex E plane for each value of l near l_0. Thus we obtain a two-dimensional submanifold of the direct product of the complex l plane with the complex E plane, with the property that each pair l_1, E_1 in this submanifold is a pair such that E_1 is a pole of $\alpha_{l_1}(E)$. These submanifolds will be finite or countable in number and we may group together resonances when they lie on the same one. One speaks of "*Regge recurrences*" of a given bound state or resonance. Unfortunately, it would take us too far afield to attempt to explain the real importance of these notions in analyzing the results of scattering experiments.

Now let us drop the assumption that our particles are spinless and see what the consequences are. We shall study only the effect on the notion of phase shift. Let our particles have spins described by the representations L and M of the rotation group R. Then $\tilde{W}_1 \otimes \tilde{W}_2$ will no longer be multiplicity free. To see what its structure actually is we analyze it as follows. If we replace L by the identity repeated $\dim \mathcal{H}(L) = n_1$ times and M by the identity repeated $\dim \mathcal{H}(M) = n_2$ times, then $\tilde{W}_1 \otimes \tilde{W}_2$ is simply $\int B^{E,l}$, where $B^{E,l}$ is $A^{E,l}$ repeated $n_1 n_2$ times. Taking account of what L and M really are amounts to replacing D^l in the definition of $B^{E,l} = n_1 n_2 A^{E,l}$ by $D^l \otimes L \otimes M$. The analysis is simplest when one of the particles is spinless and the other is described by an irreducible representation of the rotation

ISBN 0-8053-6702-0/0-8053-6703-9, pbk

group, that is, when $L = D^r$ and $M = D^0$. Then $D^l \otimes L \otimes M = D^l \otimes D^r = D^{|l-r|} \oplus D^{|l-r|+1} \oplus D^{|l-r|+1} \oplus \cdots \oplus D^{|l+r|}$ and $n_1 n_2 = 2r+1$. Thus instead of $(2r+1)A^{E,l}$ we have $A^{E,|l-r|} \oplus A^{E,|l-r|+1} \oplus \cdots \oplus A^{E,|l+r|}$ and since $|l-r|+k$ does not determine l and k uniquely, each $A^{E,j}$ will occur with multiplicity greater than one. Indeed, for $j \geqslant r$, $A^{E,j}$ will occur with multiplicity $2r+1$ and for smaller values of j with smaller multiplicity. Our operators $S^{E,j}$ that define the scattering operator S will now be defined not by a complex number but by a unitary operator in a finite-dimensional vector space whose dimension is $\leqslant 2r+1$ and is equal to $2r+1$ for $j \geqslant r$. More generally we shall have $L = D^r$, $M = D^s$,

$$D^l \otimes L \otimes M = D^l \otimes \left[D^{|r-s|} \oplus D^{(r-s)+1} \oplus \cdots \oplus D^{|r+s|} \right]$$

and things become considerably more complicated. But it will still be true that each $A^{E,j}$ will occur with multiplicity $\leqslant n_1 n_2 = (2r+1)(2s+1)$ and with just this multiplicity for $j \geqslant r+s$.

To understand the physical significance of the finite-dimensional unitary operators defining the operators $S^{E,j}$ and hence S, we use the fact that $n_{E,j} A^{E,j}$ has a canonical decomposition into replicas of $A^{E,j}$ with one component for each way of writing j in the form $l + j_1 + j_2$, where

$$j_1 = r, r-1, r-2, \ldots$$

$$j_2 = s, s-1, s-2, \ldots$$

$$l = 0, 1, 2, \ldots$$

Using this $S^{E,j}$ determines and is determined by an $n_{E,j} \times n_{E,j}$ unitary matrix whose rows and columns are indexed by the triples l, j_1, j_2 with $l + j_1 + j_2 = j$. If we write this matrix in the form $e^{2i\delta}$, where δ is a self-adjoint matrix, we obtain a family of $(n_{E,j})^2$ function of E and j, which play the role of the phase shifts in the spinless case. We may suppress the j by letting j_1, j_2, and l vary freely, and then we have a function

$$E_{j_1, j_2; l, j_1', j_2', l'} = \delta\left(E_{j_1, j_2, l, j_1', j_2', l'}\right)$$

which determines S and is zero except when $j_1 + j_2 + l = j_1' + j_2' + l'$. The integers or half integers j_1, j_2, j_1', j_2' are the spin angular momenta of the incident and emergent particles (divided by $2\pi\hbar$).

Notes and References

The idea of using the S "matrix" (more properly the S operator) to describe the "scattering" of particles by one another in quantum mechanics was introduced by J. A. Wheeler in 1937 and again independently by W. Heisenberg in 1943 and 1944. In two well-known papers published in 1945 and 1946, Møller developed the mathematics underlying Heisenberg's ideas, and the S matrix is sometimes called the Møller wave matrix. A rigorous proof of the existence of Hilbert space limits of the form $\lim_{t\to\pm\infty} U_t V_t^{-1}(\phi)$ was given for the first time by J. M. Cook ("Convergence to the Møller wave matrix," *J. Math. Phys.*, vol. 36 (1957), 82–87). Cook dealt with the case in which $U_t = e^{itH_0}$ and $V_t = e^{itH}$, where H_0 is the negative of the Laplacian in Euclidean three-space and $H = H_0 + v$, where v is multiplication by a square integrable function. Other rigorous mathematical work soon followed; for example, J. M. Jauch, "On the theory of the scattering operator" (*Helv. Phys. Acta*, vol. 31 (1958), 127–158) and T. Kato and K. Kuroda, "On existence and unitary property of the scattering operator" (*Nuova Cimento*, vol. 12 (1959), 431–454).

A very readable account of the analyticity properties of the functions describing the S operator will be found in the book by R. Omnés and M. Fróssart, *Mandelstam Theory and Regge Poles*, (Benjamin-Cummings, Reading, Mass. (1963)). The book of G. F. Chew *S Matrix Theory of Strong Interactions* (Benjamin-Cummings, Reading, Mass. (1961)) is in part a reprint collection and includes copies of the basic papers of Mandelstam and Regge alluded to in the text.

An extensive treatise dealing in detail with all aspects of scattering theory was published in 1966 by R. G. Newton. (*Scattering Theory of Waves and Particles*, McGraw-Hill, New York). Newton points out that quantum mechanical scattering has many features in common with the classical theory of the scattering of electromagnetic waves by an obstacle and begins his book with a lengthy treatment of this classical theory. The parallelism is hardly surprising since (cf. Section 22) the propagation of electromagnetic waves is also governed by a one-parameter unitary group in a Hilbert space, in this case, the Hilbert space of all states of the electromagnetic field that have finite total energy. In either case, one is studying the relationship between two unitary group representations, one of which is (in a certain sense) asymptotically equivalent to a subrepresentation of the other.

In work on the scattering of sound waves (cf. *Scattering Theory*, Academic, New York (1967)), Lax and Phillips deal with their problem in just the abstract setting described in the preceding paragraph. In addition to making contact with the quantum mechanical developments, they present a rather interesting new approach to the abstract problem that in particular provides a useful alternative method for establishing the analyticity properties of the function describing the S operator. The basic idea may be explained briefly as follows: Given a concrete unitary representation $t \to U_t$ of the additive group of the real line, physical considerations may permit one to discover closed subspaces D_+ and D_- of $\mathcal{H}(U)$ such

ISBN 0-8053-6702-0/0-8053-6703-9, pbk

that (1) $U_t(D_+)\subseteq D_+$ and $U_{-t}(D_-)\subseteq D_-$ for all $t>0$; (2) $\cap_{t>0}U_t(D_+)=\cap_{t<0}U_t(D_-)=0$; and (3) $\cup_{t>0}U_t(D_+)$ and $\cap_{t<0}U_t(D_-)$ are dense in $\mathcal{H}(U)$. One verifies at once that the ranges of the $U_t(D_+)$ as t varies over all real numbers define a system of imprimitivity P^+ for U based on the natural action of the real line on itself. Similarly, the $U_t(D_-)$ lead to another such system of imprimitivity P^-. By the imprimitivity theorem, the pair U,P^+ (resp. U,P^-) is determined to within unitary equivalence by the choice of a cardinal number d^+ (resp. d^-) which may be a positive integer or $\aleph_i$. The concrete circumstances often enable one to prove that $d^+=d^-$ and hence that there is an essentially unique unitary operator S that commutes with all U_t and maps P^+ onto P^-. It turns out that S may be identified with the scattering operator and that the concrete circumstances often allow one to prove that D_+ and D_- are orthogonal. The analytical properties of S follow easily using ideas going back to the work of Paley and Wiener on Fourier analysis in the complex domain (cf. also Y. Foures and I. E. Segal, "Causality and Analyticity," *Trans. Am. Math. Soc.*, vol. 78 (1955), 385–405).

The applications of the theory of unitary group representations to the theory of automorphic forms described in Section 30 are connected with the use of self-adjoint operators closely related to those that occur in the quantum mechanics of interacting particles. Thus, bizarre as it may seem to some, the methods developed for the analysis of quantum mechanical scattering have found important uses in the theory of automorphic forms. The reader will find further details in the notes and references to Section 30.

When more than two particles are involved, scattering theory becomes much more complicated because the particles may clump together to form various bound states which then act like particles and scatter one another. One speaks of "multi-channel scattering" as opposed to "single-channel scattering." Each way of grouping the particles into subsets that form bound states corresponds to a distinct "channel." For further details see Chapters 16 and 17 of the book of Newton cited above. An important beginning on the rigorous mathematics of multi-channel scattering was made by L. D. Faddeev in 1963 ("Mathematical Aspects of the Three-Body Problem in Quantum Scattering Theory," *Steklov Math. Inst. Leningrad Publ.*, vol. 69 (1963)). A brief account of Faddeev's work will be found in the last section of Chapter 17 of Newton's book.

The material in the text on applying the tensor product theorem for induced representations to two-particle scattering was worked out by the author in the fall of 1965 without knowledge of the parallel and slightly earlier investigations of Moussa and Stora and of Voisin. For further details and for references to this work and the later work of W. H. Klink, see page 351 of the published version of the author's Chicago lecture notes.

ISBN 0 8053 6703 0/0 8053 6702 9 pbk

22. *Radiation, special relativity, and field theory*

Although it suffices for many purposes to assume the interaction between charged particles to be as described in Section 19, this description is only a first approximation to a much more complicated interaction in which "magnetic forces" are involved as well. Until 1819 electricity and magnetism were regarded as similar but quite independent phenomena. However, in that year Oersted discovered that an electric current in a wire could deflect a compass needle, and further work by Ampere and others led to the conclusion that such a current produces a "magnetic field" indistinguishable from that produced by a suitable distribution of "magnetic poles." Somewhat later Faraday discovered that conversely a current will be induced in a wire moving with respect to a magnetic field. Developing these discoveries into a more or less complete unified theory of electromagnetism was one of the great tasks and triumphs of nineteenth century physics.

As developed by Faraday and his successor Maxwell, this theory took the "field" rather than the "action at a distance" point of view. When a charged particle moves under the influence of other charged particles, one can take the view that the first particle does not respond directly to the position of the others but instead to some change in the space in its immediate neighborhood caused by the presence of the other particles. One says that the particles have produced a "field of force" in space and that each particle moves as a result of the nature of the field in its immediate neighborhood. Up to a point it is simply a matter of taste

George W. Mackey, Unitary Group Representations in Physics, Probability, and Number Theory

ISBN 0-8053-6702-0/0-8053-6703-9, pbk

whether one thinks in terms of fields or particles. One can define the field produced by a system of point charges and the action of this field on a charge in such a way that one obtains exactly the same predicted motion as in the Newtonian action at a distance theory. However, one can also have a field theory in which changes in the field are propagated with a finite velocity and such a theory is not equivalent to an action at a distance theory in any familar sense. Faraday and Maxwell found that electric charges interact through a field of the latter sort. The celebrated Maxwell equations state that in parts of space not occupied by charges the electric and magnetic fields caused by any system of fixed or moving charges and magnets change with time and are related to one another in such a way that the following partial differential equations are satisfied:

$$\frac{\partial \vec{E}}{\partial t} = c \operatorname{curl} H, \qquad \frac{\partial \vec{H}}{\partial t} = -c \operatorname{curl} E$$
$$\operatorname{div} \vec{H} = 0, \qquad \operatorname{div} \vec{E} = 0$$

Here $\vec{E}$ and $\vec{H}$ are the vector-valued functions on space that describe the value of the electric and magnetic fields at each point and c is a universal constant. The constant c has the dimensions of a velocity and turns out to be numerically equal to the velocity of light.

If we differentiate the first equation above with respect to t and use the second to eliminate $\partial \vec{H}/\partial t$, we find that $\partial^2 \vec{E}/\partial t^2 = -c^2 \operatorname{curl}(\operatorname{curl} \vec{E})$, and since $\operatorname{div} \vec{E} = 0$, this may be simplified to

$$\frac{1}{c^2}\frac{\partial^2 \vec{E}}{\partial t^2} = \frac{\partial^2 \vec{E}}{\partial x^2} + \frac{\partial^2 \vec{E}}{\partial y^2} + \frac{\partial^2 \vec{E}}{\partial z^2}$$

Similarly, one shows that

$$\frac{1}{c^2}\frac{\partial^2 \vec{H}}{\partial t^2} = \frac{\partial^2 \vec{H}}{\partial x^2} + \frac{\partial^2 \vec{H}}{\partial y^2} + \frac{\partial^2 \vec{H}}{\partial z^2}$$

Now the partial differential equation

$$\frac{\partial^2 \phi}{\partial x^2} + \frac{\partial^2 \phi}{\partial y^2} + \frac{\partial^2 \phi}{\partial z^2} = \frac{1}{v^2}\frac{\partial^2 \phi}{\partial t^2}$$

is known as the "wave equation" and, in a sense that can easily be made precise, its solutions are "waves" propagated with velocity v. We see then that electric and magnetic fields change in space and time according to the

ISBN 0-8053-6702-0/0-8053-6703-9, pbk

laws of wave motion and that these waves are propagated with the velocity of light.

This observation suggested to Maxwell the hypothesis that light itself might consist of such electromagnetic waves and the subsequent confirmation of this hypothesis settled one of the great mysteries of physics. The long controversy over whether light consisted of waves or particles had been more or less settled in favor of waves by the work of Young and Fresnel in the first quarter of the nineteenth century. However, a question remained as to what was "waving." A comparison of the mathematical theory necessary to explain optical phenomena with the theory of wave propagation in elastic bodies had led to rather implausible hypotheses concerning the hypothetical "ether" whose vibrations were supposed to produce light.

One important consequence of Maxwell's equations is that charged particles under their mutual interactions do not satisfy the law of conservation of energy in the usual sense. This law can be recovered only by assigning energy to the electromagnetic field itself. This can be done by supposing the energy to be spread throughout space with a density equal to

$$\frac{|\vec{E}|^2+|\vec{H}|^2}{8\pi}$$

When a charged body oscillates back and forth, it gradually loses its energy and this is then "radiated" out into space as energy in the electromagnetic field.

A highly significant fact about Maxwell's equations is that they are not invariant under the Galilean group. If they hold for one observer, they will not hold for an observer moving uniformly with respect to him. In particular, the velocity of propagation of electromagnetic waves will appear different to the two observers. Indeed, for the moving observer this propagation velocity will be different in different directions.

For these reasons it was supposed that there must be such a thing as absolute space and the famous Michelson–Morely experiment was designed to measure the velocity of the earth relative to the "ether" by delicate measurements of the velocity of light in various directions. To everyone's consternation and amazement, the results of the experiment were negative. It seemed that the earth was always at rest in the ether no matter where it was in its orbit around the sun. Explaining this apparently absurd result was a major challenge to physicists which was met by the celebrated special theory of relativity of Poincaré, Lorentz, and Einstein. The basic idea is that Maxwell's equations, while not invariant under the Galilean group, are invariant under a closely related group of one-to-one

ISBN 0-8053-6702-0/0-8053-6703-9, pbk

transformations of space–time onto itself. This is the group generated by the rotations, space–time translations, and transformations of the form

$$x,y,z,t \to \frac{x-v_1t}{(1-v^2/c^2)^{1/2}}, \frac{y-v_2t}{(1-v^2/c^2)^{1/2}}, \frac{z-v_3t}{(1-v^2/c^2)^{1/2}}, \frac{t-\dfrac{(xv_1+yv_2+zv_3)}{c^2}}{(1-v^2/c^2)^{1/2}}$$

where $v^2=v_1^2+v_2^2+v_3^2$ and v_1, v_2, v_3 are arbitrary real numbers. It is a semi-direct product of the space–time translations and the group generated by the rotations and the transformations described above. This six-parameter Lie group may also be described as the connected component of the identity in the group of all linear transformations of space–time into itself that leave invariant the fundamental form $x^2+y^2+z^2-c^2t^2$. It is called the (proper) homogeneous Lorentz group and the semi-direct product of it with the group of space–time translations is called the inhomogeneous Lorentz group or the Poincaré group. If this were the true symmetry group of space–time, then electromagnetic phenomena would look the same to all observers and the Michelson–Morely experiment would be explained. Einstein's great contribution was to give convincing a priori reasons for supposing that the Poincaré group should replace the Galilean group. In broad terms he argued as follows: If we are to question the absoluteness of space, as we do in Galilean relativity, then we should also question the absoluteness of time. Einstein's analysis of the meaning, if any, to be attached to absolute time led to the conclusion that time could well depend on the observer, and, that if it did, the proper equations relating one observer to another moving with uniform velocity $v=v_1,v_2,v_3$ with respect to him were not the obvious ones $x'=x-v_1t$, $y'=y-v_2t$, $z'=z-v_3t$, and $t'=t$, but were, instead,

$$x'=\frac{x-v_1t}{(1-v^2/c^2)^{1/2}}, \quad y'=\frac{y-v_2t}{(1-v^2/c^2)^{1/2}}, \quad z'=\frac{z-v_3t}{(1-v^2/c^2)^{1/2}}$$

$$t'=\frac{t-\dfrac{(xv_1+yv_2+zv_3)}{c^2}}{(1-v^2/c^2)^{1/2}}, \quad \text{where } v^2=v_1^2+v_2^2+v_3^2$$

and c is a universal constant having the dimensions of a velocity. On this basis, the symmetry group of space–time is just the Poincaré group with c

ISBN 0-8053-6702-0/0-8053-6703-9, pbk

the velocity of light and the result of the Michelson–Morely experiment becomes intelligible.

However, if the Poincaré group is really a symmetry group for space–time and there is no absolute time, then we must re-think all the work we have done so far that is based on Galilean relativity and see if it can be modified so as to be invariant under the Poincaré group. This turns out to be rather a large order and physicists are still far from having achieved a consistent theory that includes classical mechanics as a limiting case and is invariant under the Poincaré group. An approach to the problem that has dominated work on the subject ever since was suggested by an epoch-making paper of Dirac published in 1927.

Let the configuration space of a classical Hamiltonian system be an n-dimensional vector space and let $\phi_1,\ldots,\phi_n$ be a basis. Then the general member ϕ may be written uniquely in the form $q_1\phi_1 + \cdots + q_n\phi_n$ and the equations of motion in Hamiltonian form are

$$\frac{dq_i}{dt} = \frac{\partial H}{\partial p_i}, \quad \frac{dp_i}{dt} = -\frac{\partial H}{\partial q_i}$$

where $p_1,\ldots,p_n$ are suitable auxiliary variables and H is a function of the p's and q's called the Hamiltonian. In the particular case in which H is of the form

$$\sum_{i,j=1,2,\ldots n} \frac{\mu_{ij}}{2} p_i p_j + \sum_{i,j=1,2,\ldots n} \frac{a_{ij}}{2} q_i q_j$$

that is, when the kinetic and potential energies are both quadratic forms we see that

$$\frac{\partial H}{\partial p_i} = \sum_{j=1,2,\ldots n} \mu_{ij} p_j, \quad \frac{\partial H}{\partial q_i} = \sum_{j=1,2,\ldots n} a_{ij} q_j$$

so that we may write $d\phi/dt = T(\theta)$, $d\theta/dt = -V(\phi)$, where $\phi = \sum_{i=1,2,\ldots n} q_i\phi_i$, $\theta = \sum_{i=1,2,\ldots n} p_i\phi_i$, and T and V are linear transformations. These equations are analogous to Maxwell's equations

$$\frac{1}{c}\frac{\partial \vec{E}}{\partial t} = \operatorname{curl} \vec{H}, \quad \frac{\partial \vec{H}}{\partial t} = -\frac{1}{c}\operatorname{curl} \vec{E}$$

with $\vec{E}$ playing the role of the configuration vector. As a matter of fact, it is possible to define the notion of dynamical system in abstract terms in such a way that an electromagnetic field is a dynamical system whose total

ISBN 0-8053-6702-0/0-8053-6703-9, pbk

energy is

$$\frac{1}{8\pi}\iiint\left(|\vec{E}(x,y,z)|^2+|\vec{H}(x,y,z)|^2\right)dx\,dy\,dz$$

at any given instant of time, whose configuration is given by the vector field $x,y,z\to\vec{E}(x,y,z)$ and whose rate of change is given by the vector field $x,y,z\to\vec{H}(x,y,z)$ via the relationship $\partial\vec{E}/\partial t=c\,\mathrm{curl}\,\vec{H}$. Dirac showed that it was possible to "quantize" the electromagnetic field regarded as a dynamical system, that is, to find a quantum mechanical system with self-adjoint operators corresponding to the field observables whose relationship to the original dynamical system is the same as that of a quantum mechanical system defining n interacting particles, to its classical limit.

Quite generally, given a classical dynamical system whose Hamiltonian is at most quadratic in the p's, there is a canonical way of writing down a canonical quantum mechanical refinement or "quantization." We have spelled this out in detail in an earlier section for the special case in which the quadratic term is in diagonal form. Unfortunately, this general procedure breaks down when configuration space is not finite-dimensional as we shall explain in greater detail below. However, the case in which H is the sum of quadratic forms in the position and momentum vectors as illustrated above has special features that allow one to circumvent these difficulties. One special feature is that the classical equations of motion are linear. Another is that the system can be analyzed into independent one-dimensional systems. Suppose that we have such a system in which configuration space is finite-dimensional. As shown on pages 29–34 of the author's book *Mathematical Foundations of Quantum Mechanics*, one can convert the $2n$-dimensional real vector space that is the phase space of this system into an n-dimensional complex Hilbert space $\mathcal{H}_c$ in such a fashion that the classical dynamical group becomes a one-parameter unitary group $t\to V_t^c$. Now let $\mathcal{H}_q$ denote the Hilbert space of this system quantized according to the rules alluded to above and let $t\to V_t^q$ denote the corresponding quantum dynamical group. It can be shown that the two one-parameter unitary groups are related in the following simple and significant fashion:

$$V^q\simeq I\oplus V^c\oplus V^c \circledS V^c \oplus V^c \circledS V^c \circledS V^c+\cdots$$

Here I is the one-dimensional identity representation of the additive group of the real line and $V^c \circledS V^c \circledS \cdots \circledS V^c$ (with n factors) is the subrepresentation of the n-fold tensor product of V^c with itself defined by the in-

ISBN 0-8053-6702-0/0-8053-6703-9, pbk

variant subspace of all *symmetric* tensors. Not only can we thus read off the dynamical group of the quantized system from that of the unquantized one; in addition, there are simple rules (which we shall not state) describing the position and momentum observables in terms of vectors in $\mathcal{H}(V^c)$.

One interesting fact about this representation of the quantization is that it makes no use of finite dimensionality. All aspects of the construction work equally well when $\mathcal{H}(V^c)$ is infinite-dimensional. This is interesting because one can also form a V^c for the infinite-dimensional dynamical system whose phase space is the set of all electric and magnetic fields of finite total energy. One can convert this phase space into a complex Hilbert space and find a one-parameter unitary group $t \rightarrow V_t$ that gives the same changes with time of the electromagnetic field as one obtains by integrating Maxwell's equations. It thus suggests itself that we should quantize Maxwell's equations by applying the finite-dimensional construion with V^c replaced by the V that integrates these equations. Although his approach and language were different, this is in effect what Dirac did. The electromagnetic field in quantum mechanics is the quantum system whose Hilbert space and dynamical group are constructed from the V described above by the familar

$$I \oplus V \oplus V \textcircled{s} V \oplus V \textcircled{s} V \textcircled{s} V \oplus \cdots$$

Perhaps the single most interesting consequence of writing the quantization in the indicated form is that it shows that the quantized electromagnetic field may also be regarded as a system consisting of an indefinite number of quantized particles. To see how this comes about and what we mean by it, let us first notice that the representation $t \rightarrow V_t$ of the real line may be extended in a natural way to be an irreducible unitary representation of the Poincaré group. We shall not give details but suggest only that one is led to a natural proof by considering the Lorentz invariance of Maxwell's equations. Next recall that our analysis of the possible one-particle systems led to a one-to-one correspondence between such systems and certain irreducible projective unitary representations of the Galilean group. A straightforward adaptation of the argument given there shows that if we replace the hypothesis of Galilean invariance by that of invariance under the Poincaré group, we obtain a similar correspondence between oneparticle systems and certain irreducible representations of the Poincaré group. Just as before the representations that arise are parametrized by two parameters—a continuous one that is identified with the mass of the particle and a discrete one that defines its spin. The principle difference between the two cases is that with the Poincaré group there are only two distinct multipliers (corresponding to integral and half integral

ISBN 0-8053-6702-0/0-8053-6703-9, pbk

spin); the mass parameter parametrizes different representations with the same multiplier rather than different multipliers. Suppose for the moment (contrary to fact) that the representation V of the Poincaré group which arises in the Dirac quantization of the electromagnetic field is one of those arising from what we have called a one-particle system. Then if we confine our attention to the state vectors in $\mathcal{H}(V)$, we may think of our quantum system as *being* a particle with a definite mass and spin. Of course, $\mathcal{H}(V)$ is a fairly small subspace of the whole system. How are we to interpret the rest of it? Well, if our particles are bosons, then the Hilbert space for two identical ones is $\mathcal{H}(V ⓢ V)$, that for three identical ones in $\mathcal{H}(V ⓢ V ⓢ V)$, and so forth. Thus except for $\mathcal{H}(I)$ we may interpret each summand as being the Hilbert space for an n-particle system where the particles are identical bosons with a certain definite mass and spin. $\mathcal{H}(I)$ itself is one dimensional and may be thought of as the unique state in which there are no particles—the vacuum state so to speak. Finally, then, by concentrating our attention on different sets of basic observables, we may think of our quantum system as simultaneously representating:

(a) a quantum version of the electromagnetic field;
(b) a system with an indefinite number of identical particles.

This is all very suggestive, but we must not forget that V is *not* one of the representations of the Poincaré group arising from what we have called a one-particle system. However, it is in a certain sense a limit of a sequence of such representations with the mass parameter approaching zero. The quantum system corresponding to this representation is like a one-particle system in many respects; in particular, it has a well-defined spin and particlelike momentum and angular momentum observables. However, it fails to have position observables. The exact meaning of this lack is hard to pin down. All we can say is that the quantum system in question has enough particlelike properties so that the physicists have decided to accord it particle status. They call it the photon. The dual wave–particle nature of light (and electromagnetic radiation in general) that is described so paradoxically by popular writers on quantum theory is quite satisfactorily and consistently explained by the fact that the quantized electromagnetic field may be reinterpreted as a system of photons. It is misleading to say that light is sometimes a wave and sometimes a particle. It is always something a little more complicated than either and this more complicated something may be looked at from two different points of view. From one point of view it is a quantized electromagnetic field; from another it is a system of an indefinite number of independently moving photons.

In an earlier section we learned that electrons are "indistinguishable" in the precise sense that in considering a system of several of them we must

ISBN 0-8053-6702-0/0-8053-6703-9, pbk

replace the tensor product of the one-particle Hilbert spaces by a certain subspace and this eliminates all observables that would permit us to distinguish one permutation of the electrons from another. Now we see that "photons" are indistinguishable in the same sense. Moreover, we have obtained this indistinguishability "automatically" so to speak as a consequence of the fact that photons come to us as the "quanta" of a field. We do not have to postulate it ad hoc so as to obtain agreement with experiment. This suggests the possibility that perhaps electrons and other elementary particles can be obtained by "quantizing" suitable analogs of the electromagnetic field.

As far as bosons are concerned, the truth of this supposition is quite easy to establish. Let $\mathcal{H}$ be the Hilbert space of the one particle system and let $t \rightarrow V_t$ be the dynamical group. Recall that $\mathcal{H}$ has a canonical realization as $\mathcal{L}^2(S, \mathcal{H}_c)$, where S is physical space and $\mathcal{H}_c$ is a finite-dimensional Hilbert space. Thus each element of $\mathcal{H}$ is a "vector" field on space and our dynamical group tells us how these fields change with time. Proceeding by analogy with Dirac's treatment of the electromagnetic field, one shows that $\mathcal{H}$ may be regarded as the phase space of a dynamical system and that this dynamical system may be quantized. Just as in the electromagnetic field case one finds that the dynamical group of the resulting quantum system is

$$I \oplus V \oplus V \circledS V \oplus V \circledS V \circledS V + \cdots$$

Since one obtains the system $I \oplus V \oplus V \circledS V \cdots$ by first quantizing a classical one-particle system, viewing the result as a classical field and quantizing again one speaks of "second quantization."

Electrons and other fermions may also be obtained as field quanta but only in an indirect and somewhat artificial manner. One regards the one-particle Hilbert space as the phase space of a classical dynamical system, as for bosons, and one finds that the effective Hamiltonian may be expressed as a quadratic form in any linear coordinate system of p's and q's. However, in passing to a corresponding quantum system one replaces the Heisenberg commutation rules

$$Q_j Q_k - Q_k Q_j = P_j P_k - P_k P_j = 0, \quad Q_j P_k - P_k Q_j = -i\delta_k^j \hbar$$

by the so-called "anticommutation rules":

$$Q_j Q_k + Q_k Q_j = P_j P_k + P_k P_j = 2\delta_k^j, \quad Q_j P_k + P_k Q_j = 0$$

Moreover, in forming the quantum Hamiltonian one writes the classical one in the form

$$\sum_{j,k=1,2} A_{jk}(q_j + ip_j)\overline{(q_k + ip_k)}$$

ISBN 0-8053-6702-0/0-8053-6703-9, pbk

and replaces each $q_j \pm ip_j$ by $Q_j \pm iP_j$. We notice in this connection that

$$p^2 + q^2 = (p + iq)(p - iq)$$

but that

$$P^2 + Q^2 = (P + iQ)(P - iQ)$$

if and only if P and Q commute.

Just as with the commutation relations, there are serious ambiguities in actually carrying out this procedure whenever there are infinitely many P_j and Q_j. In particular, the anticommutation relations do not determine the P_j and Q_j to within unitary equivalence except when there are only finitely many. On the other hand, if one takes a finite-dimensional dynamical system with a linear phase space and a quadratic Hamiltonian, the indicated procedure is well defined and leads to a "quantization" whose dynamical group may be obtained from the classical dynamical group just as in quantization through the commutation rules except that now the symmetrized products are replaced by antisymmetrized ones. We now argue, by analogy with the finite-dimensional case, to the conclusion that the result of quantizing the field defined by the quantum mechanics of a single fermion is $I + V + V \textcircled{A} V + \cdots$, where V is the dynamical group for the one-particle system.

Unfortunately, there seems to be no analog for the anticommutation relations of the argument that allows us to deduce the commutation relations from Euclidean invariance. The anticommutation relations must simply be written down ad hoc as a plausible formal modification of the commutation relations; a modification that happens to lead to useful procedures and points of view.

Now obtaining particles as field quanta is not just a way of understanding indistinguishability. It is interesting and important because it seems to offer an approach to dealing with particle interactions in a relativistically invariant manner. The idea is that one should think of fields rather than particles as the basic entities of the universe and that one should study interactions between fields rather than interactions between particles.

From this point of view one thinks of the interaction between electrons as arising from the interactions of the individual electrons with the electromagnetic field and this latter interaction as arising from the interaction of the "electron field" with the electromagnetic field. More precisely, let W^{Ph} be the irreducible representation of the inhomogeneous Lorentz group L that defines the quantum mechanics of a single photon, and let W^e be that

ISBN 0-8053-6702-0/0-8053-6703-9, pbk

for a single electron. Then

$$W^{Ph}=I\oplus W^{Ph}\oplus\left(W^{Ph}\circledS W^{Ph}\right)\oplus\left(W^{Ph}\circledS W^{Ph}\circledS W^{Ph}\right)\oplus\cdots$$

defines the quantized electromagnetic field and

$$W^{e}=I\oplus W^{e}\oplus\left(W^{e}\textcircled{A} W^{e}\right)\oplus\left(W^{e}\textcircled{A} W^{e}\textcircled{A} W^{e}\right)\oplus\cdots$$

defines the quantized electron field. The system describing all electrons and photons in the universe would then be described by the tensor product $W^{Ph}\otimes W^{e}$ of these two quantized fields. Using the distributive law, the tensor product may be written as a direct sum of terms of the form

$$\overbrace{\left(W^{e}\textcircled{A} W^{e}\textcircled{A}\cdots\textcircled{A} W^{e}\right)}^{f\text{ times}}\otimes\overbrace{\left(W^{Ph}\circledS W^{Ph}\circledS\cdots\circledS W^{Ph}\right)}^{k\text{ times}}$$

State vectors in this subspace are supposed to define states of the system in which there are f electrons and k photons. Clearly the unit vector of the subspace corresponding to I describes the state in which $f=k=0$. This is called the *vacuum state*. The interaction between electrons and electrons and between photons and electrons is obtained by postulating an interaction energy term in the direct product of the classical fields and attempting to "quantize" the infinite-dimensional dynamical system which results. More generally, one postulates a field for each type of particle and a sum of "interaction terms" in the product of the phase spaces of these fields. One hopes to obtain the quantum system of all the particles in the universe by quantizing this infinite-dimensional dynamical system.

Our description of the fields used to discuss the interactions of electrons is incomplete in that we have left out the "positron." Given any unitary representation L of a group G, we may define a further representation $\overline{L}$ of G by setting $\overline{L}_x=L^0_{x^{-1}}$, where $L^0_{x^{-1}}$ is the Banach space adjoint of $L_{x^{-1}}$, that is, the operator in the Banach space dual of $\mathcal{H}(L)$ that satisfies $L^0_{x^{-1}}(l)(\phi)=l(L_{x^{-1}}(\phi))$ for all ϕ in $\mathcal{H}(L)$ and all l in $\overline{\mathcal{H}(L)}$. The Hilbert space adjoint $L^*_{x^{-1}}=L_x$ is $JL^0_{x^{-1}}J^{-1}$, where $J(\psi)$ is the linear functional $\phi\to(\phi,\psi)$. J is an anti unitary map and does not set up an equivalence between L and $\overline{L}$ as unitary representation of C. In fact, they can be inequivalent. Now let L be the representation of the inhomogeneous Lorentz group L describing a single free particle. While $\overline{L}$ and L will be inequivalent as representations of L (except for mass zero particles like the photon), the particles they define will have exactly the same properties as free particles because our anti-unitary operator defines a quantum mechanical isomorphism. On the other hand, it is easy to show that the representations $\overline{L}\otimes L$ and $L\otimes L$ are

ISBN 0-8053-6702-0/0-8053-6703-9, pbk

neither unitarily equivalent nor anti-unitarily equivalent to one another. Thus as soon as one considers systems of more than one particle, one has to make a distinction between L and $\overline{L}$. One says that the particle defined by $\overline{L}$ is the *anti-particle* of the defined by L. Particles and their anti-particles behave in the same way when taken in isolation, but there is a difference between a system consisting of two particles and the system obtained by replacing first one of them by its anti-particle. The anti-particle of the electron is called the positron and the interaction between electrons and the electromagnetic field cannot be described without introducing the positron field as well as the electron field. The dynamical group of the interacting electron, positron, and electromagnetic fields does not leave invariant the subspace in which there are no positrons. The tensor product of the electron and positron fields is sometimes called the electron-positron field or the Dirac field.

The program of obtaining relativistic particle interactions by "quantizing" a system of interacting classical fields is by no means easy to carry out and has in fact met with only limited success. The basic difficulty seems to be that "quantization" is a much less clearly defined procedure for systems with an infinite number of degrees of freedom than it is for those with only a finite number. For example, when $\mathfrak{M}$ is a finite-dimensional vector space, it admits an essentially unique translation invariant measure μ so that we have a canonically associated Hilbert space $L^2(\mathfrak{M},\mu)$. However, when $\mathfrak{M}$ is infinite-dimensional, such a measure no longer exists and there is no substitute for it that has anything like the same uniqueness properties. What then is to be the Hilbert space and its relationship to the configuration observables when we are dealing with a system whose configuration space is infinite-dimensional? From another point of view the difficulty presents itself as the existence of a great number of inequivalent irreducible solutions of the Heisenberg commutation relations with no way of deciding which is appropriate to a particular problem. There are other difficulties as well, and it must be confessed that from the point of view of rigorous mathematics we simply do not know how to quantize a dynamical system with an infinite-dimensional configuration space—except when the classical equations of motion are linear. In that special case (e.g., the free electromagnetic field) we were able to recast the finite-dimensional quantization in a form that made sense in the infinite-dimensional case as well. However, nothing similar seems to be possible more generally and the classical equation of motion for a system of interacting fields is not linear for any of the interesting interactions. One may attribute our ability to deal with the linear case to the fact that in that case the system is in fact equivalent to an infinite set of *independent* finite-dimensional systems.

ISBN 0-8053-6702-0/0-8053-6703-9, pbk

The physicists have approached the problem of quantizing the infinite-dimensional dynamical system represented by a set of interacting fields by way of formal procedures that mimic those which work in finite dimensions, but invoke expressions with no well-defined mathematical meaning. In applying these procedures to the solution of actual problems, some dubious auxiliary hypotheses are made that lead finally to power-series expansions for the quantities desired. The power-series expansions have coefficients that are multiple integrals, but, except for the first term or so, these integrals diverge. Using great imagination and ingenuity, the physicists have succeeded in finding rules for replacing these divergent integrals by finite numbers, and giving semi-plausible arguments justifying the replacements. Moreover, using the first few terms of the resulting "renormalized" series, they obtain remarkable agreement with experiment, at least in dealing with interactions between electrons and photons. On the other hand, there is reason to believe that the series does not actually converge, and for the interaction between mesons, protons, and neutrons—so important for the theory of the nucleus—one does not even get reasonable approximations from the partial sums. The situation may perhaps be compared to what would happen if one had a function of the form $g(x) \equiv \phi(x)e^{-1/x^2} + P(x)$, where P is a power series and ϕ is completely unknown. Assuming the whole function to be a power series converging near 0, and expanding it, one would arrive at the false result $g(x) \equiv P(x)$, but if $P(x)$ were the dominant term near zero, this false result would yield good approximations to g for x near zero. However, it would be almost worthless for values of x near which $\phi(x)e^{-1/x^2}$ is the dominant term. The renormalized power series may be only the analog of $P(x)$, and $\phi(x)e^{-1/x^2}$ may still be completely eluding us. It seems fair to say, in spite of all the work and some remarkable success, that the physicists have told us neither how to quantize "nonlinear" infinite-dimensional dynamical systems in general, nor interacting fields in particular.

To see what the difficulties are, and how far we can go, let us approach the problem of quantizing a field in the same way as we approached (and solved) the problem of quantizing a free particle. Let G be a separable locally compact group, and let $\mathfrak{M}$ be a standard Borel G-space with an invariant measure ν. For concreteness one should think of $\mathfrak{M}$ as physical space, and G as the group of all rigid motions of $\mathfrak{M}$, but our argument will in no way depend upon such a special choice, and in a sense we shall be discussing what might be called "abstract field theory." We wish to consider the quantum mechanics of a system whose configuration space is a subspace $\mathfrak{F}$ of the space of all real-valued locally summable Borel functions on $\mathfrak{M}$. We identify functions that are equal almost everywhere, and suppose that $\mathfrak{F}$ is invariant under the linear transformations $f \rightarrow V_x f$,

ISBN 0-8053-6702-0/0-8053-6703-9, pbk

where $V_x f(m) = f(mx)$. Let $\mathcal{L}^0(\mathfrak{M},\nu)$ be the subspace of $\mathcal{L}^1(\mathfrak{M},\nu)$ consisting of those functions that take on only a finite number of values. Then each $\phi \in \mathcal{L}^0(\mathfrak{M},\nu)$ defines a linear functional on $\mathfrak{F}$ via

$$l_\phi(f) = \int_{\mathfrak{M}} f(m)\phi(m)\,d\nu(m)$$

By a (static) quantization of our system, we shall mean an assignment of a self-adjoint operator A_ϕ to each ϕ in such a way that the following conditions are satisfied:

(1) the A_ϕ commute with one another;

(2) $\phi \to A_\phi$ is linear.

We shall say that the quantization is G invariant if there exists a multiplier σ for G and a unitary σ-representation U of G such that $A_{\phi x} = U_x^{-1} A_\phi U_x$ for all x and ϕ. Here $(\phi x)(m) = \phi(mx)$.

Given any pair A, U subject to the above restrictions, let B_A denote the complete Boolean algebra of projections generated by the ranges of the projection valued measures associated to the A_ϕ by the spectral theorem. Then since $A_{\phi x} = U_x^{-1} A_\phi U_x$, it follows that $U_x^{-1} B_A U_x = B_A$. Suppose for the moment that B_A is an atomic Boolean algebra, and let S be the set of all of its atoms. Then $U_x^{-1} s U_x \in S$ for all s, so S is a G-space. Moreover, every $Q \in B_A$ will be of the form $\Sigma_{s \in E} s$ for some subset E of S. Thus there will exist a countable G-space S and a system of imprimitivity P for U based on S such that B_A is the range of P. Moreover, S and P are unique. Now it turns out to be possible to prove a continuous analog of this result. For any B_A there exists a standard Borel G-space S and a system of imprimitivity P for U based on S such that B_A is the range of P. Moreover, S and P are unique in the sense that if S' and P' have the same properties then $\exists$ a Borel isomorphism θ from an invariant Borel subset S_0 of S to an invariant Borel subset S_0' of S' such that:

(1) $\theta(sx) = \theta(s)x$ for all s and x;

(2) $P'_{\theta(E)} = P_E$ for all E;

(3) $S - S_0$ and $S' - S_0'$ have measure zero, in the sense that $P_{S-S_0} = 0 = P'_{S'-S_0'}$.

As is well known, a projection valued measure P is always associated with a unique measure class C_P such that $P_E = 0$ if and only if E is a C_P null set. When this projection valued measure is a system of imprimitivity, the measure class is necessarily invariant.

ISBN 0-8053-6702-0/0-8053-6703-9, pbk

It follows then from the preceding that a static G-invariant quantization of our system yields an invariantly associated G-space S with an invariant measure class C. By the usual direct integral techniques, any static G-in-variant quantization may be written as a continuous sum of quantizations in which the invariant measure class C is ergodic. We shall therefore concentrate attention on the ergodic case. This is doubly justified because there is reason to believe that this is the only one that arises in systems of physical interest. For example, when one quantizes a field with a quadratic potential by the trick alluded to above, the S,C that arises is ergodic. It follows that P must be uniformly k dimensional for some k, and hence that U must be induced by some k-dimensional unitary representation L of the virtual subgroup of G defined by the ergodic action on S,C. The special case in which this unitary representation of the virtual subgroup is the one-dimensional identity is the easiest to analyze further, and is the only one to have been considered seriously in the physical literature. We are not convinced that other cases can be ruled out, but will refrain from considering them further here. Assuming that L is the one-dimensional identity and that C has a finite-invariant measure is the same as assuming that the A_ϕ form a maximal commuting set of observables, and that U admits an invariant one-dimensional subspace. We shall lump these together with ergodicity, and speak of the e.m.v. assumption (e for ergodic, m for maximal, v for vacuum).

Assuming e.m.v., let μ be the unique invariant probability measure in C. Then we may realize $\mathcal{H}(U)=\mathcal{H}(P)$ as $\mathcal{L}^2(S,\mu)$ so that $U_x f(s)=f(sx)$ and P_E is multiplication by the characteristic function of E. The function 1 is a unit vector in $\mathcal{L}^2(S,\mu)$ and is said to define the vacuum state. For each $\phi\in\mathcal{L}^0(\mathfrak{M},\nu)$ let $T(\phi)=A_\phi(1)$. Then just as in our study of the homogeneous chaos, we see that T intertwines the representation of G in $\mathcal{L}^0(\mathfrak{M},\nu)$ defined by the action of G in $\mathfrak{M}$ with the representation of G in the space of all real Borel functions on S defined by the action of G in S. Conversely, given any such intertwining operator, it defines a static quantization satisfying e.m.v. If we make the small additional requirement that the vacuum state is in the domain of A for all ϕ, then the second representation of G may be taken to be a unitary representation U in $\mathcal{L}^2(S,\mu)$, and if T is continuous in the $\mathcal{L}^2(\mathfrak{M},\nu)$ topology, it may be extended uniquely to an intertwining operator defined in $\mathcal{L}^2(\mathfrak{M},\nu)$. Both of these conditions hold when one quantizes a field with a quadratic potential. Indeed, under these circumstances T is unitary and maps $\mathcal{L}^2(\mathfrak{M},U)$ onto the one-particle subspace of the quantized field. Since we are interested in fields chiefly for the quanta they generate, it is not unreasonable to suppose that T will exist for quantization of physical interest.

ISBN 0-8053-6702-0/0-8053-6703-9, pbk

In sum, then, excluding certain possibilities for reasons of varying cogency, we see that the possible invariant static quantizations correspond one-to-one to the systems consisting of the following:

(1) an ergodic action of G on a space with an invariant probability measure;

(2) a "real" intertwining operator for the unitary representation of G induced by the action in (1) and the unitary representation of G induced by the action of G on $\mathfrak{M}$.

Note that an invariant static quantization—subject to the restrictions made above—is abstractly the same thing as a homogeneous chaos. It is this relationship that "explains" the fact that both field quantization and finding the spectrum of the generalized Wiener chaos lead to the construction

$$I\oplus W\oplus(W \circledS W)\oplus(W\circledS W\circledS W)\oplus\cdots$$

One of the fundamental difficulties in quantizing an infinite-dimensional system lies in the multiplicity of possibilities for invariant static quantizations, that is, for ways of assigning self-adjoint operators to configuration observables. Even when we have made restrictions that in the finite case amount to ruling out spin, and assuming that the configuration observables are a maximal commuting set, we find a range of possibilities that so far has defied exhaustive analysis. How shall we select from this multiplicity the one appropriate to a given physical situation? If we look for a clue by examining the case of a quadratic potential energy, we find that the ergodic action of G that arises is not uniquely determined by $\mathfrak{F}$ or even by $\mathfrak{F}$ and the kinetic energy of the system. Simply multiplying the potential energy by a constant forces a change from one ergodic action to another, and it is not at all obvious what one should replace these ergodic actions by when one adds a cubic or biquadratic term.

Even after we have solved the static quantization problem and found the proper ergodic action of G etc., the problem still remains of finding the dynamical group $t\to V_t$. Of course, as we have already indicated, the solution of the static problem can depend upon the dynamics, and it is conceivable that its solution, once found, will also suggest the solution to the dynamical problem. A different approach is to ignore (for the time being at least) the dynamics of the classical system we are attempting to quantize, and seek the most general quantum field, that is, the most general pair consisting of a static quantization and a dynamical group with whatever restriction on the relationship between them seems appropriate.

ISBN 0-8053-6702-0/0-8053-6703-9, pbk

Having found this, one can then look among such pairs for those having the desired "classical limits." Of course, it is conceivable that the quantum field describing the particles that actually exist has no classical limit. Then it would be futile to start with a classical field to be quantized, but one could easily adapt the program described above. Instead of looking for the classical limit of the general quantum field, one would look at the properties of its quanta and hope to find a model for the particles that exist in the world.

Of course the triple A, U, V will define a relativistically invariant system only if further restrictions are imposed. Let us see what these are. First of all, there must exist a projective representation W of the inhomogeneous Lorentz group whose restriction to $\mathcal{G}$ $(=G)$ is U and whose restriction to the time translations is V. (We are now of course definitely restricting ourselves to the case in which $\mathfrak{M}$ is physical space.) In relating W to A, we begin by observing that to be given A and V is the same as to be given V and the mappings $t,\phi \rightarrow V_t^{-1}A_\phi V_t$. Now let g^F be the characteristic function of any Borel set F in space–time whose four-dimensional volume is finite and for each fixed t let g_t^F be the function $x,y,z \rightarrow g^F(x,y,z,t)$. Then for almost all t, $A_{g_t^F}$ will be defined and we may consider the operator

$$B_F = \int_{-\infty}^{\infty} V_t^{-1} A_t^F V_t \, dt$$

More precisely, we say that $B_F(\phi_1)$ is defined and is equal to ϕ_2 if $\int_{-\infty}^{\infty}(V_t^{-1}A_t^F V_t(\phi_1)\cdot\psi)$ exists and equals $(\phi_2\cdot\psi)$ for a dense set of vectors ψ. We assume our system to be such that the B_F and the V_t determine the A_ϕ uniquely so that our system is completely described by B and W. Both B and W are defined without reference to a particular coordinate system and we may formulate our invariance condition quite simply as follows:

$$W_\beta B_F W_{\beta^{-1}} = B_{[F]\beta} \quad \text{for all } \beta \text{ in the inhomogeneous Lorentz group} \quad (*)$$

A relativistically invariant quantum field then defines and is defined by a pair W, B satisfying (*). Such a pair comes close to what a quantum field was considered to be in the naivete of the earliest days of quantum field theory. In those days, one supposed that it would be possible to assign a self-adjoint operator $G_{x,y,z}$ to each point x,y,z of space which defined the observable corresponding to measuring the value of the field at the point. If $t \rightarrow V_t$ is the dynamical group, then (in the Heisenberg picture) $V_t^{-1}G_{x,y,z}V_t$ defines this same observable t time units later and we have an assignment

$$t,x,y,z \rightarrow G'_{x,y,z,t} = V_t^{-1} G_{x,y,z} V_t$$

ISBN 0-8053-6702-0/0-8053-6703-9, pbk

of a self-adjoint operator to each point x,y,z,t of space–time. From this point of view the B_F introduced above are just the "averages" of the $G'_{x,y,z,t}$ over the sets F:

$$B_F = \int_F G'_{x,y,z,t}\, dx\, dy\, dz\, dt$$

However, ever since the appearance of a fundamental paper by Bohr and Rosenfeld in 1933, it has been realized by physicists that only the "smeared" or "averaged" operators B_F can be expected to exist. Though many authors still write in terms of the $G'_{x,y,z,t}$, they interpret the "functions" $x,y,z,t \to G'_{x,y,z,t}$ as operator-valued distributions.

The pair B, W slightly modified is the starting point in an axiomatic approach to field theory due to A. S. Wightman and developed by his students, followers and co-workers. Wightman replaces characteristic functions of sets by C^∞ functions with compact support. Moreover, he does not assume, as we have done, that B may be obtained from an A and a V and thus allows for the possibility that it may be necessary to smear over both space and time. Our axiom to the effect that the A_E commute is replaced by a (presumably weaker) condition of "local commutativity." Wightman also assumes the existence of an analog of the invariant one-dimensional subspace of H defined by 1 in $L^2(S,\mu)$, that is, the existence of a unique "vacuum state." Let ψ_0 be a basis vector for it. Wightman also assumes that $B_{g_1} \cdots B_{g_k}(\psi_0)$ is defined for all k-tuples $g_1,\ldots,g_k$ and that the set of all finite linear combinations of vectors of the form $B_{g_1} \cdots B_{g_k}(\psi_0)$ is dense in H. Let

$$F_k(g_1,\ldots,g_k) = (B_{g_1} \cdots B_{g_k}(\psi_0), \psi_0)$$

Wightman's theory is based in part on the following key observation: If the multilinear functionals $F_1, F_2, \ldots$ are all given, then it is possible to reconstruct B and W or at least the restrictions of the operators B_g and W_β to a dense domain. Moreover, it is possible to write down axioms characterizing the sequences $F_1, F_2, \ldots$ that arise in this way. We refer the reader to Wightman's paper [*Phys. Rev.* 101 (1956)] for details and remark that the reconstruction of B and W from the F_j is quite analogous to the reconstruction of a group representation from one of the "positive-definite functions" $x \to (L_x(\phi), \phi)$.

Wightman puts further restrictions on his pairs B, W, which imply that the F_k are defined by "tempered distribution" in $4k$ space and that these distributions may be extended so as to be boundary values of analytic functions of $4k$ complex variables. In other words, it follows from the

ISBN 0-8053-6702-0/0-8053-6703-9, pbk

axioms that for each k $\exists$ a unique distribution $\tilde{F}_k$ in $4k$ space such that

$$\tilde{F}_k(g_1 \times g_2 \times \cdots \times g_k) = F_k(g_1 \cdot g_2 \cdots g_k)$$

and that there exists an analytic function of the four k complex variables $x_j + iX_j$, $y_j + iY_j$, $z_j + iZ_j$, $t_j + iT_j$, $j = 1, \ldots, k$ whose limit as $X_j \to 0$, $Y_j \to 0$, $Z_j \to 0$, $T_j \to 0$ is $\tilde{F}_k$. Thus the whole system is uniquely determined by a sequence of analytic functions of several complex variables. The Wightman school has deduced many interesting consequences of the axioms by exploiting properties of analytic functions. For details see Streater and Wightman, *PCT, Spin, and Statistics and All That*, Benjamin-Cummings, Reading, Mass. (1964, 2nd printing with corrections and additions, 1978).

The passage from a group representation W to the sum

$$I \oplus W \oplus (W \circledS W) \oplus (W \circledS W \circledS W) \oplus \cdots = \tilde{W}$$

has a simple functorial property that has considerable physical significance. Namely: $\widetilde{W_1 \oplus W_2}$ is always equivalent in a natural way to $\tilde{W}_1 \otimes \tilde{W}_2$. This means that the direct product of two independent fields is the same system one would get by quantizing a single field whose quanta are "particles" defined by the direct sum of the representations defining the quanta of the separate fields. Given that particles are always the quanta of fields, this at once suggests a physical interpretation for the occurrence of *reducible* representations L in our earlier classification of particles. When our "particle" is described by a pair U^L, P^L, where L is some *reducible* representation of the rotation group ($L = L^1 \oplus L^2 \oplus \cdots \oplus L^k$, say), we may think of our Hilbert space as being the total one-particle space in some system of interacting fields and of the individual components as corresponding to the different particle types. When we are in the one-particle subspace so that only one particle is present, this particle may be an electron, a proton, a photon, or something else. Which it is will depend upon which invariant subspace one is in.

This interpretation of reducibility is closely connected with the much studied "symmetries" of elementary particles. Our analysis of one-particle systems was based on an assumed invariance under the group of rigid motions of space. Suppose now that there is a larger symmetry group $G' \supseteq G$, and that the subgroup G'_{m_0} leaving a point m_0 of $\mathfrak{M}$ fixed is of the form $\mathfrak{R} \times \mathfrak{R}'$, where $\mathfrak{R}$ is the rotation group and $\mathfrak{R}'$ is some other compact group. Then each one-particle system will have a representation of $\mathfrak{R} \times \mathfrak{R}'$ invariantly attached to it, and if this is irreducible it will be of the form $L \otimes L'$, where L is an irreducible representation of $\mathfrak{R}$ and L' is an irreducible representation of $\mathfrak{R}'$. Now this representation restricted to $\mathfrak{R}$ will be the direct sum of k replicas of L, where k is the dimension of L'.

ISBN 0-8053-6702-0/0-8053-6703-9, pbk

From our old point of view we have k different particles, but the existence of the symmetry group G' links them together and forces them to have closely related properties. One speaks of a *particle multiplet*. In practice, such symmetry groups exist only approximately. Roughly speaking, one writes the $\mathcal{H}$ for the whole universe as a sum of terms of different orders of magnitude and finds that one has symmetry under a $G' \supseteq G$ if the lower order terms are neglected. The simplest example is that in which one ignores electromagnetic forces and weaker ones. Then there is a G' such that R' is SU(2). Now SU(2) has $\mathfrak{R}$ as a quotient group, and has one irreducible representation of every dimension. When L is $2k+1$ dimensional, one speaks of a particle of spin k. By analogy, when L' is $2k+1$ dimensional, one speaks of the members of the multiplet as having *isotopic spin k*. For example, the proton and neutron belong to a particle doublet and are said to have isotopic spin $\frac{1}{2}$. The three π mesons π^+, π^0, and π^- belong to a triplet and are said to have isotopic spin 1. Among the relatively recently discovered "strange" heavy particles there is a triple $\Sigma^+, \Sigma^0, \Sigma^-$, a doublet Ξ^-, Ξ^0, and a singlet Λ. The celebrated *eightfold way* of Gell-Mann and Ne'eman is based on the observation that the proton, neutron, and the heavy strange particles may be regarded as a part of a supermultiplet. There is a larger (and more approximate) symmetry group $G'' \supseteq G' \supseteq G$ such that $G''_{m_0} = \mathfrak{R} \times \mathrm{SU}(3)$ and SU(3) has an irreducible eight-dimensional irreducible representation whose restriction to SU(2) is the direct sum of the one-dimensional representation, two replicas of the two-dimensional representation, and the three-dimensional representation.

To the extent that this representation of SU(3) occurs, it links particles into "weakly similar" octets that are "more tightly" linked (via isotopic spin) in four subsets consisting of a singlet, two doublets, and a triplet. The eight particles consisting of the neutron, the proton, the $\Sigma^+, \Sigma^0, \Sigma^-, \Xi^-, \Xi^0$, and the Λ fit nicely into the scheme.

To see the real significance of these symmetries and groupings into multiplets, one must re-think the material on particle interactions and the S operator for particles with spin—replacing the Euclidean group by its direct product with SU(2) or SU(3). The theory predicts important relationships between the interactions of the particles in a multiplet that may be verified by experiment. The notion of isotopic spin was in fact introduced in an attempt to codify and explain similarities in the interactions of neutrons and protons.

Notes and References

The standard work on the Wightman axioms and their consequences is the book by R. F. Streater and Wightman (*PCT*: *Spin and Statistics and*

ISBN 0-8053-6702-0/0-8053-6703-9, pbk

All That, Benjamin-Cummings, Reading, Mass. (1964)). In the past decade or so, the axiomatic approach has been dominated by the extensive work of J. Glimm and A. Jaffe (and their collaborators) on what has come to be known as "constructive quantum field theory." Its goal is to produce examples of rigorously defined models that satisfy the Wightman axioms and that imply the existence of particles which interact in a nontrivial way. Its main successes so far have been achieved under the physically unrealistic hypothesis that space is one or two dimensional. However, its development has led to much interesting mathematics, new insights, and more recently to important new connections with statistical mechanics. Among the many who have contributed to this program in addition to Glimm and Jaffe, the names of K. Friedrichs, E. Nelson, I. E. Segal, and K. Symanzik should perhaps have special mention. An excellent ten–page historical summary will be found in the appendix to the very recent second printing, with corrections and addenda, of the book of Streater and Wightman cited above. The appendix to the book of Bogolubov, Logunov and Todorov, Benjamin/Cummings, Reading, Mass. (1975) also gives a very good, if less recent, account. For a lengthy survey of other aspects of axiomatic field theory the reader is referred to "Outline of axiomatic relativistic quantum field theory" by R. F. Streater in part 3 vol. 35 (1975) of *Reports on Progress in Physics*. See also the latter part of the article of Wightman in the book on the Hilbert problems cited in the notes and references to Section 17. The book of B. Simon *The $P(\phi)_2$ Euclidean (Quantum) Field Theory*, Princeton Univ. Press, Princeton, N. J. (1964), emphasizes the so-called Euclidean approach in which one replaces the Wightman functions by their analytic continuations to imaginary time and takes advantage of certain connections with probability theory. The work of Fröhlich, Guerra, Nelson, Rosen, Schrader, Simon and Spencer has been especially significant in developing this approach. A good general reference for all aspects of the subject is the report on the 1973 "Ettore Majorana" summer school (*Lecture Notes in Physics*, vol. 25, Springer-Verlag (1971)).

The approach to quantum field theory axiomatics given in the text differs from that of Wightman in making an explicit connection with ergodic theory. This point of view is somewhat further developed in the author's article "Ergodic actions of the Euclidean group and the canonical commutation relations," in *Proc.* 1967 *Int. Conf. on Particles and Fields*, Wiley-Interscience, New York (1967), 265–272. The author has long felt that it would be interesting to explore the connections between this work and Nelson's applications of probability to quantum field theory.

The study of relationships between particles and their interactions that manifest themselves via (approximate) invariance under groups such as SU(3) has advanced far beyond the SU(3) theory briefly described in the

ISBN 0-8053-6702-0/0-8053-6703-9, pbk

text. At the very time these lectures were being given there was a wave of interest in using the group SU(6) to blend rotational symmetry with SU(3) symmetry, and while this never quite worked, its partial successes were preserved in a new development known as the theory of "current algebra." Further details will be found in the book of F. J. Dyson, *Symmetry Groups in Nuclear and Particle Physics*, Benjamin-Cummings, Reading, Mass. (1966). Mainly a reprint collection, Dyson's book contains copies of the basic papers on SU(6) as well as on the beginning of current algebra. For a more integrated account of these matters, including some later developments, the reader may consult the surveys by L. Michel and L. O'Raifeartaigh in the published proceedings of the 1969 Battelle Rencontres (*Lecture Notes in Physics*, vol. 6, Springer-Verlag (1970), 36–143 and 144–236).

In 1974 a new particle was independently discovered in two different laboratories and variously named the J and the ψ. Attempts to correlate it with field theory models led to a revival of interest in a 1964 proposal of Glashow and Björken that, roughly speaking, amounts to replacing the group SU(3) by SU(4). Certain features of SU(3) symmetry had suggested from the first that the particles to which it applied (the so-called strongly interacting particles or hadrons) might all be bound states (cf. Section 20) of three unobserved and more fundamental particles that are referred to as "quarks." Postulating the existence of these hypothetical particles makes sense of a number of observed phenomena and also lets one hope for a simplified theory of hadron interactions. Just as the complicated laws governing the ways in which atoms combine to form molecules are strict logical consequences of the quantum mechanical formalism combined with the Coulomb law governing the interactions of nuclei and electrons, so might a simple interaction law between quarks imply all the complexities observed in hadron interactions. From the quark point of view, the passage from SU(3) to SU(4) amounts to postulating a fourth quark. Certain difficulties in the quark point of view are gotten around by supposing that each of the four quarks is in reality a triple of slightly different particles that are picturesquely distinguished by supposing them to have different "colors." An informative popular account of all of this will be found in an article by Glashow, "Quarks with color and flavor," *Scientific American*, October (1975).

Another extremely interesting development of the past decade or so has grown out of the 1964 discovery that two more or less discarded types of field theoretic models for the particles of nature could be combined in a way that eliminated a major drawback of each. Taking advantage of this discovery, S. Weinberg and A. Salam in 1967 indicated a way of unifying

ISBN 0-8053-6702-0/0-8053-6703-9, pbk

the weak and electromagnetic interactions, and a major step toward the implementation of this program was taken by G. 't. Hooft in 1971. Since the details are complicated and the literature of the subject is now rather large, we shall say no more but will content ourselves with referring the reader to the excellent review article of Sidney Coleman ("Secret Symmetry: An Introduction to Spontaneous Symmetry Breakdown and Gauge Fields," in 1973 *Ettore Majorana Summer School Proceedings*, Academic, New York) and to the popular account of S. Weinberg, "Unified theories of elementary particles," *Scientific American*, July (1974).

ISBN 0-8053-6702-0/0-8053-6703-9, pbk

23. *Temperature, heat, and thermodynamics*

So far our description of the physical world has completely ignored the important phenomena associated with the words "heat" and "temperature." To fill in this gap we must explain what it means in terms of our basic concepts for one material body to be "hotter" than another and derive from these concepts the basic laws and notions of the classical theory of heat—both in isolation and in their interaction with other parts of physics.

Let us begin by describing the notions of heat and temperature as we find them in everyday experience. Our basic observation is that material bodies are simply ordered by a relation that we sense directly. We touch two bodies A and B and have sensory experiences that we have learned to describe by saying that "A is hotter than B" or that "A has a higher temperature than B". Concerning this ordering, experience shows that a given body does not maintain a constant temperature and in fact that if two bodies of different temperature are placed in contact the temperature of the hotter one decreases and that of the other increases until the two have the same temperature. This equilibrium temperature depends not only upon the initial temperatures of the two bodies but also upon what they are made of and upon how large each body is. For bodies made of given materials at given initial temperatures, the equilibrium temperature increases and approaches that of the hotter body the larger this hotter body is. Similarly, it decreases and approaches that of the cooler body the larger the cooler body is.

In order to make a quantitative theory out of these qualitative facts, one

George W. Mackey, Unitary Group Representations in Physics, Probability, and Number Theory

ISBN 0-8053-6702-0/0-8053-6703-9, pbk

needs first of all a way of deciding when two substances are at the same temperature which is independent of ones relatively crude sense of touch. This is found in the fact that most substances expand in volume (though not in mass) when they become hotter and contract when they become cooler. If one chooses a substance that expands and contracts rather a lot, one can measure these expansions and contractions with a ruler and thus set up a one-to-one correspondence between the ordered set of all possible temperatures and the points on a line. The ordinary mercury thermometer contains a thin column of the liquid metal mercury in a tube attached to a reservoir. When placed in contact with a body whose temperature is to be measured and which is large compared to the amount of mercury in the reservoir, the temperature of the mercury changes to approximately that of the body in question and the length of the column adjusts accordingly. The error caused by the fact that contact with the thermometer changes the temperature to be measured can be made as small as one wishes by using small thermometers and large amounts of material.

It must be emphasized that the use of thermometers allows us to define temperature quite precisely as a point in a certain ordered set, but does *not* permit us to put the points of this ordered set into one-to-one correspondence with the real numbers in any canonical way. If we use thermometers made of different substances, then the temperature T_1 as measured by the first will be a certain monotone function f of the measure T_2 as measured by the other. If f were always of the form $x \rightarrow ax+b$, where $a>0$, then choosing a particular substance and a particular design of the thermometer would be simply a matter of choosing an origin and a unit of temperature difference and one could make an arbitrary choice with little or no loss in generality. However, f is in general not linear and no canonical numerical parametrization of temperatures is available at this stage of the development of the theory. We must think of the set $\mathfrak{T}$ of all possible temperatures as an abstract ordered set.

Let A and B be two bodies. For each τ_1 and τ_2 in $\mathfrak{T}$ let $E_{A,B}(\tau_1,\tau_2)$ be the equilibrium temperature that is reached when body A at temperature τ_1 is placed in contact with body B at temperature τ_2. From all that we have said at the moment the functions $E_{A,B}$ from $\mathfrak{T}\times\mathfrak{T}$ to $\mathfrak{T}$ are completely independent of one another except that we have indicated a qualitative relationship between $E_{A,B}$ and $E_{A',B}$ when A and A' are bodies of the same material but different sizes. However, experiment shows that these functions are related in a very fundamental way that may be described in mathematical terms as follows: There exists for each body A a function H_A from $\mathfrak{T}\times\mathfrak{T}$ to the real numbers that has the following properties:

(1) $H_A(\tau_1,\tau_2)+H_A(\tau_2,\tau_3)=H_A(\tau_1,\tau_3)$ for all τ_1, τ_2, and τ_3.

ISBN 0-8053-6702-0/0-8053-6703-9, pbk

(2) For each A, B, τ_1 and $\tau_2, E_{A,B}(\tau_1,\tau_2)$ is the unique $\tau_3 \in \mathfrak{T}$ such that $H_A(\tau_1,\tau_3) = H_B(\tau_3,\tau_2)$.

(3) If A_1 and A_2 are made of the same substance, then $H_{A_2}(\tau_1,\tau_2) = \lambda H_{A_1}(\tau_1,\tau_2)$, where λ is the ratio of the mass of A_2 to that of A_1.

The family of functions H_A is clearly not unique since we may replace each H_A by H_A', where $H_A' = \mu H_A$ and μ is any fixed nonzero real number. However, this is the extent of the nonuniqueness. If we declare that $H_A(\tau_1,\tau_2)$ shall be 1 for any particular choice of τ_1, τ_2, A, then H_A is uniquely determined for all A; that is, the experimental functions $E_{A,B}$ are consistent with only one such choice.

One speaks of $H_A(\tau_1,\tau_2)$ as the *quantity of heat* required to change the temperature of A from τ_1 to τ_2 and one chooses the arbitrary constant so that $H_A(\tau_1,\tau_2) > 0$ whenever $\tau_2 > \tau_1$. Because of (1) the function H_A of two variables may be written in the form $H_A(\tau_1,\tau_2) = H_A^0(\tau_2) - H_A^0(\tau_1)$, where H_A^0 is a function from $\mathfrak{T}$ to the real numbers that is uniquely determined up to an additive constant. One thinks of H_A^0 as the heat content or total quantity of heat contained in A and one can then think of temperature changes as taking place because of the motion of heat from one body to the other. The existence and uniqueness of the H_A allows us to measure these heat exchanges in numerical terms and tells us that in any complex series of heat exchanges the total amount of heat lost by bodies that become cooler is equal to the total amount gained by those that become warmer.

The function H_A normalized by choosing A to be a body of unit mass is a fundamental function of τ attached to each material substance. A priori it must be determined by experiment for each substance. However, one of our aims in the next section is describe a theory from which it may be computed. Once a temperature scale has been chosen, H_A^0 is just a real function of one real variable determined up to an additive constant. The derivative of this function is independent of the constant and determines H_A^0 up to this constant. It is called the *specific heat* of the given substance. In many cases it approaches a constant for high values of the temperature, and for many common substances everyday temperatures are high enough for this limiting constant to be a good approximation to the function. Thus one often talks of the specific heat of a substance as though it were a number. Knowing this number for all substances involved we can compute the equilibrium temperature when various amounts of various substances are brought together at different temperatures. Of course, whether or not the specific heat tends to a constant depends upon the temperature scale. Our statement about this refers to the still undefined "absolute" scale.

ISBN 0-8053-6702-0/0-8053-6703-9, pbk

The statements made above about conservation of quantity of heat are only valid when there is no accompanying gain or loss of mechanical energy. It is a familiar fact that the law of conservation of energy in its purely mechanical form seems to fail when there is "friction" between moving parts. For example, when a weight slides down an inclined plane that is not perfectly smooth a certain amount of mechanical energy disappears. On the other hand, whenever this happens the moving parts increase in temperature without causing a corresponding decrease in the temperature of other bodies. Similarly, when a compressed gas expands and lifts a weight, the temperature of the gas falls although nothing else is heated up in consequence. At the same time the mechanical energy of the system increases. It seems that mechanical energy can disappear provided that heat is created in the process and that heat can disappear if mechanical energy is created in the process. This suggests the possibility that the amount of heat created or destroyed is directly proportional to the amount of mechanical energy simultaneously destroyed or created and that one may recover a more universally valid conservation law by thinking of heat as a form of energy. Careful experiments carried out by Joule and others around 1850 showed that this possibility was a reality. Heat and mechanical energy are mutually convertible into one another and conservation holds only for the total obtained by adding the mechanical energy to the total amount of heat where the latter is measured in energy units. This relationship between heat and mechanical energy is known as the *first law of thermodynamics*.

Because of the intraconvertibility of heat and mechanical energy, we must formulate our statement about the existence of the fundamental functions H_A^0 with more care. Whenever the heated substance expands and moves other material bodies, some of its heat content turns into mechanical energy and it is only if we apply external constraints to keep the volume constant that the heat content is preserved. Moreover, the application of this constraint changes the internal structure of the body so as to make it effectively into a different body. In other words, for a given mass of a given substance H_A^0 is only well defined when the volume V, which A is constrained to occupy, is given. Thus H_A^0 is really a function of two variables T and V and is determined by what has been said only up to an additive function of V. Once we have chosen a temperature scale this dependence on T and V can be expressed quite simply by saying that the specific heat C_A is a well-defined function of T and V. H_A^0 is then an arbitrary solution of $\partial H/\partial T = C_A$ and as such is determined only up to an additive function of V.

Of course, knowledge of the function C_A also tells us nothing about the mechanical effects of volume changes or the extent to which heat energy is

ISBN 0-8053-6702-0/0-8053-6703-9, pbk

transformed into mechanical energy. It must be supplemented by another function that we obtain by considering so-called "adiabatic" changes. By the use of heat insulators we may allow expansions and contractions to take place in such a fashion that no heat is gained or lost to surrounding bodies. Such expansions and contractions are said to be adiabatic and when they take place the resultant gain or loss in mechanical energy must be compensated for by a consequent fall or rise in the temperature of the body A itself. If the body is kept at a given volume by external constraints and these are changed slightly, the volume will change and so will the temperature. The change in temperature will depend on factors other than the change in volume unless the change is carried out under "equilibrium conditions." In crude terms, this means that we make a series of very small volume changes and wait for the temperature to settle after each change before making the next. More precisely, we consider the limiting result of making the small volume changes even smaller. The resulting relationship between volume and temperature depends only upon the initial volume and temperature and its graph is called the adiabatic curve through the given point of the V,T plane. If V_1,T_1 and V_2,T_2 are two different points on the same adiabatic, let $-W_A(V_1,T_1,V_2,T_2)$ denote the total mechanical energy communicated to the constraining system in going along this adiabatic from V_1,T_1 to V_2,T_2. It is now an immediate consequence of the first law of thermodynamics that there exists a function U_A defined in the V,T plane such that

$$U_A(V_1,T_1)-U_A(V_2,T_2)=W_A(V_1,T_1,V_2,T_2) \tag{1}$$

whenever V_1,T_1 and V_2,T_2 lie on the same adiabatic and

$$U_A(V,T_1)-U_A(V,T_2)=H_A^0(V,T_1)-H_A^0(V,T_2) \tag{2}$$

The function U_A is clearly determined up to an additive constant by conditions (1) and (2). It is called the internal energy function for the body A and plays the same role for arbitrary changes in V and T that H_A^0 plays for changes at constant volume. For each fixed V it differs from H_A^0 by a constant depending upon V and we may regard it as obtained from H_A^0 by choosing the arbitrary function of V in a manner consistent with (2). That such a choice is possible is the content of the first law of thermodynamics as far as single bodies are concerned.

Knowing the function U_A of course gives us the functions H_A^0 and C_A but does not tell us which curves in the $V-T$ plane are adiabatic. It tells us the total energy but does not tell us how it divides into heat energy and mechanical energy when changes are made. If we have introduced a

ISBN 0-8053-6702-0/0-8053-6703-9 pbk

particular parametrization for temperature, then at each point V,T we may compute the slope $\lambda_A(V,T)=dT/dV$ of the adiabatic at that point, and given this function we may recover the adiabatics. The two functions λ_A and U_A together tell us all the thermodynamics functions of our substance and it is the job of the experimental thermodynamist to find these functions for each body A. Of course λ_A depends only upon the substance of which A is made and U_A/m_A does likewise where m_A is the mass of A. Thus it is really for each (homogeneous) substance rather than each body that these determinations must be made.

Actually one usually does not work with λ itself but a certain other function p_A called the *state function.* ($p_A(v,T)$ is called the *pressure* at volume v and temperature T and the equation $p=p_A(v,T)$ is called the "equation of state.") p_A has the advantage of having a direct intuitive physical meaning. By definition

$$p_A=-\left(\frac{\partial U_A}{\partial v}+\frac{\partial U_A}{\partial T}\lambda_A\right)$$

$$=-\left(\frac{\partial U_A}{\partial v}+\lambda_A C_A\right)$$

that is, p_A is the rate of change of U_A with respect to v as v and T vary along an adiabatic. Clearly the pair U_A,p_A determines the pair U_A,λ_A and conversely.

It is also worth noting that U_A is determined up to a multiplicative constant as soon as we know λ_A, p_A, and C_A and that these three functions may all be defined without introducing the first law. Specifically we have

$$\frac{\partial U_A}{\partial T}=C_A,\quad \frac{\partial U_A}{\partial v}=-p_A+\lambda_A C_A$$

On the other hand, it is certainly not true that a U exists for every mathematically possible triple C_A,λ_A,p_A. U_A will exist if and only if

$$\frac{\partial C_A}{\partial v}=\frac{\partial}{\partial T}(\lambda_A C_A-p_A)$$

that is, if and only if $C_A\,dT+(\lambda_A C_A-p_A)dv$ is an exact differential. We have already remarked that for a single homogeneous substance the first law is equivalent to the existence of the function U_A. Similarly, it is equivalent to the solvability of the equations

$$\frac{\partial U}{\partial T}=C_A,\quad \frac{\partial U}{\partial v}=\lambda_A C_A-p_A$$

ISBN 0-8053-6702-0/0-8053-6703-9, pbk

for U, that is, to the exactness of the differential $C_A\,dT+(\lambda_A C_A-p_A)\,dv$.

An experimental scientist might have been led to the first law without any conceptions concerning a joint heat, mechanical energy conservation law. Painstakingly measuring C_A, λ_A, and p_A for each of a variety of bodies or substances, he might have found that these were not three independent functions but that one had the universal relationship

$$\frac{\partial C_A}{\partial v}=\frac{\partial}{\partial T}(\lambda_A C_A-p_A)$$

and that this made it possible to replace the pair C_A,λ_A by the single function U_A.

The celebrated second law of thermodynamics tells us that p_A and U_A are also not independent and that they can both be obtained from a single function by differentiation. Like the first, this relationship not only can be verified by experiment, but also can be derived from a simple general principle about the relationship between heat and mechanical energy. This second principle involves a certain limitation in the *extent* to which one can convert heat energy into mechanical energy. It may be formulated in several equivalent ways. One of them says that it is impossible to transfer a nonzero quantity of heat from a body at one temperature to a body at a higher temperature without effecting changes in other bodies. Another says that it is impossible to turn a nonzero quantity of heat into mechanical energy without effecting changes in other bodies. In proving the equivalence of these formulations as well as in deriving their consequences for the function U_A and p_A, one considers what happens to these functions for a given body A as one goes around a Carnot cycle, that is, a closed curve made up of two adiabatics and two lines of constant temperature. If one computes the heat emitted or absorbed along each component and the mechanical work done on or by the substance, one sees that the net effect is to take a quantity of heat Q_1 at a temperature T_1, transfer Q_2 of it to a body of lower temperature T_2, and turn the difference into mechanical energy. The ratio $(Q_1-Q_2)/Q_1$ is called the *efficiency* of the given substance between the temperatures T_1 and T_2. Using two different bodies and going around the Carnot cycle in one direction with one substance and in the other direction with the other, one shows easily that the general principle enunciated above implies that the efficiency is independent of Q_1,Q_2 and the substance and depends only upon the temperatures T_1 and T_2. Let $e(T_1,T_2)$ denote this efficiency. It is easy to prove that to within a multiplicative constant there is one and only one way of parametrizing temperature so that

$$e(T_1,T_2)=\frac{T_1-T_2}{T_1}$$

ISBN 0-8053-6702-0/0-8053-6703-9 pbk

When the constant is chosen so that the freezing point of water is 273.1°, we get the celebrated absolute temperature scale of Kelvin.

On the other hand, we may easily compute $e(T_1, T_2)$ for any given substance using U_A and p_A and work out the condition that

$$e(T_1, T_2) = \frac{T_1 - T_2}{T_1}$$

It is that $(dU_A + p_A\,dV)/T$ be an exact differential where T is the absolute temperature. In other words, for any actual substance U_A and p_A must be so related that

$$\frac{\partial}{\partial T}\left[\frac{\left(\frac{\partial U_A}{\partial v} + p\right)}{T}\right] = \frac{\partial}{\partial v}\left[\frac{\left(\frac{\partial U_A}{\partial T}\right)}{T}\right]$$

Let S_A be the function (determined up to an additive constant) of which $(U_A + p_A\,dv)/T$ is the differential. Then S_A is called the *entropy* of the substance. Consider the function

$$F_A = U_A - TS_A$$

$$\frac{\partial F_A}{\partial T} = -S_A$$

$$\frac{\partial F_A}{\partial v} = \frac{\partial U_A}{\partial v} - T\frac{\partial S_A}{\partial v} = -p_A$$

$$\frac{\partial S_A}{\partial v} = \frac{\left(p_A + \frac{\partial U_A}{\partial v}\right)}{T}$$

$$\frac{\partial S_A}{\partial T} = \frac{\left(\frac{\partial U_A}{\partial T}\right)}{T}$$

Hence

$$U_A = F_A - T\left(\frac{\partial F_A}{\partial T}\right)$$

$$p_A = -\frac{\partial F_A}{\partial v}$$

Thus once we know the single function F_A, called the Helmholtz free energy, we may compute U_A and p_A and hence all thermodynamical functions.

In all of this we have assumed that A is a single homogeneous body that does not undergo "chemical changes" or changes of state such as melting, evaporating, and the like. When these possibilities exist one has in effect a continuum of different bodies and one must study the variability of F_A, U_A, p_A, etc. with these parameters. We shall not have time to embark upon this study and will remark only that one applies the first and second laws as before and obtains exact relationships of great importance for chemistry.

ISBN 0-8053-6702-0/0-8053-6703-9, pbk

24. Statistical mechanics

We have described the notions of temperature and quantity of heat and the laws of thermodynamics at the so-called "phenomenological" level. That is, we have described them in terms of what we actually observe. However, if we take seriously the view that all phenomena can be explained in terms of interacting elementary particles, we must seek an "explanation" of the notions and laws of thermodynamics in these same terms. Given a system of particles and given the interactions between these particles, what will we observe at the phenomenological level as the "temperature" of the system, and how are we to compute the functions p_A, U_A, etc. for the body made up of these particles? This is one of the major questions that the subject known as statistical mechanics sets out to answer.

Statistical mechanics was originally developed within the framework of classical mechanics and we shall depart from the policy followed up to now in these lectures by beginning with a treatment of the subject in its classical form. We shall derive the quantum form independently, but it is interesting for several reasons to compare the classical and quantum forms.

Let Ω be the phase space of a system of interacting particles confined within a finite volume v. If we assume the usual Euclidean model for space, this means that phase space consists of all $6n$ tuples $q_1 \cdots q_{3n}$, $p_1 \cdots p_{3n}$, where for each $j = 1 \cdots n$, $q_{3j+1}, q_{3j+2}, q_{3j+3}$ lies in a certain subset M of space of volume v, and the p_j are arbitrary. Here the q_k are the coordinates and the p_j the momenta of our n particles. We make Ω into a

ISBN 0-8053-6702-0/0-8053-6703-9, pbk

George W. Mackey, Unitary Group Representations in Physics, Probability, and Number Theory

measure space by means of the "Liouville measure" ζ, where

$$\zeta(E)=\int_E dq\cdots dq_{3n}\,dp\cdots dp_{3n}$$

Given a material body made up of n material particles, where n is a very large number ($>10^{20}$ for most familiar objects), there are an enormous number of points of phase space consistent with any given phenomenological state. Observable differences are not nearly fine enough to distinguish between points of phase space, and observable properties must somehow be related to averages of actual observables over large subsets of phase space. What we actually do is to suppose that each phenomenological state is associated with a certain probability measure α in Ω and that this association has the following significance: If we were suddenly given supernatural powers and could determine which point in Ω actually represented the state of our system, we would get different answers at different trials for systems in the given phenomenological state and these different answers would be distributed statistically in a manner described by the measure α. In other words, we would find the point to lie in the set E a fraction $\alpha(E)$ of the time. Now the phenomenological states we have been discussing are "equilibrium states" that are constant in time. Thus the α describing such a state must be a probability measure that is invariant under the action of the dynamical group of the system. Our problem then is to study the possible invariant probability measures in Ω and for each temperature T decide which one describes our body at temperature T.

To begin with, let us observe that the function H on Ω which determines the dynamical group is a constant on the orbits of our group in Ω. This is because the differential equations of motion are

$$\frac{dq_i}{dt}=-\frac{\partial H}{\partial p_i},\quad \frac{dp_i}{dt}=\frac{\partial H}{\partial q_i}$$

The Hamiltonian function H of course is the total energy of the system as a function of the q's and p's. For each fixed real number E let Ω_E be the *constant energy hypersurface* Ω_E consisting of all points ω with $H(\omega)=E$. Because of the constancy of H on the orbits of the dynamical group, it follows that each Ω_E is invariant under this group. Hence by a general theorem in measure theory any invariant probability measure α may be decomposed essentially uniquely as an integral over the reals of invariant probability measures α_E one in each Ω_E. This same theorem may be used to decompose the Liouville measure ζ and in this way we get a canonical invariant measure ζ_E in each Ω_E. It can be shown to follow from our

ISBN 0-8053-6702-0/0-8053-6703-9, pbk

hypothesis about the finite volume that for any H likely to arise in physics the ζ_E all give finite measure to Ω_E and can be normalized so as to be probability measures. Thus there is a canonical invariant probability measure in each Ω_E. Strictly speaking, it is only defined for almost all E but supplementary considerations involving the differentiable structure of Ω_E make it possible to remove this qualification. A very plausible hypothesis now presents itself: *a system whose total energy is E is described by the invariant probability measure* ζ_E.

Let us tentatively make this hypothesis and investigate some of the consequences. First of all, suppose that we have two systems with phase spaces Ω^1 and Ω^2 and Hamiltonians H^1 and H^2 in energy states E_1 and E_2. The combined system will have phase space $\Omega^1 \times \Omega^2$ and the Hamiltonian $H^1 + H^2$, where

$$(H^1 + H^2)(\omega_1, \omega_2) = H^1(\omega_1) + H^2(\omega_2)$$

Of course $\Omega_{E_1+E_2}$ will not be $\Omega_{E_1} \times \Omega_{E_2}$, but the union of all sets of the form $\Omega_F \times \Omega_C$ such that $F + C = E_1 + E_2$. Moreover, in the state described by $\zeta_{E_1+E_2}$, H^1 will have an expected value $\overline{H^1}$ that in general will not equal E_1, and H^2 will have an expected value $\overline{H^2}$ that in general will not equal E_2. Indeed, $\zeta_{E_1+E_2} = \zeta_{F_1+F_2}$ whenever $E_1 + E_2 = F_1 + F_2$ and $\overline{H^1}$ cannot equal both E_1 and F_1. On the other hand, there is a unique number ΔE such that if we take an amount of energy ΔE from the first system and add it to the second, we will change their states in such a way that $\overline{H^1}$ and $\overline{H^2}$ in the combined system are precisely equal to the energy values defining the original systems. This suggests that we think of the first system as being at a higher or lower temperature than the second according as ΔE is positive or negative, that is, according as $E_1 - \overline{H^1}$ is positive or negative and that we think of ΔE as the amount of heat energy that must move from the first system to the second in order to equalize the temperature.

If we pursue this idea we find that it "works." We do get a model for heat and temperature that explains the laws of thermodynamics and gives quantitatively correct predictions—except in so far as quantum effects are important. However, the necessary computations are complicated and awkward, and, in addition, one is suspicious of a statistical theory in which the energy E is given a sharp value. Indeed, when our combined system is in the state defined by $\zeta_{E_1+E_2}$, neither H^1 nor H^2 does have a sharp value, but each has a certain statistical distribution. Fortunately, there is another approach that leads to substantially the same results and is much easier to work with. Moreover, it replaces ζ_E by a measure that is absolutely continuous with respect to ζ.

ISBN 0-8053-6702-0/0-8053-6703-9, pbk

In this alternative approach one proceeds as follows. In a phenomenological state in which we observe the total energy to be E, our probability measure α must be such that $\int_\Omega H\,d\alpha = E$. There are many such α's including ζ_E and some represent more precise information about just where in Ω we are than others. We now take into account the fact that we know nothing about the system except for E and ask for that α which gives us the "greatest degree of randomness" or the "least amount of information" consistent with the requirement that $\int_\Omega H\,d\alpha = E$. Of course, asking for such an α makes no sense unless we have some way of measuring "randomness" or "information."

Fortunately, the recently developed subject known as information theory provides such a way. Suppose that we have a discrete set of possibilities with probabilities $\alpha_1,\ldots,\alpha_k$, $\alpha_i \geqslant 0$, $\alpha_1 + \cdots + \alpha_k = 1$. Clearly, when one of the α_j is 1 and all others are zero, we have minimal randomness and maximal information. Moreover, when $\alpha_1 = \alpha_2 = \cdots = \alpha_k = 1/k$, we have maximal randomness and minimal information. However, it is not a priori clear how to compare the amount of randomness in such a set of probabilities as $\frac{1}{4}, \frac{1}{2}, \frac{1}{4}$ with that in $\frac{1}{9}, \frac{1}{3}, \frac{1}{9}, \frac{1}{9}, \frac{1}{3}$ or indeed whether one can do so in a consistent fashion. Nevertheless, it turns out that there is a set of plausible axioms that one can write down concerning the properties that a "good" measure of randomness should have and that there is one and only one way of defining randomness so that these axioms are satisfied. This definition assigns the number $\alpha_1 \log(1/\alpha_1) + \alpha_2 \log(1/\alpha_2) + \cdots$ to the probability distribution $\alpha_1, \alpha_2, \ldots, \alpha_n$ with the understanding that $0\log(1/0) = 0$. Proceeding by analogy, we measure randomness in the continuous case by the expression $\int \rho \log(1/\rho)\,d\zeta$, where $\rho = d\alpha/d\zeta$ and we restrict attention to those α's for which $d\alpha/d\zeta$ exists. Arguments can be made justifying these steps once one has agreed that ζ is the proper "reference measure." Why this reference measure should be ζ is less clear. At this stage one cannot say much more than that it is the "obvious choice." However, more cogent arguments can be given based on quantum mechanics.

If we accept $\int_\Omega \rho \log(1/\rho)\,d\zeta$ as the appropriate measure of randomness, our problem now is to find that ρ which maximizes $\int_\Omega \rho \log(1/\rho)\,d\zeta$ subject to the auxiliary requirement that $\int_\Omega H\rho\,d\zeta$ be equal to a fixed constant E. Standard arguments from the calculus of variations lead to the conclusion that ρ must be of the form $Ae^{-H/B}$, where A and B are constants. The fact that ρ must be a probability measure at once gives us a means of determining A. We have $A\int_\Omega e^{-H/B}\,d\zeta = 1$ so that we must have

$$A = \frac{1}{\displaystyle\int_\Omega e^{-H/B}\,d\zeta}$$

ISBN 0-8053-6702-0/0-8053-6703-9, pbk

Thus ρ must be of the form

$$\frac{e^{-H/B}}{\int_{\Omega} e^{-H/B}\,d\zeta}$$

and we have only a one-parameter family of possibilities for ρ. For determining B we have the condition that $\int_{\Omega} H\rho\,d\zeta = E$. In other words, B must be a solution of the equation

$$E = \frac{\int He^{-H/B}\,d\zeta}{\int e^{-H/B}\,d\zeta}$$

and in most problems of physical interest this equation has a unique solution. Indeed, in such problems H is bounded below and may be normalized so as to be $\geqslant 0$. When this is done the function

$$B \rightarrow \frac{\int He^{-H/B}\,d\zeta}{\int e^{-H/B}\,d\zeta}$$

turns out to be an unbounded monotone increasing function defined for all $B > 0$ and 0 for $B = 0$. The measure

$$\frac{e^{-H/B}\,d\zeta}{\int_{\Omega} e^{-H/B}\,d\zeta}$$

actually differs but little from the measure ζ_E where

$$E = \frac{\int_{\Omega} He^{-H/B}\,d\zeta}{\int_{\Omega} e^{-H/B}\,d\zeta}$$

It is a "weighted average" of many ζ_F's with the F's near to E being given by far the greatest weights. The great advantage of using it is that the parameter B may be interpreted as proportional to the temperature. To see this, let us consider two systems with phase spaces Ω^1 and Ω^2 and

ISBN 0-8053-6702-0/0-8053-6703-9, pbk

Hamiltonians H^1 and H^2 in states of maximal randomness described by parameter values B_1 and B_2. Let

$$E_1=\frac{\int H^1 e^{-H^1/B_1}\,d\zeta_1}{\int e^{-H^1/B_1}\,d\zeta_1},\quad E_2=\frac{\int H^2 e^{-H^2/B_2}\,d\zeta_2}{\int e^{-H^2/B_2}\,d\zeta_2}$$

and let B be the unique value such that

$$E_1+E_2=\frac{\int (H^1+H^2)e^{-(H^1+H^2)/B}\,d\zeta_1\times\zeta_2}{\int e^{-(H^1+H^2)/B}\,d\zeta_1\times\zeta_2}$$

Let

$$E_1^0=\frac{\int H^1 e^{-H^1/B}\,d\zeta_1}{\int e^{-H^1/B}\,d\zeta_1},\quad E_2^0=\frac{\int H^2 e^{-H^2/B}\,d\zeta_2}{\int e^{-H^2/B}\,d\zeta_2}$$

Then $E_1^0+E_2^0=E_1+E_2$ and we see that if an amount of energy $E_1-E_1^0$ goes from system 1 to system 2, then the combined system will be in a state of maximal randomness with parameter B. Clearly $E_1-E_1^0\geqslant 0$ if and only if $B_1\geqslant B_2$ and we may think of B as describing the temperature of the system at least as far as ordering is concerned. Actually the analysis to follow shows that B must actually be proportional to the absolute temperature. $B=kT$, where k is a universal constant known as Boltzmann's constant.

We have now obtained an algorithm for computing the internal energy function U_A of a material body A as soon as we know its dynamical constitution, that is, as soon as we know H and Ω. It is

$$U_A(T)=\frac{\int H_A e^{-H_A/kT}\,d\zeta}{\int e^{-H_A/kT}\,d\zeta}$$

At first it seems to be a function of T alone. However, we recall that H_A contains the volume v as a parameter and hence $U_A(T)$ depends upon v since it depends upon H. Similar considerations lead to a closely related

ISBN 0-8053-6702-0/0-8053-6703-9, pbk

algorithm for the state function p_A. It is

$$p_A(T) = \frac{\int \frac{\partial H_A}{\partial v} e^{-H_A/kT} d\zeta}{\int e^{-H_A/kT} d\zeta}$$

What ever doubts we may have about the derivations of these algorithms, the algorithms themselves seem to be correct (except for the changes necessitated by quantum mechanics) and we have a more or less universally accepted way of computing the thermodynamic properties of material bodies from hypotheses about their dynamical structure. As one check on their validity, let us show that they imply the second law of thermodynamics (at least in the special case of a single homogeneous body not capable of chemical change). For this we must only show that $(dU_A + p_A dv)/T$ is an exact differential. But let

$$P_A(T) = \int e^{-H_A/kT} d\zeta$$

Then

$$\frac{\partial P_A(T)}{\partial T} = \frac{1}{kT^2} \int H_A e^{-H_A/kT} d\zeta$$

and

$$\frac{\partial P_A(T)}{\partial v} = -\frac{1}{kT} \int \frac{\partial H_A}{\partial v} e^{-H_A/kT} d\zeta$$

Hence

$$U_A(T) = kT^2 \frac{\frac{\partial P_A(T)}{\partial T}}{P_A(T)} = kT^2 \frac{\partial}{\partial T} \log P_A(T)$$

$$p_A(T) = \frac{kT \frac{\partial P_A(T)}{\partial v}}{P_A(T)} = kT \frac{\partial}{\partial v} \log P_A(T)$$

Hence

$$\frac{\partial U_A(T)}{\partial v} = kT^2 \frac{\partial^2}{\partial T \partial v} \log P_A(T)$$

ISBN 0-8053-6702-0/0-8053-6703-9, pbk

so

$$\frac{\dfrac{\partial U_A(T)}{\partial v}+p_A}{T}=kT\frac{\partial}{\partial v}\left(\frac{\partial}{\partial T}\log P_A(T)+\frac{\log P_A(T)}{T}\right)$$

$$=\frac{\partial}{\partial v}\left(kT\frac{\partial}{\partial T}\log P_A(T)+k\log P_A(T)\right)$$

and

$$\frac{1}{T}\frac{\partial U_A(T)}{\partial T}=2k\frac{\partial}{\partial T}\log P_A(T)+kT\frac{\partial^2}{\partial T^2}\log P_A(T)$$

$$=\frac{\partial}{\partial T}\Bigl(kT\frac{\partial}{\partial T}\log P_A(T)+k\log P_A(T)$$

Thus

$$\frac{dU_A+p_A\,dv}{T}=dS_A$$

where

$$S_A(T)=kT\frac{\partial}{\partial T}\log P_A(T)+k\log P_A(T)$$

and the second law is established. S_A is of course the entropy function of our system. Using the definition of $P_A(T)$, we obtain the following formula for computing S_A.

$$S_A(T)=k\log\int_\Omega e^{-H_A/kT}\,d\zeta+\frac{\dfrac{1}{T}\displaystyle\int H_A e^{-H_A/kT}\,d\zeta}{\displaystyle\int e^{-H_A/kT}\,d\zeta}$$

It is interesting to compare this expression with the expression $\int\rho\log(1/\rho)\,d\zeta$ for the amount of randomness when

$$\rho=\frac{e^{-H_A/kT}}{\displaystyle\int e^{-H_A/kT}\,d\zeta}=\frac{e^{-H_A/kT}}{P_A(T)}$$

ISBN 0-8053-6702-0/0-8053-6703-9, pbk

The latter expression reduces to

$$\log P_A(T) + \frac{1}{kT}\frac{\int H_A e^{-H_A/kT}}{P_A(T)} = kS_A(T)$$

In other words, the entropy is precisely the constant $1/k$ multiplied by the amount of randomness.

Of course, all three functions p_A, U_A, S_A are computable from the single function F_A as described in the section on thermodynamics and we compute without difficulty that

$$\begin{aligned} F_A &= U_A - S_A T \\ &= kT^2\frac{\partial}{\partial T}\log P_A(T) - kT^2\frac{\partial}{\partial T}\log P_A(T) - kT\log P_A(T) \\ &= -kT\log P_A(T) \end{aligned}$$

In other words, $e^{-F_A(T)/kT}$ is precisely the function

$$P_A(T) = \int_\Omega e^{-H_A/kT}\,d\zeta$$

in terms of which we have found it convenient to express the mean values of H_A and $\partial H_A/\partial v$. This function P_A is called the *partition function* for the system. Although it is defined as an integral over Ω, it is convenient to represent it as an integral of a slightly different sort. For each Borel subset E of the real line, let $\beta_A(E) = \zeta(H_A^{-1}(E))$. Then the measure β_A is a measure in the real line and it is a simple exercise in real variable theory to show that

$$\begin{aligned} P_A(T) &= \int_\Omega e^{-H_A/kT}\,d\zeta \\ &= \int e^{-x/kT}\,d\beta_A(x) \end{aligned}$$

In other words, the thermodynamic properties of a body A are completely determined by a certain measure β_A in the real line. The partition function is just $\hat{\beta}_A(1/kT)$, where $\hat{\beta}_A(y) = \int e^{-xy}\,d\beta_A(x)$ is the Laplace transform of β_A. Strictly speaking, of course, β_A is not just a measure but a function that assigns a measure to every volume v. However, it is an important fact that

ISBN 0-8053-6702-0/0-8053-6703-9, pbk

for each fixed v the partition function as a function of $1/kT$ is the Laplace transform of a positive measure.

As a first example let us consider the system consisting of n particles that do not interact with one another and move freely except for being constrained to stay in a subset R of space of finite volume v. We note that such a system is in fact a product of n independent systems where each independent system consists of just one particle. Quite generally a simple argument shows that the partition function for a product of independent systems is the product of the separate partition functions so that we need only compute β for the case $n=1$. If m is the mass of the particle, then

$$H(q_1,q_2,q_3,p_1,p_2,p_3)=\frac{p_1^2+p_2^2+p_3^2}{2m}$$

if $q_1,q_2,q_3\in R$. Thus $H(q_1,\cdots,p_3)\leqslant E$ if and only if $q_1,q_2,q_3\in R$ and $p_1^2+p_2^2+p_3^2\leqslant 2mE$. Thus $\beta(0,E)=v$ (volume of sphere of radius $\sqrt{2mE}$) $=\frac{4}{3}\pi v(2mE)^{3/2}=cm^{3/2}vE^{3/2}$, where $c=\frac{8}{3}\sqrt{2}\,\pi$. Hence

$$P(T)=cvm^{3/2}\int_0^{\infty}e^{-x/kT}\tfrac{3}{2}x^{1/2}\,dx$$
$$=\tfrac{3}{2}cvm^{3/2}\int_0^{\infty}e^{-x/kT}x^{1/2}\,dx$$

Setting $y=x/kT$ this becomes

$$\tfrac{3}{2}cvm^{3/2}\int_0^{\infty}e^{-y}(kT)^{1/2}y^{1/2}kT\,dy=\tfrac{3}{2}cvm^{3/2}(kT)^{3/2}\int_0^{\infty}y^{1/2}e^{-y}\,dy$$
$$=c^1m^{3/2}v(kT)^{3/2}$$

where c^1 is a universal constant. Finally, then P for the n-particle system is $c_0v^n(kT)^{3n/2}$, where c_0 is a constant depending only an n and the masses of the particles. The free energy function then is

$$F(T,v)=-kT\log c_0-nkT\log v-\tfrac{3}{2}nkT\log kT$$

Thus

$$U(T,v)=\tfrac{3}{2}nkT \quad\text{and}\quad p(T,v)=\frac{nkT}{v}$$

Thus for a "gas" made up of n independent particles (of arbitrary masses) we find that pv/T is a constant equal to nk and that the specific heat (at

ISBN 0-8053-6702-0/0-8053-6703-9, pbk

constant volume) is also a constant equal to $\frac{3}{2}nk$. We have derived the laws governing the behavior of a so-called "perfect gas." It is found that most actual gases approximate this behavior when p/T is small or equivalently when n/v is small. By measuring pv/T we can obtain the number nk experimentally. Thus we not only verify the atomic hypothesis for gases, but can compare the numbers of atoms contained in different gases. Actually, to find n, of course, we must have some way of measuring k. Of course, actual gases differ from perfect ones in that the atoms do interact and in principle one can test hypotheses about the nature of this interaction by computing the corresponding β and hence P, F, U, and p. This program has been carried out in some cases but there are formidable mathematical difficulties in general.

As a second example, let us consider the problem of computing the specific heat of a solid. We conceive of the solid as a set of n particles that interact in such a fashion that there is a configuration of stable equilibrium in which the particles remain fixed at definite distances from one another. When displaced from their equilibrium positions, the particles vibrate back and forth about these positions and as long as the vibrations are of small amplitude the motion is very approximately described by a Hamiltonian function H, which is a positive-definite quadratic form in both positions and momenta. It then follows from standard linear algebra that there exists a linear change of variable $q_j^0 = \Sigma a_{jk} q_k$ of such a character that the q_j^0 move independently of one another with Hamiltonians of the form $p^2/2m + (\mu/2)q^2$. The partition function of our system is consequently a product of the partition functions of "harmonic oscillators." To compute the partition function of the harmonic oscillator with Hamiltonian $p^2/2m + (\mu/2)q^2$, we observe that the set of all p,q with

$$\frac{p^2}{2m} + \frac{\mu q^2}{2} \leqslant E$$

is the interior of the ellipse

$$\frac{p^2}{(\sqrt{2mE}\,)^2} + \frac{q^2}{\left(\sqrt{\frac{2}{\nu}E}\right)^2} = 1$$

and therefore has measure $\pi 2E\sqrt{\mu/m}$. Thus $\beta(0,x) = 2\pi x\sqrt{m/\mu} = x/\nu$, where $\nu = (1/2\pi)\sqrt{m/\mu}$ is the frequency with respect to which the oscillator oscillates. The corresponding partition function is

$$\frac{1}{\nu}\int_0^\infty e^{-x/kT}\,dx = \frac{kT}{\nu}$$

ISBN 0-8053-6702-0/0-8053-6703-9, pbk

Thus the partition function for our solid is $(kT)^{3n}/\nu_1\nu_2\cdots\nu_{3n}$, where the ν_j are the frequencies of the harmonic oscillators into which its motion may be decomposed. From this we conclude that the free energy function F is given by

$$F(T) = -kT\log(kT)^{3n} - kT\log\nu_1\cdots\nu_{3n}$$
$$-=3nkT\log kT - kT\log\nu_1\cdots\nu_{3n}$$

and that $U(T)=3nkT$. Thus the specific heat of a solid is a constant and this constant is $3k$ times the number of atoms or fundamental particles that it contains. This result agrees with experiment for most substances at room temperatures and above but fails for low temperatures. Indeed, for some substances it fails even at room temperatures.

The failure to explain the specific heats of solids at low temperatures is only one of several ways in which classical statistical mechanics failed to live up to expectations. In addition, its predictions about the specific heats of gases turned out to be in disagreement with experiment for polyatomic gases except for rather high temperatures. It predicted a specific heat proportional to the number of degrees of freedom but some of these seemed "frozen" at lower temperatures. With the discovery of the electron, these discrepencies became much worse because this meant that each atom had many more degrees of freedom than had previously been supposed—and according to the theory each degree of freedom contributed the same amount to the specific heat. Indeed, the theory fell into considerable disrepute and its originators became quite discouraged.

Actually, all that was wrong was that quantum mechanics had not yet been discovered. Quantum mechanics changes statistical mechanics in a very interesting way and leads to formulas that are in agreement with those of the classical theory for high temperatures but not for low. The low-temperature quantum formulas do agree with experiment and fully vindicate the theory.

The so-called "old quantum theory" that preceded quantum mechanics proper originated in 1900 with Max Planck's successful attempt to explain still another anomaly in the predictions of classical statistical mechanics. As we saw in Section 22 the electromagnetic field may be regarded as a classical dynamical system whose Hamiltonian is quadratic in the coordinates and momenta. When the field is enclosed in a finite volume, the resulting infinite-dimensional dynamical system decomposes into a product of independent harmonic oscillators with frequencies $\nu_1, \nu_2, \nu_3, \ldots$ depending upon the geometrical properties of the enclosure. When the enclosure is rectangular, an easy application of Fourier analysis allows one

ISBN 0-8053-6702-0/0-8053-6703-9, pbk

to determine that the number of ν_j between ν and $\nu+\Delta\nu$ is approximately $(8\pi V\nu^2/c^3)\Delta\nu$, where the volume V is large and $\Delta\nu$ is small compared to ν but large enough to contain a representative number of ν_j. A deeper analysis shows that the shape is irrelevant and that the formula is valid for any enclosure of volume V. Now according to classical statistical mechanics each oscillator should contribute kT to the total internal energy at temperature T and accordingly the energy for unit volume in the frequency range between ν and $\nu+\Delta\nu$ should be $(8\pi\nu^2 kT/c^3)\Delta\nu$. This formula was put forward by Rayleigh to explain the energy distribution in "black-body radiation" at temperature T. It agrees with experiment when ν/T is small but is quite wrong for ν/T large. Indeed, for large ν/T the data agree with a formula (due to Wien) of the form $A\nu^3 e^{-b\nu/T}\Delta\nu$, where A and b are constants to be determined by experiment. Planck obtained a synthesis of these two formulas by reasoning rather different from but analogous to that which follows. Suppose that we did not like to integrate but preferred to take discrete sums and pass to the limit. Then in computing $P(T)$ for a harmonic oscillator of frequency ν, we might first replace the measure $(1/\nu)dx$ by a discrete measure approximating it. If we assign a measure h to the points $0, a, 2a, 3a$, the interval $[0, na]$ will be assigned the measure $(n+1)ha$ and this will approximate na/ν for large n if and only if $a = h\nu$. Thus we may compute $P(T)$ by first forming $\sum_{n=0}^{\infty} e^{-nh\nu/kT}$ and then taking the limit as h tends to zero. Now this expression is a geometric series whose sum is

$$\frac{1}{1-e^{-nh\nu/kT}}$$

and for $h \neq 0$ the corresponding U is

$$\frac{h\nu}{e^{h\nu/kT}-1}$$

Now as $h \to 0$ the denominator becomes very close to

$$1+\frac{h\nu}{kT}-1=\frac{h\nu}{kT}$$

and the whole expression approaches kT as expected. However, whenever $h \neq 0$ and kT is much smaller than $h\nu$, we may neglect 1 and get $h\nu e^{-h\nu/kT}$. If we substitute this value instead of kT in deriving Rayleigh's formula the latter becomes

$$\frac{8\pi h\nu^3}{c^3} e^{-h\nu/kT}\Delta\nu$$

ISBN 0-8053-6702-0/0-8053-6703-9, pbk

and this is just Wien's formula if we take $A=8\pi h/c^3$, $b=h/k$. In other words, if we choose just the right nonzero value for h, we obtain a formula that reconciles those of Wien and Rayleigh and agrees with experiment over the whole temperature range. This is in essence Planck's great discovery and the constant h is the famous Planck's constant. Note that experiments allow us to determine A and b and hence to find numerical values for h and k through the formulas $h=Ac^3/8\pi$, $k=Ac^3/8\pi b$. Since specific heat data give us nk, we now can determine the number n of atoms in a material body from experiment.

Of course Planck's idea applies equally well to formulas for the specific heats of solids. The formula for the internal energy now becomes

$$U(T)=\sum_{j=1}^{3n}\frac{h\nu_j}{e^{h\nu_j/kT}-1}$$

Far from being independent of T and the ν_j the specific heat depends upon all of them and in a rather complicated way. On the other hand, when kT is large compared to *all* ν_j, then each $e^{h\nu_j/kT}$ is approximately $1+h\nu_j/kT$ and $U(T)$ becomes approximately $3nkT$. Thus Planck's formulas accounts both for the applicability of the classical theory at high temperature and for its deviation from this at low temperatures. (Actually the application of Planck's ideas to specific heats were made by Einstein in 1906.) It is also worth noticing that Planck's replacement of $(1/\nu)dx$ by a discrete measure removes another anomaly. According to the classical theory, an infinite number of oscillators, as in the electromagnetic field, should have an infinite specific heat $U(T)=kT+kT+\cdots$ with infinitely many summands. Planck replaces this obviously divergent series by

$$\sum_{j=1}^{\infty}\frac{h\nu_j}{e^{h\nu_j/kT}-1}$$

which converges whenever the ν_j go to with sufficient rapidity.

Of course the problem remains of accounting for the mysterious replacement of $(1/\nu)dx$ by a discrete measure in some rational way and of deciding what replacement to use when dealing with problems other than the harmonic oscillator. Physicists struggled with this problem for a quarter of a century and a more or less final answer was provided by the development of quantum mechanics in the latter half of the 1920s.

Let us recall that the quantum mechanical analog of a probability measure in phase space is a probability measure on the set of all projections in our Hilbert space and that the most general such is of the form

ISBN 0-8053-6702-0/0-8053-6703-9, pbk

$F \rightarrow \mathrm{Tr}(F\Lambda)$, where Λ is a positive self-adjoint operator whose trace exists and is equal to one. If H is the Hamiltonian operator of our system, then its expected value in the state described by Λ is $\mathrm{Tr}(\Lambda H)$ and as in the classical case we may seek that Λ of maximal randomness for which $\mathrm{Tr}(\Lambda H)$ has a fixed value E. To do this we require a way of measuring randomness for measures of the form $F \rightarrow \mathrm{Tr}(F\Lambda)$. This is not difficult to find. It is easy to convince oneself that pure states correspond to minimal randomness. Moreover, each Λ is defined by a discrete probability distribution on the pure states. Indeed if $\phi_1, \phi_2, \phi_3, \ldots$ is a basis of eigenvectors for Λ and $\gamma_1, \gamma_2, \ldots$ are the corresponding eigenvalues, then the state defined by Λ may be equally well described by saying that the pure state defined by ϕ_j occurs with probability γ_j. The arguments from information theory listed above now lead us at once to the hypothesis that the amount of randomness in the state described by Λ should be measured by the number $\Sigma \gamma_j \log 1/\gamma_j$. This is just $-\mathrm{Tr}(\Lambda \log \Lambda)$. Thus our problem is to find the positive-definite self-adjoint trace operator Λ that maximizes $-\mathrm{Tr}(\Lambda \log \Lambda)$ subject to the restrictive conditions $\mathrm{Tr}\,\Lambda = 1$ and $\mathrm{Tr}(\Lambda H) = E$. Arguments analogous to those used in the classical case show that Λ must be of the form $Ae^{-H/B}$, where A and B are constants. As before the condition $\mathrm{Tr}\,\Lambda = 1$ tells us that $A = 1/\mathrm{Tr}(e^{-H/B})$ so that

$$H = \frac{e^{-H/B}}{\mathrm{Tr}(e^{-H/B})}$$

and the parameter B is the (in general unique) solution of the equation

$$E = \frac{\mathrm{Tr}(He^{-H/B})}{\mathrm{Tr}(e^{-H/B})}$$

Moreover, just as before we identify B with kT, where k is Boltzmann's constant and T is the absolute temperature. It is now easy to check that the formulas of classical statistical mechanics go over into the formulas of quantum statistical mechanics if we make just one change. This change consists in replacing the formula for the partition function

$$P(T) = \int e^{-x/kT} d\beta(x)$$

by the formula

$$P(T) = \mathrm{Tr}(e^{-H/kT})$$

ISBN 0-8053-6702-0/0-8053-6703-9, pbk

All other thermodynamic functions are expressible in terms of P and the mode of expression is the same in the two cases. Actually, the change is even less drastic than it appears to be. If $\phi_1, \phi_2, \ldots$ is a basis of eigenvectors for H and $H(\phi_j) = E_j\phi_j$ then

$$e^{-H/kT}\phi_j = e^{-E_j/kT}\phi_j$$

and

$$P(T) = \sum_{j=1,2,\ldots} e^{-E_j/kT} = \int e^{-x/kT}\, d\beta_q(x)$$

where β_q is the measure that assigns to each Borel set the number of H eigenvalues contained in it. In other words, everything is the same including the formula for $P(T)$ if we replace the continuous measure β derived from the classical Hamiltonian by the discrete measure β_q obtained as above by counting the eigenvalues of the quantum Hamiltonian. In the case of a single harmonic oscillator of frequency ν, it is not difficult to verify that the quantum Hamiltonian has the eigenvalues $(k+\frac{1}{2})2\pi\hbar\nu$, where $k = 1,2,\ldots$ and if we subtract an inessential constant we get $2\pi\hbar\nu, 4\pi\hbar\nu, 6\pi\hbar\nu, \ldots$. Thus β_q is just the measure introduced above as a discrete approximation to $(1/\nu)dx$ provided that $\hbar = h/2\pi$. In other words, Planck's discrete approximation that mysteriously gives right answers is actually the measure which quantum mechanics tells us we must use.

The question now arises as to how β_q and β compare more generally and the above example suggests the conjecture that perhaps in "all" cases β_q is a discrete approximation to β in the sense that

$$\lim_{x\to\infty} \frac{\beta([0,x])}{\beta_q([0,x])} = 1$$

Now $\beta([0,x])$ is usually a continuous function of x more or less explicitly describable in terms of the function occurring in the classical Hamiltonian and hence in the associated quantum mechanical differential operator. Thus this conjecture if true would give us explicit asymptotic formulas for the eigenvalues of certain partial differential operators. The literature in differential equations both ordinary and partial contains theorems providing such formulas and they have just the form suggested by the above considerations. In other words, it is true (for a wide variety of dynamical systems) that

$$\lim_{x\to\infty} \frac{\beta([0,x])}{\beta_q([0,x])} = 1$$

ISBN 0-8053-6702-0/0-8053-6703-9, pbk

and we may regard the comparison of quantum and classical statistical mechanics as providing a possible motivation for conjecturing many known results about the asymptotic properties of eigenvalues of differential operators.

The above described relationship between β and β_q may be expected to imply some sort of relationship between the partition functions

$$P_q(T)=\int e^{-x/kT}\,d\beta_q(x)$$

and

$$P(T)=\int e^{-x/kT}\,d\beta(x)$$

In this connection we note that $P(T)=\hat{\beta}(1/kT)$, where $\hat{\beta}(s)=\int e^{-sx}\,d\beta(x)$ and similarly for $P_q(T)$. In other words, except for the change of variable $s=1/kT$ the partition function is just the Laplace transform of the measure defining it. Now the relationship between measures and their Laplace transforms has been extensively studied on an abstract mathematical level and one has results such as the following: Not only do β and $\hat{\beta}$ determine one another uniquely, but the asymptotic behavior of $\hat{\beta}$ for small s is directly related to the asymptotic behavior of $\beta([0,x])$ for large x. For example, one has a theorem asserting that if β is concentrated in the positive axis and

$$\lim_{x\to\infty}\frac{\beta([0,x])}{x^{\gamma}}=A$$

then

$$\lim_{s\to 0}\hat{\beta}(s)s^{\gamma}=A\Gamma(\gamma+1)$$

Of course $s\to 0$ if and only if $T\to\infty$. Thus the theorem just cited implies that the behavior of P for large T is determined by $\beta([0,x])$ for large x at least when $\beta([0,x])\sim Ax^{\gamma}$. This suggests that a more refined study of the relationship between β and β_q on the one hand and between $\beta(\text{resp}\,\beta_q)$ and $\hat{\beta}(\text{resp}\,\hat{\beta}_q)$ on the other should make it possible to prove quite generally that classical and quantum statistical mechanics must agree at high temperatures. We shall not attempt such a study here. We simply wish to emphasize that statistical mechanics provides a motivation for a rather detailed study of the relationship between a measure and its Laplace transform and provides physical interpretations for theorems in the subject. This is especially interesting because number theory also provides

ISBN 0-8053-6702-0/0-8053-6703-9, pbk

such a motivation—indeed much of the theory of the Laplce transform (especially for discrete β) was developed in response to the needs of number theory. Combining these two connections, one is led to physical interpretations of results in number theory as well as applications of number theoretical results to physics. One interesting question about the Laplace transform that is suggested by the needs of physics is this. We measure the specific heat directly. If we know it exactly for all T, then we know U and hence know P up to a multiplicative constant. Hence we know β up to a multiplicative constant. Hence we know all eigenvalues of the Hamiltionian. In practice, of course, we can measure the specific heat with only limited accuracy and only over a finite interval. The question then arises as to how this inaccuracy affects our knowledge of the eigenvalues. In other words, what kind of variation in β is consistent with a variation in $\hat{\beta}$ that changes $\hat{\beta}$ only slightly in a "long" finite interval.

We close this section with two examples in which the quantum mechanical partition function is especially closely related to the analytic functions which occur in number theory. First let us consider quantum mechanical systems of the sort that arise when we quantize a field as in Dirac's theory of the electromagnetic field. We begin with a classical mechanical system for which both the kinetic and potential energies are quadratic and in which phase space is an infinite-dimensional vector space. Then, as we have seen, we may convert the classical phase space into a complex Hilbert space in such a fashion that the classical dynamical group becomes a one-parameter unitary group $t \to W_t$. Moreover, as we have also seen, the corresponding quantum dynamical group $t \to U_t$ has an especially simple relationship to $t \to W_t$. In fact,

$$U = I + W + W \circledS W + W \circledS W \circledS W + \cdots$$

where $W \circledS W \circledS \cdots \circledS W$ means the symmetrized n-fold tensor product of W with itself. Now let $W = e^{-itH_0}$, $U = e^{-itH}$ and let H_0 have a pure point spectrum with eigenvalues $E_1, E_2, \ldots$. Then the eigenvalues of H will belong to eigenvectors in one of the spaces $\mathcal{H}(W \circledS W \circledS \cdots \circledS W)$ and will consist of all numbers $l_1 E_1 + l_2 E_2 + \cdots + l_\nu E_\nu$, where $l_1 \cdots l_\nu$ vary over the nonnegative integers and ν is arbitrary. Thus the partition function will be the function

$$\sum_{\substack{l_i \geq 0 \\ l_i \text{ integral}}} \exp\left(\frac{-(l_1 E_1 + \cdots + l_\nu E_\nu)}{kT} \right)$$

ISBN 0-8053-6702-0/0-8053-6703-9, pbk

Now

$$\exp -\left(\frac{l_1 E_1}{kT} + \cdots + \frac{l_\nu E_\nu}{kT}\right) = \frac{1}{(e^{E_1/kT})^{l_1} \cdots e^{(E_\nu/kT)l_\nu}}$$

$$= \frac{1}{\left[(e^{E_1})^{l_1}(e^{E_2})^{l_2}\cdots(e^{E_\nu})^{l_\nu}\right]^{1/kT}}$$

Thus as a function of $s = 1/kT$ the partition function is $\Sigma 1/N^s$ where N varies over all possible products

$$(e^{E_1})^{l_1}(e^{E_2})^{l_2}\cdots(e^{E_\nu})^{l_\nu}$$

Notice the parallel with the celebrated Riemann zeta function of number theory. In fact, if the E_j are chosen so that the $e^{E_j} = p_j$ the jth prime then n varies over the positive integers and the partition function is $\zeta(1/kT)$, where ζ is the Riemann zeta function.

Like the Riemann zeta function, the partition function for any free boson field admits an "Euler product" decomposition. Indeed recall that the classical system from which we start can be regarded as composed of independent oscillators and by a general principle the partition function for the whole system is the product of the partition functions for these separate oscillators. The partition function for the oscillator of energy E_j (we are taking units such that $\hbar = 1$) is

$$\sum_{l=0}^{\infty} e^{-lE_j/kT} = \frac{1}{1 - 1/(e^{E_j})^{1/kT}}$$

Thus our partition function may be written as the infinite product

$$\prod_{j=1}^{\infty} \frac{1}{\left(1 - (e^{E_j})^{\frac{1}{kT}}\right)}$$

Of course when e^{E_j} is the jth prime and we set $s = 1/kT$, we get the usual Euler product for the Riemann zeta function.

Note that the partition function for the oscillator has the property that its reciprocal is also the Laplace transform of a (signed) measure. In fact, this reciprocal is

$$1 - \frac{1}{(e^{E_j})^{1/kT}}$$

ISBN 0-8053-6702-0/0-8053-6703-9, pbk

and as a function of $1/kT = s$ this is

$$1 - \frac{1}{(e^{E_j})^s} = 1 - e^{-E_j s} = \int e^{-xs}\, d\beta(x)$$

where β is the point measure that assigns the measure 1 to the point 0 and -1 to the point E_j. It follows that the entire partition function has this property and hence that the internal energy function does too. In fact, we compute in a routine way that the latter has the form

$$U(T) = \sum_{\substack{j=1\\n=1}}^{\infty} E_j e^{-nE_j/kT}$$

$$= \int e^{-x/kT}\, d\rho(x)$$

where

$$\rho([0,x]) = \sum_j g(x,j) E_j$$

where $g(x,j)$ = largest positive integer l such that $lE_j \leqslant x$. Thus the asymptotic behavior of $U(T)$ for large T is connected with the distribution of the numbers e^{E_j} and hence with the distribution of the E_j themselves. When the E_j are the logarithms of the primes, the function $U(T)$ with $s = 1/kT$ is just $-\zeta'(s)/\zeta(s)$. Knowing the distribution of the integers we can draw conslusions about the behavior of $\zeta(s)$ for s near 1 (1 instead of 0 because the series for ζ converges only for $s > 1$). From this and certain auxilary information about ζ, one finds the behavior of $-\zeta'/\zeta$ near 1, and from this one gets information about the distribution of the primes—in particular, the celebrated prime number theorem of Hadamard and de la Vallee Poussin. We remark that it is less easy to go from information about $\hat{\beta}$ to information about β than vice versa. One needs auxilary results called Tauberian theorems. Our main point here is that one could have been led to the main outline of the proof of the prime number theorem by using the physical interpretation of Laplace transforms provided by statistical mechanics. In particular, the function $-\zeta'/\zeta$ whose representation as a Dirichlet series (Laplace transform with discrete measure) plays a central role in the proof has a direct physical interpretation as the internal energy function.

Our other example is that of a quantum mechanical perfect gas made up of spinless identical particles. Since the particles are independent, we can

ISBN 0-8053-6702-0/0-8053-6703-9, pbk

deduce the eigenvalues of the n-particle case from that of the one-particle case. The eigenvalue spectrum depends to some extent on the shape of the container but may be computed exactly when the shape is cubical. We shall confine attention to this case. One computes easily that when the cube has side of length a the Hamiltonian

$$-\frac{\hbar}{2m}\left(\frac{\partial^2}{\partial x^2}+\frac{\partial^2}{\partial y^2}+\frac{\partial^2}{\partial z^2}\right)$$

with the appropriate boundary conditions has as eigenvalues all numbers of the form

$$-\frac{\hbar}{2m}\frac{\pi^2}{a^2}\left(l_1^2+l_2^2+l_3^2\right)$$

where l_1, l_2, and l_3 vary independently over the nonnegative integers. Thus the partition function for the one-particle gas is

$$\sum_{l_1,l_2,l_3=0}^{\infty}\exp\left(-\frac{\hbar\pi^2}{2mkTa^2}\left(l_1^2+l_2^2+l_3^2\right)\right)$$

$$=\sum_{n=0}^{\infty}\rho(n)\exp\left(-\frac{\hbar\pi^2}{2mkTa^2}n\right)$$

where $\rho(n)$ is the number of ways of representating n as a sum of three squares. In number theory one studies functions like $\rho(n)$ via the "θ series"

$$\theta_\rho(z)=\sum_{n=1}^{\infty}\rho(n)e^{\pi izn}$$

It is thus interesting to note that the partition function for our one-particle gas is just $\theta_\rho(i\hbar\pi/2mkTa^2)$ especially since the study of θ series in number theory and the theory of automorphic functions has provided a wealth of information about the behavior of θ_ρ. To study the n-particle gas it will not be sufficient as in the classical case to replace the partition function of the one-particle gas by its nth power; we must also take account of the "identity" of the particles and the passage from $\mathcal{H}^n$ to its symmetric subspace. Thus if $E_1, E_2, \ldots$ are the eigenvalues in the one-particle problem those in the n-particle problem will be all sums $k_1E_1+k_2E_2+\cdots+k_vE_v$ for which $k_1+k_2+\cdots+k_v=n$.

ISBN 0-8053-6702-0/0-8053-6703-9, pbk

Notes and References

A new approach to statistical mechanics was inaugurated in 1963 with the publication of a fundamental paper by H. Araki and E. J. Woods in the *Journal of Mathematical Physics*. It was continued beginning in 1965 by D. Ruelle, R. L. Dobrushin, and others. The key idea is to get rid of "edge effects" and study the so-called "thermodynamic limit" by supposing the system under consideration to fill all of space. Among other advantages one can then exploit the existence of symmetries that otherwise are only approximately present. In particular, ergodic theory enters into statistical mechanics in a quite new way. A standard work on this new development is the book of Ruelle, *Statistical Mechanics*—Rigorous Results, Benjamin-Cummings, Reading, Mass. (1968). A brief account of some of the main ideas will be found in part 10 of the author's article "Ergodic theory and its significance for probability theory and statistical mechanics," *Advan. Math.*, vol. 12 (1974), 178–268 (mentioned in the Preface as containing a further developement of the material of Sections 15 and 16 of this book). Part 9 of this article is also relevant to Section 24. In particular, pages 240 and 241 contain the author's views on the role of ergodicity and the ergodic theorem in understanding the "approach to equilibrium" and the statistical nature of the laws of thermodynamics.

Another new development began in 1966 with the publication of "Covariance algebras in field theory and statistical mechanics" (*Comm. Math. Phys.*, vol. 3, 1–28) by S. Doplicher, D. Kastler, and D. Robinson. A covariance algebra is the system consisting of a C^* algebra $\mathcal{A}$, a separable locally compact group G, and a homomorphism α of G into the group of automorphisms of $\mathcal{A}$. Following ideas introduced by I. E. Segal in the 1940s, one thinks of the self-adjoint elements of $\mathcal{A}$ as the observables of a physical system and of G as the time translation group, a group of symmetries, or as a combination of the two. The replacement of the algebra of all bounded linear operators in the Hilbert space of state vectors by a more general C^* algebra is important both in dealing with quantum field theory and with the quantum version of the infinite system limit of statistical mechanics developed by Ruelle et al. Segal had long advocated such a replacement in quantum field theory. For further details the reader is referred to pages 241 through 246 of the published version of the author's Chicago lecture notes and the treatise of G. Emch (*Algebraic Methods in Statistical Mechanics and Quantum Field Theory*, Wiley, New York (1972)).

ISBN 0-8053-6702-0/0-8053-6703-9, pbk

25. Fourier analysis and analytic functions

We have spoken of a relationship between statistical mechanics and number theory arising from their common use of certain kinds of Dirichlet series. These Dirichlet series appeared naturally as the result of an analysis of a physical situation. However, they also have a mathematical significance that relates them closely to Fourier analysis and hence to the abstract general theory developed in the first 15 sections of these lectures. Indeed if

$$\sum_{j=1}^{\infty} c_j e^{-\lambda_j y}$$

is a Dirichlet series that converges at all, it will converge for all $y > a$ for some a. Moreover, the series

$$\sum_{j=1}^{\infty} c_j e^{i\lambda_j z} = \sum_{j=1}^{\infty} c_j e^{i\lambda_j (x+iy)}$$

$$= \sum_{j=1}^{\infty} c_j e^{\lambda_j ix} e^{-\lambda_j y}$$

will converge for all $z = x + iy$ in the half-plane; $y > a$ and will define an analytic function there. Now for fixed y the function defined by the series is

$$x \to \sum_{j=1}^{\infty} a_j e^{i\lambda_j x}$$

George W. Mackey, Unitary Group Representations in Physics, Probability, and Number Theory

ISBN 0-8053-6702-0/0-8053-6703-9, pbk

where $a_j = c_j e^{-\lambda_j y}$ and thus is just the Bohr–Fourier series expansion for a certain almost periodic function. This function has a unique extension that is analytic in a half-plane and the function defined by our original (real variable) Dirichlet series is just the restriction of this analytic continuation to a line parallel to the imaginary axis. The fact that certain Bohr–Fourier series may be extended to be analytic functions is a special case of a rather general theorem that brings Fourier analysis into close touch with the theory of analytic functions even at the abstract level of Section 6. This connection between Fourier analysis and analytic functions might well have been explained as part of the general theory of the first 15 sections. Since it was not, it seems appropriate to do so here as a sort of preface to the remaining sections on the connections of the general theory with number theory and the theory of automorphic functions.

Let us recall that the Fourier transform of a suitably restricted complex-valued function f on the separable locally compact commutative group G is the function f defined on the dual group by the formula

$$\hat{f}(\chi) = \int f(x)\chi(x)\,d\mu(x)$$

where μ is Haar measure on G. Now if we were to drop the restriction to unitary representations that we have made up to now, we would obtain a larger class of irreducible representations of G by considering characters that are not necessarily unitary, that is, continuous homomorphisms of G into the multiplicative group of *all* nonzero complex numbers. Moreover, the formula

$$\hat{f}(\chi) = \int f(x)\chi(x)\,d\mu(x)$$

makes sense whether or not χ is unitary and the Fourier transform $\hat{f}$ of any summable function f has a natural extension to a function $\tilde{f}$ defined on all generalized characters for which $|f\chi|$ is summable. Let us look at this extended Fourier transform a little more closely. First of all, it is clear that every generalized character is uniquely a product of an ordinary character and a generalized character taking on only positive real values. Secondly, every generalized character taking on only positive real values can be written uniquely in the form $x \rightarrow e^{l(x)}$, where l is a continuous homomorphism of G into the group of all real numbers under addition. Thus the group of all generalized characters is isomorphic to $\hat{G} \times \overline{G}$, where $\hat{G}$ is the ordinary character group and $\overline{G}$ is the additive group of the real *vector space* of all homomorphisms l. It is now easy to see that for each f the

ISBN 0-8053-6702-0/0-8053-6703-9, pbk

domain of $\tilde{f}$ is $\hat{G}\times K$, where K is some convex subset of the vector space $\overline{G}$. To see what sort of thing this extended Fourier transform is, let us look at some simple examples. Suppose first that G is the additive group of the real line. Then the most general member of $\overline{G}$ is $x\rightarrow -\sigma x$ where σ is a real number, the most general member of $\hat{G}$ is $x\rightarrow e^{-i\tau x}$, where τ is a real number, and

$$\tilde{f}(\sigma,\tau)=\int_{-\infty}^{\infty} f(x)e^{-(\sigma+i\tau)x}\,dx$$

The right-hand side exists as an absolutely convergent integral for σ in some finite or infinite interval. Thus $\tilde{f}(\sigma,\tau)$ is defined in the whole σ,τ plane, in a half-plane with a vertical boundary or in a vertical strip. Moreover, it coincides with the two-sided Laplace transform of f and as such is an analytic function of $\sigma+i\tau$ in its domain of definition. In the special case in which f is zero on the left half-plane, $\tilde{f}(\sigma,\tau)$ will be defined at least in a right half-plane and reduces to the usual one-sided Laplace transform of f:

$$\begin{aligned}\tilde{f}(\sigma,\tau)&=\int_{0}^{\infty} e^{-(\sigma+i\tau)x}f(x)\,dx\\&=\int_{0}^{\infty} e^{-sx}f(x)\,dx\end{aligned}$$

where $s=\sigma+i\tau$.

Another simple example is that in which G is the additive group of all integers. In this case it is more illuminating to write the general real-valued character as $n\rightarrow r^{n}$ instead of as $e^{(\log r)n}$. The general ordinary character is, of course, $n\rightarrow e^{in\theta}$ so $\tilde{f}$ is defined on a subset of the set of all pairs r,θ, where $r>0$, θ is a real number, and we do not distinguish between θ and $\theta+2\pi$. But such a pair can be thought of as defining a point in the polar complex plane: $z=re^{i\theta}$. Moreover,

$$\begin{aligned}\tilde{f}(r,\theta)&=\sum_{n=-\infty}^{\infty} f(n)r^{n}e^{in\theta}\\&=\sum_{n=-\infty}^{\infty} f(n)(re^{i\theta})^{n}\end{aligned}$$

so

$$\tilde{f}(z)=\sum_{n=-\infty}^{\infty} f(n)z^{n}$$

ISBN 0-8053-6702-0/0-8053-6703-9, pbk

$\tilde{f}$ is defined in a subset given by $r_1 < r < r_2$, where $0 \leqslant r_1 \leqslant r_2 \leqslant \infty$, that is, either in an annulus centered at 0, the exterior of a circle with center at 0, the interior of a circle with center at 0, and 0 removed or the entire finite plane with 0 removed. Moreover, $\tilde{f}$ is the analytic function whose Laurent expansion coefficients are the numbers $f(n)$. In the special case in which $f(n)=0$ for $n<0$, $f(r,\theta)$ is defined at least for all $r \leqslant 1$ and $f(z)$ is the analytic function whose power-series expansion coefficients about the origin are the $f(n)$:

$$\tilde{f}(z) = \sum_{n=0}^{\infty} f(n) z^n$$

In both cases the domain of $\tilde{f}$ is identifiable in a natural way with a subset of the complex plane and $\tilde{f}$ is an analytic function there. Moreover, in the general mapping $f \to \tilde{f}$ from functions on G to functions on $\hat{G} \times K$ we have a unification of the one-sided (two-sided) Laplace transform on the one hand, with the transformation from a sequence of coefficients to the function defined by the corresponding power series (Laurent series) on the other. Clearly this unification extends to Laplace transforms for functions of several variables and to power and Laurent series in several complex variables. We need only replace G by the additive group of a higher dimensional vector space or by the additive group of all n-tuples of integers. We can even get power series in infinitely many variables by letting G be the additive group of all infinite sequences of integers each of which has only finitely many nonzero terms.

The question now arises as to whether or not there is a sense in which the extended Fourier transform $\tilde{f}$ is always "analytic" in its domain of definition. We shall see that the answer is "yes" whenever $\tilde{f}$ is really an extension of $\hat{f}$. Note that $\overline{G}$ may consist only of zero. This is so, for example, when G is compact or when it is discrete and has only elements of finite order. Accordingly, we shall restrict attention to those groups for which $\overline{G}$ is big enough to separate points in G, that is, those for which $x \neq 0$ implies $l(x) \neq 0$ for some $l \in \overline{G}$. It is easy to see that G has this property if and only if any one and hence all of the following conditions hold:

(i) $\hat{G}$ is connected;

(ii) $\hat{G}$ is the direct product of a finite-dimensional vector group and a compact connected group;

(iii) G is the direct product of a finite-dimensional vector group and a discrete countable group with no elements of finite order.

ISBN 0-8053-6702-0/0-8053-6703-9, pbk

Let us suppose now that $\hat{G}$ is connected and that $\overline{G}$ is finite-dimensional (the restriction on $\overline{G}$ is not really essential but relieves one of the necessity of talking about analytic functions of infinitely many complex variables). Each member l of $\overline{G}$, being a homomorphism of G into the additive group of the real line, has an adjoint l^* that is a homomorphism of the real line into $\hat{G}$. Here $l^*(t)$ is the character $x \to e^{itl(x)}$. Fixing some value of t, say $t=1$, we obtain a homomorphism $l \to e^{il(\cdot)}$ of $\overline{G}$ into $\hat{G}$ and it is easy to verify that the range of this homomorphism is dense in $\hat{G}$. (It will be all of $\hat{G}$ if and only if $\overline{G}$ is the direct product of finitely many replicas of the additive group of the real line and the integers.) The direct product of this homomorphism with the identity mapping of $\overline{G}$ into $\overline{G}$ will then be a homomorphism ϕ of $\overline{G} \times \overline{G}$ into $\hat{G} \times \overline{G}$ whose range is dense. Using this homomorphism we define an ergodic action of $\overline{G} \times \overline{G}$ on $\hat{G} \times \overline{G}$ by setting

$$(\chi, l)(l_1, l_2) = (\chi, l) \cdot \phi(l_1, l_2)$$

where $\cdot$ means group multiplication. That this is an ergodic action (with pure point spectrum) follows from the considerations of Section 15. Now $\overline{G} \times \overline{G}$ being the direct product ("=" direct sum) of two real vector spaces is in a natural way a complex vector space. One needs only define $i(l,m) = -m, l$. Thus it makes sense to say that a function on $\overline{G} \times \overline{G}$ is an analytic function. Now for each $\chi, l \in \hat{G} \times \overline{G}$ the mapping

$$l_1, l_2 \to (\chi, l)\phi(l_1, l_2)$$

maps $\overline{G} \times \overline{G}$ onto this orbit and allows us to regard a function on the orbit as a function on $\overline{G} \times \overline{G}$. Changing l_1, l_2 to another member of the same orbit merely translates the functions in $\overline{G} \times \overline{G}$. Thus it makes sense to say that a function defined on $\hat{G} \times \overline{G}$ (or an open subset) is analytic on each orbit. Finally, then, we may define a function g defined on some open subset of $\hat{G} \times \overline{G}$ to be analytic there if it is a Borel function whose restriction to every orbit is analytic. Note that functions which are analytic in the sense just described are quite different from functions of several variables which are analytic in only some of these variables. Indeed, suppose that the differential of our analytic function is zero. This implies that it is a constant on each orbit. *But then ergodicity implies that it is a constant almost everywhere*. Thus an analytic function in this generalized sense is determined by its "derivatives" just as completely as a function of several variables that is analytic in all of them.

With analyticity so defined, it is not difficult to show that the extended Fourier transform $\tilde{f}$ of a function f on a group G is always analytic on

ISBN 0-8053-6702-0/0-8053-6703-9, pbk

$\hat{G}\times K$ at least when G is restricted as indicated above.

From another point of view, the transform $\tilde{f}$ of f is a family of ordinary Fourier transforms. For each $l\in K$, $\chi\to f(\chi,l)=f_l(\chi)$ is the ordinary Fourier transform of $x\to f(x)e^{l(x)}$. But by the Fourier inversion formula we may compute fe^l wherever we know $\hat{f}_l$. Hence we may compute f and hence $fe^{l'}$ for any other l'. Thus the theory of the Fourier transform allows us to compute all values of $\tilde{f}$ as soon as we know $\chi\to f_l(\chi)$ for a single value of l. Let us look at this computation in the simple special case in which G is the additive group of the integers. Let

$$\tilde{f}(z)=\sum_{n=-\infty}^{\infty} a(n)z^n$$

where the series converges in the annulus $r_1<|z|<r_2$. Let $r_1<r'<r<r_2$. To determine $f(n)$ from $\tilde{f}(re^{i\theta})$, we first determine $r^n f(n)$ by computing Fourier coefficients

$$r^n f(n)=\frac{1}{2\pi}\int_0^{2\pi} f(re^{i\phi})e^{-in\phi}\,d\phi$$

Thus

$$f(n)=\frac{1}{2\pi r^n}\int_0^{2\pi} f(re^{i\phi})e^{-in\phi}\,d\phi$$

and

$$\begin{aligned}\tilde{f}(r'e^{i\theta}) &= \sum_{n=-\infty}^{\infty} f(n)(r')^n e^{in\theta}\\ &= \frac{1}{2\pi}\sum_{n=-\infty}^{\infty}\left(\frac{r'}{r}\right)^n\int_0^{2\pi}\tilde{f}(re^{i\phi})e^{in(\theta-\phi)}\,d\phi\end{aligned}$$

In the special case in which $f(n)=0$ for $n<0$, this formula reduces in a straightforward manner to the Cauchy formula

$$\tilde{f}(z')=\frac{1}{2\pi i}\int\frac{f(z)\,dz}{z-z'}$$

where the integration is taken over the circle $|z|=r$. Numerous other versions of the Cauchy integral formula in one or more complex variables may be obtained by suitable modification of the above argument, and it

ISBN 0-8053-6702-0/0-8053-6703-9, pbk

seems legitimate to regard this fundamental tool of complex analysis as belonging to the general theory that we have been expounding.

The fact that it is useful to extend Fourier analysis into the complex domain in the manner indicated above is of course familiar to analysts. Applied to Fourier transforms on the line it is the theme of the well-known book of Paley and Wiener (*Fourier transforms in the complex domain*) Amer. Math. Soc. Collquium Pub. vol. 19, New York (1934) and "complex methods" form an important chapter in the theory of Fourier series. Part of its utility may no doubt be ascribed to the flexibility that we obtain in having a whole family of Fourier transforms for a given function. Moreover, obtaining a representation for a given function by inverting we do not need to stick to one l. We can let l be a function of χ and choose this function in any way that is most convenient. This fact may be regarded, at least in part, as the basis for the utitlity of the technique of "contour integration." Also, many functions that are too unbounded to have Fourier transforms in the usual sense become bounded when multiplied by $e^{l(x)}$ for suitable l. Thus going into the complex domain permits us to extend Fourier analysis beyond its normal range.

In general, as we have seen,

$$\int f(x)\chi(x)\,d\mu(x)=\tilde{f}(\chi)$$

will not exist for all χ because $|\chi(x)|$ is not bounded when χ is not unitary. On the other hand, if f vanishes outside of a subset K of G, then $\tilde{f}(\chi)$ will exist for all χ for which $|\chi(k)|\leqslant 1$ for all k in K. Now if $\chi=\chi_0, e^{-l}$, where χ_0 is real and l is a homomorphism of G into the additive reals, then to say that $|\chi(k)|\leqslant 1$ is to say that $l(k)\geqslant 0$. For each subset K of G let K^0 be the set of all $l\in\bar{G}$ with $l(k)\geqslant 0$ for all $k\in K$ and for each subset C of $\bar{G}$ let C^0 be the set of all $k\in G$ with $l(k)\geqslant 0$ for all $l\in C$. Then we verify at once that $C^{00}\supseteq C$, $K^{00}\supseteq K$, and that $K^{000}=K^0$, $C^{000}=C^0$. Thus $C=C^{00}$ if and only if $C=K^0$ for some K and $K=K^{00}$ if and only if $K=C^0$ for some C. Moreover, if we define C (resp K) to be a *closed cone* whenever $C^{00}=C$ (resp $K^{00}=K$), we see that $K\to K^0$ and $C\to C^0$ are mutually inverse maps setting up a one-to-one inclusion inverting correspondence between the closed cones in G and $\bar{G}$, respectively. Let us call K^0 the *dual* of K and vice versa. Then it follows from what we have said above that whenever f vanishes outside of a closed cone K, then $\tilde{f}$ is defined inside of $\bar{G}\times K^0$ and in particular inside of $l\times K^0$.

When G is the additive group of the real line, the only possible closed cones are the sets $x\geqslant 0$ and $x\leqslant 0$ and their duals, respectively, are $y\geqslant 0$

and $y \leqslant 0$. Restricted to functions that vanish outside of $x \geqslant 0$, our Laplace transform becomes the familiar one-sided Laplace transform $f(s) = \int_0^\infty e^{-st}\,dt$. Similarly, when G is the additive group of the integers we get the power-series transformation.

When $\overline{G}$ is more than one dimensional, there are many more possibilities for K and K^0. An especially interesting special case is that in which G is the additive group of a finite-dimensional vector space over the reals and we demand in addition that $T(K) = K^0$ for some isomorphism T of G with $\overline{G}$ such that $T(x,x) > 0$ for $x \neq 0$ and $T(x)(y) = T(y)(x)$ for all x and y. M. Koecher calls the interior of such a K a *domain of positivity*.

The example that inspired Koecher's definition is that in which G is the additive group of all $n \times n$ real symmetric matrices and K is the subgroup of those whose eigenvalues are all nonnegative. For each $A \in G$, $B \rightarrow \mathrm{Tr}\,AB$ is a member l_B of $\overline{G}$ and $B \rightarrow l_B$ is an isomorphism of $\overline{G}$ with $\overline{G}$ that maps K onto K^0 and has the other required properties. $\hat{G}$ may of course be identified with $\overline{G}$ also and the interior of $\hat{G} \times K^0 = \overline{G} \times K$ is C. L. Siegel's famous generalization of the upper half-plane (to which it reduces when $n = 1$). Let Γ be a discrete subgroup of G such that G/Γ is compact. Then $\overline{\Gamma} = \overline{G}$ and $\hat{\Gamma} = \hat{G}/\Gamma^{\perp}$. Thus we may identify functions on $\hat{\Gamma} \times \overline{G}$ with functions on $\hat{G} \times \overline{G}$ that are constant on the $\Gamma^{\perp} \times e$ cosets. Siegel's "automorphic forms" on $\hat{G} \times \overline{G}$ are "periodic" in the sense of being constant on the $\Gamma^{\perp} \times e$ cosets, where Γ is the subgroup of all symmetric matrices with integer coefficients. Thus they may be regarded as functions on $\Gamma^{\perp} \times \overline{G}$ and turn out to be Laplace transforms of functions on Γ. We shall have more to say about these automorphic forms in later sections.

For any domain of positivity Y let Σ_Y denote the group of all linear transformations of $\overline{G}$ into $\overline{G}$ that carry Y onto Y. If Σ_Y acts transitively on Y, Koecher says that Y is homogeneous. Siegel's Y is homogeneous and Koecher shows that part at least of the theory of automorphic forms on Siegel generalized upper half-planes may be extended to the case in which K is the closure of an arbitrary homogeneous domain of positivity. Identifying $\hat{G}$ with $\overline{G}$ as we may, each element α of Σ_Y defines an automorphism of $\hat{G} \times \overline{G}$, namely, the one taking l_1, l_2 into $\alpha(l_1), \alpha(l_2)$. These automorphisms preserve the complex structure of $\hat{G} \times \overline{G}$ and so do those of the form $\chi, l \rightarrow \chi\chi_0, l$ defined by the members of $\hat{G}$. Together these two groups of automorphisms of $G \times \overline{G}$ generate a group that is isomorphic to a semi-direct product of $\hat{G}$ and Σ_Y and is transitive whenever Y is homogeneous. In the special case of the ordinary upper half-plane $(n = 1)$, this group is the group of all transformations of the form $z \rightarrow az + z_0$, where a is > 0 and z_0 is an arbitrary complex number. The group of all analytic automorphisms of the upper half-plane (with the point at ∞) is generated by this group and the single transformation $z \rightarrow -1/z$ that leaves the imaginary axis fixed. Koecher defines an analog of $z \rightarrow -1/z$ for any homogeneous positivity domain. It is the unique analytic extension of an involutory map $y \rightarrow y^*$ of Y onto Y. In various examples including the Siegel upper half-planes, this transformation and those of the form $\chi, l \rightarrow \alpha(\chi)\chi_0, \alpha(l)$ generate the group of all analytic automorphisms.

There is an interesting relationship between the Laplace transform for functions in G and the regular representation of $\hat{G}$. We have already noted that the Laplace transform amy be regarded as a family of Fourier transforms parameterized by members of $\overline{G}$. If we use the Fourier transform in the $\mathcal{L}^2$ sense, then each member $\hat{f}$ of $\mathcal{L}^2(\hat{G})$ is associated with a family of other members parametrized by those members l of $\overline{G}$ for which $fe^{-l} \in \mathcal{L}^2(G)$. Let $A_l(f)$ be the member of $\mathcal{L}^2(\hat{G})$ so associated with f and l and let U be the regular representation of $\hat{G}$. Then for each l, A_l is a linear transformation of a subspace of $H(U)$ into itself and we have $A_l U_\chi = U_\chi A_l$ as well as $A_{l_1+l_2} = A_{l_1} \cdot A_{l_2}$ whenever both sides are defined. We have described the Laplace transform (for square summable functions, as a sort of extension of U from G to part of $\hat{G} \times \overline{G}$. The operators $A_l U_\chi$ are of course not unitary. On the other hand, for each fixed vector $\hat{f}$ in $\mathcal{L}^2(\hat{G})$, the function $\chi, l \rightarrow A_l U_\chi f$ can be shown to be analytic function from $\hat{G} \times K$ to $\mathcal{L}^2(\hat{G})$ for some $K \subseteq \overline{G}$ and A_l can be shown to be uniquely defined by this property.

More generally, let V be any unitary representation of $\hat{G}$ and let P be the associated projection valued measure on G. For each $g \in \mathcal{H}(V)$ let K_g be the set of all $l \in \overline{G}$ such that e^{-l} is square integrable with respect to the measure $F \rightarrow (P_F(g) \cdot g)$. Then we may define $A_l(g)$ for each $l \in K_g$ as the unique vector such that

$$A_l(g) \cdot g' = \int e^{-l(x)} d(P_F(g) \cdot g')$$

for all g' in $\mathcal{H}(V)$. We find that $A_l V_\chi(g)$ is analytic on $G \times K_g$ and that this definition of A_l agrees with the one above when V is the regular representation of $\hat{G}$.

ISBN 0-8053-6702-0/0-8053-6703-9, pbk

26. Number theory

In the remaining sections of these notes we propose to describe some of the many interrelationships that exist between the theory of infinite-dimensional group representations, number theory, and the theory of automorphic functions of one or more complex variables. We begin with a semi-historical summary account of the nature of number theory with heavy emphasis on those parts of the subject with which we shall be chiefly concerned.

Number theory has a long and complex history going back to the work of Euclid (around 300 B.C.) and Diophantus (around 300 A.D.) and including important contributions by Fermat in the seventeenth century and by Euler in the eighteenth. However, most of the early work consists of loosely related fragments and special methods for solving special Diophantine problems, and it is probably not too misleading to think of the systematic theory as having begun in 1773–1775 with Lagrange's two memoirs on binary quadratic forms with integer coefficients. These were followed in 1798 by Legendre's *Théorie des nombres*, a book containing Lagrange's results in simplified and extended form as well as a systematic attack on ternary forms. Only three years later in 1801 another book appeared that was destined to become one of the great classics of mathematics. This was the *Disquisitiones arithmeticae* of Gauss, then a young man of 24. In it Gauss put the work of Lagrange and Legendre in what remained for a long time its final form. Actually, he had rediscovered much of their work before becoming aware of its existence. In addition, he added many new ideas, gave the first proof of the celebrated quadratic

George W. Mackey, Unitary Group Representations in Physics, Probability, and Number Theory

reciprocity law, and formalized and systematically used the notion of congruence.

This work of the end of the eighteenth century on quadratic forms may be regarded as containing the germs of a large share of the later developments. One large and important branch of number theory is the theory of n-ary quadratic forms. Another quasi-independent branch is the theory of algebraic number fields—culminating in the so-called "class field theory." This subject in part at least owes its origin to attempts to generalize the quadratic reciprocity law to powers higher than the second. Moreover, the theory of the simplest algebraic number fields—those that are quadratic extensions of the rationals—turns out to be equivalent to the theory of binary quadratic forms. Indeed, some of the more important theorems about algebraic number fields were first proved as theorems about binary quadratic forms. These do not extend to n-ary forms but find their natural generalization as theorems about number fields.

A few words are in order about the nature of these theories. Of course, one of the central problems of number theory is the solution of Diophantine equations, that is, algebraic equations in which one insists that the solutions be integers and compensates by having fewer equations than unknowns. Single linear equations in two unknowns are easily dealt with in an elementary way (via the theory of congruences) and one of the simplest nontrivial problems is that of a single quadratic equation in two unknowns with no linear terms. This is the problem of finding all solutions of the equation $An^2+Bmn+Cm^2=D$, where A, B, C, and D are given integers and n and m are unknown integers. Of course, there need not exist any solutions at all and another form of the problem is this: which integers D can be "represented" by the quadratic form $An^2+Bmn+Cm^2$ and in how many different ways.

Of course, the answer will depend upon A, B, and C but one does not have a completely new problem every time one of A, B, and C is changed. Let $J\times J$ be the group under addition of all pairs of integers and let Γ be the group of all automorphisms of $J\times J$. The most general such takes n,m into $an+bm, cn+dm$ where a, b, c, and d are integers such that $\begin{vmatrix} a & b \\ c & d \end{vmatrix} = \pm 1$. That such matrices $\begin{pmatrix} a & b \\ c & d \end{pmatrix}$ exist in abundance is clear since we may choose a and d arbitrarily and then choose b to be any factor of $ad-1$. If $\gamma\in\Gamma$ and $Q(n,m)=An^2+Bmn+Cm^2$, then $n,m\rightarrow Q(\gamma(n,m))$ is also a quadratic form that we may denote by Q^γ. Clearly Q and Q^γ represent exactly the same integers and represent each integer that they represent at all in exactly the same number of ways. For the purposes of the problem under discussion, Q and Q^γ do not need to be distinguished and one of the

ISBN 0-8053-6702-0/0-8053-6703-9, pbk

central problems that arises, the "reduction theory" problem, is that of finding a convenient way of choosing one Q from each equivalence class, that is, from each Γ orbit. The representation problem then need only be studied for these especially chosen and especially simple Q's. That there are an infinite number of inequivalent Q's follows from the observation that the "discriminant" B^2-4AC of the form is invariant under $Q \to Q^\gamma$ and may be any integer. It also follows from this observation that one may break the problem down by considering separately each possible value of the discriminant. It is a fundamental theorem in the subject that there are only finitely many inequivalent Q's having any given discriminant. Their number is called the *class number* for this discriminant. Some of the deeper theorems in number theory concern ways of computing the class number and relationships between the class numbers for different discriminants.

An important approach to the problem of finding the solutions of $An^2+Bmn+Cm^2=D$ is conveniently explained in the special case $A=C=1$, $B=0$. Let us introduce $i=\sqrt{-1}$ and write n^2+m^2 in the form $(n+im)(n-im)$. The complex numbers of the form $n+im$, where n and m are integers, are called the *Gaussian integers*. They form a subring of the ring of all complex numbers and this ring is like the ring of ordinary arithmetic in that "unique factorization" holds. One defines an element to be a *unit* if its reciprocal is also in the ring and a *prime* if it cannot be written as a product of two other elements neither of which is a unit. To say that unique factorization holds means that every nonzero element is a product of primes and that these primes (except of course for order) are uniquely determined up to multiplication by units. In the ring of Gaussian integers the units are clearly ± 1 and $\pm i$. Moreover, it can be shown that every prime is either a unit times an ordinary prime or else of the form $n+im$, where $n^2+m^2=(n+im)(n-im)$ is a prime. In other words, every ordinary prime is either a Gaussian prime or the product of two complex conjugate Gaussian primes and we obtain all Gaussian primes by factoring the ordinary ones. The ordinary primes that are also Gaussian primes turn out to be precisely those primes p such that $p+1$ is a multiple of 4. It is clear that one can get a complete overview of the integral solutions of $n^2+m^2=D$ by factoring D into Gaussian primes.

More generally, any binary quadratic form can be associated in a similar fashion with a ring of "algebraic integers" and the problem of solving $Ax^2+Bxy+Cy^2=D$ reduced to a factorization problem. The ring that arises depends only upon the discriminant B^2-4AC of the form. There is a serious complication in that the relevant ring need not be one in which unique factorization holds. This difficulty and its elucidation led to the theory of ideals in algebraic number fields but only much later. We shall give details below.

ISBN 0-8053-6702-0/0-8053-6703-9, pbk

The law of quadratic reciprocity in its most elementary form is a theorem about the solvability of a certain class of quadratic Diophantine equations. Let a and b be odd primes and consider the equation $x^2 = ay + b$, where x and y are unknown integers. The theorem in question allows us to tell whether $x^2 = ay + b$ has a solution as soon as we know whether or not $x^2 = by + a$ has a solution. Specifically the first equation has a solution if and only if the other does whenever $\frac{1}{2}(a-1)\frac{1}{2}(b-1)$ is even and the first equation has a solution if and only if the other does *not* whenever $\frac{1}{2}(a-1)\frac{1}{2}(b-1)$ is odd. One usually writes this in the form

$$\left(\frac{b}{a}\right) = \left(\frac{a}{b}\right)(-1)^{((a-1)/2)((b-1)/2)}$$

where (b/a) is 1 or -1 according as $x^2 = ay + b$ does or does not have a solution. The symbol (b/a) is called the Legendre symbol after Legendre who discovered the truth of the theorem (though his proof was incomplete). This theorem in spite of its elementary statement and apparently rather special character is of fundamental importance in number theory. It is both a useful practical tool and deep fact whose real meaning only begins to appear when it is suitably reformulated and generalized. We remark that its more sophisticated reformulations are equivalent to the main statement plus two "supplements":

(1) $(-1/a) = (-1)^{(a-1)/2}$ for a odd and positive;

(2) $(2/a) = (-1)^{(a^2-1)/8}$ for a odd.

The first really big advance in number theory beyond the contents of Gauss' classic of 1801 was made in Dirichlet's 1839 memoir "Recherches sur diverse applications de l'analyse infinitesimal à la théorie des nombres." In this paper Dirichlet introduced the systematic use of tools from analysis including densities of infinite sets of integers and infinite series of the form

$$\sum_{n=1}^{\infty} \frac{a(n)}{n^s}$$

which are now called Dirichlet series. Using these tools Dirichlet was able to show, among other things, that every arithmetic progression of the form $an + b$ $(n = 1, 2, \ldots)$ with $(a, b) = 1$ contains an infinite number of primes and that the class number h of a binary quadratic form with discriminant $d < -4$ is given by the formula

$$h = -\frac{1}{|d|}\sum_{n=1}^{|d|-1} n\left(\frac{d}{n}\right)$$

ISBN 0-8053-6702-0/0-8053-6703-9, pbk

It is remarkable that this formula of Dirichlet has never been proved in complete generality without using analysis in spite of its purely arithmetical character.

Although Dirichlet is appropriately regarded as the founder of analytic number theory, he was not the first to obtain arithmetic results using analytic tools. In 1829 his contemporary, Jacobi, published a celebrated pioneering work on elliptic functions. In this work Jacobi obtained a formula for the number of representations of an integer as a sum of four squares as an almost accidental consequence of general identities involving theta functions. Moreover, similar formulas for 2, 6, and 8 squares were implicit in other parts of the work. However, it was not until well into the twentieth century that these results were properly understood as part of a general theory connecting quadratic forms with the theory of automorphic functions.

Dirichlet was also not the first to use Dirichlet series in investigating the primes. Euler had noted the identity

$$\sum_{n=1}^{\infty} \frac{1}{n^s} = \prod_{p=1}^{\infty} \left(1 - \frac{1}{p^s}\right)^{-1}$$

the famous "Euler product" formula for the zeta function and had used it to prove that the sum of the reciprocals of the primes diverges.

We have said that two of the main branches of number theory grew out of the work of Lagrange, Legendre, and Gauss in binary quadratic forms. A third branch is concerned with the use of Dirichlet series to study the distribution of the prime numbers amongst the integers and may be thought of as beginning with the work of Euler and Dirichlet already referred to. However, it received its real impetus in a fundamental paper of Riemann published in 1859 in which Riemann showed that the behavior of the "zeta function"

$$s \to \sum_{n=1}^{\infty} \frac{1}{n^s} = \zeta(s)$$

for *complex* s was intimately related to questions about the distribution of the primes. (cf. Section 24) In particular, he showed that although the series converges only in the half-plane $\sigma > 1$ ($s = \sigma + it$), the zeta function has an analytic continuation that satisfies the functional equation

$$\frac{\zeta(s)\Gamma\left(\frac{s}{2}\right)}{\pi^{s/2}} = \frac{\zeta(1-s)\Gamma\left(\frac{1-s}{2}\right)}{\pi^{(1-s)/2}}$$

and is analytic in the entire s plane except for a simple pole at $s = 1$.

ISBN 0-8053-6702-0/0-8053-6703-9, pbk

The three branches of number theory that we have singled out for special mention developed more or less independently during the rest of the nineteenth century. The generation of mathematicians succeeding that of Dirichlet (1805–1859) and Jacobi (1804–1851) included Riemann (1826–1866), Hermite (1822–1901), Eisenstein (1823–1852), Kronecker (1823–1891), and Dedekind (1831–1916). We have already mentioned the pioneering work of Riemann. Hermite and Eisenstein were active in generalizing the known facts about binary quadratic forms to general n-ary forms. In 1847 Eisenstein obtained purely arithmetic proofs of Jacobi's results on representing integers by 4, 6, and 8 squares and at the same time found formulas (previously unknown) for sums of 5 and 7 squares. Hermite studied reduction theory in the general case and in 1851 published an important paper on the subject in which analytical methods were used. Kronecker followed Jacobi in obtaining number theoretical results from the theory of elliptic functions and in particular found various remarkable identities involving class numbers.

Dedekind's principal contribution was in developing the theory of ideals in algebraic number theory to clarify the problems arising from nonunique factorization in various generalizations of the Gaussian integers. We have seen how one can solve Diophantine problems by factoring "integers" in extension fields of the rationals. However, the fact that this factorization need not be unique was not clearly recognized at first and in 1849 Kummer (1810–1893) achieved what he thought was a proof of the celebrated Fermat conjecture on the equation $x^n+y^n=z^n$. He generalized the Gaussian integers by replacing $\pm 1, \pm i$ by more general nth roots of unity, factored Fermat's equation, and argued without taking account of the nonuniqueness of his factorization. In Dedekind's theory developed in the early 1870s the basic idea is to shift attention from the integers themselves to the "principal ideals" that they generate. Given two ring elements x and y it is easy to see that x is a unit times y if and only if the set Rx of all multiples of x is equal to the set Ry of all multiples of y. Rx is called the *principal ideal* generated by x. When unique factorization holds the factorization of an element x is reflected in the following (strictly unique) factorization of the principal ideal that it generates. If $x = x_1 x_2 \cdots x_k$, where $x_1 \cdots x_k$ are primes, then $Rx=(Rx_1)(Rx_2)\cdots(Rx_n)$, where AB means the set of all sums of products $a_1b_1+a_2b_2+\cdots+a_kb_k$ with $a_j \in A$, $b_j \in B$. One can generalize the notion of principal ideal by allowing more generators and defining an *ideal* to be any subset of R of the form $Rx_1+Rx_2+\cdots$, that is, any subset closed under addition and subtraction as well as under multiplication by arbitrary members of R. An ideal that cannot be written as a product of two others is said to be *prime* and it turns out that in an algebraic number ring every ideal is uniquely a

ISBN 0-8053-6702-0/0-8053-6703-9, pbk

product of prime ideals. For those particular rings in which every ideal is principal, the usual unique factorization theorem is an immediate consequence. More generally, one says that two ideals I_1 and I_2 belong to the same *class* if there exist principal ideals Rx_1 and Rx_2 such that $I_1Rx_1 = I_2Rx_2$ and it is one of the fundamental theorems of the subject that there are only a finite number of distinct classes. Thus one has a notion of class number for algebraic number fields as well as for quadratic forms. A different but equivalent approach to that of Dedekind was discovered by Kronecker and published in 1881. Kummer himself also had an approach to the problem.

Another important impetus to the study of algebraic number fields was the problem of generalizing the quadratic reciprocity law to higher powers than the second. Gauss around 1830 had obtained a reciprocity law for fourth powers by studying the factorization of Gaussian integers and a large part of Kummer's work was devoted to this problem of "higher reciprocity laws."

Although various advances of interest were made in the 1880s and early 1890s especially in the theory of n-ary quadratic forms the next really big steps were taken at the very end of the century by men born forty years later than Riemann and his contemporaries. In 1896 Hadamard (1865–1963) and de la Vallé Poussin (1866–1962) carried out a part of Riemann's program by proving the celebrated prime number theorem and in 1897 Hilbert (1862–1943) published his famous "report" on algebraic number fields. In it he not only simplified and unified much earlier work, but also founded the subject known as "class field theory." Let the algebraic number field k_1 contain another k_2 as a subfield and let G be the group of all automorphisms of k_1 that leave the elements of k_2 fixed. The field k_1 is said to be a normal extension of k_2 if k_2 is exactly the set of all elements left fixed by every member of G and to be an abelian extension if in addition G is abelian. Class field theory may be regarded as the study of the possible abelian normal extensions of a given algebraic number field and its theorems include all the higher reciprocity laws as special cases. We shall say more about its actual content in a later section. While Hilbert laid the foundations for the theory and formulated its main theorems, he was able to prove them only in special cases and then only by complicated analytical methods using Dirichlet series, etc. Completing and simplifying Hilbert's work was one of the main tasks of the number theorists of the twentieth century. A certain degree of completion was achieved in 1922 when Takagi proved the last of Hilbert's main conjectures but simplifying the proofs and reformulating and refining the results is still going on.

An important technique for studying Diophantine equations involves reducing them modp^k for various prime powers p^k. If $An^2 + Bmn + Cm^2 =$

ISBN 0-8053-6702-0/0-8053-6703-9, pbk

D has a solution in integers, then it will certainly have one in the ring of integers modp^k for each p and k. A similar statement holds for other Diophantine equations as well as for more complex Diophantine problems such as deciding when two forms are equivalent. Of course, solvability modp^k for all p and k is in general only a necessary condition for the solvability of the original problem and one has the interesting question of deciding when it is sufficient. This technique was given considerable flexibility and power with the invention of "p-adic numbers" by K. Hensel (1861–1941) in 1901. For a given prime p the p-adic numbers form a locally compact field in which the rational numbers are dense and the "p-adic integers" form a compact open subring in which the rational integers are dense. The importance of the p-adic numbers lies in part in the fact that problems about solutions mod p^k for all k can be reduced to problems about solutions in the field of p-adic numbers. Via the duality theory for locally compact abelian groups, they form one of the main links between number theory and the theory of infinite-dimensional group representations.

The principal names in the next generation of number theorists are Hardy (1877–1942), Littlewood (1885–1977), Ramanujan (1887–1920), and Hecke (1887–1947). One could also mention others, but these are the ones of most relevance for us.

Let k be an algebraic number field and for each ideal d in the corresponding ring of integers let $N(d)$ be the number of elements in the residue class ring R_k/d. $N(d)$ is always finite and we verify that $N(d_1d_2)=N(d_1)N(d_2)$ for all ideals d_1 and d_2. Now let χ be any complex-valued function on the set of all ideals such that $\chi(d_1d_2)=\chi(d_1)\chi(d_2)$ for all d_1 and d_2 and such that certain other conditions are satisfied. We may form the Dirichlet series $\Sigma_d\chi(d)/N(d)^s$. It can be shown to converge for all s with real part greater than 1 and to define an analytic function there. This function is called the *L function* with character χ for the field, and one writes

$$L(\chi,s)=\sum_d \frac{\chi(d)}{N(d)^s}$$

In the particular case in which $\chi(d)\equiv 1$, it is called the *zeta function* of the field and when k is the rational field it obviously reduces to the classical Riemann zeta function. It is easy to see that the L function of an arbitrary k has an Euler product decomposition,

$$L(\chi,s)=\prod_{\mathfrak{p}} \frac{1}{\left(1-\dfrac{\chi(\mathfrak{p})}{N(\mathfrak{p})^s}\right)}$$

ISBN 0-8053-6702-0/0-8053-6703-9, pbk

where $\mathfrak{p}$ varies over the *prime* ideals of R_k. In 1917 and 1918 Hecke wrote a series of papers in which he proved that the zeta and L functions of an arbitrary algebraic number field have meromorphic continuations to the entire s plane and satisfy functional equations closely analogous to that found by Riemann for the original zeta function. Moreover, he introduced certain generalized L functions in which χ is replaced by something more general called a "Grössencharacter" and showed that these generalized L functions also have meromorphic continuations and satisfy Riemann-type functional equations.

A whole new field of investigation was opened up in 1917 by Hardy and Ramanujan with their introduction of the "circle method" for the study of the asymptotic properties of number theoretical functions. Let $P(n)$ be the number of "partitions" of n, that is, the number of ways of representing n as a sum of smaller positive integers. The so-called "generating function" $\sum_{n=0}^{\infty}P(n)z^n$ is analytic in the unit circle $|z|<1$ and has $|z|=1$ as a natural boundary. Applying the general methods of the theory of functions of a complex variable to study the behavior of this analytic function near the boundary, Hardy and Ramanujan obtained an asymptotic formula for $P(n)$ of amazing accuracy—the value given by the formula differing from the true value by less than an integer. Then in 1920 Hardy and Littlewood began a series of papers in which they applied the same method to obtain asymptotic formulas for $S_k^r(n)$ the number of ways of representing n as a sum of r kth powers and, in particular, obtained a new solution to Waring's problem. This is the problem of showing that for every k there exists an r such that $S_k^r(n)>0$ for all n. The first solution was obtained by Hilbert in 1909.

In the special case in which $k=2$, the work of Hardy and Littlewood makes contact with the work of Jacobi, Eisenstein, etc. on exact formulas for the number of representations of an integers as a sum of r squares for $r\leqslant 8$. Indeed, for $5\leqslant r\leqslant 8$ Hardy showed that the asymptotic formulas were exact and could be reduced to the previously known exact formulas. The generating function $\sum_{n=1}^{\infty}z^nS_k^r(n)$ becomes a "modular form" under the change of variable $z\rightarrow e^{2\pi i\tau}$, that is, it transforms in a certain simple way (to be described below) when τ is replaced by $(a\tau+b)/(c\tau+d)$, where $\begin{pmatrix} a & b \\ c & d \end{pmatrix}$ is a member of a subgroup of finite index the modular group of all 2×2 integer matrices with determinant one, and Hardy used the theory of modular forms in establishing the exactness of his asymptotic formulas for $5\leqslant r\leqslant 8$.

The connection between modular forms and asymptotic formulas for sums of squares became greatly clarified by work of Hecke in 1926. Hecke showed that by making proper use of the theory of modular forms one

ISBN 0-8053-6702-0/0-8053-6703-9, pbk

could do much more than show that certain asymptotic formulas were exact. One could in fact obtain the asymptotic results themselves for all r (but with $k=2$) and by simpler and more elegant arguments.

We now have two apparently quite different analytic methods for studying the asymptotic properties of number theoretical functions. On the one hand, we have the use of Dirichlet series such as zeta and L functions for the study of the primes, and on the other we have generating functions $\Sigma f(n)z^n$ and the closely related periodic functions obtained by replacing z by $e^{2\pi i\tau}$ that are used in studying Waring's problem, the partition problem, etc. These methods are in fact closely connected and this connection was clearly brought out and exploited by Hecke and his student Petersson in a series of papers beginning in 1935. The basic observation is the following: Let $a_1, a_2, \ldots$ be a sequence of complex numbers. Then one can form the Dirichlet series $\Sigma_{n=1}^{\infty} a_n/n^s$ and the transformed generating function $\Sigma_{n=1}^{\infty} a_n e^{2\pi inz}$. Now if in $e^{2\pi inz}$ we set $z=iy$, where y is real and positive, and take the Mellin transform of the resulting $e^{-2\pi ny}$, we get $\Gamma(s)/(2\pi n)^s$. Thus the Mellin transform of $\Sigma_{n=1}^{\infty} a_n e^{-2\pi ny}$ (when suitable convergence conditions are satisfied) is

$$\frac{\Gamma(s)}{(2\pi)^s} \sum_{n=1}^{\infty} \frac{a_n}{n^s}$$

From this it is easy to deduce that $\Sigma_{n=1}^{\infty} a_n e^{2\pi inz}$ is a modular form as a function of z if and only if $\Sigma_{n=1}^{\infty} a_n/n^s$ satisfies a functional equation analogous to that of Riemann. Using the theory of modular forms Hecke was able to show that any Dirichlet series satisfying a suitable functional equation and appropriate auxiliary hypotheses can be written as a sum of a finite number of other series each of which has an infinite product expansion analogous to the Euler product for the Riemann zeta function. We note that our three branches of number theory, which developed more or less separately in the nineteenth century, became intimately related during the twentieth.

The work of Hardy–Littlewood and Hecke on sums of squares was considerably, generalized, clarified and deepened by Carl Ludwig Siegel (1896–) in a series of papers beginning in 1935.

Siegel did not restrict himself to sums of squares, but considered arbitrary quadratic forms both definite and indefinite with coefficients in an algebraic number field. Moreover, he replaced the problem of finding the number of representations of an integer by a form by the more general one of finding the number of representations of one form by another. Finally, by considering certain averages of representation numbers, he was

ISBN 0-8053-6702-0/0-8053-6703-9, pbk

able to replace asymptotic formulas by exact ones and to show that his results were entirely equivalent to certain identities in the theory of modular forms. The exact formulas in earlier work turned out to correspond to cases in which Siegel's averages were over a one element set. In dealing with representations of one form by another, Siegel had to pass from modular forms in one complex variable to modular forms in several complex variables, and the needs of the theory of quadratic forms thus inspired considerable progress in the theory of automorphic functions of several complex variables.

A third important advance was made in 1935 by C. Chevalley with his introduction of the notion of the "idèle group" of an algebraic number field. The notion of p-adic number has a generalization in which p is replaced by a prime ideal $\mathfrak{p}$ in a ring of algebraic integers. The $\mathfrak{p}$-adic numbers form a locally compact field that contains the given algebraic number field as a dense subfield. This field is also always a finite extension of the field of p-adic numbers for that rational prime p whose decomposition into prime ideals includes $\mathfrak{p}$. Let $k_{\mathfrak{p}}$ be the field of $\mathfrak{p}$-adic numbers for the algebraic number field k. Then each $x \neq 0$ in $k_{\mathfrak{p}}$ defines an automorphism of the additive group of $k_{\mathfrak{p}}$ via $x' \to xx'$. If this automorphism does not leave Haar measure fixed, it multiplies it by a constant $\rho(x)$ and $x \to \rho(x)$ is a homomorphism of the multiplicative group $k_{\mathfrak{p}}^*$ of the nonzero elements of $k_{\mathfrak{p}}$ into the multiplicative group of positive real numbers. The subgroup $k_{\mathfrak{p}}^0$ of all x with $\rho(x) = 1$ can be shown to be compact, open, and totally disconnected. It is called the group of units of $k_{\mathfrak{p}}^*$. One also has dense imbeddings of k into connected locally compact fields, that is, into the real and complex number fields (there are no others). However, there are only finitely many of these. Denote them by $\phi_1 \cdots \phi_r$ and let F_j^* denote the multiplicative group of the field in which ϕ_j imbeds k. The *idèle group* I_k of k is the direct product of $\Pi_{j=1}^r F_j^*$ with the subgroup $\Pi'_{\mathfrak{p}} k_{\mathfrak{p}}^*$ of $\Pi_{\mathfrak{p}} k_{\mathfrak{p}}^*$ consisting of all $\mathfrak{p} \to x_{\mathfrak{p}}$ such that $x_{\mathfrak{p}}$ is a unit for all but a finite number of prime ideals $\mathfrak{p}$. $\Pi_{\mathfrak{p}} k_{\mathfrak{p}}^0$ is a compact in the product topology and there is a unique way of topologizing $\Pi'_{\mathfrak{p}} k_{\mathfrak{p}}^*$ so that its subgroup $\Pi_{\mathfrak{p}} k_{\mathfrak{p}}^0$ is open and has the product topology. In the resulting topology $\Pi'_{\mathfrak{p}} k_{\mathfrak{p}}^*$ is locally compact. Thus I_k as the product of a finite number of locally compact groups has a natural locally compact topology. The natural imbedding of k^* in each $k_{\mathfrak{p}}^*$ and each F_j^* defines an imbedding of k^* in I_k. In contradistinction to the other imbeddings, this one has a closed range. Thus k^* appears in a natural way as a closed subgroup of I_k. It is called the subgroup of *principal idèles*. Moreover,

$$I_k \Big/ \left(\prod_{\mathfrak{p}} k_{\mathfrak{p}}^0 \times \prod_{j=1}^{r} F_j^* \right)$$

ISBN 0-8053-6702-0/0-8053-6703-9, pbk

is canonically isomorphic to the group of all fractional ideals, that is, to the group generated by the semi group of all ideals in R_k. The image of k^* in this quotient is the subgroup generated by the principal ideals so that the subgroup of I_k consisting of all characters that are one in both k^* and $\Pi_{\mathfrak{p}} k_{\mathfrak{p}}^0 \times \Pi_{j=1}^r F_j^*$ may be identified with the character on the group of ideal classes. Those that are one just on k^* may be identified with Hecke's Grössencharacter and this is certainly the easiest way to formulate the definition of the latter.

Using the idèle group as a tool, Chevalley was able to free class field theory from its dependence on the theory of analytic functions. He did so in the sense that he replaced arguments involving zeta functions and other Dirichlet series by arguments involving the topology of locally compact groups. In particular, he showed in 1940 that a fundamental theorem of class field theory could be reformulated as the assertion that a certain mapping sets up a one-to-one correspondence between the subgroups of finite index of I_k/k^* and the possible normal abelian extensions of k.

Closely related to the notion of idèle group is that of "adèle group" or group of "valuation vectors." This notion was introduced by Artin and Whaples in 1945. It is defined in just the same way as idèle group except that the additive group of $k_{\mathfrak{p}}$ replaces the multiplicative group and $k_{\mathfrak{p}}^0$ is replaced by the group of all $\mathfrak{p}$-adic integers. It was used to considerable effect by Tate in his 1950 Princeton thesis. Following a suggestion of Artin he showed that one could use Fourier analysis on adèle groups to obtain simple natural proofs of Hecke's functional equations for general zeta and L functions.

In the late 1950s Ono and Tamagawa generalized the adèle-idèle notion still further and showed how to attach a locally compact group to any "algebraic group" in a way that reduces to the adèle group when the algebraic group is the additive group of an algebraic number field and to the idèle group when the algebraic group is the multiplicative group of this field. Certain theorems in algebraic number theory that can be formulated as about the idèle group of the field make sense for the adèle groups of more general algebraic groups, and these general results become theorems about quadratic forms when one specializes to suitable noncommutative "algebraic groups." Thus one can partially reunify the two branches of number theory that grew out of the theory of binary quadratic forms.

The adèle group G_A of an "algebraic group" G over the rationals always contains G canonically imbedded as a discrete subgroup. In certain cases for example, when G is "semi-simple," there is a canonical way of choosing the Haar measure for G_A and one can then ask for the measure of G_A/G with respect to this canonical measure. (Since G is discrete, there is a canonical choice for the Haar measure in G_A/G.) Tamagawa (and

ISBN 0-8053-6702-0/0-8053-6703-9, pbk

independently Kneser) observed that some of the main results of C.L. Siegel on representation numbers of quadratic forms are completely equivalent to the statement that G_A/G has measure two whenever G is a certain kind of orthogonal group. Accordingly, it has become customary to call the measure of G_A/G the *Tamagawa number* of G. To the extent that one can obtain the Tamagawa numbers of other algebraic groups, one obtains significant extensions of the work of Siegel.

An important advance was made in rather a different direction in Selberg's paper given at the International Colloquium on Zeta Functions in Bombay in 1956 (published the same year in the Journal of the Indian Mathematical Society). Selberg discovered that the so called "Poisson summation formula" of classical Fourier analysis had a noncommutative generalization that could be applied to obtain an array of important identities in number theory and the theory of automorphic functions. It is now referred to as the Selberg trace formula. Let f be a suitably restricted continuous complex-valued function on the real line and let Γ be the subgroup of the additive group of the real line consisting of the integers. Then

$$\tilde{f}(x) = \sum_{\gamma \in \Gamma} f(x+\gamma)$$

is a periodic function of period 1 and may be expanded in a Fourier series $\Sigma_{n=-\infty}^{\infty} C_n e^{2\pi i n x}$. Here

$$C_n = \int_0^1 \tilde{f}(x) e^{-2\pi i n x\, dx}\, dx = \hat{f}(-2\pi n)$$

where $\hat{f}$ is the Fourier transform of f. Since $\tilde{f}(0) = \Sigma_{n=-\infty}^{\infty} f(n)$, we conclude that

$$\sum_{n=-\infty}^{\infty} f(n) = \sum_{n=-\infty}^{\infty} \hat{f}(2\pi n)$$

and this is the classical Poisson summation formula. It has an immediate generalization in which the additive group of the reals is replaced by the additive group G of all k-tuples of real numbers and Γ is replaced by the additive group of all k-tuples of integers. This formula plays a key role in number theory. Applied to functions of the form $e^{-Q(x_1 \cdots x_n)}$, where Q is a positive-definite quadratic form, it yields the fact that the periodic generating functions arising in the analytic theory of quadratic forms are modular forms; via the connection between generating functions and Dirichlet

ISBN 0-8053-6702-0/0-8053-6703-9, pbk

series it also yields a proof of the functional equation for the Riemann zeta function and its various generalizations. Again applied to a function of the form $e^{-Q(x_1\cdots x_n)}$, it occurs as an important step in Hecke's proof of a generalization of the quadratic reciprocity law due Hilbert.

There is a further more or less obvious generalization of the Poisson summation formula in which the group G is an arbitrary separable locally compact commutative group and Γ is an arbitrary closed subgroup. The formula then becomes

$$\int_{\Gamma} f(\gamma)\,d\gamma = \int_{\Gamma^{\perp}} \hat{f}(\sigma)\,d\sigma$$

where $\hat{f}$ is the Fourier transform of f and $\Gamma^{\perp}$ is the "annhilator" of Γ in G, that is, the group of all characters χ such that $\chi(\gamma)=1$ for all $\gamma\in\Gamma$. In his proof of the functional equation for the Hecke generalized L functions, Tate uses the generalized Poisson summation formula in the special case in which G is the adèle group of an algebraic number field and Γ is the canonical image of the additive group of field itself in the adèle group.

In the noncommutative version of the Poisson summation formula, Γ is an arbitrary closed subgroup of a separable locally compact group G, but the formulation involves heavy technicalities unless G/Γ is compact. When this is the case G/Γ has a finite invariant measure μ and the unitary representation U of G defined in $L^2(G/\Gamma, \mu)$ by setting $U_x(f)(s)=f(sx)$ has a discrete decomposition into irreducibles. These irreducibles play the role of $\Gamma^{\perp}$ in the commutative theorem. The actual statement is a little complicated even when G/Γ is compact and we shall postpone giving it to a later section. Selberg considered essentially only the case in which Γ is a discrete subgroup of a semi-simple Lie group, but generalized in another direction by (in effect) replacing U by the representation of G induced by a finite-dimensional unitary representation of Γ. As we shall see later the study of automorphic functions and modular forms in one variable is intimately related to the properties of induced representations of the form U^L, where L is a finite-dimensional unitary representation of a discrete subgroup Γ of $\mathrm{SL}(2,R)$. Actually Selberg's work was inspired by the work of Maass and others on various generalizations of automorphic functions and modular forms and some of his most striking results were obtained in the special case $G=\mathrm{SL}(2,R)$.

A remarkable synthesis of some of the major twentieth century contributions to number theory occurs in a pair of long memoirs published by A. Weil in Acta Mathematica in 1964 and 1965. The main result of this work is a formula from which one may compute the Tamagawa numbers of

ISBN 0-8053-6702-0/0-8053-6703-9, pbk

most semi-simple Lie groups. Since these semi-simple Lie groups include the orthogonal groups involved in the Tamagawa number formulation of Siegel's results on quadratic forms, Weil's formula may be considered to be a generalization of Siegel's. Now in one formulation Siegel's main result takes the form of an identity between a "θ series" and an "Eisenstein series." Given a quadratic form one may associate with it a function defined in the upper half-plane by a certain infinite series (called a θ series) and show that it satisfies a certain identity. One may also associate with it another infinite series (called an Eisenstein series) that more or less obviously satisfies the identity in question. A refinement of showing that the sum of the θ series satisfies the identity in question is to show that it is identical to the sum of the Eisenstein series. Weil's formula also takes the form of asserting that a θ series is equal to an Eisenstein series, but his θ series and Eisenstein series are abstract generalizations of the usual objects. Taking advantage of the close connections between group representations and the theory of automorphic functions (to be explained in a later section), Weil translates the classical notions into group theoretical ones and carries out his arguments in the framework of the theory of locally compact groups. He also exploits the technical advantages of the adèle notion in a fashion quite analogous to that used by Tate in proving the functional equations for the Hecke zeta functions. Indeed his θ series and his Eisenstein series are both constructions made starting with a function on an adèle group.

In closing this brief resume of the number theory of the past two centuries, we remind the reader that we have been heavily selective and have omitted altogether many important topics. Among these are various refinements of the prime number theorem, the work of Vinogradov and the Russian school on the Goldbach hypothesis and on refinements and simplifications of the work of Hardy and Littlewood, connections with algebraic geometry such as the Mordell–Weil theorem, the work of Artin and Artin–Tate on class field theory, Diophantine approximation, and the geometry of numbers, etc. We have aimed only at giving the reader an idea of those parts of the subject most easily and naturally relatable to our central theme.

Notes and References

The approach of Kummer to the problem of nonunique factorization was prior to that of Kronecker and Dedekind and was quite satisfactory as far as it went; however, it only applied to rather special cases.

Siegel's work on quadratic forms did *not* include that of Hecke, as suggested by the text. Siegel was content to derive formulas for the *average*

ISBN 0-8053-6702-0/0-8053-6703-9, pbk

number of representations as the forms in question varied over a "genus." Hecke, on the other hand, made considerable progress in getting results for the individual forms themselves. Siegel did his work on averages in a more general context, but did not attempt to deal with individual forms.

Even with the indicated limitations, this brief history should have mentioned the 1923 thesis of Artin as well as the paper published in the same year in which he assigned an "*L* series" to each character of the Galois group of an algebraic number field. In doing the latter, he not only made what is probably the first application of (noncommutative) group representation theory to number theory, but, in addition, he took an important step toward a nonabelian class field theory and paved the way for his own beautiful reformulation of abelian class field theory a few years later. In his thesis Artin inaugurated a variant of algebraic number theory that has developed in parallel fashion ever since and plays an important role in the arithmetic theory of algebraic curves.

Recent investigations of Harold Edwards have discredited the widely believed story cited in the text concerning Kummer's "belief" that he had proved the Fermat conjecture. It seems that Kummer was protected from making this mistake by having made an analogous one in dealing with another problem. Other mathematicians of the time, such as Lamé, did make the mistake.

ISBN 0-8053-6702-0/0-8053-6703-9, pbk

27. Diophantine equations and Fourier analysis

Without doubt the fundamental problem of number theory is the solution of Diophantine equations and systems of such. Not only did the subject begin in this way, but many of its most sophisticated problems reduce in the end to answering question about the solutions of systems of Diophantine equations. Given such a system the first question that arises is as to whether it has any solutions. More generally, one can ask how many solutions it has. More generally, still, one can consider a family of systems of equations depending upon one or more parameters and ask for the number of solutions as a function of these parameters. In fact, to imbed a problem in a family of different but related problems is often the key to its solution. In any event, once we have functions to deal with, even if they are integer-valued functions defined on the set of all k-tuples of integers, it is natural to study them with the methods of analysis. We shall see in this section that applying Fourier analysis to these functions in various ways leads more or less directly to the main analytical tools of number theory: generating functions, modular forms, Dirichlet series, p-adic numbers, and adèles. A concrete example, which we shall study in detail further on, is given by Waring's problem. Here we have a single equation

$$n = x_1^k + x_2^k + \cdots + x_r^k$$

with integer unknowns $x_1, \cdots, x_r$ and we wish to study the number $S_r^k(n)$ of solutions of this equation as a function of the parameter n.

George W. Mackey, Unitary Group Representations in Physics, Probability, and Number Theory

ISBN 0-8053-6702-0/0-8053-6703-9, pbk

Quite generally we will have before us an integer-valued function of m integers that is to be examined and if possible expressed in terms of simpler functions. Now the set Z^m of all m-tuples of integers is a group under addition and, as we know, functions on groups may be Fourier analyzed. Unfortunately, the functions that arise in number theoretical problems are usually unbounded and are seldom if ever square summable. Thus the theory we have been developing does not apply immediately. However, there is a way out; in fact, there are several. For example, we may be interested in our function only for positive values of the variables or at least only for values lying in some closed cone K in Z^m (in the sense of Section 25). Then we may replace it by a function that is zero outside of K and consider the Laplace transform $\tilde{f}$ on $\hat{Z}^m \times K^0$ instead of the Fourier transform on $\hat{Z}^m$. $\tilde{f}$ will exist throughout the interior of $\hat{Z}^m \times K^0$ whenever f does not have too large a rate of growth. In the important special case in which m is one and K consists of the positive integers, $\tilde{f}$ will be defined by the power series $\Sigma_{n=1}^{\infty} f(n) z^n$ and this is just the so-called generating function used by Hardy and Ramanujan in discussing the partition function and by Hardy and Littlewood in their work on Waring's problem. Their asymptotic formulas for $f(n)$ result from an ingenious choice of the contour in reobtaining f from $\tilde{f}$.

As mentioned in Section 25, it is often convenient to regard $\hat{Z}^m$ as the quotient of the vector space $\overline{Z^m}$ by a discrete subgroup so that $\tilde{f}$ becomes a periodic function on the interior of $\overline{Z^m} \times K^0$. When m and K are as above, this leads to the periodic function $\Sigma_{n=1}^{\infty} f(n) e^{2\pi i n z}$ obtained from the generating function by making the exponential change of variable $z \to e^{2\pi i n z}$. This shift in viewpoint is useful because it is these periodic functions that turn out to be modular forms or "nearly" modular forms for certain important number theoretical functions f.

The introduction of zeta functions and other Dirichlet series can be motivated in several ways. We shall discuss two. First of all, when we have converted $\tilde{f}$ into a periodic function on the interior of $\overline{Z_m} \times K^0$, it follows from the analyticity of $\tilde{f}$ that $\tilde{f}$ is completely defined by its values on $e \times int(K^0)$. Let $\tilde{f}^0$ denote the function on $int(K^0)$ obtained by restricting $\tilde{f}$ to $e \times int(K^0)$. As observed in Section 25, $int(K^0)$ will often be transitive under the group A_K of all automorphisms of $\overline{Z_m}$ leaving K^0 fixed. In some cases A_K will be commutative or at least have a transitive commutative subgroup. When it does, $\tilde{f}^0$ may be regarded as a function on a commutative group and again Fourier analyzed. In any case, one may attempt to apply the methods of noncommutative Fourier analysis to the study of $\tilde{f}^0$. We shall look here only at the very special but important case in which m is one and K is the positive integers. Then K^0 is a half line and A_K is the

ISBN 0-8053-6702-0/0-8053-6703-9, pbk

group of all transformations $y \to ay$ where a varies over the positive real numbers. Thus A_K is transitive and commutative and in fact isomorphic to the additive group of the real line. We see that

$$\tilde{f}^0(y) = \sum_{n=1}^{\infty} e^{-2\pi n y} f(n)$$

and choosing $y=1$ as reference point we may identify the points of K^0 with those of A_K. The general character is $y \to y^s$ for s real and Haar measure is dy/y. Thus the Fourier transform of $\tilde{f}^0$ is

$$s \to \int_0^{\infty} \tilde{f}^0(y) y^s \frac{dy}{y} = \int_0^{\infty} \tilde{f}^0(y) y^{s-1} dy$$

that is, the so-called Mellin transform of $\tilde{f}^0$. Assuming suitable convergence we may take this transform term by term and find that the resulting function is

$$s \to \sum_{n=1}^{\infty} f(n) \int_0^{\infty} e^{-2\pi n y} y^{s-1} dy$$

If we make the change of variable $y = t/2\pi n$, the integral becomes

$$\int_0^{\infty} e^{-t} \left(\frac{t}{2\pi n} \right)^{s-1} \frac{dt}{2\pi n} = \frac{1}{(2\pi)^s n^s} \int_0^{\infty} e^{-t} t^{s-1} dt = \frac{\Gamma(s)}{(2\pi)^s n^s}$$

where Γ is the classical gamma function. Thus our transformed function is

$$s \to \frac{\Gamma(s)}{(2\pi)^s} \sum_{n=1}^{\infty} \frac{f(n)}{n^s}$$

that is, the known function $\Gamma(s)/(2\pi)^s$ multiplied by the Dirichlet series

$$s \to D(s) = \sum_{n=1}^{\infty} \frac{f(n)}{n^s}$$

We have remarked that in a certain important class of problems the function

$$\sum_{n=1}^{\infty} f(n) e^{2\pi i n z}$$

ISBN 0-8053-6702-0/0-8053-6703-9, pbk

is a modular form. It follows from the definitions concerned (to be given in Section 29) that a function f on the upper half-plane that is already periodic with period one will be what is called a modular form of dimension $-k$ if and only if

$$\tilde{f}\left(\frac{i}{y}\right) = (iy)^k \tilde{f}(iy)$$

Now replacing y by $1/y$ replaces the (multiplicative) Fourier transform $\tilde{f}$ by $s \to \tilde{f}(-s)$. Moreover, multiplication of $\tilde{f}$ by y^k is just multiplication by a character and replaces $\tilde{f}$ by $s \to \tilde{f}(s+k)$. Thus $\tilde{f}$ will be a modular form of dimension $-k$ if and only if $\tilde{f}(s) = i^k \tilde{f}(k-s)$. Accordingly, we arrive at the observation whose exploitation by Hecke we alluded to in the last section:

$$\sum_{n=1}^{\infty} f(n) e^{2\pi i n z}$$

is a modular form of dimension $-k$ if and only if

$$\sum_{n=1}^{n=\infty} \frac{f(n)}{n^s} = D(s)$$

satisfies the functional equation

$$\frac{D(s)\Gamma(s)}{(2\pi)^s} = i^k \frac{\Gamma(k-s)D(k-s)}{(2\pi)^{k-s}}$$

One is led to the Dirichlet series $D(s)$ quite directly if one thinks of the positive integers as a subset of the multiplicative group of all positive rational numbers rather than as a subset of the additive group of all integers. The multiplicative group G of all positive rational numbers is isomorphic to the group of all finite sequences of integers under addition via the mapping $r \to p_1^{m_1} p_2^{m_2} \ldots p_k^{m_k}$, where the m_j are integers and the p_j are distinct primes. For this G the $\overline{G}$ of Section 25 is infinite-dimensional and each $\tilde{f}$ is really an analytic function of infinitely many complex variables. However, every element of $\overline{G}$ generates a one-dimensional subspace of $\overline{G}$ and a complex one-dimensional subspace of $\overline{G} \times \overline{G}$. Using this subspace, $\tilde{f}$ becomes a function of one complex variable. The most general element of $\overline{G}$ has the form

$$r \to l(p_1)^{m_1} l(p_2)^{m_2} \cdots l(p_k)^{m_k}$$

ISBN 0-8053-6702-0/0-8053-6703-9, pbk

where $r=p_1^{m_1}p_2^{m_2}\ldots p_k^{m_k}$ the p_j are primes, the m_j are integers, and $p\to l(p)$ is an arbitrary function from the primes to the real numbers. Now in the special case in which $l(p_j)=p_j$, we may compute $l(p_1)^{m_1}\cdots l(p_k)^{m_k}$ without factoring since it is just $p_1^{m_1}\cdots p_k^{m_k}=r$ itself. Using this special element of $\overline{G}$ to reduce $\tilde{f}$ to a function of one complex variable, we find that

$$\tilde{f}(s)=\sum_{n=1}^{\infty}\frac{f(n)}{n^s}$$

The fact that $\tilde{f}$ is "really" a function of infinitely many complex variables suggests that it might be useful to look at other choices for l and even to make $\tilde{f}$ into a function of several l's simultaneously. The special case

$$\tilde{f}'(s,t)=\sum_{n=1}^{\infty}\frac{f(n)}{n^s}e^{-(m_1+m_2+\cdots+m_k)t}$$

where $n=p_1^{m_1},p_2^{m_2},\ldots,p_k^{m_k}$ is a function of two complex variables whose consideration is suggested by certain analogous concepts in quantum statistical mechanics. Specifically, $\tilde{f}'$ is related to $\tilde{f}$ just as the so-called "grand partition function" is related to the ordinary one. In the "grand partition function" $m_1+m_2+\cdots+m_k$ is the total number of particles for the eigenstate in question (cf. Section 24).

There is still a third approach to the problem of Fourier analyzing an unbounded function on a closed cone K in J^m. It may be possible to convert it into a bounded function by dividing by a suitable polynomial in members of $\overline{J^m}$ and to do so in such a way that the resulting bounded function has a "mean value." As above it will be convenient to confine our discussion to the case in which $m=1$ although it will not be difficult to see how to generalize much of what we say. In that case we need only consider polynomials that are powers of a single member of $\overline{J^m}$ and our supposition takes the form

$$f(n)=n^{\beta}f_0(n)$$

where

$$\lim_{n\to\infty}\frac{f_0(1)+f_0(2)+\cdots+f_0(n)}{n}$$

exists. For example, if $f(n)$ is the number of integral solutions of $n=x_1^k+x_2^k+\cdots+x_r^k$, then quite elementary arguments suffice to show that if

$$f_0(n)=\frac{f(n)}{n^{r/k-1}}$$

ISBN 0-8053-6702-0/0-8053-6703-9, pbk

then f_0 has a mean value in the sense that

$$\frac{f_0(1)+\cdots+f_0(n)}{n}$$

has a finite limit as $n\to\infty$. Assuming that β exists so that f has the property in question, we may shift our attention to f_0. The existence of a mean value in the indicated sense reminds us of ergodic theory and of Wiener's generalized harmonic analysis and suggests that f_0 might be defined for negative integers so as to be a nonexceptional path function in a stationary stochastic process (cf. Section 17). If so, the autocorrelation coefficient

$$m\to\lim_{n\to\infty}\frac{f_0(1+n)f_0(1)+f_0(2+m)f_0(2)+\cdots+f_0(n+m)f_0(n)}{n}$$

will exist and will be of the form

$$m\to\int_0^{2\pi}e^{im\theta}\,d\alpha(\theta)$$

for some finite measure α. If this measure α is Lebesgue measure or close enough to it to be in the "imperfectly predictable case" of the Wiener prediction theory, this means that f_0 has a "strongly random character" and we cannot hope to do much in the way of obtaining a simple explicit expression for it. In particular, the extension to the negative integers will *not* be unique. At the opposite extreme α might be discrete. In this case, the extension to the negative integers is unique and f_0 itself will be expressible in a suitable sense as a sum of constant multiples of characters in the support of α. This is almost but not quite the same thing as saying that f_0 is almost periodic (we do not require *uniform* approximation).

By definition an almost periodic function on a locally compact commutative group G is a continuous complex-valued function whose translates have a compact closure in the topology of uniform convergence. It is a fundamental theorem that the almost periodic functions are precisely the uniform limits of finite linear combinations of characters. According to another theorem there is a unique translation invariant linear functional M on the space of all almost periodic functions which is such that $M(1)=1$. $M(f)$ is called the *mean value* of f and indeed when G is the group of integers

$$M(f)=\lim_{n\to\infty}\frac{f(1)+\cdots+f(n)}{n}$$

ISBN 0-8053-6702-0/0-8053-6703-9, pbk

for all almost periodic f. To each almost periodic function f one assigns a formal Fourier series $\Sigma_{\chi\in\hat{G}} C_\chi \chi$, where $C_\chi = M(f\bar{\chi})$. It can be shown that for each f, $C_\chi = 0$ for all but countably many values of χ. This series need not actually converge but it determines f uniquely and in fact bears just the same relation to f that the ordinary Fourier series of a continuous function bears to G. Indeed A. Weil has shown that one may reduce the theory of almost periodic functions to the theory of continuous functions on a compact group by the following device. Given a countable dense subgroup, Γ of $\hat{G}$ give it the discrete topology and let ϕ be the natural one-to-one map of Γ into $\hat{G}$. Then ϕ^*, the dual of ϕ, maps G in a one-to-one way onto a dense subgroup of the separable compact group $\hat{\Gamma}$. Let us identify G with its image in $\hat{\Gamma}$. Then is can be shown that the restrictions to G of the continuous functions on $\hat{\Gamma}$ are precisely the almost periodic functions whose nonzero Fourier coefficients are those belonging to members of Γ. Moreover, for each such almost periodic f, $M(f)$ is equal to the Haar integral of the unique extension of f to $\hat{\Gamma}$. (Actually one can find a nonseparable compact group—the dual of the discretization of $\hat{G}$—which works for all almost periodic functions at once.)

With the notions of the two preceding paragraphs in mind, it is natural to see whether our function f_0, having a mean value, also admits Fourier coefficients $C_\chi = M(f_0\bar{\chi})$ for the various characters of our underlying group, and if so to study the relationship of f_0 to the formal Fourier series $\Sigma C_\chi \chi$. Here we must bear in mind that a function which tends to zero for large n will necessarily have all its Fourier coefficients equal to zero. Thus even if $\Sigma C_\chi \chi$ converges uniformly the resulting function can differ from f_0.

The main results of Hardy and Littlewood on Waring's problem may be understood in these terms and may be formulated as follows: Let $f(n)$ be the number of integer solutions of $n = x_1^k + \cdots + x_r^k$. Assume that $r \geqslant 2^k(2k+1)$. Then

$$f(n) = n^{(r/k-1)} f_0(n)$$

where f_0 is such that all Fourier coefficients $C_\chi = M(f_0\bar{\chi})$ exist. Let $\chi_a(n) = e^{2\pi i a n}$ and let us abbreviate C_{χ_a} by C_a. Then $C_a = 0$ except when a is rational and the series $\Sigma C_a e^{2\pi i a n}$ converges uniformly to a function S which is such that

$$\lim_{n\to\infty} (f_0(n) - S(n)) = 0$$

It follows that for large n, $f(n)$ is asymptotic to $n^{(r/k-1)} S(n)$, where

$$S(n) = \sum_a C_a e^{2\pi i a n}$$

and a varies over all rational numbers in the unit interval.

ISBN 0-8053-6702-0/0-8053-6703-9, pbk

Hardy and Littlewood also obtain simple explicit formulas for the Fourier coefficients C_a. Let $\gamma_0 = \gamma(r,k) = M(f_0) = C_0$. Let $C_a = \gamma_0 C_a'$. Then when $a = p/q$, where p and q are relatively prime, they show that

$$C_a' = \frac{1}{q^r}\left(\sum_{m=0}^{q-1} e^{2\pi i(m^k)p/q}\right)^r$$

To see what these formulas mean let us note that

$$\left(\sum_{m=0}^{q-1} e^{2\pi i(m^k)p/q}\right)^r = \sum_{m_1,\ldots,m_r=0}^{q-1} \exp\left(\frac{2\pi i p}{q}(m_1^k + \cdots + m_r^k)\right)$$
$$= \sum_{t=0}^{q-1} \rho(t)\exp\left(2\pi i \frac{p}{q} t\right)$$

where $\rho(t)$ is the number of solutions of the equation $m_1^k + \cdots + m_r^k = t$ and where $m_1 \cdots m_r$ and t are now regarded as members of the ring obtained from the ring of integers by factoring out the ideal of multiples of q. Thus $C_{p/q}'$ is just the pth Fourier coefficient of the function $l \to \rho(l)/qr-1$ where l is considered as an integer mod q. Of course the term $C_a' e^{2\pi i a n}$, where $a = p/q$, depends on n only through the residue of $n \bmod q$. Thus the above observation tells us that we can compute it as though we had a different problem—that of solving a congruence mod q. The series

$$\frac{S(n)}{\gamma_0} = \sum_a C_a' e^{2\pi i a n}$$

with C_a' as indicated is known as the *singular series*.

The function S/γ_0 being the sum of a *uniformly convergent* sequence of finite linear combinations of characters is almost periodic. Hence it is the restriction to the integers of a continuous function on a "compactification" $\hat{\Gamma}$ of the additive group Z of the integers. In this case, Γ is the subgroup of $\hat{Z}$ consisting of all characters of the form $n \to e^{2\pi i a n}$, where a is rational. More significantly perhaps, Γ may be described as the subgroup of $\hat{Z}$ consisting of all elements of finite order. Hence it is a "torsion group," that is, a group in which every element has finite order and as such is uniquely decomposable as a direct sum over the primes. If G is any torsion group, then for each prime p we may consider the subgroup G_p of all elements whose order is a power of p and prove without difficulty that every element of x is uniquely of the form $x_1, x_2, \ldots, x_j$, where each x_i belongs to one of the G_p and no G_p occurs more than once. In other words, G is

ISBN 0-8053-6702-0/0-8053-6703-9, pbk

isomorphic to the so-called direct sum of the subgroups G_p. Now the dual of a direct sum is the direct product of the duals of the factors. Thus the compact group $\hat{\Gamma}$ in which we have imbedded the integers and to which the function S/γ_0 has a unique continuous extension S^1 is canonically a direct product over the primes of certain compact groups $\hat{\Gamma}_p$. This factorization is of particular interest to us because it is easily shown that S^1 has a corresponding factorization. For each p ∃ a continuous function S_p defined on $\hat{\Gamma}_p$ such that

$$S^1(x_{p_1},x_{p_2},\ldots)\equiv\prod_p S_p(x_p)\cdots$$

To see this it suffices to consider the term $C'_a e^{2\pi ian}$ in the singular series more closely. We may write the rational number a uniquely in the form

$$\frac{q_1}{p_1^{m_1}}+\frac{q_2}{p_2^{m_2}}+\cdots+\frac{q_t}{p_t^{m_t}}$$

where $p_1,p_2,\ldots,p_t$ are distinct primes. Correspondingly, the group of integers mod$p_1^{m_1}\cdots p_t^{m_t}$ is a direct product of the groups obtained by reducing the integers mod$p_1^{m_1},p_2^{m_2},\ldots,p_t^{m_t}$, respectively. It follows then that the number of solutions of the congruence problem involved in computing C'_a is a product over the primes $p_1\cdots p_t$ of the corresponding problem with a replaced in turn by

$$\frac{q_1}{p_1^{m_1}},\frac{q_2}{p_2^{m_2}},\ldots,\frac{q_t}{p_t^{m_t}}$$

It follows easily that S^1 is a product as indicated if we take S_p to be the unique continuous extension to $\hat{\Gamma}_p$ of

$$n\rightarrow\sum_{a\in A_p} C'_r e^{2\pi ian}$$

where A_p is the set of all a with $\chi_a\in\Gamma_p$. Since Γ_p is itself dense in $\hat{Z}$, the group Z is dense in $\hat{\Gamma}_p$ as well as in $\hat{\Gamma}$.

Consider the group $\hat{\Gamma}_p$ a little more closely. We have noted that it contains Z as a dense subgroup. Moreover, it is a ring as well as a group and it is not difficult to verify that the multiplication operation has a unique continuous extension from $Z\times Z$ to $\hat{\Gamma}_p\times\hat{\Gamma}_p$. In this way $\hat{\Gamma}_p$ becomes a compact topological ring that is in fact an integral domain. This compact ring is known as the ring of p-adic integers. The additive group $\hat{\Gamma}_p$ has no elements of finite order and therefore may be imbedded in a unique,

ISBN 0-8053-6702-0/0-8053-6703-9, pbk

unique, minimal infinitely divisible extension $\tilde{\hat{\Gamma}}_p$. We give $\tilde{\hat{\Gamma}}_p$ the unique topology that makes the natural imbedding of $\hat{\Gamma}_p$ in $\hat{\Gamma}_p$ a homeomorphism with an open range. Ith this topology $\hat{\Gamma}_p$ is locally compact. Moreover, the multiplication in $\hat{\Gamma}_p$ extends in a natural way to $\hat{\Gamma}_p$ in such a way that it becomes a topological field. This locally compact topological field contains the field of rational numbers as a dense subfield and is known as the field of p-adic numbers.

The functions S_p defined on the p-adic integers can be interpreted as giving a measure of the number of solutions of $n = x_1^k + \cdots + x_r^k$ in p-adic integers. For each $l = 1,2,\ldots$ the ring Z_p^l of integers modp^l is in a natural way a quotient ring of the ring of p-adic integers and the entire ring is a "limit" in a certain sense of these quotient rings. Let $a_l^p(n)$ be the number of solutions of our equation in Z_{p^l}. Then $\Sigma_{n=1}^{p^l} a_l^p(n)$ is equal to p^{lr} the number of elements in $(Z_{p^l})^r$ and so the "average" over all possible n of $a_l^p(n)$ is $p^{lr}/p^l = p^{l(r-1)}$. Let

$$l_l^p(n) = \frac{a_l^p(n)}{p^{l(r-1)}}$$

Then

$$\frac{l_l^p(n_1)}{l_l^p(n_2)} = \frac{a_l^p(n_1)}{a_l^p(n_2)}$$

so that l_l^p measures the "relative number" of solutions and is the unique relative measure whose average with respect to n is one. The significant thing is that $\lim_{l\to\infty} l_l^p(n)$ exists and is equal to $S_p(n)$. Thus $S_p(n)$ can be thought of as a (relative) measure of the "number" of p-adic solutions.

The factor $\gamma_0 n^{(r/k-1)}$ has a similar interpretation. Consider the solutions of $n = x_1^k + \cdots + x_r^k$ in the field of all real numbers and suppose that k is even. Let $\phi(x_1 \cdots x_r) = x_1^k + \cdots + x_r^k$ and for each Borel subset E of the positive real numbers let $\tilde{\mu}(E)$ be the r-dimensional Lebesgue measure of $\phi^{-1}(E)$. Then $\tilde{\mu}$ is absolutely continuous with respect to one-dimensional Lebesgue measure and the Radon Nikodym derivative is $\gamma_0 n^{(r/k-1)}$. In other words, $\gamma_0 n^{(r/k-1)}$ is the $(r-1)$-dimensional measure of the hypersurface $n = x_1^k + \cdots + x_r^k$ in Euclidean r-space. As such it is a reasonable (relative) measure of the "number" of solutions even though the actual cardinal number is always infinite.

In the works of Siegel on quadratic forms, this point of view toward the results is adopted systematically. In fact, Siegel shows more or less directly that his means of representation numbers are infinite products of "solution measures."

ISBN 0-8053-6702-0/0-8053-6703-9, pbk

Consider now the compact group $\hat{\Gamma} = \Pi_p \hat{\Gamma}_p$. It is a group without elements of finite order and has correspondingly a unique, minimal infinitely divisible extension $\hat{\hat{\Gamma}}$ that may be topologized just as we did $\hat{\hat{\Gamma}}_p$ so that $\hat{\Gamma}$ is a compact open subgroup. It is easy to see that it is just the subgroup of $\Pi_p \hat{\hat{\Gamma}}_p$ consisting of those elements $\{x_p\}$ such that $x_p \in \hat{\Gamma}_p$ for all but finitely many p. Moreover, since we now know how to extend S_p to $\hat{\hat{\Gamma}}_p$ and the product of the S_p is defined on $\hat{\Gamma} = \Pi_p \hat{\Gamma}_p$, it is clear that we have a natural way of extending $S^1 = \Pi_p S_p$ to be a function defined and continuous on all of $\hat{\hat{\Gamma}}$. Because of the parallel between the S_p and $\gamma_0 n^{(r/k-1)}$, it is natural to form the product of $\hat{\hat{\Gamma}}$ and the reals and on this product to consider the product of S^1 extended to $\hat{\hat{\Gamma}}$ and $\gamma_0 x^{(r/k-1)}$. But $\hat{\hat{\Gamma}}$ times the reals is just the adèle group of the rationals as defined in Section 26. Thus F is a continuous function on the adèle group whose restriction to the integer subgroup of the discrete subgroup of all "principal adèles" is the function to which f is asymptotic.

We have gone into so much detail about the various ways of looking at the Hardy–Littlewood results because the concepts involved are central to much of number theory. In particular, there are many number theoretical problems in which one discusses what happens for the integers or rationals in terms of what happens for the p-adics for each prime p and in terms of what happens for the reals and possibly the complex numbers. Consider, for example, reduction theory for quadratic forms. The theory classifying quadratic forms over the reals or complexes is quite simple and is familiar to every student of elementary abstract algebra. There is a corresponding theory for quadratic forms over the p-adics that is somewhat more involved but fairly simple and elegant. Now given a quadratic form over the rationals, it may be regarded as a quadratic form over the reals and also as one over the p-adics for each prime p. Clearly two forms cannot be equivalent over the rationals unless they are equivalent over the reals and over the p-adics for each prime p. In this case the converse holds (i.e., the so-called Hasse principle is valid) and two forms that are equivalent over the p-adics for all p and also over the reals are equivalent over the rationals. Thus to classify forms over the rationals we have only to determine a complete set of invariants for forms over the reals and p-adics and then decide which combinations of these occur for forms over the rationals. This problem has been given a complete and elegant solution. For forms over the integers the situation is less straightforward, but the problem may be approached and solved in the same spirit.

As another example, let us look at class field theory regarded as the problem of classifying the possible finite normal abelian extensions of a given algebraic number field. Moreover, for simplicity's sake, let us just

ISBN 0-8053-6702-0/0-8053-6703-9, pbk

consider the normal abelian extensions of the rationals. Given a finite extension F^1 of a field F, for each element $x \neq 0$, let $N(x)$ be the product of all the transforms of x by automorphisms of F^1 leaving the elements of F fixed. Then N is a homomorphism of the multiplicative group of F^1 into that of F. Now let F be either the field of all real numbers or the field of all p-adic numbers for some prime p. In the so-called "local class field theory" one proves that the range of N is closed has finite index in the multiplicative group F^* of F and that this subgroup of F^* of finite index determines the extension to within isomorphism. One proves also that every closed subgroup of F^* of finite index occurs and hence that the finite normal abelian extensions of F correspond one-to-one to the closed subgroups of finite index of F^*. When F is real, these facts are quite well known and trivial. F^* then has only one closed subgroup of finite index and this corresponds to the field obtained by adjoining $\sqrt{-1}$. In this case, $N(x+iy)=x^2+y^2$ and the closed subgroup of finite index is the group of all positive real numbers.

A finite normal abelian extension of the rationals defines such an extension for the p-adic numbers for each p and also for the real numbers. Hence it defines a closed subgroup of finite index of the multiplicative group of each of these fields and hence a subgroup of their product. The main theorem of global class field theory in this case tells us that this subgroup of the product determines the extension up to isomorphism and also tells us which subgroups occur. The subgroups that occur are just those which are inside the idèle group, contain the principal idèles, and are closed and have finite index in the idèle group. Thus there is a natural one-to-one correspondence between the closed subgroups of finite index of the group of "idèle classes," on the one hand, and the finite normal abelian extensions of the rationals on the other.

There is a further statement identifying the Galois groups of the extensions with the quotient groups of the subgroups of finite index. It implies, in particular, that the quadratic extensions correspond to the subgroups of index 2. The quadratic reciprocity law with its two supplements turns out to be completely equivalent to the statement that the subgroups of index 2 which occur contain the group of principal idèles.

Notes and References

The connection between almost periodicity and the work of Hardy and Littlewood seems to have been pointed out for the first time by M. Kac ("Almost periodicity and the representations of integers as sums of squares," *Am. J. Math* vol. 62 (1940)).

ISBN 0-8053-6702-0/0-8053-6703-9, pbk

28. Semi-direct products, algebraic number fields, and diophantine equations

In the last section we attempted to demonstrate the relevance of Fourier analysis on commutative groups to questions in number theory. In this we shall try to do the same for the representation theory of the simplest noncommutative groups—those that are semi-direct products of two commutative groups. Let F be a field. Let F^a and F^* respectively, denote the additive group of F and the multiplicative group of all nonzero elements of F. Each $y \in F^*$ defines an automorphism of F^a; namely, the mapping $x \rightarrow yx$. Moreover, if α_y denotes the automorphism defined by y, then $y \rightarrow \alpha_y$ is a homomorphism of F^* into the group of all automorphisms of F^a and we may form the noncommutative semi-direct product $F^a \circledS F^*$ of our two groups. If F is an algebraic number field, F^a will have a distinguished subgroup—the group J of all algebraic integers and F^a will be the unique minimal divisible extension of J. It turns out that much of the theory of algebraic number fields may be developed for much more general systems N, H, J, α, where J is a subgroup of the commutative group N, and α is a homomorphism of the commutative group H into the group of automorphisms of N. We shall give details below.

An obvious way of generalizing the construction of the semi-direct product $F^a \circledS F^*$ consists in replacing α_y by β_y, where $\beta_y(x) = y^k x$ for some positive integer k. More generally, still, one can replace F_a by its direct "product" with itself s_1 times, F^* by its direct product with itself s_2 times,

George W. Mackey, Unitary Group Representations in Physics, Probability, and Number Theory

ISBN 0-8053-6702-0/0-8053-6703-9, pbk

and α_y by $\beta_{y_1,\ldots,y_{s_2}}$ where

$$\beta_{y_1,y_2,\ldots,y_{s_2}}(x_1,x_2,\ldots,x_{s_1})=y_1^{m_{11}}y_2^{m_{12}}\cdots y_{s_2}^{m_{1s_2}}x_1,$$

$$y_1^{m_{21}}y_2^{m_{22}}\cdots y_{s_2}^{m_{2s_2}}x_2,\ldots,y_1^{m_{s_11}}y_2^{m_{s_12}}\cdots y_{s_2}^{m_{s_1s_2}}x_{s_1}$$

and (m_{jk}) is a matrix of nonnegative integers. Semi-direct products of this form are especially interesting when taken in conjunction with certain subgroups of the additive factors. Let N be the direct "product" of F^a with itself s_1 times and let H be the direct product of F^* with itself s_2 times and let β be as above. Let $s_1=t_1+t_2+\cdots t_r$, where each t_j is a positive integer and let N_0 be the subgroup of N consisting of all members $x_1,x_2,\ldots,x_{s_1}$ such that

$$x_1+\cdots+x_{t_1}=0$$

$$x_{t_1+1}+\cdots x_{t_1+t_2}=0$$

$$\vdots$$

$$x_{t_{r-1}+1}+\cdots+x_{s_1}=0$$

Now suppose that F is countable or finite and consider the dual $\hat{N}$ of N and the semi-direct product of $\hat{N}$ with H defined by $h\rightarrow\beta^*_{h^{-1}}$, where $\beta^*(\chi)(n)=\chi(\beta(n))$. The annihilator $N_0^\perp$ of N_0 in $\hat{N}$ is a closed subgroup of $G=\hat{N}Ⓢ H$. From the point of view of the theory of group representations, one of the fundamental problems about the pair $N_0^\perp$, G is the determination of the structure of the representations of G induced by the characters of $N_0^\perp$. In the important special case in which the subgroups H_n (see below) are all finite, it is easy to see that for each $c\in\hat{N}_0^\perp$ the induced representation U^c of G is a discrete direct sum of irreducibles. Thus we may describe U^c completely by specifying the multiplicity with which each irreducible occurs. The interesting thing is that this multiplicity is equal to the number of solutions of a system of r algebraic equations in s_1 unknowns. Moreover, this system can be made to coincide with an arbitrary such system chosen in advance by suitably choosing the m_{ij}, the character c, and the irreducible representation of G.

To see this, let us begin by recalling the complete and explicit description of the irreducible unitary representations of G that follows from the theory expounded in Section 10. For each $n\in N$, let H_n be the subgroup consisting of all $h\in H$ for which $\beta_h(n)=n$ and let σ be any character of H_n. Then χ, $h\rightarrow\chi(n)\sigma(h)I$ is a one-dimensional irreducible unitary representation $n\sigma$ of the subgroup $\hat{N}H_n$ of G and we may form the induced representation $U^{n\sigma}$ of G. It follows from the theory in question that $U^{n\sigma}$ is irreducible and that every irreducible unitary representation of G is equivalent to some $U^{n\sigma}$. (Of course the $U^{n\sigma}$ may be equivalent among themselves—the conditions for this being explained in Section 10.)

ISBN 0-8053-6702-0/0-8053-6703-9, pbk

Our question then is this: Given $n \in N$ and a character σ of H_n, with what multiplicity is $U^{n\sigma}$ contained in the induced representation U^c, where $c \in N_0^{\perp}$? To answer it, let us compute U^c using the "chain rule" described in Section 9 with $\hat{N}$ as the intermediate subgroup. We have $U^c = U^W$, where W is the representation of $\hat{N}$ induced by c and hence is the direct sum of all characters $\chi \to \chi(n)$, where n lies in a certain N_0 coset, namely, that "defined by c." By way of explanation, we recall that $N_0^{\perp}$ is canonically isomorphic to N/N_0 so that c defines (in a certain sense "is") an N_0 coset. If we choose a member n_0 of N such that $\chi(n_0) \equiv c(\chi)$ for all $\chi \in N_0^{\perp}$, then W is just the direct sum of all $n_0\nu$ for ν in N_0 and of course n_0 may be replaced by $n_0\nu'$ for any ν' in N_0. It follows that $U^c = U^W = \Sigma_{\nu \in J} U^{n_0\nu}$. Now for any $n \in N$, U^n is easily seen to be equivalent to U^{nR}, where R is the regular representation of H_n and nR is the representation $\chi, h \to \chi(n)R_h$ of $\hat{N}H_n$. Thus U^n is a direct sum or direct integral of terms of the form $U^{n\sigma}$, where σ varies over the characters of H_n. Let us now assume that H_n is finite with order $o(H_n)$. Thus U^c will be a discrete direct sum and will contain $U^{n\sigma}$ a number of times that is independent of σ. Applying the conditions for equivalence of two $U^{n,\sigma}$ explained in Section 10, we reach the conclusion that this number is $i(N_0, n, n_0)/o(H_n)$, where $i(N_0, n, n_0)$ is the number of elements h of H for which $\beta_h(n)n_0^{-1} \in N_0$.

To make the connection with systems of equations, we note that when N N_0, H, and β are constructed from the additive and multiplicative groups of a field, as indicated above, and $h = y_1, \cdots y_{s_2}$, $n = a_1, \cdots a_{s_1}$, $n_o = b_1, \cdots b_{s_1}$, then $\beta_h(n)n_0^{-1} \in N_0$ if and only if $y_1, \ldots, y_{s_2}$ satisfies the following system of equations:

$$
\begin{aligned}
& a_1 y_1^{m_{11}} y_2^{m_{12}} \cdots y_{s_2}^{m_{1s_2}} \\
& \qquad + a_2 y_1^{m_{21}} \cdots y_{s_2}^{m_{2s_2}} + \cdots + a_{t_1} y^{m_{t_11}} \cdots y^{m_{t_1 s_2}} = b_1 + b_2 + \cdots + b_{t_1} \\
& a_{t_1+1} y_1^{m_{t_1+11}} y_2^{m_{t_1+12}} + \cdots + a_{t_1+2} y_1^{m_{t_1+2}} + \cdots + \cdots = b_{t_1+1} + \cdots + b_{t_1+t_2} \\
& \qquad \vdots \\
& a_{t_{r-1}+1} y_1^{m_{t_{r-1}+11}} y_2 + \cdots + \cdots = b_{t_{r-1}} + \cdots + b_{s_1}
\end{aligned}
$$

Thus the number of solutions of this system of equations in the given field is equal to $i(N_o, n, n_o)$ and, except for the factor $o(H_n)$, is equal to the multiplicity with which U^L contains the irreducible representation $U^{n,0}$.

It is clear that any system of r polynomial equations in s_2 unknowns can be thrown into the indicated form. Thus we may reduce the study of the number of solutions of such equations to the study of the number $i(N_o, n, n_o)$ for certain semi-direct products or (almost equivalently) to the study of the structure of certain induced representations. By suitably

ISBN 0-8053-6702-0/0-8053-6703-9, pbk

modifying the construction, one can obtain an entirely similar result in which integer solutions are involved rather than rational ones.

Although we are far indeed from such a goal at the present time, if we are ever to obtain theorems about semi-direct products analogous to those of Hardy and Littlewood for Waring's problem, we shall need to generalize the technique of factoring over the primes involved in writing

$$f(n) \sim \gamma_0 n^{(r/k-1)^{-1}} \prod_p S_p(n)$$

We shall find a reasonable candidate for this in the idèle–adèle groups of algebraic number theory—formulated in terms of the semi-direct products F^a Ⓢ F^* and then extended to more general semi-direct products. We turn our attention now to a description of these notions.

We begin with some general (and somewhat digressive) remarks about the structure of countable discrete commutative groups. Given any such group G, we may consider the subgroup G^f of all elements of finite order. Then G^f is a "torsion group" in the sense that all elements are of finite order and G/G^f is "torsion free" in the sense that only the identity has finite order. In this way the study of the structure of a general G is reduced to the separate study of the torsion and torsion-free cases and to an extension problem. Now countable torsion groups have been completely classified by Ulm and a simplified account of this classification will be found in Kaplansky's monograph *Infinite Abelian Groups*. Much less is known about torsion-free groups. However, in a sense which we shall now explain, each such has a structure closely related to that of an associated torsion group. It is not hard to show that the discrete commutative group G is torsion free if and only if its compact dual $\hat{G}$ is "infinitely divisible" in the sense that $x^n = y$ is always solvable for x. Conversely, $\hat{G}$ is torsion free if and only if G is infinitely divisible and it follows that G has *no* infinitely divisible elements if and only if the set $\hat{G}^f$ of all elements of finite order in $\hat{G}$ is *dense* in $\hat{G}$. Now it is easy to show that every torsion-free discrete group is the direct product of the subgroup of all infinitely divisible elements and a subgroup in which only the identity is infinitely divisible. Since a commutative group that is both torsion free and infinitely divisible is a vector space over the rationals (and hence completely determined by its dimension), one need only consider those torsion-free groups in which no element but the identity is completely divisible. By what has gone before these are just the groups whose compact duals admit a dense set of elements of finite order. But if $\hat{G}^f$ is dense in $\hat{G}$, we have a continuous isomorphism ϕ of the torsion group $\hat{G}^f$ (made discrete) into a dense subgroup of $\hat{G}$. The dual of ϕ will then be a continuous isomorphism of G onto a dense subgroup of $G^{\wedge f \wedge}$. In other words, G has a canonical dense imbedding in the compact dual of a torsion group. While this torsion

ISBN 0-8053-6702-0/0-8053-6703-9, pbk

group need not be countable, it will be in many important cases. We shall be interested here only in the rather trivial case in which G is isomorphic to the group of all n-tuples of integers for some n, for example, the group of all integers in an algebraic number field. However, it seems worth noticing that the construction of the dense imbedding $G^{\wedge f \wedge}$ has a rather general significance.

Now let J be a commutative discrete group that is free and finitely generated (i.e., is isomorphic to the additive group of all n-tuples of integers for some n) and let $J^{\sim}$ be its unique, minimal completely divisible extension. Then $\tilde{J}$ will be an n-dimensional vector space over the rationals. We propose to study the system consisting of J and a subgroup H of the group of all automorphisms of $\tilde{J}$. Our motivating example is that in which J is the group of all integers in an algebraic number field and H is the group of all automorphisms of $\tilde{J}$ of the form $x \to x \cdot y$, where y is a nonzero element of the field and $\cdot$ denotes field multiplication. In this case, of course, $\tilde{J}$ is just the additive group of the field. We define H_0 the *unit group* of H to be the subgroup consisting of all $h \in H$ such that $h(J) = J$. H is of course a group of linear transformations in the rational vector space $\tilde{J}$. With respect to a basis it will be a group of matrices. If this basis is chosen in J, as it may be, the members of H_0 will all have integral elements.

We now introduce the group $J^{\wedge f}$ of all elements of finite order in the dual $J^{\wedge}$ of J as above and note in this case that $J^{\wedge f}$ is countable so that $J^{\wedge f \wedge}$ is a separable compact group. As above we have a canonical dense imbedding of J in $J^{\wedge f \wedge}$ and it is clear that this has a unique extension to a dense imbedding of $\tilde{J}$ in $J^{\wedge f \wedge \sim}$. We give $J^{\wedge f \wedge \sim}$ the unique topology that makes $J^{\wedge f \wedge}$ a compact open subgroup with its original topology. We denote this imbedding by the symbol ϕ_0 and call $J^{\wedge f \wedge \sim}$ the non-Archimedean component of the *adèle group* of $\tilde{J}$ with respect to J. Now let $\bar{J}$ denote the vector space of all real characters of J in the sense of Section 25. Then each $x \in J$ defines a member of the vector space dual of $\bar{J}$ and we have a natural imbedding of J in $\bar{J}^*$. As a real vector space, $\bar{J}^*$ is infinitely divisible and the imbedding just mentioned has an obvious unique extension to an isomorphism ϕ_A of $\tilde{J}$ onto a dense subset of $\bar{J}^*$. We call $\bar{J}^*$ the Archimedean component of the adèle group of $\tilde{J}$ with respect to J. ϕ_0 and ϕ_A together give a dense imbedding of $\tilde{J} \times \tilde{J}$ in $J^{\wedge f \wedge \sim} \otimes \bar{J}^*$. We call the latter group the adèle group of $\tilde{J}$ with respect to J. Let ϕ be the restriction of $\phi_0 \times \phi_A$ to the diagonal of $\tilde{J} \times \tilde{J}$. It is quite easy to show that the range of ϕ is closed and has a compact quotient. Thus J has a natural imbedding as a closed subgroup with a compact quotient. It can be shown moreover that $J^{\wedge f \wedge \sim} \times \bar{J}^*$ is isomorphic to its own dual in such a way as to map $\tilde{J}$ onto its own annihilator $\tilde{J}^{\perp}$ in $(J^{\wedge f \wedge \sim} \times \bar{J}^*)$. As in the special case considered

ISBN 0-8053-6702-0/0-8053-6703-9 pbk

in Section 27, $J^{\wedge f \wedge}$ has a natural decomposition as a direct product over the primes. For each prime p let $\hat{J}^p$ denote the subgroup of $\hat{J}^f$ consisting of all elements whose order is a power of the prime p. Then each element of $\hat{J}^f$ is uniquely a product of elements, one from each of a finite number of the $\hat{J}^p$.

Correspondingly, $J^{\wedge f \wedge}$ is the complete direct product of subgroups isomorphic to the $J^{\wedge p \wedge}$, the subgroup $J^{\wedge p \wedge'}$ isomorphic to $J^{\wedge p \wedge}$ being the intersection of the $J^{\wedge q \perp}$ for $q \neq p$. It is easy to see that $J^{\wedge f \wedge \sim}$ is the subgroup of the complete direct product of all of the $J^{\wedge p \wedge' \sim}$ consisting of those members whose $J^{\wedge p \wedge' \sim}$ component is in $J^{\wedge p \wedge' \sim}$ is isomorphic to the additive group of a finite-dimensional vector space over the field of all p-adic numbers.

It is not difficult to show that every automorphism $h \in H$ extends uniquely to an automorphism of $\hat{J}^{f \wedge \sim}$ and to an automorphism of $\bar{J}^*$. The automorphism of $J^{\wedge f \wedge \sim}$ maps each $J^{\wedge p \wedge' \sim}$ into itself and is a linear transformation with respect to the p-adic vector space structure of the latter. Let us denote this linear transformation by T_h^p. Similarly, the automorphism of $\bar{J}^*$ defined by h is a real linear transformation that we denote by T_h^∞.

For $\alpha = \infty$ or a prime p, the mapping $h \to T_h^\alpha$ is a representation of H by linear transformations in a finite-dimensional vector space. Although this representation is not unitary (or even defined in a Hilbert space), the notions of primarity and disjointness are still meaningful and we may write T^α uniquely as a direct sum of disjoint primary representations. In the special case in which J and H come from an algebraic number field as described earlier (and α is a prime p), these primary components correspond one-to-one to the prime ideals into which the natural prime p decomposes. In the general case, accordingly, we refer to these components as the *primes of the system which lie over p*. We refer to the components of T^∞ as the *infinite primes* of the system. Actually, it will be convenient to distinguish between a primary component of T^α and the space in which it acts and to call the latter a prime of the system. If $\mathfrak{p}$ is such a prime, we shall denote the corresponding subrepresentation of T^α by $T^{\mathfrak{p}}$.

Each $T^{\mathfrak{p}}$ is a representation of H in a finite-dimensional vector space over the reals or over the p-adic numbers for some p. Let $H_{\mathfrak{p}}$ denote the closure of the range of $h \to T_h^{\mathfrak{p}}$ using the natural topology in the set of all invertible linear transformations in a finite-dimensional vector space over a locally compact field. Let $H_{\mathfrak{p}}^0$ be the closure of the group of all elements of the form $T_h^{\mathfrak{p}}$, where $h \in H_0$ and preserves Haar measure in the space of $T^{\mathfrak{p}}$.

ISBN 0-8053-6702-0/0-8053-6703-9, pbk

The subgroup $H_{\mathfrak{p}}^{0}$ is compact and open when $\mathfrak{p}$ is a finite prime. We now define the adèle group of H_J (with respect to J) to be the subgroup of the direct product of all $H_{\mathfrak{p}}$ consisting of those sequences $x \rightarrow x_{\mathfrak{p}}$ such that $x_{\mathfrak{p}} \in H_{\mathfrak{p}}^{0}$ for all but a finite number of $\mathfrak{p}$. The direct product of all the $H_{\mathfrak{p}}^{0}$ for $\mathfrak{p}$ finite is compact in the product topology. Since there are only a finite number of infinite $\mathfrak{p}$'s, we may form

$$\prod_{\mathfrak{p}\text{ finite}} H_{\mathfrak{p}}^{0} \times \prod_{\mathfrak{p}\text{ infinite}} H_{\mathfrak{p}}$$

and obtain a locally compact group K contained in the adèle group H_J and having countable index therein. We give H_J the unique topology that restricts in K to the topology just described and makes K an open subgroup.

When J and H come from an algebraic number field, as above, the adèle group H_J is just the idèle group of the field as defined by Chevalley. When H is an "algebraic" subgroup of the group of all automorphisms of $\tilde{J}$ and conditions are such that certain density theorems hold, then H_J coincides with the adèle group introduced by Ono and Tamagawa. Ono observed that if we take H to be the group of all automorphisms of $\tilde{J}$ that leave a quadratic form fixed, then the properties of H_J and its related structures reflect the number theoretical properties of this quadratic form in much the same way that Chevalley's idèle group reflects the properties of the underlying algebraic number field. In this way the notion of adèle group effects a certain unification of algebraic number theory and the theory of quadratic forms. In the general case we have not only an analog of the idèle group, but also of the ideal group and of the group of ideal classes. The former is the subgroup of the direct product of the ranges of the $T^{\mathfrak{p}}$ taken over finite primes and defined by the requirement that all but a finite number of components should come from an h that lies in H_0 and preserves Haar measure. One has a natural imbedding of H in this analog of the ideal group, and when H is commutative one may form the quotient group. This quotient group is the analog of the group of ideal classes and the number of elements in it is the class number. When H is not commutative, the ideal classes need not form a group, but one still has the notion of class number.

It would be interesting to investigate the meaning of the various concepts of algebraic number theory when our semi-direct product comes from a system of algebraic equations as described earlier. Of course, in this event we have an added element of structure in the subgroup N_0. From it we can form its closure $\overline{N}_0$ in the adèle group and the projection $\overline{N_0^p}$ of this closure on each of the p-adic components of the adèle group of N with

ISBN 0-8053-6702-0/0-8053-6703-9, pbk

respect to J. In this way we get for each prime p a semi-direct product of a finite-dimensional vector space over the p-adic numbers with a group of linear transformations in this space and in addition a well-defined subgroup $\overline{N_0^p}$ of the vector space part. Presumably the analog of the function S_p in the Hardy–Littlewood solution of Waring's problem will be derived from a study of this triple of groups.

Recall that in setting up the relationship between solution numbers and multiplicities in induced representations, we had to consider the semi-direct product of H with the compact dual of N. We can obtain similar results in a more symmetrical way by using adèle groups instead of N and $\hat{N}$. Since the adèle group of N with respect to J is its own dual and in such a way that $N^{\perp}=N$, we see that the representation of the adèle group induced by the identity representating N is the direct sum of all the characters in $N(=N^{\perp})$. If we now replace the first N by the annihilator of N_0 in the adèle group, we get a subgroup N_{00} that contains N and is such that the representation of the adèle group induced by the identity of N_{00} is the direct sum of all members of N_0.

If we are to hope to obtain any sort of product formula for $i(N_0,n,n_0)$, we shall need to generalize the connection between $i(N_0,n,n_0)$ and the structure of U^c to cases in which N and H need not be discrete and the U^c need not be a discrete direct sum. This is because the local semi-direct products into which $J\wedge f\wedge^{\sim}\times\bar{J}^{*}$ ⓈH_J factors are not of this character and neither are the semi-direct products that we obtain by replacing each normal subgroup by its dual. We do this through an extension of the notion of character from finite- to infinite-dimensional representations and, in particular, a consideration of the relationship between the character of an induced representation and the character of the representation that induces it.

To begin with let Γ be a closed subgroup of finite index of the separable locally compact group G and let L be a finite-dimensional unitary representation of Γ. Then U^L will also be finite-dimensional and by definition the character χ_{U^L} of this finite-dimensional representation will be the function $x\rightarrow\mathrm{Tr}(U_s^L)$. Since the character of U^L depends only on the equivalence class of L, it depends in fact only on the character χ_L of L and we expect a formula expressing χ_{U^L} in terms of χ_L. Such a formula is easy to establish and may be stated as follows. Let χ_L^0 be the function on G that is zero outside of Γ and agrees on Γ with χ_L. Then for each x, $\chi_0^L(yxy^{-1})$ depends only on the right Γ coset $\tilde{y}$ to which y belongs and so may be regarded as a function on G/Γ. Finally

$$\mathrm{Tr}(U_x^L)=\chi_{U^L}(x)=\frac{1}{o(G/\Gamma)}\sum_{\tilde{y}\in G/\Gamma}\chi_0^L(yxy^{-1})$$

ISBN 0-8053-6702-0/0-8053-6703-9, pbk

When G/Γ is infinite, U^L is infinite-dimensional and $\mathrm{Tr}(U_s^L)$ will not in general exist. On the other hand, for each complex-valued function f on G that is summable with respect to Haar measure dx, one can form U_f^L, where

$$\left(U_f^L(\phi)\cdot\psi\right)=\int_G f(x)\left(U_x^L(\phi)\cdot\psi\right)dx$$

for all ϕ and ψ in the Hilbert space and in favorable cases $\mathrm{Tr}(U_f^L)$ will exist for all f in some dense subspace of $\mathfrak{L}^1(G)$. We may then regard the linear functional $f\to\mathrm{Tr}(U_f^L)$ as "being" the character of U^L. Indeed, when U^L is finite-dimensional so that χ_{U^L} exists in the classical sense, the linear functional $f\to\mathrm{Tr}(U_f^L)$ coincides with $f\to\int\chi_{U^L}(x)f(x)\,dx$. It is interesting to express $\mathrm{Tr}(U_f^L)$ in terms of χ_L in this case; the answer is readily computed to be

$$\frac{1}{o(G/\Gamma)}\sum_{\tilde{y}\in G/\Gamma}\left[\int_\Gamma f(y^{-1}\gamma y)\chi(\gamma)\,d\gamma\right]$$

This answer is interesting because it is an expression that can be written so as to make sense whenever G/Γ has a finite invariant measure ν. We need only normalize ν so that $\nu(G/\Gamma)=1$ and replace $\Sigma_{y\in G/\Gamma}$ by $\int_{G/\Gamma}d\nu(\tilde{y})$. In other words, whenever G/Γ has a finite invariant measure ν, we may regard the linear functional

$$f\to\int_{G/\Gamma}\left[\int_\Gamma f(y^{-1}\gamma y)\chi_L(\gamma)\,d\gamma\right]d\nu(\tilde{y})$$

as the "character" of the induced representation U^L and seek to express it in terms of irreducible characters. If U^L decomposes discretely $=m_1M^1+m_2M^2+\cdots$, where each M^j is irreducible and each m_j is a finite positive integer, then

$$\mathrm{Tr}\,U_f^L=\sum m_j\,\mathrm{Tr}\,M_f^j$$

whenever the relevant traces exist. If U^L does not decompose discretely but G is a Type I group, one can write

$$U^L=\int\nu(M)M\,d\omega(M)$$

ISBN 0-8053-6702-0/0-8053-6703-9, pbk

where $\nu(M)$ is a positive integer for all M and ω is a measure in $\hat{G}$. Of course only the class of ω will be uniquely determined by U^L. On the other hand, it is natural to conjecture the existence of a unique choice for ω such that

$$\operatorname{Tr} U_f^L = \int_{\hat{G}} \operatorname{Tr} M_f \, d\omega(M)$$

for all f in some suitably large subset of $L^1(G)$. To the extent that such a conjecture is true, one can think of the measure ω in $\hat{G}$ as defining the character of U^L and as being a natural generalization of the set of multiplicities m_j. Indeed, when U^L does decompose discretely, ω will be the discrete measure that assigns m_j to M^j and zero to the complement of the set of M^j.

It is interesting to look at the special case of the above in which G is commutative and L is the one-dimensional identity. Then G/Γ will have a finite invariant measure if and only if it is compact and then U^I will be the direct sum of all characters in the discrete subgroup $\Gamma^\perp$ of $\hat{G}$. Since G is commutative, $f(y^{-1}xy)=f(x)$ and $\operatorname{Tr} U_f^L$ becomes just $\int_\Gamma f(\gamma)\,d\gamma$. On the other hand, if $M_x^j = \chi(x)I$, then

$$\operatorname{Tr} M_f^j = \int_G \chi(x) f(x)\,dx = \hat{f}(\chi)$$

Thus the formula

$$\operatorname{Tr}\left(U_f^L\right) = \Sigma m_j \operatorname{Tr} M_f^j$$

reduces to

$$\int_\Gamma f(\gamma)\,d\gamma = \sum_{\chi \in \Gamma^\perp} \hat{f}(\chi)$$

and this in turn reduces to the classical Poisson summation formula when Γ is a discrete subgroup of a finite-dimensional real vector group. Because of this one can think of the formula

$$\int_{G/\Gamma} \left[\int_\Gamma f(y^{-1}\gamma y) \chi_L(\gamma)\,d\gamma \right] d\nu(\tilde{y}) = \int_{\hat{G}} \operatorname{Tr} M_f \, d\omega(M)$$

as a noncommutative generalization of the Poisson summation formula. The Selberg trace formula referred to in Section 26 is essentially the special case in which G is a semi-simple Lie group and Γ is a discrete subgroup.

ISBN 0-8053-6702-0/0-8053-6703-9, pbk

However, Selberg does not introduce the measure ω as such except when it is discrete. Instead, he attempts to find and subtract off the continuous part of U^L. Moreover, he restricts himself to functions f that are constants on the right K cosets for a maximal compact subgroup K. This makes it possible to write $\operatorname{Tr} M_f$ in the form $\int_{\hat{G}} f(x)s_M(x)\,dx$, where s_M is a so-called "spherical function" and exists whether or not M has a character that can be described by a function. It is natural for him to make this restriction because he was motivated by the work of Maass and others on automorphic functions and these are functions defined on G/K.

Actually one can make sense of the left-hand side of the generalized Poisson summation formula even when G/Γ does not admit a finite invariant measure. All that we really need is that it be possible to speak of the mean value of a function on G/Γ. Suppose then that $\tilde{\Gamma}$ is the normalizer of Γ in G and that $\tilde{\Gamma}_L$ is the subgroup of $\tilde{\Gamma}$ consisting of all y which define automorphisms of Γ leaving χ_L fixed. Then

$$\int_\Gamma f(y^{-1}\gamma y)\chi_L(\gamma)\,d\gamma$$

will be constant not only on the right cosets but on the right $\tilde{\Gamma}_L$ cosets as well. Thus if $G/\tilde{\Gamma}_L$ has a finite invariant measure (and we normalize so that the measure of $G/\tilde{\Gamma}_L$ is one), the integral of $\int_\Gamma f(y^{-1}\gamma y)\chi_L(\gamma)\,d\gamma$ over $G/\tilde{\Gamma}_L$ with respect to this measure may be interpreted as the mean value on G/Γ.

That this is a significant generalization may be seen by looking at the case in which G is commutative. Then $\tilde{\Gamma}_L = G$ for all Γ and L and for any closed subgroup Γ our expression makes sense (even though $\operatorname{Tr} U_f^L$ exists only when G/Γ is compact). It takes the form: $\int_\Gamma f(\gamma)\chi_L(\gamma)\,d\gamma$. Now it is not difficult to show that the classical Poisson summation formula has the following generalization:

$$\int_\Gamma f(\gamma)\,d\gamma = \int_{\Gamma^\perp} \hat{f}(\chi)\,d\chi$$

Here Γ is an arbitrary closed subgroup of the separable locally compact commutative group G, $d\gamma$ and $d\chi$ are suitably chosen Haar measures in Γ and $\Gamma^\perp$, and f is any member of a large class of functions including the continuous functions with compact support. If we replace f by $f\chi$, where χ is any member of $\hat{G}$, this reduces to

$$\int_\Gamma f(\gamma)\chi(\gamma)\,d\gamma = \int_{\Gamma^\perp\chi} \hat{f}(\chi)\,d\chi$$

ISBN 0-8053-6702-0/0-8053-6703-9, pbk

Thus

$$\int_{\Gamma} f(\gamma)\chi_L(\gamma)\,d\gamma=\int_{\Gamma^{\perp}\chi_0}\hat{f}(\chi)\,d\chi$$

where χ_0 is any member of $\hat{G}$ reducing on Γ to χ_L. In other words, our measure ω does exist in this case and is a Haar measure on $\Gamma^{\perp}$ translated by χ_0.

An interesting noncommutative example is obtained by letting $\Gamma=\{e\}$ so that $\tilde{\Gamma}=G$. Then our formula reads:

$$f(e)=\int_{\hat{G}}\operatorname{Tr}(M_f)\,d\omega(M)$$

and this (modulo regularity hypotheses on f) is just the statement of the Plancherel theorem for a unimodular Type I group—ω being the so called "Plancherel measure."

Now let us consider the special case in which G is a semi-direct product of two commutative groups $\hat{N}$ and H with respect to a homomorphism β of H into the group of automorphisms of $\hat{N}$, and Γ is the annihilator $N_0^{\perp}$ of a closed subgroup N_0 of N. When N and H are discrete and a certain auxiliary condition is satisfied, the representations U^L are discretely decomposable and ω exists as a discrete measure. We saw earlier in this section how to compute ω (i.e., the multiplicities of occurrence of the irreducible constituents of U^L) in terms of the action of H on N. We shall show now that when N and H are not discrete, ω can exist as a nondiscrete measure that can be computed from the action of H on N in a manner generalizing our earlier result.

Let f be a suitably restricted complex-valued function on $G=\hat{N}\circledS H$. Then for each h_1,h_2 in H we may consider the function on $\hat{N}$ which takes χ to $f(\beta_{h_1}(\chi),h_2)$ and its Fourier transform

$$\begin{aligned}n\to\int f(\beta_{h_1}(\chi),h_2)\chi(n)\,d\chi&=\Delta(h_1)\int f(\chi,h_2)\beta_{h_1^{-1}}(\chi)(n)\,d\chi\\&=\Delta(h_1)\int f(\chi,h_2)\chi(\alpha_{h_1}(n))\,d\chi\\&=\Delta(h_1)F(\alpha_{h_1}(n),h_2)\end{aligned}$$

where $F(n,h_2)$ is the Fourier transform of $\chi\to f(\chi,h_2)$ and $\Delta(h_1)$ is the factor by which Haar measure is changed by the automorphism $\chi\to$

ISBN 0-8053-6702-0/0-8053-6703-9, pbk

$\beta_{h_1^{-1}}(\chi)$. Now by the general commutative version of the Poisson summation formula we have

$$\int_{N_0^\perp} f(\beta_{h_1}(\chi), h_2)\, d\chi = \Delta(h_1) \int_{N_0} F(\alpha_{h_1}(n), h_2)\, dn \qquad (*)$$

Next let us consider the irreducible unitary representation $U^{n,\sigma}$ of G induced by the character $\chi, \xi \to \chi(n)\sigma(\xi)$ of the subgroup $\hat{N} \circledS H_n$, where $n \in N$ and $\sigma \in \hat{H}_n$. A calculation that we leave to the reader shows that

$$\mathrm{Tr}(U_f^{n,\sigma}) = \int_H \Delta(h_1) \left[\int_{H_n} F(\alpha_{h_1}(n), \xi)\sigma(\xi)\, d\xi \right] dh_1$$

Suppose for the moment that $H_n = \{e\}$ for $n \neq e$. Then our formula simplifies to

$$\mathrm{Tr}(U_f^n) = \int_H \Delta(h_1) F(\alpha_{h_1}(n), e)\, dh_1$$

Let $\theta(n)$ be the member of $\hat{G}$ to which U^n belongs and for each Borel subset E of $\hat{G}$ let $\omega(E)$ be the Haar measure in N_0 of $N_0 \cap \theta^{-1}(E)$. Then

$$\int_{\hat{G}} \mathrm{Tr}(M_f)\, d\omega(M) = \int \mathrm{Tr}(U_f^\gamma)\, d\gamma$$

Hence

$$\int_{\hat{G}} \mathrm{Tr}(M_f)\, d\omega(M) = \int_{N_0} \int_H \Delta(h_1) F(\alpha_{h_1}(n), e)\, dh_1\, d\gamma$$

Now if our function f is such that we may interchange the orders of integration, we get

$$\int_{\hat{G}} \mathrm{Tr}(M_f)\, d\omega(M) = \int_H \Delta(h_1) \int_{N_0} F(\alpha_{h_1}(n), e)\, d\gamma\, dh_1$$

and we see that the right-hand side is just the right-hand side of (*) integrated with respect to h_1 and with $h_2 = e$. Thus by

$$\int_{\hat{G}} \mathrm{Tr}(M_f)\, d\omega(M) = \int_H \int_{J^\perp} f(\beta_{h_1}(\chi), e)\, d\chi\, dh$$

and the right-hand side of this is just the left-hand side of our noncommutative generalization of the Poisson summation formula except that we

ISBN 0-8053-6702-0/0-8053-6703-9, pbk

have an integral over H instead of a "mean value" over H. Of course, the fact that the integral exists implies that the mean value is zero. Clearly we must be somewhat more general and permit ourselves to use any invariant linear functional on the functions on G/Γ that happens to exist whether it be a mean value or not. With this agreement we do indeed have a well-defined measure ω on $\hat{G}$ generalizing the multiplicity function in the discrete case. So far, we have only actually exhibited it in the special case in which σ is the identity character and the H_n are trivial. However, there is no great difficulty in removing these restrictions. To remove the first we have only to replace our function $\chi \rightarrow f(\beta_{h_1}(\chi), e)$ by its product with the character $\chi \rightarrow \chi(n_0)$ for some $n_0 \in N$ in applying the commutative Poisson summation formula. We find then that $\omega(E)$ is the Haar measure in $N_0\Gamma$ of $n_0\Gamma \cap \theta^{-1}(E)$.

In closing this rather speculative section we warn the reader that putting the various pieces together (adèle groups of semi-direct products, induced characters, solution numbers as induced representation multiplicities) to form a genuine theory of Diophantine equations is a much more difficult task than it appears at first sight. The writer thought for awhile that he could at least formulate some reasonable conjectures. However, his attempts in this direction showed him that he still had much to learn.

ISBN 0-8053-6702-0/0-8053-6703-9, pbk

29. Quadratic forms and automorphic forms

We have discussed the results of Hardy and Littlewood from several points of view, but have given no hint as to how such results might be proved. Hardy and Littlewood themselves used Cauchy's theorem to express $f(n)$ in terms of $\tilde{f}(z)$ choosing the path of integration so as to make use of the peculiarities of $\tilde{f}$, in particular, the fact that $\tilde{f}(x+iy)$ tends rapidly to ∞ whenever x is rational. Their analysis was long and complicated and involved many delicate estimates. However, in the important special case in which $k=2$, there is an easier method that is especially interesting because of its contacts with other parts of mathematics and, in particular, with the representation theory of certain semi-simple Lie groups. This method applies not only to sums of squares but to positive-definite quadratic forms in general. It may be generalized so as to apply to the representations of one quadratic form by another and suitably modified may be applied to corresponding problems involving indefinite forms.

Let $f_Q(n)$ denote the number of ways of representing the integer n by the positive-definite quadratic form Q and let

$$\tilde{f}_Q(z)=\sum_{n=0}^{\infty} e^{2\pi inz}f_Q(n)$$

The basic idea is to show that $\tilde{f}_Q$ is a "modular form" in the sense that it satisfies a certain functional equation and has certain boundedness properties, and then to apply the methods of the theory of automorphic functions to find an explicit formula for the most general possible modular form.

George W. Mackey, Unitary Group Representations in Physics, Probability, and Number Theory

ISBN 0-8053-6702-0/0-8053-6703-9, pbk

This explicit formula is of such a character that one can almost "read off" the desired facts about f.

The functional equation is obtained by applying the Poisson summation formula to the function

$$x_1, x_2, \ldots, x_s \rightarrow e^{2\pi z i Q(x_1, \cdots, x_s)}$$

where z is a complex parameter whose imaginary part is positive. The details are rather complicated in the general case and we shall confine ourselves (for the present at least) to the special case in which

$$Q(x_1, x_2, \ldots, x_s) = x_1^2 + x_2^2 + \cdots + x_s^2$$

and $s = 4k$ for some positive integer k. The Fourier transform of $x \rightarrow e^{2\pi i z x^2}$ is

$$y \rightarrow \int_{-\infty}^{\infty} e^{2\pi i z x^2} e^{2\pi i x y}\, dx = \frac{e^{-(i\pi/2z)y^2}}{\sqrt{-2iz}}$$

Thus that of

$$x_1, \cdots, x_n \rightarrow e^{-2\pi z(x_1^2 + \cdots + x_s^2)}$$

is

$$y_1, y_2, \ldots, y_s \rightarrow \left(\frac{1}{\sqrt{-2iz}} \right)^s e^{-(i\pi/2z)(y_1^2 + \cdots + y_s^2)}$$

The Poisson summation formula in this case is then

$$\sum_{n_1, n_2, \ldots, n_s = -\infty}^{\infty} e^{2\pi i z(n_1^2 + \cdots + n_s^2)} = \left(\frac{1}{\sqrt{-2iz}} \right)^s \sum_{n_1, n_2, \ldots = -\infty}^{\infty} e^{-(i\pi/2z)(n_1^2 + \cdots + n_s^2)}$$

Now the term $e^{2\pi i z n}$ occurs on the left-hand side $f_Q(n)$ times and a similar statement may be made about terms on the right. Thus our relationship becomes

$$\sum_{n=0}^{\infty} f_Q(n) e^{2\pi i n z} = \left(\frac{1}{-2iz} \right)^{s/2} \sum_{n=0}^{\infty} f_Q(n) e^{-(i\pi/2z)n}$$

or

$$\tilde{f}_Q(z) = \left(\frac{1}{-2iz} \right)^{s/2} \tilde{f}_Q\left(-\frac{1}{4z} \right)$$

ISBN 0-8053-6702-0/0-8053-6703-9, pbk

and when $s=4k$ the ambiguous factor $(1/iz)^{s/2}$ becomes $(-1)^k/(2z)^{2k}$. If we replace z by $z/2$ throughout, we obtain the relevant functional equation

$$\tilde{f}_Q\left(-\frac{1}{2z}\right)=(-1)^k z^{2k}\tilde{f}_Q\left(\frac{z}{2}\right)$$

In examining the implications of this equation for f and $\tilde{f}$, it is convenient to replace $\tilde{f}$ by θ, where $\theta(z)=f(z/2)$ so that the functional equation becomes

$$\theta\left(-\frac{1}{z}\right)=(-1)^k z^{2k}\theta(z)$$

By definition $\tilde{f}$ is periodic of period 1 so that θ is periodic of period 2. Moreover, the transformations $z\to z+2$ and $z\to -1/z$ are both of the form

$$z\to\frac{az+b}{cz+d}$$

where $ad-bc=1$ and a, b, c, and d are integers. The group of all such is a discrete subgroup Γ_0 of $SL(2,R)$ called the modular group. Let Γ denote the subgroup of Γ_0 generated by $z\to z+2$ and $z\to -1/z$. Moreover, for each prime q let $Q_q(n)$ be the canonical image of n in the field of integers mod q. Then

$$\begin{pmatrix} a & b \\ c & d \end{pmatrix}\to\begin{pmatrix} Q_q(a) & Q_q(b) \\ Q_q(c) & Q_q(d) \end{pmatrix}$$

is a homomorphism of Γ_0 onto a finite group and so its kernel is a normal subgroup of Γ_0 of finite index. We denote it by the symbol Γ_q. (The subgroups Γ_q are called the congruence subgroups of Γ_0.)

A simple calculation now shows that Γ contains Γ_2 and another shows that our functional equation and the periodity together for θ are equivalent to the statement that

$$\theta\left(\frac{az+b}{cz+d}\right)=\pm(cz+d)^{2k}\theta(z)$$

for all

$$\begin{pmatrix} a & b \\ c & d \end{pmatrix}\in\Gamma$$

ISBN 0-8053-6702-0/0-8053-6703-9, pbk

the + sign occurring if and only if

$$\begin{pmatrix} a & b \\ c & d \end{pmatrix} \in \Gamma_2$$

Thus finally we conclude that

$$\theta\left(\frac{az+b}{cz+d}\right) = (cz+d)^{2k}\theta(z) \quad \text{for all} \begin{pmatrix} a & b \\ c & d \end{pmatrix} \in \Gamma_2$$

Analytic functions satisfying identities of this character for the discrete subgroups of $SL(2,R)$ play a central role in the theory of automorphic functions. They are known as automorphic forms (or as modular forms when the discrete subgroup is a subgroup of finite index of the modular group). We digress to say a few words about this theory. Let R be any Riemann surface and let $\tilde{R}$ be its simply connected covering surface. Let Γ_R be the fundamental group of R. Then Γ_R is a countable group of automorphisms of $\tilde{R}$ whose space of orbits may be identified with R. Now it turns out that to within isomorphsim there are only three possibilities for $\tilde{R}$. It must be either the set of all complex numbers, the set of all complex numbers with the point at ∞ added—the so-called Riemann sphere or the set of all complex numbers in the interior of the unit circle. Accordingly, one says that R is *parabolic*, *elliptic* or *hyperbolic*. Of course $|z|<1$ is isomorphic to the so-called "upper half-plane"—the set of all complex numbers with positive imaginary part and it will be convenient for us to consider the hyperbolic case in this presentation. In each case the group G of all automorphisms of $\tilde{R}$ is transitive so that $\tilde{R}$ may be looked upon as a coset space G/K, where K is a closed subgroup of G. When R is hyperbolic the most general automorphism of $\tilde{R}$ is

$$z \to \frac{az+b}{cz+d}$$

where a, b, c, and d are real and $ad-bc=1$. Thus G is isomorphic to $SL(2,R)/Z_2$, where Z_2 is the two element center of $SL(2,R)$. The subgroup K_i leaving i fixed is the set of all matrices of the form

$$\begin{pmatrix} \cos\theta & \sin\theta \\ -\sin\theta & \cos\theta \end{pmatrix}$$

for θ real. It is a maximal compact subgroup. To find the most general hyperbolic Riemann surface we must find the most general countable discrete subgroup Γ of G such that the orbits of the natural action of Γ on

ISBN 0-8053-6703-9, pbk
ISBN 0-8053-6702-0/0-8053-6703-9, pbk

G/K_i constitute a Riemann surface in the quotient complex structure. Similar statements may be made in the other two cases, but we shall concentrate our attention on this one.

Given a hyperbolic Riemann surface defined by the discrete subgroup Γ of G, it is easy to see that the meromorphic functions f on this Riemann surface correspond one-to-one to the meromorphic functions on the upper half-plane $= G/K_i$ which are invariant under the action of Γ. Meromorphic functions on the upper half-plane invariant under the action of Γ are called *automorphic functions* (relative to Γ). Thus to determine the most general automorphic function in the upper half-plane (relative to Γ) is to determine the most general meromorphic function on the hyperbolic Riemann surface $G/K_i \diagdown \Gamma$.

Automorphic forms were introduced as an aid in the construction of automorphic functions. By definition, an (unrestricted) automorphic form of weight k for the group Γ is a meromorphic function g on the upper half-plane such that

$$g\left(\frac{az+b}{cz+d}\right)=(cz+d)^{2k}g(z)$$

for all z in the upper half-plane and all

$$\begin{pmatrix} a & b \\ c & d \end{pmatrix} \in \Gamma$$

Clearly the ratio of any two automorphic forms of the same weight for the same Γ is an automorphic function relative to Γ. Poincaré in his work on automorphic functions in the 1880s developed methods for the construction of automorphic forms and then used the forms to construct the functions.

An automorphic form is an unrestricted automorphic form that is analytic everywhere in the upper half-plane and in addition is finite in a certain sense on the "compactification" of the Riemann surface $G/K_i \diagdown \Gamma$. This compactification can be described in group theoretical terms. The group G actually acts on the real axis plus the point at ∞ as well as on the upper half-plane, and the former is a "boundary" of the latter. It is thus natural to think of the orbit space of Γ acting on the line as being a "boundary" of $G/K_i \diagdown \Gamma$ and to compactify the latter by adjoining points of the former. On the other hand, we do not want to add all orbits on the boundary of G/K_i. To see why, it is convenient to introduce the notion of a fundamental domain for Γ. This is a subset of the upper half-plane that meets each Γ orbit once and only once and consists of an open set together with part of its boundary. If a fundamental domain exists whose closure is

ISBN 0-8053-6702-0/0-8053-6703-9, pbk

bounded and does not meet the real axis then $G/K_i\backslash\Gamma$ is already compact and we need adjoin no boundary points. More generally, there might exist a fundamental domain whose closure meets the boundary in only a finite number of points. In that case, it will suffice to adjoin the orbits of this finite point set. We note that the graph of a fundamental domain meeting the real axis in a finite point must look something like the figure below near that point:

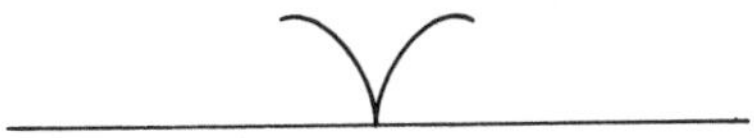

For this reason it has become customary to refer to the added points as *cusps*. Since G acts transitively on the real line plus the point at ∞, we can realize the boundary of G/K_i as G/T, where T is the subgroup leaving ∞ fixed. Then each cusp will be defined by a right coset Tx—or more invariantly by a double coset $Tx\Gamma$. Note that T is the group of all $\begin{pmatrix} a & b \\ 0 & 1/a \end{pmatrix}$ and that the subgroup N of all $\begin{pmatrix} 1 & b \\ 0 & 1 \end{pmatrix}$ is normal in T and is its maximal nilpotent subgroup. It turns out that the double cosets $Tx\Gamma$ that define cusps and have to be added to $G/K_i\backslash\Gamma$ in order to compactify it are just those for which $x\Gamma x^{-1}\cap N$ is not the identity so that the index of $x\Gamma x^{-1}\cap N$ in N is compact. This property seems to depend upon the choice of x in $Tx\Gamma$, but replacing x by $tx\gamma$ with $t\in T$, $\gamma\in\Gamma$ replaces

$$x\Gamma x^{-1}\cap N \quad \text{by} \quad t(x\Gamma x^{-1})t^{-1}\cap N = t(x\Gamma x^{-1}\cap N)t^{-1}$$

and so the property holds for all x in $Tx\Gamma$ if it holds for any.

Let $Tx\Gamma$ be a cusp and let $\gamma\neq e$ be an element of $\Gamma\cap x^{-1}Nx$. Then $\gamma\in x^{-1}Tx$ so $x\gamma\in Tx$ so $Tx\gamma^{-1}=Tx$ and Tx is a fixed point of a so-called parabolic element of Γ, that is, an element that is conjugate to a member of N. The converse also holds and one can also describe the cusps as the Γ orbits of the fixed points of the parabolic elements. Clearly there can be at most a countable number of these. Actually, we shall be concerned only with the case in which the number of cusps is finite. This is always so for example when Γ is a subgroup of Γ_0 of finite index.

To complete our definition of automorphic form we must specify in what sense it is to be finite on the compactification of $G/K_i\backslash\Gamma$. Clearly we must demand that it be finite at each cusp and it only remains to say what we mean by its value at a cusp. Let $T\alpha\Gamma$ be a cusp and let g be an unrestricted modular form of weight k for Γ. If $\alpha=\begin{pmatrix} a & b \\ c & d \end{pmatrix}$, let

$$J(z,\alpha)=\frac{1}{(cz+d)^2}$$

ISBN 0-8053-6702-0/0-8053-6703-9, pbk

and let

$$h(z)=g(\alpha(z))J(\alpha,z)^k$$

Then h is an unrestricted automorphic form of weight k for $\alpha^{-1}\Gamma\alpha$. Let b_0 be the smallest positive real number such that

$$\begin{pmatrix} 1 & b_0 \\ 0 & 1 \end{pmatrix} \in \alpha^{-1}\Gamma\alpha$$

Then h is periodic with period b_0. Hence h may be written as $h_0(e^{2\pi i z/b_0})$, where h_0 is meromorphic at $z=0$. We say that g is finite at the cusp $T\alpha\Gamma$ if h_0 is analytic at $z=0$. One is tempted to define $h_0(0)$ to be the value of h at the cusp, but this value depends to some extent on α. On the other hand, one can always choose $\alpha=\begin{pmatrix} a & b \\ c & d \end{pmatrix}$ so that $a=d=1$ and $b=0$, and with this restriction on α, $h_0(0)$ is uniquely determined. We call this uniquely determined complex number the *value* of g at the cusp in question.

Returning to our number theoretical problem about sums of squares, we recognize first that our function θ, where $\theta(z)=\tilde{f}(z/2)$ is an unrestricted automorphic form of weight k for Γ_2. Actually, it is easy to show that $G/K_i\diagdown\Gamma_2$ has just three cusps and that θ in fact is an automorphic form that takes on the values 0, 1, and $(-1)^k$ at these cusps. Now the general theory of automorphic forms not only tells us that there can be only finitely many linearly independent such forms for a given k and Γ, but provides a method of constructing them all from certain explicit infinite series involving the function $J(z,\alpha)$ introduced above. It is from a study of these series that we may "read off" the results of Hardy and Littlewood.

In terms of J the identity defining an unrestricted modular form is just

$$g(\gamma(z))J(z,\gamma)^k\equiv g(z)$$

Now for each fixed γ, $g\to g'$, where

$$g'(z)=g(\gamma(z))J(z,\gamma)^k$$

is a linear operator T_γ and $\gamma\to T_\gamma$ is a homomorphism. If Γ were a finite group, the most general solution of our identity would clearly be $\Sigma_{\gamma\in\Gamma}T_\gamma(g_0)$, where g_0 is arbitrary. More generally, we can seek solutions by trying to find g_0's for which the series

$$\sum_{\gamma\in\Gamma} T_\gamma(g_0)=\sum_{\gamma\in\Gamma} g_0(\gamma(z))J(z,\gamma)^k$$

ISBN 0-8053-6702-0/0-8053-6703-9, pbk

converges. Actually there is little hope of convergence because $J(z,\gamma)\equiv 1$ whenever $\gamma=\begin{pmatrix}1 & b\\ 0 & 1\end{pmatrix}$ and there are infinitely many such whenever the point at ∞ is a cusp—in particular, whenever Γ has finite index in Γ_0. On the other hand, the set of all such γ is a subgroup $\Gamma\cap N$ of Γ and if we choose g_0 to be invariant under the action of this subgroup, we may replace our sum over Γ by one over the cosets $\operatorname{mod}\Gamma\cap N$. The formula

$$\sum_{\gamma\in\Gamma/\Gamma\cap N} g_0(\gamma(z))J(z,\gamma)^k$$

will make sense and be a formal solution of our identity whenever $g_0(\gamma(z))\equiv g_0(\gamma)$ for $\gamma\in\Gamma\cap N$. Actually this series can be shown to converge whenever $k\geqslant 2$ and $g_0(z)\equiv 1$. The sum is an automorphic form of weight k for Γ that takes the value 1 at the cusp $T\Gamma$ (i.e., the cusp at ∞) and is zero at all the other cusps. It is called the *Eisenstein series* for the cusp $T\Gamma$ and may clearly be written in the form

$$\sum \frac{1}{(mz+n)^{2k}}$$

where m and n range over a certain infinite subset of the set of all pairs of integers. Transforming by α one obtains for each cusp $T\alpha\Gamma$ an automorphic form of weight k that is one at the cusp $T\alpha\Gamma$ and zero at all other cusps. It also has a series expansion of the form

$$\sum \frac{1}{(mz+n)^{2k}}$$

where m and n now range over some other infinite subset of the set of all pairs of positive integers. This series is called the Eisenstein series for the cusp in question.

Given any automorphic form of weight k for Γ, let $a_1,\cdots,a_i$ be its values at the cusps. Clearly we may write it in the form $a_1\phi_1+\cdots+a_i\phi_i+\psi$, where the ϕ_j are the Eisenstein series for the cusps and ψ is an automorphic form of weight k that is zero at all the cusps, that is, is what is called a *cusp form*. Thus to obtain an explicit expression for the most general possible automorphic form of weight k for a subgroup Γ_q of Γ, it remains only to find such an expression for the most general possible cusp form. Now in the series

$$\sum_{\gamma\in\Gamma/\Gamma\cap N} g_0(\gamma(z))J(z,\gamma)^k$$

we may take g_0 to be any function of the form $z\rightarrow e^{2\pi i r z/\lambda}$, where $r=1,2,\ldots$ and λ is chosen so that g_0 is invariant under $\Gamma\cap N$ and otherwise is as large as possible. As with $r=0$, the series can be shown to converge for $k\geqslant 2$.

ISBN 0-8053-6702-0/0-8053-6703-9, pbk

However, the sum ψ_r is now an automorphic form that vanishes at ∞ as well as at the other cusps; that is, it is a cusp form. Moreover, one chooses $r_1, r_2, \ldots, r_i$ so that the ψ_r form a basis for the finite-dimensional space of all cusp forms.

To obtain the results of Hardy and Littlewood, one now simply works backwards. Given an Eisenstein series or a cusp form ψ_r, one knows that it has period 2 and hence may be written in the form $\Sigma_{n=0}^{\infty} F(n)e^{\pi inz}$. (The negative coefficients are zero since our form is finite at ∞.) Using the explicit representation as an infinite series described above and an identity resulting from the Poisson summation formula, one is able to compute the coefficients $F(n)$. The function f_Q that we desire is a linear combination of these $F(n)$, the coefficients of the Eisenstein series being explicitly known. It turns out that the part of f_Q arising in this way from the Eisenstein series is just the product of the singular series with the appropriate power of n while the part arising from the cusp form is the lower order term that prevents our asymptotic formula from being an exact one. It follows that the formula will in fact be exact whenever $G/K_i \diagdown \Gamma$ is such that there are no cusp forms and this is so for Γ_2 when $k=2$ and $k=4$. For further details, the reader is referred to the notes of Gunning: "Lectures on Modular Forms," No. 48, in the Annals of Mathematics Studies.

One of the contributions of C. L. Siegel was to show that if one replaces $f_Q(n)$ by its average f_Q over all forms in a genus, then the cusp form component is always zero and $f_Q(n)$ is exactly equal to an explicit linear combination of the F's arising from Eisenstein series. In this way his main result could be shown to be equivalent to an identity in the theory of automorphic forms—the identity between $\Sigma f_Q(n)e^{\pi inz}$ and the corresponding linear combination of Eisenstein series. Of course, he considered quite general forms. Moreover, he considered the more general problem of representing one form by another. This problem can be connected with the theory of automorphic forms in a strictly analogous way. However, one must use automorphic forms in several variables defined in the so called "Siegel upper half-plane" (cf. Section 25).

Notes and References

The theory described in this section is largely due to Hecke. The account given here is far from doing justice to what Hecke accomplished in using the theory of modular forms to study the number theory of quadratic forms. In particular, it omits the material alluded to on page 321. A more complete summary will be found in sections 19 and 22 of an article of the author's entitled "Harmonic Analysis as the Exploitation of Symmetry" mentioned in the preface.

ISBN 0-8053-6702-0/0-8053-6703-9, pbk

30. Automorphic forms and group representations

The central role played in Section 29 by coset spaces and group actions suggests the existence of a close relationship between the theory developed there and the theory of induced representations of groups. We shall now describe this relationship and at the same time indicate ways in which the theory of Section 29 can be generalized.

We begin by showing that the relationship between the upper half-plane G/K_i and its "boundary" G/T may be thrown into an abstract form that is capable of extensive generalization. Notice that there is only one $T:K_i$ double coset and hence that the compact group K_i acts transitively on G/T. It follows that G/T admits a unique probability measure β that is invariant under the K_i action. This measure β will not of course be invariant under G, but we may consider the set M of all transforms of β by G. M is then a transitive G-space and the subgroup leaving β fixed clearly contains K_i. Actually it is equal to K_i and we may thus identify G/K_i with the set M of probability measures on G/T. Now it is natural to say that a sequence $\mu_1, \mu_2, \ldots$ of probability measures in G/T converges to a point x_0 of G/T if $\int \phi(x)\, d\mu_n(x)$ converges to $\phi(x_0)$ for all continuous complex-valued functions ϕ on G/T. In this way we can think of points of G/T as being limits of sequences of points of $M = G/K_i$ and it turns out this notion coincides with the more concrete one which we get by identifying G/T and G/K_i with the real line and upper half-plane, respectively. We

George W. Mackey, Unitary Group Representations in Physics, Probability, and Number Theory

ISBN 0-8053-6702-0/0-8053-6703-9, pbk

note incidentally that forming

$$\int \phi(x)\, d\mu(x) = \tilde{\phi}(\mu)$$

for μ in M assigns a function on $M = G/K_i$ to each function ϕ on G/T and that $\tilde{\phi}$ is a function whose "boundary values" are given by ϕ. The mapping $\phi \rightarrow \tilde{\phi}$ coincides with the classical Poisson integral mapping a continous function on the unit circle into the unique harmonic function in the unit disk whose boundary values are given by ϕ.

More generally let G be a noncompact semi-simple Lie group with a finite center and let K be a maximal compact subgroup. By the Iwasawa structure theorem (cf. Section 14) there exists a closed nilpotent subgroup N of G and a commutative closed subgroup D isomorphic to a finite-dimensional real vector space such that every element of G is uniquely of the form ndk, where n is in N, d is in D, and k is in K. Let T be the normalizer of N in G. Then $T \supseteq ND$ and $TK = G$. Thus K acts transitively on G/T and G/T admits a unique K invariant probability measure β whose transforms by G constitute a transitive G-space M that may be identified with G/K. Just as above we can define a point x of G/T to be a limit of a sequence of points of G/K if

$$\lim_{n\to\infty} \int \phi(x)\, d\mu_n(x) = \phi(x)$$

for all continuous ϕ on G/T where the measures μ_n are the translates of β defining the points of $M = G/K$.

Unlike the simple special case discussed above in which $G = \mathrm{SL}(2, R)$, it is *not* true in general that $G/T \cup G/K$ is a compact space and hence a "compactification" of G/K. For example, if $G = \mathrm{SL}(2, R) \times \mathrm{SL}(2, R)$, then G/K has a natural compactification isomorphic to the direct product of two replicas of the unit disk $|z| \leqslant 1$. G/K appears in this as the set of all z_1, z_2 with $|z_1| < 1$ and $|z_2| < 1$ and G/T is the two-diemsnional subset of all z_1, z_2 with $|z_1| = |z_2| = 1$ and not the three-dimensional set obtained by subtracting G/K from the compactification. To obtain a compactification of G/K including G/T, we consider the closure $\overline{M}$ of M in the compact space of *all* probability measures on G/T. $\overline{M}$ will then be a compact nontransitive G-space containing M as a dense subset and "containing" G/T as one transitive component of the action of G in $\overline{M} - M$.

In the general case there will also be proper closed subgroups T' of G which properly contain T and these may be used in exactly the same way to construct other compactifications of G/K. However, for some choices of T', the canonical map of G/K on M need not be one-to-one. Excluding these, it can be shown that one obtains in this way all possible compactifications of G/K having certain natural properties. For further details

ISBN 0-8053-6702-0/0-8053-6703-9, pbk

including an account of which subgroups T' lead to one-to-one maps, we refer the reader to the original literature (H. Furstenberg, Annals of Mathematics vol. 77 (1963), 335–386 and C. C. Moore, American Journal of Mathematics vol. 86 (1964), 201–218 and 258–278).

For each triple G, K, T' and each continuous complex-valued function ϕ on G/T', we may form the function $\tilde{\phi}(\mu)=\int\phi(x)\,d\mu(x)$ for $\mu\in M$ and thus obtain a linear map $\phi\to\tilde{\phi}$ from functions on G/T' to functions on G/K that generalizes the Poisson integral. The extent to which it shares the properties of the Poisson integral is a complicated question that we shall not discuss here. Various special cases have been examined in recent years in works of D. Lowdenslager and R. Hermann, and the general case is studied in the above cited work of Furstenberg.

Our next goal is to relate the decomposition of automorphic forms into sums of Eisenstein series and cusp forms to certain general facts about intertwining operators for induced representations. Preparatory to this we introduce and discuss a general method for constructing such intertwining operators. Let H_1 and H_2 be closed subgroups of the separable locally compact group G and let χ_1 and χ_2 be one-dimensional characters of H_1 and H_2, respectively. We wish to discuss the possible intertwining operators for the induced representations U^{χ_1} and U^{χ_2}. (The restriction to one-dimensional representations is made for simplicity only. More general cases are needed and can be dealt with along analogous lines.) In the very special case in which G is a finite group, it is easy to write down the most general possible intertwining operator. It is of the form $f\to S_A(f)$, where

$$S_A(f)(x)=\sum_{y\in G/H_1} A(xy^{-1})f(y)$$

and A is any complex-valued function on G that satisfies the identity

$$A(\xi x\eta)=\chi_2(\xi)A(x)\chi_1(\eta)\quad\text{for all } \xi\in H_2,\ \eta\in H_1 \tag{*}$$

That this is so is not difficult to establish directly. It is also an easy corollary of the theorem in Section 14 about tensor products of induced representations.

To find the most general solution of (*) we observe first that every solution is uniquely determined throughout any $H_2:H_1$ double coset by its value at any point therein. Moreover, given any solution and any x in G, the solution remains a solution if we reduce it to zero outside of H_2xH_1 and leave it unchanged in H_2xH_1. Hence there is a basis for the vector space of all solutions consisting entirely of solutions that vanish outside of some fixed double coset. Given $z\in G$, there will exist a nontrivial solution of (*) vanishing outside of H_2zH_1 if and only if $\chi_2(\xi)=\chi_1(z^{-1}\xi z)$ whenever

ISBN 0-8053-6702-0/0-8053-6703-9, pbk

$\xi \in H_2 \cap zH_1z^{-1}$ and then it will be a constant multiple of the function $\xi z\eta \to \chi_1(\xi)\chi_2(\eta)$. The formula

$$S_A(f)(x) = \sum_{y \in G/H_1} A(xy^{-1})f(y)$$

reduces in this case to

$$S_A(f)(x) = \sum_{\xi \in H_2/H_2 \cap zH_1z^{-1}} f(z^{-1}\xi x)\overline{\chi_2(\xi)}$$

In other words, for each $z \in G$ such that $\chi_2(\xi) = \chi_1(z^{-1}\xi z)$ for all $\xi \in H_2 \cap zH_1z^{-1}$, we obtain an intertwining operator S_z for U^{χ_1} and U^{χ_2} by setting

$$S_z(f)(x) = \sum_{\xi \in H_2/H_2 \cap zH_1z^{-1}} f(z^{-1}\xi x)\overline{\chi_2(\xi)}$$

Changing z to z_1 in the same $H_2:H_1$ double coset changes S_z only by a multiplicative constant and choosing one z from each admissible $H_2:H_1$ double coset gives us a basis for the space of all intertwining operators.

Now while no such complete general theorem is available when G is infinite we can at least define generalizations of the S_z and hope that when H_1 and H_2 are not too wild these operators will in some sense generate the space of all possible intertwining operators. Indeed suppose that G, H_1, and H_2 are all unimodular and that $\chi_2(\xi) \equiv \chi_1(z^{-1}\xi z)$ for all $\xi \in H_2 \cap zH_1z^{-1}$. Then $\dfrac{H_2}{H_2 \cap zH_1z^{-1}}$ admits an invariant measure ν and we may define

$$S_z(f)(x) = \int_{H_2/H_2 \cap zH_1z^{-1}} f(z^{-1}\xi x)\overline{\chi_2(\xi)}\, d\nu(\xi)$$

for suitably restricted f in $\mathcal{H}(U^{\chi_1})$ and obtain an intertwining operator for U^{χ_1} and U^{χ_2} whenever S_z is defined and bounded for all f in a dense subspace of $\mathcal{H}(U^{\chi_1})$. It is easy to see that the existence of S_z depends only upon the double coset to which z belongs and that changing from one z to another in the same double coset just multiplies S_z by a constant.

In order to deal with cases in which H_1 and H_2 need not be unimodular it is convenient to introduce a slight modification in the definition of induced representation. Let H be a closed non unimodular subgroup of the separable locally compact group G and let δ be the homomorphism of H into the positive real numbers such that $\mu(hE) = \delta(h)\mu(E)$ for all h and E

ISBN 0-8053-6702-0/0-8053-6703-9, pbk

where μ is any right invariant Haar measure in H. Let ν be a quasi invariant measure in G/H and let P_x be the Radon Nikodym derivative of the x translate of ν with respect to ν. We can look at P_x as a function on G which is constant on the right H cosets and then it may be written in the form

$$y \to \frac{\rho(yx)}{\rho(y)}$$

where ρ is a real valued Borel function on G which satisfies the identity $\rho(hy)=\delta(h)\rho(y)$ for all h in H and all y in G. Let χ be a character of H and for each $f\in\mathcal{H}(U^{\chi})$ as defined in Section 9 let $\tilde{f}(y)=f(y)\sqrt{\rho(y)}$. Then if $g=U_x^{\chi}(f)$ so that $g(y)=f(yx)\sqrt{\rho(yx)/\rho(y)}$ then $\tilde{g}(y)=f(yx)\sqrt{\rho(yx)}=\tilde{f}(yx)$. In other words, if we replace $\mathcal{H}(U^{\chi})$ by the set of all functions of the form $\tilde{f}$ for $f\in\mathcal{H}(U^{\chi})$, then the expression for U^{χ} becomes

$$U_x^{\chi}(\tilde{f})(y)=\tilde{f}(yx)$$

Now the set of all $\tilde{f}$ for $f\in\mathcal{H}(U^{\chi})$ may be described as the set of all functions ψ such that $\psi/\sqrt{\rho}$ is in $\mathcal{H}(U^{\chi})$. Thus we arrive at the following equivalent definition of U^{χ}. Let $\mathcal{H}'(U^{\chi})$ consist of all complex-valued Borel functions ψ on G such that

(1) $\psi(hy)=\chi(h)\delta(h)^{-1/2}\psi(y)$

(2) $\int_{G/H}|(\psi(y)\rho(y))|^2\,d\nu(y)<\infty$

Then $U_x^{\chi}(\psi)(y)=\psi(yx)$.

Returning now to the definition of S_z, let us choose quasi-invariant measures ν_1 and ν_2 in G/H_1 and G/H_2, respectively, and let ρ_1 and ρ_2 be the functions related to ν_1 and ν_2 as ρ was related to ν above; δ_1 and δ_2 playing the role of δ. Now consider the coset space $H_2/H_2\cap zH_1z^{-1}$ and let $\delta_1^z(zhz^{-1})=\delta_1(z)$ for all $h\in H_1$. For all ξ in $H_2\cap zH_1z^{-1}$ let $\delta_3^z(\xi)=\delta_1^z(\xi)/\delta_2(\xi)$. Let ρ_3^z be a Borel function on H_2 that satisfies the identity $\rho_3^z(\xi h)=\delta_3^z(\xi)\rho_3^z(h)$ for all $h\in H_2$ and all $\xi\in H_2\cap zH_1z^{-1}$. It follows from the general theory of quasi-invariant measures that there exists a unique quasi-invariant measure ω in $H_2/H_2\cap zH_1z^{-1}$ such that $\rho_3^z(h_1h_2)\rho_3^z(h_1)$ defines the Radon Nikodym derivatives of ω with respect to its translates as above. Finally, given $f\in\mathcal{H}^{\pi}(U^{\chi_1})$, we may define $S_z(f)$ as follows:

$$S_z(f)(x)=\int_{\xi\in H_2/H_2\cap zH_1z^{-1}} f(z^{-1}\xi x)\,\overline{\chi_2(\xi)}\,\delta_2(\xi)^{1/2}\rho_3^z(\xi)^{1/2}\,d\omega(\xi)$$

ISBN 0-8053-6702-0/0-8053-6703-9, pbk

and verify that it has the properties announced in the unimodular case. Note that the definition makes sense even for representations induced by nonunitary characters.

As a first example let $G = \mathrm{SL}(2,R)$ and let H_1 and H_2 be the T and K_i introduced above. There is just one double coset and hence at most one intertwining operator. We may take $z=e$ and then $H_2 \cap zH_1z^{-1} = T \cap K_i$ so that we may choose ω invariant and $\delta_2 = 1$. Thus

$$S_e(f)(x) = \int_{K_i} f(\xi x)\overline{\chi_2(\xi)}\, d\omega(\xi)$$

Now let us choose $\chi_1 = \delta_1^{1/2}$. Then $\mathcal{H}'(U^{\chi_1})$ will consist of functions that are constant on the T cosets and may be regarded as functions on G/T. If we take χ_2 to be the identity, then S_e maps functions on G/T into functions on G/K and turns out to be the Poisson integral map discussed at the beginning of this section. In other words, that map may be looked upon as an example of an intertwining operator of the form S_z. Actually, we have seen in our study of the representation theory of $\mathrm{SL}(2,R)$ that the representations of the form U^{χ_1} with χ_1 unitary are almost all irreducible and that none of them is contained discretely in the regular representation. Since U^{χ_2} is a subrepresentation of the regular representation, there are no nontirvial intertwining operators for U^{χ_1} and U^{χ_2} except when χ_1 is one of the nonunitary characters for which U^{χ_1} exists as a unitary representation. When χ_1 is one of these, we can discover the inner product in U^{χ_1} by looking at the range of S_e in the Hilbert space of U^{χ_2}.

The abstract version of the theory of Eisenstein series and cusp forms to which we alluded in the beginning of our discussion of the operators S_z results from considering the S_z for U^{χ_1} and U^{χ_2}, where χ_1 and χ_2 are the identity characters, H_1 is the subgroup N of $G = \mathrm{SL}(2,R)$, and H_2 is a discrete subgroup Γ. Let us denote χ_1 by I_N, χ_2 by I_Γ. We wish to study the intertwining operators S_z for U^{I_N} and U^{I_Γ}. Now U^{I_N} has a simply, easily discovered structure. If we induce only up to T, we get the regular representation of T/N lifted to T. This follows from the normality of N in T. Hence by the inducing in stages theorem $U^{I_N} = \int_{\hat{D}} U^{\chi'}$, where $D = T/N \simeq$ all $\begin{pmatrix} a & 0 \\ 0 & a^{-1} \end{pmatrix}$, χ' is the character χ of D lifted up to T, and χ varies over $\hat{D}$. In other words, U^{I_N} is the direct integral of all members of the principal series—each one occurring twice—with respect to the same measure class that occurs in the decomposition of the regular representation. Given an arbitrary unitary representation V of G, it may be split uniquely as a direct sum of two subrepresentations $V \simeq V^c \oplus V^d$, where V^d is disjoint from the continuous part of the regular representation and V^c is

ISBN 0-8053-6702-0/0-8053-6703-9, pbk

quasi-equivalent to a subrepresentation of the continuous part of the regular representation. From what we have said about U^{I_N} it is clear that $\mathcal{H}(V^c)$ must coincide with the closed linear span of the ranges of all intertwining operators for U^{I_N} and V and that $\mathcal{H}(V^d)$ must coincide with the common null space of all intertwining operators for V and U^{I_N}.

Now let $V=U^{I_N}$. If the intertwining operators S_z span the space of all intertwining operators in a suitable sense, then $\mathcal{H}(U^{I_\Gamma,c})$ must coincide with the closed linear span of the ranges of all the S_z and $\mathcal{H}(U^{I_\Gamma,d})$ must coincide with the common null space of their adjoints. Here all groups are unimodular and we may use the simple formula for S_z that here reduces even further to

$$S_z(f)(x)=\sum_{\gamma\in\Gamma/\Gamma\cap zNz^{-1}} f(z^{-1}\gamma x)$$

The adjoint of this is easily computed to be

$$S_z^*(g)(x)=\int_{N/z^{-1}\Gamma z\cap N} f(znx)\,dv(n)=S'_{z^{-1}}$$

Notice that there is a sharp difference depending upon whether $N\cap z^{-1}\Gamma z$ is trivial or not. N is isomorphic to the additive group of the real line and if $N\cap z^{-1}\Gamma z$ is nontrivial, the integration involved in the definition of S_z^* is over a compact group. Otherwise it is over the whole real line and S_z^* will certainly not be a bounded operator defined throughout $\mathcal{H}(U^{I_\Gamma})$. Accordingly, we restrict attention to those S_z's for which $N\cap z^{-1}\Gamma z$ is nontrivial. Of course, S_z depends to within a multiplicative constant only upon the $\Gamma:N$ coset to which z belongs. Moreover, the range of S_z depends only upon the $\Gamma:T$ double coset to which z belongs. This is because T normalizes N. Quite generally if $tH_1t^{-1}=H_1$ and

$$S_zf(x)=\int_{\xi\in H_2/H_2\cap zH_1z^{-1}} f(z^{-1}\xi x)\,dv(\xi)$$

then

$$\begin{aligned} S_{zt}f(x)&=\int_{\xi\in H_2/H_2\cap zH_1z^{-1}} f(t^{-1}z^{-1}\xi x)\,dv(\xi)\\ &=S_zf_t(x)\end{aligned}$$

where $f_t(x)=f(t^{-1}x)$. In other words, if we pick just one z from each $\Gamma:T$ double coset, we will obtain a set of intertwining operators whose ranges

ISBN 0-8053-6702-0/0-8053-6703-9, pbk

may be hoped to have $\mathcal{H}(U^{I_{\Gamma,c}})$ as their closed linear space. Moreover, we need only consider those $\Gamma : T$ double cosets $\Gamma z T$ for which $N \cap z^{-1}\Gamma z$ is nontrivial. But these correspond one-to-one to the cusps as defined in the last section. Thus we have one family of essentially equivalent intertwining operators S_z for each "cusp" on G/T relative to Γ.

For a wide class of discrete subgroups Γ of $G = \mathrm{SL}(2,R)$, one can show that the S_z for the cusps $Tz\Gamma$ have ranges that actually do span the continuous part of U^{I_Γ} and in addition that the orthogonal complement of the span of the ranges of the S_z is a discrete direct sum of irreducibles. This decomposition of U^{I_Γ} into continuous and discrete pieces is closely related to the decomposition of automorphic forms of a given weight into Eisenstein series and cusp forms. Each cusp contributes a subrepresentation of the continuous part and the intertwining operator associated with it may be explicitly constructed using a corresponding "Eisenstein series."

On the other hand, for each positive integer k the dimension of the space of all cusp forms for Γ of weight k is equal to the multiplicity with which U^{I_Γ} contains the member of the discrete series of irreducible unitary representations of G denoted by D_{2k}^{+} in the original notation of Bargmann (cf. Section 14). This is the unique irreducible unitary representation (to within equivalence) whose restriction to the subgroup K of all

$$\begin{pmatrix} \cos\theta & \sin\theta \\ -\sin\theta & \cos\theta \end{pmatrix} = x_\theta$$

is the direct sum of all characters of the form $x_\theta \to e^{i(2k+2j)\theta}$, where $j = 0, 1, 2, \ldots$. If Bargmann's description of D_{2k}^{+} on a Hilbert space of analytic functions in the unit circle is transformed to the upper half-plane H^{+}, it may be formulated as follows. $\mathcal{H}(D_{2k}^{+})$ is the space of all analytic functions on H^{+} that are square summable with respect to the measure $y^{2k-2}\,dx\,dy$. Moreover,

$$\left((D_{2k}^{+})\begin{pmatrix} a & b \\ c & d \end{pmatrix} f\right)(z) = f\left(\frac{az+b}{cz+d}\right)(cz+d)^{-2k}$$

That there might be a connection between the multiplicity of D_{2k}^{+} in U^{I_Γ} and the dimension of the space of cusp forms for Γ of weight k suggests itself at once if one confronts the above realization of D_{2k}^{+} with the identity

$$f\left(\frac{az+b}{cz+d}\right) = (cz+d)^{2k} f(z) \quad \text{all} \quad \begin{pmatrix} a & b \\ c & d \end{pmatrix} \in \Gamma \qquad (*)$$

used in defining an automorphic form of weight k for Γ. Indeed, the subspace of $\mathcal{H}(D_{2k}^{+})$ on which the restriction of D_{2k}^{+} to Γ reduces to the

ISBN 0-8053-6702-0/0-8053-6703-9, pbk

identity is just exactly the space of square summable analytic functions that satisfy (*). *If* the Frobenius reciprocity theorem in its naive form held for the pair Γ, G and *if* square summability were the same as being a cusp form, the statement above equating the multiplicity of D_{2k} in U^{I_Γ} to the dimension of the space of cusp forms would follow at once. While neither "if" is valid, their failures cancel and one can prove the statement in question. Indeed one can use cusp forms to define intertwining operators between U^{I_Γ} and D_{2k}^+ and so obtain a canonical isomorphism of $R(U^{I_\Gamma}, D_{2k}^+)$ with the space of cusp forms. In other words, cusp forms may be identified with intertwining operators.

This relationship between automorphic forms and discrete components of U^{I_Γ} seems to have been noticed for the first time by Gelfand and Fomin who published a paper in 1952 applying it to the study of the ergodicity of flows on compact surfaces of constant negative curvature. The surfaces in question are orbit spaces H^+/Γ, where Γ is a discrete subgroup of $G = \mathrm{SL}(2, R)$ such that G/Γ is compact. In this case there are no cusps and U^{I_Γ} decomposes discretely.

The interpretation of cusp forms as intertwining operators of the D_{2k}^+ with U^{I_Γ} is useful in understanding the abstract significance of the so-called "Hecke operators" used in obtaining Euler product decompositions for the Dirichlet series associated with certain automorphic forms. Since the product of a self-intertwining operator for U^{I_Γ} with an intertwining operator for D_{2k}^+ and U^{I_Γ} is again an intertwining operator for D_{2k}^+ and U^{I_Γ}, we see that the space of all intertwining operators for D_{2k}^+ and U^{I_Γ} is an $R(U^{I_\Gamma}, U^{I_\Gamma})$ module. Hence, via the above identification, we see that it always makes sense to "multiply" a cusp form by a member of the ring $R(U^{I_\Gamma}, U^{I_\Gamma})$ of intertwining operators for U^{I_Γ} and U^{I_Γ} and that the space of cusp forms of a given weight is thus converted into a module or representation space for this ring. Now we know how to associate a (formal) intertwining operator with each $\Gamma : \Gamma$ double coset, the construction having been explained in detail earlier in this section. Suppose we choose a double coset $\Gamma z \Gamma$ in which there are only a finite number of right and left cosets. Then $\Gamma / \Gamma \cap z\Gamma z^{-1}$ will be finite and the integral that occurs in the general case reduces to a finite sum and the formal intertwining operator becomes an actual one. Hence each $\Gamma z \Gamma$ double coset with the indicated finiteness property defines a linear operator in the space of cusp forms and the vector space generated by these operators is a ring that may be called the ring of *Hecke operators*. The original operators introduced by Hecke in a special case all belong to this ring and generate a commutative subring of it. In that case, one can decompose the module of cusp forms into one-dimensional irreducible submodules and so obtain a natural basis for the space of all cusp forms of a given weight. It is the members of this

ISBN 0-8053-6702-0/0-8053-6703-9, pbk

basis whose corresponding Dirichlet series (cf. Section 27) have Euler products. Since each Hecke operator is associated naturally with an operator A in the space of U^{I_Γ}, one can look at $\mathrm{Tr}(AU_f^{I_\Gamma})$ instead of at $\mathrm{Tr}(U_f^{I_\Gamma})$ in deriving the Selberg trace formula. Selberg did this and obtained thereby an interesting generalization of the trace formula.

Now U^{I_Γ} can contain discrete components which are not of the form D_{2k}^+ and it is natural to ask whether there is some generalization of the concept of automorphic form that is related to these as automorphic forms of positive integral weight are to the D_{2k}^+. This question is easy to answer for the other discrete series members. Those of the form D_j^+ for j an odd positive integer correspond to automorphic forms of half integer weight and those of the form D_j^- for j any positive integer are just the complex conjugates of the D_j^+ and occur in U^{I_Γ} with the same multiplicity since $\overline{U^{I_\Gamma}} \simeq U^{I_\Gamma}$. Much more interesting is the case in which some member of the principal or complementary series occurs discretely in U^{I_Γ}. The necessary generalization of the concept of automorphic form was introduced by H. Maass in 1949 three years before the appearance of the paper of Gelfand and Fomin and with a motivation having nothing to do with the unitary representation theory of $\mathrm{SL}(2,R)$.

Maass was motivated rather by the fact that Hecke's theory connecting modular forms with Dirichlet series satisfying a functional equation applies to too limited a class of functional equations. In particular, Hecke's theory applies to the zeta function of a quadratic extension of the rational field only when the extension contains nonreal numbers. Maass found that he could construct a theory analogous to that of Hecke (and applying in a similar fashion to the zeta functions of real quadratic extensions) by replacing the classical automorphic forms in the upper half-plane by a new class of functions that he called nonanalytic automorphic forms. In defining these he replaced the condition of analyticity by that of being an eigenfunction of the "Laplacian"

$$y^2\left(\frac{\partial^2}{\partial x^2}+\frac{\partial^2}{\partial y^2}\right)$$

and the condition that

$$f\left(\frac{az+b}{cz+d}\right)=(cz+d)^{2k}f(z)$$

for all $\begin{pmatrix} a & b \\ c & d \end{pmatrix}\in\Gamma$ by the condition that

$$f\left(\frac{az+b}{cz+d}\right)=f(z)$$

ISBN 0-8053-6702-0/0-8053-6703-9, pbk

for all $\begin{pmatrix} a & b \\ c & d \end{pmatrix} \in \Gamma$. The eigenvalue λ plays the role of the weight of a classical automorphic form in that one considers the (usually finite-dimensional) vector space $V_{\Gamma,\lambda}$ of all nonanalytic automorphic forms for a given Γ and λ.

When G/Γ is compact, the space $V_{\Gamma,\lambda}$ is always finite-dimensional and is in fact zero-dimensional except when λ belongs to a certain countable subset $\lambda_1, \lambda_2, \ldots$ of real numbers. These real numbers, of course, are just the members of the point spectrum of the self-adjoint operator in $\mathcal{L}^2(H^+/\Gamma)$ defined by $y^2(\partial^2/\partial x^2 + \partial^2/\partial y^2)$. It is these eigenvalues and their multiplicties that describe the components of U^{I_Γ} which belong to the *principal* series of irreducible unitary representations of G. To see how this comes about let us consider the induced representation U^{I_K} of G. It is easily seen to be multiplicity free and to be a direct integral of the principal series members associated with those characters of the diagonal subgroup that are 1 on the center of G. Since G/K may be identified with H^+, the operator $y^2(\partial^2/\partial x^2 + \partial^2/\partial y^2)$ defines an operator D in $\mathcal{H}(U^{I_K}) = \mathcal{L}^2(G/K)$, which is self-adjoint and commutes with all $U_x^{I_K}$. The decomposition of U^{I_K} as a direct integral of irreducibles thus decomposes D as a direct integral of real constants and maps the corresponding members of the principal series onto points of the spectrum of D. Let s be any real number different from zero and let W^s denote the member of the principal series induced by the character $\begin{pmatrix} a & \mu \\ 0 & 1/a \end{pmatrix} \to |a|^{is}$. A computation shows that the corresponding eigenvalue of D is $-(s^2 + \frac{1}{4})$. Finally, it is not difficult to show that the correspondence $W^s \to -(s^2 + \frac{1}{4})$ persists when one compares U^{I_Γ} to the spectrum of $y^2(\partial^2/\partial x^2 + \partial^2/\partial y^2)$ in H^+/Γ. Thus (when G/Γ is compact), the dimension of $V_{\Gamma, -(s^2+1/4)}$ is equal to the multiplicity with which U contains W^s. It is rather remarkable reflection on the coherence of the whole subject that the generalized automorphic forms introduced by Maass in 1949 for number theoretical purposes were precisely what were needed to complete the 1952 observations of Gelfand and Fomin.

That Maass's nonanalytic automorphic forms have the indicated relationship to the structure of U^{I_Γ} was first pointed out by Gelfand and Pyatetzki Shapiro in a paper published in Uspehi Mat. Nauk. vol. 14 (1959), 172–194. In the same paper they indicate that the complementary series may be taken care of in a similar manner. Notice that $-(s^2 + \frac{1}{4})$ is real and positive not only when s is real but also when s is pure imaginary and $0 < |s| \leqslant \frac{1}{2}$. The supplementary series of irreducible unitary representations of G may be obtained by analytic continuation of W^s for values of s of the form ti, where $i^2 = -1$ and $0 < t \leqslant \frac{1}{2}$, and for $-\frac{1}{4} < \lambda < 0$ the dimension of $V_{\Gamma,\lambda}$ is equal to the multiplicity in U^{I_Γ} of the complementary series member defined by $i\sqrt{\lambda - \frac{1}{4}}$.

ISBN 0-8053-6702-0/0-8053-6703-9, pbk

In their 1959 paper Gelfand and Pyatetzki–Shapiro refer not only to Maass' 1949 paper, but also to the now celebrated paper that Selberg published in 1956 and which contains his famous "trace formula." While Selberg's paper preceded and stimulated that of Gelfand and Pyatetzki–Shapiro, its content is much easier to appreciate after one has understood the connection between the structure of U^{I_Γ} and Maass's nonanalytic automorphic forms as explained in the later paper. This is because Selberg states and discusses his trace formula without any reference to infinite-dimensional unitary representations—induced or otherwise. He is able to do this by using the spectrum of $y^2(\partial^2/\partial x^2+\partial^2/\partial y^2)$ in H^+/Γ as a substitute for the list of principle and supplementary series members contained discretely in U^{I_Γ}. As explained in Sections 26 and 28, the Selberg trace formula may be regarded as a special case of a noncommutative generalization of the classical Poisson summation formula. The real line in the classical formula is replaced by a connected semi-simple Lie group G and the discrete subgroup of all integers by a discrete subgroup Γ of G such that G/Γ is compact (or more generally has a finite invariant measure). The left-hand side $\Sigma_{n=-\infty}^{\infty} f(n)$ of the Poisson formula becomes a sum over those conjugate classes in G that contain elements of Γ, each summand being a multiple (independent of f but depending on Γ) of the integral of f over the conjugate class in question. In the case in which G/Γ is compact so that U^{I_Γ} decomposes discretely the right-hand side, $\Sigma_{n=-\infty}^{\infty}\hat{f}(n)$ is replaced by a sum over the irreducible constituents of U^{I_Γ} the contribution of the irreducible V^n being $\operatorname{Tr} V_f^n$, where

$$V_f^n=\int_c f(x)V^n\,dx$$

To see the analogy between

$$\hat{f}(n)=\int_{-\infty}^{\infty} e^{2\pi inx}f(x)\,dx$$

in the classical case and $\operatorname{Tr} V_f^n$ one has only to remember that whenever V^n has a character χ^n which is a function then

$$\operatorname{Tr}(V_f^n)=\int_c f(x)\chi^n(x)\,dx$$

In the special case in which $G=\mathrm{SL}(2,R)$, one can obtain the trace formula just described in the form in which Selberg gave it by using the connection between irreducible unitary representations of G and the spectrum of $y^2(\partial^2/\partial x^2+\partial^2/\partial y^2)$ to convert $\sum_n \operatorname{Tr}(V_f^n)$ into a sum over the eigenvalues

ISBN 0-8053-6703-9, pbk ISBN 0-8053-6702-0/0-8053-6703-9, pbk

$y^2(\partial^2/\partial x^2+\partial^2/\partial y^2)$ regarded as an operator in $\mathfrak{L}^2(H^+/\Gamma)$ of a certain integral transform g of f. By restricting himself to functions f on G that are constant on the right K cosets, Selberg avoided the contribution of the discrete series members.

When G/Γ is not compact, the right-hand side of the trace formula has to be supplemented by one or more integrals corresponding to the continuous part of U^{I_Γ}. There is one such for every cusp and these integrals relate in an interesting way both to the part of Maass's 1949 paper in which he discusses the analog of Eisenstein series for his nonanalytic automorphic forms and to the double coset intertwining operators for U^{I_Γ} and U^{I_N} discussed above. In particular, we put the words "Eisenstein series" in quotation marks in stating above that these series may be used to construct the intertwining operators because it is the Maass–Eisenstein series rather than the classical ones that must be used.

To describe Maass' analog of Eisenstein series, consider the function on the upper half-plane that takes z into $y^s=|y|^s$, where $z=x+iy$ and s is a fixed complex number. Evidently it is an eigenfunction of the operator $y^2(\partial^2/\partial x^2+\partial^2/\partial y^2)$ and the corresponding eigenvalue is $s(s-1)$. Hence its transform by $\begin{pmatrix} a & b \\ c & d \end{pmatrix}$ in $\mathrm{SL}(2,R)$ is an eigenfunction with the same eigenvalue. One easily computes that this transform is $z\rightarrow y^s/|cz+d|^{2s}$. Now, for any discrete subgroup Γ of $\mathrm{SL}(2,R)$, the sum of the series

$$\sum_{\begin{pmatrix} a & b \\ c & d \end{pmatrix}\in\Gamma} \frac{y^s}{|cz+d|^{2s}}$$

is formally Γ invariant and by the above each term is an eigenfunction of $y^2(\partial^2/\partial x^2+\partial^2/\partial y^2)$ with eigenvalue $s(s-1)$. Thus if the series converged in a suitable manner, its sum would be a Maass nonanalytic automorphic form for Γ with eigenvalue $s(s-1)$. By omitting certain terms just as with the classical Eisenstein series (more exactly, their generalizations by Hecke to discrete subgroups other than $\mathrm{SL}(2,Z)$), Maass was able (for the cases that concerned him) to assign to each cusp a series of terms of the form $y^s/|cz+d|^{2s}$ which converges if $\mathrm{Re}(s)>2$. Moreover, it converges in such a manner that the sum is a Γ invariant eigenfunction of $y^2(\partial^2/\partial x^2+\partial^2/\partial y^2)$ with eigenvalue $s(s-1)$. This convergent subsequence of

$$\sum_{\begin{pmatrix} a & b \\ c & d \end{pmatrix}\in\Gamma} \frac{y^s}{|cz+d|^s}$$

is Maass' analog of the Eisenstein series for the cusp in question.

ISBN 0-8053-6702-0/0-8053-6703-9, pbk

There is a difficulty in that $s(s-1)$ is real and less than or equal to $-\frac{1}{4}$ only when $\text{Re}(s)=\frac{1}{2}$. Thus for the most interesting values of s the Eisenstein series does *not* converge. Maass overcame this difficulty by an analytic continuation in s. For each fixed value of z the Eisenstein series described above is a Dirichlet series in s. Maass showed that it is like the more familiar Dirichlet series of number theory in admitting a continuation to a function that is meromorphic in the entire s plan and satisfies a functional equation. These analytic continuations as functions of z are Γ invariant eigenfunctions of $y^2(\partial^2/\partial x^2+\partial^2/\partial y^2)$. It is the continuation of the convergent Eisenstein series to values of s with $\text{Re}(s)=\frac{1}{2}$ which Maass needed for his extension of Hecke's theory and which Selberg uses to compute the contribution to his trace formula of the continuous spectrum of $y^2(\partial^2/\partial x^2+\partial^2/\partial y^2)$ in H^+/Γ (equivalently of the continuous part of U^{I_Γ}).

It is natural to attempt to study the structure of U^{I_Γ} in a similar spirit when $\text{SL}(2,R)$ is replaced by a more general semi-simple Lie group G, K_i is replaced by a maximal compact subgroup K, and Γ is a suitably restricted discrete subgroup of G. If we are to relate such a study to a theory of automorphic forms, we must replace functions on the upper half-plane by functions on the "symmetric space" G/K and the operator

$$y^2\left(\frac{\partial^2}{\partial x^2}+\frac{\partial^2}{\partial y^2}\right)$$

by suitable "generalized Laplacians." The case $G=\text{SL}(2,R)$ is exceptional in that the ring of invariant differential operators on G/K_i is generated by the single operator

$$y^2\left(\frac{\partial^2}{\partial x^2}+\frac{\partial^2}{\partial y^2}\right)$$

In general, the structure of this ring is more complicated and its study is a nontrivial problem. An elegant solution has been given by Helgason (Acta Mathematica vol. 102 (1960), 239–299). These automorphic forms must of course include nonanalytic ones. Indeed, it is only for certain symmetric spaces G/K that it makes sense to speak of analytic functions at all. There are formidable difficulties in the way of carrying out the program described at the beginning of this paragraph with any degree of completeness, but a good start has been made in the cited 1956 paper of Selberg as well as in subsequent work of Selberg, Gelfand, Pyatetzki–Shapiro,

ISBN 0-8053-6702-0/0-8053-6703-9, pbk

Harish–Chandra, Langlands, and others. (See especially the 1962 Stockholm Congress address of Selberg and Gelfand for further details.)

A major difficulty is that the structure of U^{I_Γ} can be much more complicated. In addition to a discrete part and a part equivalent to a subrepresentation of the regular representation, one has to expect continuous components of "intermediate dimension." When $G = \mathrm{SL}(2,R)$, the dual $\hat{G}$ is essentially a one-dimensional manifold plus a countable discrete set and there are no "intermediate dimensions." Correspondingly, it is less straightforward to compactify the space $G/K\backslash\Gamma$. In addition to adding points, one must also add higher dimensional "compactifying components" and the whole business becomes much more involved and is by no means clearly understood. Selberg's Stockholm address gives a detailed account of some of the attempts he has made to study the structure of these "compactifying components" in certain manageable special cases.

Selberg has approached these questions from a direct geometric point of view. Gelfand, on the other hand, has shown that one can be led in a systematic way to the appropriate generalization of the notion of cusp by using general concepts from the theory of semi-simple Lie groups. Let G be a semi-simple Lie group with a finite center and let K be a maximal compact subgroup. Let Γ be a discrete subgroup of G satisfying regularity hypotheses too complicated to state here. Let T' be any closed subgroup of G defining a compactification of G/K as described at the beginning of this section. Let N' be its unique maximal normal nilpotent subgroup. Then N' is what Gelfand calls a *horospherical subgroup*, although he states the definition rather differently. A *horosphere* is an orbit in G/Γ of a horospherical subgroup N'. Notice that the horosphere containing $\Gamma z \in G/\Gamma$ is the homogenous space of N' defined by the subgroup of all $n' \in N'$ with $\Gamma z n' = \Gamma z$, that is, $z n' z^{-1} \in \Gamma$, that is, by the subgroup $N' \cap z^{-1}\Gamma z$. In other words, the compact horospheres correspond one-to-one to the $N' z \Gamma$ double cosets such that $N'/N' \cap z^{-1}\Gamma z$ is compact. In view of the identification of each compact horosphere with a compact homogeneous space we can speak of the integral over a compact horosphere of a function on G/Γ and consider the subspace $\mathcal{H}^0$ of $\mathcal{H}(U^{I_\Gamma})$ consisting of those members whose integrals over *every* compact horosphere are equal to zero. The word "every" here implies that we consider all compact orbits of all horospherical subgroups. Gelfand and Pyatetzki–Shapiro have shown that the subrepresentation defined by $\mathcal{H}^0$ is a discrete direct sum of irreducibles.

To see the significance of this result, notice that the double coset $T' z^{-1}\Gamma$ that contains $N' z^{-1}\Gamma = (\Gamma z N')^{-1}$ is a point in the Γ orbit of the compactification of G/K defined by T' and that to say that $N'/(N' \cap z^{-1}\Gamma z)$ is compact reduces in the case in which $G = \mathrm{SL}(2,R)$ to saying that this Γ

ISBN 0-8053-6702-0/0-8053-6703-9, pbk

orbit belongs to the compactification of $G/K\setminus\Gamma$ and is therefore a cusp. H^0 therefore is the common null space of the intertwining operators S_z^* associated with the various cusps where the definition of cusp is now obvious. Consequently, $H^{0\perp}$ is the closed linear span of the intertwining operators S_z. In other words, U^{I_Γ} can be decomposed into a discrete direct sum of irreducibles and a discrete direct sum of subrepresentations of representations of the form $U^{I_{N'}}$ with one summand for each generalized cusp.

The trick of studying the structure of an induced representation such as U^{I_Γ} by means of the operators S_z intertwining it with induced representations of "known" structure was used earlier by Gelfand and Graev to study the structure of U^{I_K} in certain cases. They called it the horospherical method and related it to certain transforms occurring in integral geometry. Let G be the group generated by the translations and rotations in Euclidean n space and let K be the subgroup of all rotations about some fixed point. Then G/K may be identified with Euclidean n space and in it we may consider the family S of all $(n-1)$-dimensional hyperplanes. For each continuous function f on G/K having compact support and each $s \in S$, let $\tilde{f}(s)$ be the integral of f over the hyperplane s. Then $\tilde{f}$ is a complex-valued function in S that uniquely determines f. For $n=2$ and 3 the transform $f \to \tilde{f}$ was studied by J. Radon who in particular obtained an "inversion formula" allowing one to compute f from $\tilde{f}$. Accordingly, it has become customary to call it the *Radon transform*. Notice that G acts in a natural way on S and that S is in fact a transitive G-space. Thus we may find a closed subgroup H of G such that S can be identified with G/H and $f \to \tilde{f}$ becomes a mapping from functions on G/K to functions on G/H. Defined for square summable functions the Radon transform is an intertwining operator (perhaps unbounded) for U^{I_K} and U^{I_H} and in fact coincides with S_e the intertwining operator associated with the double coset KH. If one tries to generalize Radon's results from Euclidean space to the symmetric space G/K, it is natural to use horospheres instead of hyperplanes and one is led to the transformation S_z occurring in the Gelfand–Graev analysis of U^{I_K}. Helgason has studied a quite general and abstract Radon transform that turns out to be essentially an intertwining operator S_z for a pair of induced representations $U^{I_{H_1}}$ and $U^{I_{H_2}}$. We note by the way that by changing one of the subgroups to a suitable conjugate we may always suppose without loss of generality that $z=e$.

In those special cases in which G/K has a complex analytic structure one can hope to connect the theory of analytic automorphic forms with the occurrence of square integrable representations of G in the discrete part of U^{I_Γ} as we have already done for $G=\mathrm{SL}(2,R)$. Some work has been done in this direction, but we shall not attempt to describe it.

ISBN 0-8053-6702-0/0-8053-6703-9, pbk

In our discussion of automorphic forms in the upper half-plane we considered only forms of integral weight. Because of this, in our applications to number theory we had to restrict ourselves to those sums of s squares in which s is a multiple of 4. Actually it is possible to develop a theory of automorphic forms of weight r, where r is any real number, and we conclude this section with a few indications as to how one proceeds. Because of the fact that the possible weights are parametrized by the characters of K, it is rather surprising at first sight that it is possible to have anything but integral weights. However, the difficulty is easily met by introducing projective representations. It turns out that $\mathrm{SL}(2,R)$ admits a continuum of dissimilar projective multipliers and that each is associated with a theory of automorphic forms. Let $J(z,\alpha)=(cz+d)^{-2}$, where $\alpha=\begin{pmatrix} a & b \\ c & d \end{pmatrix}$ as in earlier discussions. Then

$$J(z,\alpha_1\alpha_2)=J(z,\alpha_1)J(\alpha_1(z),\alpha_2)$$

for all α_1 and α_2 and it follows that we obtain a (nonunitary) representation of $G=\mathrm{SL}(2,R)$ if we set $(V_\alpha f)(z)=f(\alpha(z))J(z,\alpha)$ for all continuous complex-valued functions f on the upper half-plane. If r is a real number, then $J(z,\alpha)^r=e^{r\log J(z,\alpha)}$, which in general has countably many different values. For each fixed α let us choose a particular branch of the analytic function $z\rightarrow J(z,\alpha)^r$ and henceforth interpret $J(z,\alpha)^r$ as the value given by this branch. Then $J^r(z,\alpha_1\alpha_2)/J^r(z,\alpha_1)$ will be some branch of $z\rightarrow J^r(\alpha_1(z),\alpha_2)$ but not necessarily the one we have chosen. Hence

$$\frac{J^r(z,\alpha_1\alpha_2)}{J^r(z,\alpha_1)}=e^{2\pi ikr}J^r(\alpha_1(z),\alpha_2)$$

where k is some integer depending upon α_1 and α_2. Thus

$$J^r(z,\alpha_1\alpha_2)\equiv J^r(z,\alpha_1)J^r(\alpha_1(z),\alpha_2)\sigma_r(\alpha_1,\alpha_2)$$

where $\sigma_r(\alpha_1,\alpha_2)$ is a complex number of modulus unity and it follows that if we define $V^r_\alpha f(z)=f(\alpha(z))J^r(z,\alpha)$, then $V^r_{\alpha_1\alpha_2}=V^r_{\alpha_1}V^r_{\alpha_2}\sigma_r(\alpha_1,\alpha_2)$ so that V^r is a projective representation of $G=\mathrm{SL}(2,R)$ with multiplier σ_r.

By analogy with what has gone before, we would expect to define an (unrestricted not necessarily analytic) automorphic form of weight r for the discrete group Γ to be a member of the space of V^r invariant under the restriction of V^r to Γ. Such of course cannot exist unless σ_r reduces to the identity on Γ and in general it does not. However, it can be shown that σ_r is always a *trivial* multiplier on the modular group Γ_0 and hence on all of its subgroups. If Γ is one of these subgroups, this means that there exists a

ISBN 0-8053-6702-0/0-8053-6703-9, pbk

complex valued function v_r of modulus one defined on Γ such that

$$\sigma_r(\gamma_1, \gamma_2) = \frac{v_r(\gamma_1\gamma_2)}{v_r(\gamma_1)v_r(\gamma_2)}$$

for all γ_1, γ_2 in Γ. Now let $W^r_\alpha = V^r_\alpha / v_r(\alpha)$. Then

$$W^r_{\alpha_1\alpha_2} = t_r(\alpha_1, \alpha_2) W^r_{\alpha_1} W^r_{\alpha_2}$$

where

$$t_r(\alpha_1, \alpha_2) = \frac{\sigma_r(\alpha_1, \alpha_2)}{\dfrac{v_r(\alpha_1\alpha_2)}{v_r(\alpha_1)v_r(\alpha_2)}}$$

and hence reduces to one on Γ. Since W restricted to Γ is an ordinary representation, we can have invariant elements and these by definition are our automorphic forms of weight r for Γ. Written out, the defining identity involves the function v_r and takes the form

$$f(\gamma(z)) = v_r(\gamma) J^{-r}(z, \gamma) f(z)$$

The function v_r is called a *multiplier system* for Γ and r. One speaks of an automorphic form with a given weight, group Γ, and multiplier system.

The multiplier system v_r is of course a one-dimensional σ_r representation of Γ and we may construct the σ_r representation of G induced by v_r. This will play the role for automorphic forms of nonintegral weight played by U^{I_Γ} for those of integral weight. That the multipliers σ_r are not trivial on G may be seen by restricting them to K. K is a homomorphic image of the additive group of the real line and any character of the real line defines a one-dimensional projective representation of K that will be ordinary if and only if the character vanishes on the kernel of the natural map of the real line on K. The σ_r restricted to K define multipliers in natural one-to-one correspondence with those defined by the characters of the real line. Of course, they are not all inequivalent. There is a constant c such that σ_{r_1} and σ_{r_2} are similar multipliers if and only if $r_1 - r_2$ is an integer multiple of c.

Notes and References

An important advance made shortly before these lectures were given (but not mentioned in the text) consists in the observation that it is useful

ISBN 0-8053-6702-0/0-8053-6703-9, pbk

to introduce the noncommutative adèle groups of Ono and Tamagawa into the study of automorphic forms and into the study of the interplay between their theory and the theory of unitary group representations. In particular, the infinite product structure of the adèle groups helps to "explain" the Euler products found by Hecke in the 1930s. For further details and references to the literature see pages 304 through 308 of the published version of the author's Chicago lecture notes. Among the references we single out for special mention is the book of Gelfand, Graev, and Pjatetskii–Shapiro, *Theory of Representations and Automorphic Functions, Generalized Functions VI*, Saunders, (1969) (original Russian edition, 1966). It contains a detailed exposition of much of the material in the text as well as the omitted discussion of adèle groups.

In 1967 A. Weil published an article ("Uber die Bestimmung Dirichletscher Reihen durch Funktional Gleichungen," *Math Ann.*, vol. 168, 149–156) in which he generalized somewhat the classical Hecke theory connecting modular forms with Dirichlet series and then went on to suggest a way in which one might use this connection to study the arithmetic of elliptic curves much as Hecke had used it to study n-ary quadratic forms. At about the same time, R. P. Langlands gave the J. K. Whittemore lectures at Yale University (published as *Euler Products*, Yale Univ. Press, New Haven (1971)) and announced some far reaching ideas connecting Hecke's theory of Euler products with unitary group representations and generalizing it to arbitrary semi-simple Lie groups. In particular, he suggested the possibility of identifying Artin's new L series with the generalized Hecke Dirichlet series associated with the group $GL(n,R)$— thus applying Hecke's methods for quadratic forms to non-Abelian class field theory. The program suggested by these closely related ideas of Weil and Langlands is simultaneously very promising and apparently very difficult to work out in detail. It has attracted the attention of a large number of mathematicians and generated a large and rapidly growing literature. The reader will find a few more details and some references on pages 313 through 316 of the Chicago book cited above. Two useful expository articles not cited are: the Bourbaki seminar talk of Borel "Formes automorphes et series de Dirichlet," *Sem. Bourbaki*, vol. 466, (1975) (in *Springer Lecture Notes*, vol. 514 (1976)) and "Elliptic Curves and Automorphic Representations" by S. Gelbart, *Advan. Math*, vol. 21 (1976), 235–292.

The address by Gelfand in the Proceedings of the 1962 International Mathematical Congress in Stockholm contains a provocative but rather mysterious remark suggesting an analogy between quantum mechanical scattering theory and the use of intertwining operators defined by Eisenstein series in removing the continuous part of U^{I_Γ}. In partial explanation

ISBN 0-8053-6702-0/0-8053-6703-9, pbk

of this remark, consider the operators S^+ and S^- defined in Section 21 and which intertwine the two unitary representations defining the dynamics of the free and interacting pairs of particles, respectively. The interacting representation is usually a direct sum of two subrepresentations: one being a discrete direct sum of irreducibles and the other being unitarily equivalent to the free particle representation. In many cases S^+ and S^- set up this unitary equivalance isolating the discrete part just as the Eisenstein series do for U^{I_Γ}. Moreover, the analytic properties of the S operator $S^+(S^-)^{-1}$ may be derived from corresponding properties of S^+ and S^- and the zeros and poles of the analytic continuation determine the components of the discrete part of the interacting representation just as the zeros and poles of the analytically continued Eisenstein series determine the discrete components of U^{I_Γ}.

This analogy noted by Gelfand was first exploited in a concrete way by the same Faddeev cited in the notes and references to Section 21 as having begun the rigorous mathematics of mutli-channel scattering. Recall that for suitable discrete subgroups of $SL(2,R)$ the decomposition of U^{I_Γ} is equivalent to the spectral resolution of the Laplace operator in H^+/Γ, and this in turn is equivalent to decomposing the unitary representation of the real line generated by this Laplace operator. In his paper "Expansion in eigenfunctions of the Laplace operator in the fundamental domain of a discrete group in the Lobachevsky plane," *Trudy Mosk Mat. Obsc.*, vol. 17 (1967), 322–356 (Russian), Faddeev applied scattering theory techniques to prove some of the important facts about these decompositions. Five years later in collaboration with B. S. Pavlov he wrote a second paper showing how the Lax–Phillips scattering theory (cf. notes and references to Section 21) could be used to advantage in the same problem and, in particular, made possible a new proof of the analytic continuability of the Eisenstein series. Still later (1973) in collaboration with V. L. Kalinin and A. B. Venkov he used the Lax–Phillips theory to give a proof of the Selberg trace formula for suitable discrete subgroups of $SL(2,R)$. In a recent book (*Scattering Theory for Automorphic Functions*, Annals of Mathematics Studies vol. 87, Princeton Univ. Press, Princeton, N. J., and University of Tokyo Press (1976)), Lax and Phillips have given a self-contained treatment of their scattering theory together with the applications to the theory of automorphic forms found by Faddeev and his collaborators. We refer to the introduction and bibliography of this book for further details and for references to the later Faddeev papers, as well as to other detailed proofs of the Selberg trace formula and the analytic continuability of the Eisenstein series.

More material relevant to section 30 will be found in the Notes and References to sections 14 and 21.

ISBN 0-8053-6702-0/0-8053-6703-9, pbk

31. *The Weil–Siegel formula*

We saw in the last section that many results in the theory of automorphic forms are equivalent to or have close analogs in the theory of unitary group representations. This suggests that the arguments of Hecke and Siegel connecting the theory of automorphic forms with the theory of quadratic forms might be converted into arguments connecting the theory of quadratic forms directly with the theory of group representations. As far as Siegel's results are concerned, this program has been carried out by A. Weil in two papers published in Acta Mathematica in 1964 and 1965 and already mentioned in Section 26. In this section we shall describe some of the notions and results occuring in these papers.

The automorphic forms in $n(n+1)/2$ variables that occur in Siegel's work are, as already remarked, defined in the Siegel generalized upper half-plane. The group of automorphisms of this (as a complex manifold) is the symplectic group $\mathrm{Sp}(n)$, that is, the group of all linear transformations of a $2n$-dimensional real vector space that leaves invariant a nondegenerate alternating bilinear form. (It is easy to verify that $\mathrm{Sp}(1)$ is isomorphic to $\mathrm{SL}(2,R)$.) Now $\mathrm{Sp}(n)$ can be conveniently realized in the following way: Let X be an n-dimensional real vector space and let X^* be its dual. Then $X \oplus X^*$ is a $2n$-dimensional real vector space that admits a "natural" nondegenerate alternating bilinear form, namely, the form B defined by the equation

$$B(x_1 l_1; x_2 l_2) = l_1(x_2) - l_2(x_1)$$

George W. Mackey, Unitary Group Representations in Physics, Probability, and Number Theory

ISBN 0-8053-6702-0/0-8053-6703-9, pbk

Thus $\mathrm{Sp}(n)$ is isomorphic to the group of all nonsingular linear transformations of $X \oplus X^*$ into itself which preserve the "natural" bilinear form B.

The real vector space X is also a separable locally compact commutative group and X^* is canonically isomorphic to its dual. This suggests studying the more general situation in which X is replaced by an arbitrary locally compact commutative group G, X^* is replaced by the dual $\hat{G}$ of G, and the bilinear form B is replaced by the function on $G \times \hat{G} \times G \times \hat{G}$ taking x_1, χ_1, x_2, χ_2 into $\chi_1(x_2)/\chi_2(x_1)$. The group of all automorphisms of $G \times \hat{G}$ that leave this function invariant may then be regarded as a generalization of $\mathrm{Sp}(n)$ which it is natural to call $\mathrm{Sp}(G)$. Of course, $\mathrm{Sp}(G)$ will not have a natural locally compact topology except in special cases. It is useful to work in this more general context because among other things one wants to study symplectic groups over the p-adic numbers and the adèle groups of the symplectic groups over the rationals.

A key role in Weil's argument is played by the observation that a group closely related to $\mathrm{Sp}(G)$ amits a "natural" projective representation. On the group $G \times \hat{G}$ let $\sigma(x_1,\chi_1;x_2,\chi_2) = \overline{\chi_2(x_1)}$. Then σ is a projective multiplier for $G \times \hat{G}$ which we have studied earlier. In particular, we have seen that to within equivalence $G \times \hat{G}$ has a unique irreducible σ representation and that a realization of it is the σ representation of $G \times \hat{G}$ induced by the identity representation of $e \times \hat{G}$. The latter, of course, has $\mathcal{L}^2(G)$ as its space and the representation may be defined by the formula

$$(U_{x_1,\chi_1} f)(x) = \chi_1(x) f(x x_1)$$

Now for each automorphism α of $G \times \hat{G}$ let

$$\sigma^\alpha(x_1,\chi_1;x_2,\chi_2) = \sigma(\alpha(x_1,\chi_1);\alpha(x_2,\chi_2))$$

Then σ^α will also be a projective multiplier for $G \times \hat{G}$. Moreover, α will be in $\mathrm{Sp}(G)$ if and only if it preserves the multiplier

$$x_1\chi_1, x_2\chi_2 \rightarrow \frac{\sigma(x_1,\chi_1;x_2,\chi_2)}{\sigma(x_2,\chi_2;x_1,\chi_1)}$$

that is, if and only if

$$\frac{\sigma^\alpha(x_1,\chi_1;x_2,\chi_2)}{\sigma(x_1,\chi_1;x_2,\chi_2)} = \frac{\sigma^\alpha(x_2,\chi_2;x_1,\chi_1)}{\sigma(x_2,\chi_2;x_1,\chi_1)}$$

In other words, α will be in $\mathrm{Sp}(G)$ if and only if σ^α/σ is symmetric. Thus the group $\mathrm{Sp}'(G)$ of all α, for which σ^α and σ are similar, is a subgroup of

ISBN 0-8053-6702-0/0-8053-6703-9, pbk

Sp(G). In the important special case in which $x \to x^2$ is an automorphism of G, the converse is true and we may define Sp(G) alternatively as the group of all automorphisms α of $G \times \hat{G}$ such that σ^α and σ are similar multipliers. Whether or not $x \to x^2$ is an automorphism of G we may consider the subgroup Sp$'(G)$ and more generally the set $B_0(G)$ of all pairs α, f, where $\alpha \in \mathrm{Sp}'(G)$ and f sets up the similarity between σ^α and σ, that is,

$$\frac{\sigma^\alpha(w_1,w_2)}{\sigma(w_1,w_2)} = \frac{f(w_1 w_2)}{f(w_1)f(w_2)}$$

for all w_1, w_2 in $G \times \hat{G} \times G \times \hat{G}$. If $\alpha, f \in B_0(G)$ and U is a σ representation of $G \times \hat{G}$, then $w \to U_{\sigma(w)}$ is a σ^α representation of $G \times \hat{G}$ and $w \to U_{\alpha(w)}/f(w)$ is again a σ representation. Thus α, f defines a permutation of the set of all σ representations of $G \times \hat{G}$. Clearly α_2, f_2 takes $w \to U_{\alpha_1(w)}/f_1(w)$ into

$$w \to \frac{U_{\alpha_1\alpha_2(w)}}{f_1(\alpha_2(w))f_2(w)}$$

Hence $B_0(G)$ is a group under the composition law

$$(\alpha_1, f_1)(\alpha_2, f_2) = (\alpha_1\alpha_2, f_1^{\alpha_2} f_2)$$

where $f_1^{\alpha_2}(w) = f_1(\alpha_2(w))$. Now as we have already remarked $G \times \hat{G}$ has a *unique* irreducible σ representation. Hence for each $\alpha, f \in B_0(G)$ there must exist a unitary operator $V_{\alpha,f}$ that sets up an equivalence between U and the representation $w \to U_{\alpha(w)}/f(w)$; and $V_{\alpha,f}$ is unique up to a multiplicative constant. It follows at once that $\alpha, f \to V_{\alpha,f}$ is a projective unitary representation of $B_0(G)$ with a multiplier whose similarity class is uniquely determined. Notice that the natural homomorphism $\alpha, f \to \alpha$ of $B_0(G)$ onto Sp$'(G)$ has as kernel the character group of $G \times \hat{G}$; that is, $\hat{G} \times G$. Thus $B_0(G)$ is an extension of Sp$'(G)$ by $\hat{G} \times G$. In the special case in which $x \to x^2$ is an automorphism of G, it is easy to show that $B_0(G)$ is a semi-direct product of $\hat{G} \times G$ and Sp(G). In this case then we need only restrict V to Sp(G) to obtain a "natural" projective representation of Sp(G).

Now let Γ be any closed subgroup of G and let $\Gamma^\perp$ be its annihilator in $\hat{G}$. Then $\Gamma \times \Gamma^\perp$ is a closed subgroup of $G \times \hat{G}$ on which σ reduces to the identity. Thus we may consider the σ representation ${}^\Gamma U$ of $G \times \hat{G}$ induced by the identity representation of $\Gamma \times \Gamma^\perp$ and the general theory of induced σ representations (c.f. Section 11) allows us to conclude that it is irreducible and thus equivalent to U, indeed, that there is a canonical unitary

ISBN 0-8053-6702-0/0-8053-6703-9, pbk

operator Z setting up the equivalence in question. Let $B_0(G,\Gamma)$ denote the subgroup of $B_0(G)$ consisting of all α,f such that f takes the value one at each point of $\Gamma\times\Gamma^{\perp}$ and $\alpha(\Gamma\times\Gamma^{\perp})=\Gamma\times\Gamma^{\perp}$. For each $\theta\in\mathcal{H}({}^{\Gamma}U)$ and each α,f in $B_0(G,\Gamma)$, let ${}^{\Gamma}R_{\alpha,f}(\theta)=\theta'$, where $\theta'(x,\chi)=\theta(\alpha(x,\chi))f(x,\chi)$. It follows from the definition of ${}^{\Gamma}U$ as a σ induced representation that $\theta'\in\mathcal{H}({}^{\Gamma}U)$ and that ${}^{\Gamma}R_{\alpha,f}$ is a unitary operator in this Hilbert space; moreover, that $\alpha,f\to{}^{\Gamma}R_{\alpha,f}$ is a unitary representation of $B_0(G,\Gamma)$. It also follows at once from the definitions concerned that ${}^{\Gamma}R_{\alpha,f}$ sets up an equivalence between ${}^{\Gamma}U$ and $w\to{}^{\Gamma}U_{\alpha(w)}/f(w)$. Hence for each $\alpha,f\in B_0(G,\Gamma)$, $Z^{-1}\,{}^{\Gamma}R_{\alpha,f}Z$ must be a constant multiple of $V_{\alpha,f}$. In other words, the arbitrary constant in $V_{\alpha,f}$ may be chosen so that V restricted to $B_0(G,\Gamma)$ is an ordinary representation equivalent to ${}^{\Gamma}R$. This means in particular that the multiplier class associated with the natural projective representation V of $B_0(G)$ becomes the trivial class when restricted to $B_0(G,\Gamma)$. In the special case in which G is the additive group of the real line and Γ is the subgroup of all integers, we may identify $\hat{G}$ with the additive group of the real line via the mapping $y\to\chi_y$, where $\chi_y(x)=e^{2\pi ixy}$. Then $\Gamma^{\perp}$ is also the subgroup of all integers and $B_0(G,\Gamma)$ becomes a semi-direct product of the additive group of all pairs of integers with the group Γ_0 of all integer matrices in $SL(2,R)$. This example and its n-dimensional generalization are of course what one has in mind in introducing the abstract groups $\Gamma\times\Gamma^{\perp}$ and $B_0(G,\Gamma)$. Recall that we have already made reference to the fact that multipliers for $SL(2,R)$ become trivial on Γ_0(Section 30).

Consider now the representation

$$\alpha,f\to Z^{-1}\,{}^{\Gamma}R_{\alpha,f}Z={}^{\Gamma}S_{\alpha,f}$$

of $B_0(G,\Gamma)$ by unitary operators in $\mathcal{L}^2(G)$. When G is a finite group it follows immediately from the definitions concerned that the linear functional $\phi\to\Sigma_{\gamma\in\Gamma}\phi(\gamma)$ is carried into itself by all of the operators ${}^{\Gamma}S_{\alpha,f}$ for $\alpha,f\in B_0(G,\Gamma)$. This linear functional is not defined on the whole Hilbert space $\mathcal{L}^2(G)$ when G is infinite, but there are various dense subspaces on which it is defined and which are invariant under ${}^{\Gamma}S_{\alpha,f}$. Weil chooses one of these and shows that for all ϕ in it

$$\int_{\Gamma}\phi(\gamma)\,d\gamma=\int_{\Gamma}\phi'(\gamma)\,d\gamma$$

whenever $\phi'={}^{\Gamma}S_{\alpha,f}(\phi)$ and $\alpha,f\in B_0(G,\Gamma)$. Formally this result is quite trivial. Indeed, if $Z(\phi)=\psi$, then one verifies at once that $\int_{\Gamma}\phi(\gamma)\,d\gamma=\psi(e,e)$ and ${}^{\Gamma}R_{\alpha,f}$ obviously preserves the functional $\psi\to\psi(e,e)$. On the other

ISBN 0-8053-6702-0/0-8053-6703-9, pbk

hand, whenever there exists an isomorphism of G with $\hat{G}$ that maps Γ on $\Gamma^{\perp}$, one can choose α,f so that ϕ' is the Fourier transform of ϕ transferred back to G by the isomorphism. In this case the identity

$$\int_{\Gamma}\phi(\gamma)\,d\gamma=\int_{\Gamma}\phi'(\gamma)\,d\gamma$$

reduces to the Poisson summation formula.

Now let $\tilde{B}_0(G)$ be the group of all unitary operators in $\mathcal{L}^2(G)$ of the form $cV_{\alpha,f}$, where $|c|=1$. Then the center of $\tilde{B}_0(G)$ is isomorphic to the multiplicative group of all complex numbers of modulus one and the quotient of $\tilde{B}_0(G)$ by its center is isomorphic to $B_0(G)$. Similarly, let $\tilde{B}_0(G,\Gamma)$ be the subgroup of $\tilde{B}_0(G)$ consisting of all ${}^{\Gamma}S_{\alpha,f}$ with $\alpha,f\in B_0(G,\Gamma)$. Clearly $\tilde{B}_0(G,\Gamma)$ is isomorphic to $B_0(G,\Gamma)$. Let ϕ be any function in the dense subspace of $\mathcal{L}^2(G)$ mentioned above. This subspace is actually invariant under all $s\in\tilde{B}_0(G)$. Thus for each such ϕ we obtain a function $\tilde{\phi}$ on $\tilde{B}_0(G)$ by setting $\tilde{\phi}(s)=\int\phi'(\gamma)\,d\gamma$, where $\phi'=s\phi$. It follows immediately from the generalization of the Poisson summation formula described in the last paragraph that for all

$$s_1\in\tilde{B}_0(G,\Gamma)$$

$$\begin{aligned}\tilde{\phi}(s_1s)&=\int_{\Gamma}s_1s\phi(\gamma)\,d\gamma\\&=\int_{\Gamma}s\phi(\gamma)\,d\gamma\\&=\tilde{\phi}(s)\end{aligned}$$

Thus $\tilde{\phi}$ is a constant on the $\tilde{B}_0(G,\Gamma)$ left cosets.

Functions on $\tilde{B}_0(G)$ constant on the left $\tilde{B}_0(G,\Gamma)$ cosets are Weil's substitute for automorphic forms. The fact that

$$\tilde{\phi}(s)=\int_{\Gamma}s\phi(\gamma)\,d\gamma$$

is such a function is Weil's abstract analog of the theorem that the "θ series" $z\rightarrow\sum e^{2\pi mz}f_Q(n)$ associated with the sequence of representation numbers $f_Q(n)$ for a quadratic form Q is an automorphic form.

In one form the basic result of Siegel is the assertion that a suitable weighted average of θ series taken over all the quadratic forms in a "genus" is equal to a certain Eisenstein series. In Weil's theory, the analog of replacing a single θ series by an average over a genus consists in

ISBN 0-8053-6702-0/0-8053-6703-9, pbk

replacing the linear functional $\phi\to\int_\Gamma\phi(\gamma)\,d\gamma$ by

$$\phi\to\int_{H/H_\Gamma}\int_\Gamma\phi(h(\gamma))\,d\gamma\,dh$$

where H is a (locally compact) group of automorphisms of G and H_Γ is the subgroup leaving Γ fixed. Let $B_0^H(G,\Gamma)$ be the subgroup of $\tilde{B}_0(G,\Gamma)$ consisting of all members that commute with $\phi\to\phi^h$, where $\phi^h(x)=\phi(h(x))$. Then it follows at once from the invariance of $\phi\to\int\phi(\gamma)\,d\gamma$ under $\tilde{B}_0(G,\Gamma)$ that

$$\phi\to\int_{H/H_\Gamma}\int_{\gamma\in\Gamma}\phi(h(\gamma))\,d\gamma\,dh$$

is invariant under the subgroup $\tilde{B}_0^H(G,\Gamma)$. An obvious way of obtaining a linear functional invariant under this subgroup and under H is to take the linear functional $\phi\to\phi(l)$ and "sum" over all of its transforms by members of the subgroup. Of course, when the subgroup is not discrete we must integrate instead of adding. On the other hand, in cases of interest Γ is a discrete group and $\phi\to\phi(l)$ is invariant under a sub-subgroup whose coset space is discrete. Thus in these cases we obtain a $\tilde{B}_0^H(G,\Gamma)$ invariant linear function by summing the transforms of $\phi\to\phi(l)$ by the members of a countable subset of $\tilde{B}_0^H(G,\Gamma)$. Such a sum is Weil's abstract analog of our Eisenstein series, and his analog of the Siegel formula equates a linear functional of the form

$$\phi\to\int_{H/H_\Gamma}\int_\Gamma\phi(h(\gamma))\,d\gamma\,dh$$

to such an Eisenstein series.

Although one can write down the Weil Siegel formula for quite general G, Γ, and H, there is no reason to expect it to be true except in special cases and Weil is interested only in the case in which G is an adèle group and Γ is the discrete subgroup of all principal adèles. Let J be a finitely generated free commutative group and let X be its divisible extension so that X may also be regarded as a finite-dimensional vector space over the rationals. Let X_A be the adèle group of X with respect to J (cf. Section 28) and identify X itself with the discrete group of all principal adèles. Looking upon X as a vector space, let J^0 denote the subgroup of the dual X^* of X consisting of all linear functionals that take integer values on J. Then X^* is the divisible extension of J^0 and the adèle group of X^* with respect to J^0 is

ISBN 0-8053-6702-0/0-8053-6703-9, pbk

canonically isomorphic to the dual $\hat{X}_A$ of X_A. Moreover, this canonical isomorphism maps X^* onto the annihilator $X^{\perp}$ of X in $\hat{X}_A$. Similarly, we may identify $X_A \otimes X_A^*$ with the adèle group $(X \otimes X^*)_A$ of $X \otimes X^*$ with respect to the free commutative subgroup $J \otimes J^0$. Hence $X_A \otimes \hat{X}_A$ may be identified with this adèle group. It follows that every automorphism β of $X \otimes X^*$ defines an automorphism $\bar{\beta}$ of $X_A \otimes \hat{X}_A$ that carries $X \otimes X^*$ onto itself and agrees there with β. In particular, every β that is "symplectic" in the sense that it preserves the alternating bilinear form taking x_1, l_1, x_2, l_2 into $l_2(x_1) - l_1(x_2)$ defines a $\bar{\beta}$ that is in $\mathrm{Sp}(X_A)$ and maps $X \otimes X^*$ onto itself. Moreover, for each such β there is a canonical way of assigning a function f on $X_A \otimes \hat{X}_A$ so that the pair $\bar{\beta}, f$ is a member of $B_0(X_A, X)$. Let $\mu(\beta) = \bar{\beta}, f$. Then μ is a homomorphism of $\mathrm{Sp}'(X)$ into $B_0(X_A, X)$ and it is easy to see that μ is one-to-one. (We use the $'$ to indicate that X is being regarded as a rational vector space in forming the symplectic group.) For each $\beta \in \mathrm{Sp}'(X)$ we may form ${}^{\Gamma}S_{\mu(\beta)}$ and obtain a unitary operator in $\mathcal{L}^2(X_A)$. We denote it by r_β. What Weil calls the Eisenstein–Siegel series is a linear functional of the form $\phi \to \Sigma r_\beta(\phi)(l)$, where the summation is over a subset of $\mathrm{Sp}'(X)$ to be described below.

Let H_0 be a group of automorphisms of X and form the adèle group H_A of H_0 with respect to J as defined in Section 28. This will contain H_0 as a discrete subgroup and will act on X_A as a group of automorphisms in such a fashion that each $h \in H_0$ leaves X fixed and acts upon it in the originally given manner. On the other hand, we observe that each $h \in H_0$ also defines an automorphism of X^* and hence an automorphism T_h of $X \otimes X^*$. Thus we may define a subgroup $\mathrm{Sp}'(X/H_0)$ of $\mathrm{Sp}'(X)$ consisting of all members of $\mathrm{Sp}'(X)$ that commute with all automorphisms of $X \otimes X^*$ of the form T_h. Needless to say, Weil does not consider the most general possible group H_0. He starts with an algebraic number field k, a finite-dimensional semi-simple algebra $\mathcal{Q}$ over k (containing k in its center), an involution i in $\mathcal{Q}$, and a finite-dimensional $\mathcal{Q}$ module X. H_0 is then the group of all elements a in $\mathcal{Q}$ for which $aa^i = 1$. To define the Eisenstein–Siegel series, let $P(X)$ denote the subgroup of $\mathrm{Sp}'(X)$ consisting of all β for which $\beta(X^*) = X^*$. Then, if ϕ is a continuous member of $\mathcal{L}^2(X_A)$, we may form $r_\beta(\phi)(l)$ and verify that as a function of β it is a constant on the $\mathrm{Sp}'(X/H_0) \cap P(X)$ cosets. Thus

$$E(\phi) = \sum_{\beta \in \frac{\mathrm{Sp}'(X/H_0)}{(\mathrm{Sp}'(X) \cap P(X))}} r_\beta(\phi)(1)$$

makes sense whenever the series converges and is what Weil calls the

ISBN 0-8053-6702-0/0-8053-6703-9, pbk

Eisenstein–Siegel series. The averaged θ series to which Weil shows $E(\phi)$ to be equal is

$$I(\phi)=\int_{H_A/H_0}\sum_{\xi\in X}\phi(h(\xi))\,dh$$

The conditions on X and H_0 sufficient to guarantee convergence and equality of E and I are too complicated to state here.

Let us recall that by Weil's generalization of the Poisson summation formula the linear functional $\phi\to\Sigma_{\xi\in X}\phi(\xi)$ is invariant under the r_β for $\beta\in\mathrm{Sp}'(X)$ and a fortiori for $\beta\in\mathrm{Sp}'(X/H_0)$. But for $\beta\in\mathrm{Sp}'(X/H_0)$, one can show that r_β commutes with $\phi\to\phi^h$. Hence the averaged functional

$$\phi\to\int_{H_A/H_0}\sum_{\xi\in X}\phi(h(\xi))=J(\phi)$$

is invariant under the r_β for $\beta\in\mathrm{Sp}'(X)$ as well as under $\phi\to\phi^h$. On the other hand, $\phi\to\phi(l)$ is clearly invariant under $\phi\to\phi^h$ and hence so is E. Moreover, by construction, E is invariant under the r_β for $\beta\in\mathrm{Sp}'(X/H_0)$. Thus both E and I are invariant under the same large group and one might hope to deduce their equality from a suitable uniqueness theorem. Weil's proof actually proceeds partly along such lines, but also uses induction on the dimension of X.

Although Weil states the identity $I(\phi)=E(\phi)$ at the end of his first paper as the main result to be proved in the second, he actually needs (and proves) a more refined theorem in order to deduce his number theoretical corollaries. In this more refined theorem I and E are exhibited as countable sums that are equal term by term. There is one summand for each orbit in the action of H on X, and as far as I is concerned the existence of such a decomposition is formally obvious. One simply writes $\Sigma_{\xi\in X}\phi(h(\xi))$ as

$$\sum_{\omega\in\Omega}\sum_{\xi\in\omega}\phi(h(\xi))$$

where Ω is the set of all orbits ω in X under the action of H and then integrates term by term to obtain

$$I(\phi)=\sum_{\omega\in\Omega}I_\omega(\phi)$$

ISBN 0-8053-6702-0/0-8053-6703-9, pbk

where

$$I_\omega(\phi)=\int_{H_A/H}\sum_{\xi\in\omega}\phi(h(\xi))\,dh$$

A less transparent argument shows that E is a sum of linear functions of the form

$$\phi\to\int_{X_A}\phi(x)\,d\mu_\omega(x)$$

where μ_ω is a measure whose support is contained in the H_A orbit generated by the H_0 orbit ω. Moreover, μ_ω is shown to be a product of measures in the p-adic components of the adèle groups X_A—the component measure being describable in terms of the action of H_p on X_p. Actually, each X_p has a further decomposition related to the decomposition of the natural prime p into prime ideals of the number field k and μ_ω is even a product measure with respect to this finer factorization.

In a long digression in the first of his two papers, Weil shows how his methods and constructions lead to a proof of the law of quadratic reciprocity via the notion of "second degree character." Just as a character of an abelian group is an analog of a linear functional on a vector space, so a second-degree character is an analog of a quadratic form. By definition a *second-degree character* on a locally compact commutative group G is a continuous complex-valued function θ on G of modulus one such that $\theta(x,y)/\theta(x)\theta(y)$ is an ordinary character in x for each fixed y (and, of course, also in y for each fixed x). If θ is a second-degree character, let $\rho_\theta(x)$ denote the unique member of $\hat{G}$ such that

$$\rho_\theta(x)(y)\equiv\frac{\theta(xy)}{\theta(x)\theta(y)}$$

Then ρ_θ is a continuous homomorphism of G into $\hat{G}$ and $\rho_\theta(x)(y)=\rho_\theta(y)(x)$. If ρ_θ is an isomorphism of G onto $\hat{G}$, the second-degree character θ is said to be *nondegenerate*. Notice that if ρ is any isomorphism of G onto $\hat{G}$, then $x\to\rho(x)(x)$ is a second-degree character. If $x\to x^2$ is an automorphism, then $x\to x^{1/2}$ is well-defined and if $\rho(x)(y)=\rho(y)(x)$, then $\rho(x)(x^{1/2})$ is a second-degree character ψ such that $\rho_\psi=\rho$. When G is finite, we may form the Fourier transform $\hat{\theta}$ of the second-degree character θ.

$$\hat{\theta}(\chi)=\sum_{x\in G}\theta(x)\chi(x)$$

ISBN 0-8053-6702-0/0-8053-6703-9, pbk

If we replace θ by θ_y, where $\theta_y(x)=\theta(xy)$ and y is arbitrary, we get

$$\hat{\theta}_y(\chi)=\sum_{x\in G}\theta(x)\chi(xy^{-1})=\sum_{x\in G}\theta(x)\chi(x)\overline{\chi(y)}=\overline{\chi(y)}\,\hat{\theta}(\chi)$$

$$\hat{\theta}_y(\chi)=\sum_{x\in G}\theta(x)\theta(y)\rho_\theta(x)(y)\chi(x)=\theta(y)\hat{\theta}(\rho_\theta(y)\chi)$$

Thus setting $\chi=e$ and equating the two right-hand members, we get

$$\hat{\theta}(e)=\theta(y)\hat{\theta}(\rho_\theta(y)) \quad \text{so} \quad \hat{\theta}(\rho_\theta(y))=\frac{\hat{\theta}(e)}{\theta(y)}$$

In other words, if we use ρ_θ to identify G and $\hat{G}$, then the Fourier transform of a nondegenerate second-degree character θ is a constant multiple of $1/\theta$. When G is infinite, the Fourier transform of θ does not exist in the usual sense since θ is not integrable. However, one can define it in an indirect manner (by considering the Fourier transforms of its convolutions with continuous functions of compact support) and show that $\hat{\theta}=c/\theta$ whenever we use ρ_θ to identify G and $\hat{G}$. The complex number $c/|c|$ is an important invariant that Weil calls $\gamma(\theta)$. His proof of the quadratic reciprocity law is based on the following lemma (which appears as Theorem 5 in his first Acta paper). Let θ be a second-degree character on the locally compact commutative group G and let Γ be a closed subgroup of G. Suppose that θ is identically 1 on Γ and that ρ_θ is an isomorphism of G on $\hat{G}$ that maps Γ onto $\Gamma^\perp$. Then $\gamma(\theta)=1$.

When G is finite, this result is a more or less immediate consequence of the Poisson summation formula. Indeed

$$\sum_{\gamma\in\Gamma}\theta(\gamma)=\sum_{\gamma\in\Gamma}1=o(\Gamma)=\sum_{\gamma\in\Gamma^\perp}\hat{\theta}(\chi)=\sum_{\gamma\in\Gamma}\hat{\theta}(\rho_\theta(\gamma))$$
$$=\sum_{\gamma\in\Gamma}\frac{\lambda\gamma(\theta)}{\theta(\gamma)}=\lambda o(\Gamma)\gamma(\theta)$$

where $\lambda>o$. Thus $\gamma(\theta)=\lambda/|\lambda|=1$. In the general case, Weil gives a short ingenious proof based upon the constructions used in his generalization of the Poisson summation formula. Let F_{ρ_θ} be the unitary operator from $\mathfrak{L}^2(G)$ onto $\mathfrak{L}^2(G)$ obtained by Fourier transforming and using ρ_θ to identify G with $\hat{G}$. Let W_θ be the result of following F_{ρ_θ} with multiplication by θ. A mildly tedious but straightforward calculation shows that W_θ is in $\tilde{B}_0(G)$ and that its canonical image in $B_0(G)$ is an element whose cube is the identity. Thus W_θ^3 must be a constant multiple of the identity and an

ISBN 0-8053-6702-0/0-8053-6703-9, pbk

examination of all the definitions shows that this constant is $\gamma(\theta)$. On the other hand, when the hypotheses of Weil's Theorem 5 hold, it follows more or less immediately that the image of W_θ in $B_0(G)$ is in $B_0(G,\Gamma)$ and hence that W_θ^3 is the identity. Thus $\gamma(\theta)=1$.

The application of this result to the proof of the quadratic reciprocity law depends upon setting up a correspondence between second-degree characters and quadratic forms. Let k be a locally compact field that is not discrete and let χ be a nontrivial character of the additive group k^+ of k. For each $y\in k$, $x\rightarrow\chi(yx)$ is also a character that we denote by χ_y. It is readily verified that $y\rightarrow\chi_y$ sets up an isomorphism between k^+ and $\hat{k}^+$. Now let X be a finite-dimensional vector space over k and let $l\in X^*$. Then $x\rightarrow\chi(l(x))$ is a character of the additive group of X that we denote by χ_l; $l\rightarrow\chi_l$ is an isomorphism of X^* on $\hat{X}$.

Suppose that f is a quadratic form on X. Then $f(x+y)-f(x)-f(y)=\rho(x)(y)$, where ρ is a symmetric linear transformation from X to X^*. Clearly $\rho(x)(x)=f(2x)-2f(x)=2f(x)$ so $f(x)=\frac{1}{2}\rho(x)(x)$ if k is not of characteristic 2. In this case, then we have a natural one-to-one correspondence between quadratic forms and symmetric linear transformations. Given a character χ and a quadratic form f, the composite function $\chi\circ f$ is clearly a character of the second degree for which we may compute Weil's invariant γ. We shall denote it by $\gamma_\chi(f)$ and remark that it is "almost" independent of χ.

For example, if k is the real field an elementary calculation shows that $\gamma_\chi(f)$ is always

$$\exp\left(\pm\frac{\pi i}{4}(a-b)\right)$$

where a is the number of positive terms and b the number of negative in the reduction of f to diagonal form. The $\pm$ depends upon χ and, in particular, upon whether $\lambda>0$ or <0 in $\chi(x)=e^{2\pi i\lambda x}$.

Now let k be any nondiscrete locally compact field except C, let a and b be nonzero members of k, and let

$$f(x,y,z,t)=x^2-ay^2-bz^2+abt^2$$

A computation shows that $\gamma_\chi(f)=1$ if a can be written in the form x^2-by^2 and $\gamma_\chi(f)=-1$ if a cannot be so written. Weil proves the quadratic reciprocity law by applying Theorem 5 to the adèle group of an algebraic number field using a second-degree character on the adèle group whose local components are of the form $\chi\circ f$, where f is as first described. The fact that $\gamma_\chi(f)=1$ then reduces to the fact that the product of the local "norm

ISBN 0-8053-6702-0/0-8053-6703-9, pbk

residue symbols" (a/b) is one and this is one form of the law in question.

Weil describes his argument as merely a new formulation of a classical one due to Cauchy and Hecke. In tracing the connection it is important to observe the fact that $\hat{\theta} = C/\theta$ is an abstract version of the theorem which says that $Ae^{-x^2/B}$ has a Fourier transform of the same form and more generally that this is so for $Ae^{-Q(x_1 \cdots x_n)}$, where Q is a positive-definite quadratic form.

On the other hand, in understanding what is going on from the point of view of the theory of group representations, it is useful to observe that a second-degree character is just a one-dimensional projective representation of the group G for the (trivial) multiplier $\rho_\theta(x)(y)$. From this point of view the invariant $\gamma(\theta)$ is a function on the so-called σ-dual of G; that is, the set of all irreducible σ-representations of G where $\sigma(x,y) = \rho_\theta(x)(y)$. Now since σ is trivial, all irreducible σ-representations of G are one dimensional and we may define a "Fourier transform" taking functions on G into functions on $\hat{G}^\sigma$ by the formula

$$\hat{f}^\sigma(\theta) = \int_G f(x)\theta(x)\,dx$$

and just as in the classical case a function is uniquely determined by its Fourier transform.

Thus the invariant $\theta \to \gamma(\theta)$ may be completely described as the σ-Fourier transform of some function on G. An easy calculation shows that when the Haar measures are suitably normalized, this is just the function that is identically one on G. It would be interesting to go further and interpret the hypotheses and conclusions of Weil's theorem in terms of σ-representations. However, we shall not attempt this here.

Notes and References

For further developments of the ideas in Weil's two *Acta* papers see various articles by J. Igusa beginning with "Harmonic analysis and theta functions" *Acta Math.* vol. 120, (1968) 187-222 and including "Complex powers and asymptotic expansions." I and II *J. Reine und Angew. Math.* 268/269 (1974) 116-130 and 278/279 (1975) 307-321. See also pages 311-313 of the published version of the author's Chicago lecture notes and the work of Cartier, Segal, Shale, and the author referred to therein.

ISBN 0-8053-6702-0/0-8053-6703-9, pbk

Index